José A. Salfity
(Ed.)

Cretaceous Tectonics of the Andes

IUGS UNESCO IGCP 242

International earth evolution sciences Monograph Series

R. A. Gayer
The Tectonic Evolution of the Caledonide-Appalachian Orogen

A. Vogel ● H. Miller ● R. O. Greiling
The Renish Massif

J. Pohl
Research in Terrestrial Impact Structures

C.-Y. King ● R. Scarpa
Modeling of Volcanic Processes

S. El-Gaby ● R. O. Greiling
The Pan-African Belt of Northeast Africa and Adjacent Areas

C. O. Ofoegbu
Groundwater and Mineral Resources of Nigeria

C. O. Ofoegbu
The Benue Trough: Structure and Evolution

S. J. Freeth ● C. O. Ofoegbu ● K. M. Onuoha
Natural Hazards in West and Central Africa

R. A. Gayer ● R. O. Greiling ● A. K. Vogel
Rhenohercynian and Subvariscan Fold Belts

J. A. Salfity
Cretaceous Tectonics of the Andes

José A. Salfity
(Ed.)

Cretaceous Tectonics of the Andes

vieweg

Die Deutsche Bibliothek – CIP-Einheitsaufnahme

Cretaceous tectonics of the Andes: [IGCP, IUGS UNESCO 242] / José A. Salfity (ed.). – Braunschweig; Wiesbaden: Vieweg, 1994
(Earth evolution sciences)
ISBN 978-3-528-06613-0 ISBN 978-3-322-85472-8 (eBook)
DOI 10.1007/978-3-322-85472-8

NE: Salfity, José A. [Hrsg.]

Softcover reprint of the hardcover 1st edition 1994

Vieweg is a subsidiary company of the Bertelsmann Publishing Group International.

Produced by W. Langelüddecke, Braunschweig
Printed on acid-free paper

ISBN 978-3-528-06613-0

Contents

Editorial
J.A. Salfity VI

The Mesozoic Passive Margin of Northern South America
J. L. Pindell / J. P. Erikson 1

The Colombian Andes During Cretaceous Times
J. F. Toussaint / J. J. Restrepo 61

Kimmeridgian to Paleocene Tectonic and Geodynamic Evolution of the Peruvian (and Ecuadorian) Margin.
E. Jaillard 101

Kimmeridgian? to Paleocene Tectonic Evolution of Bolivia
T. Sempere 168

Tectono-Sedimentary Evolution of the Cretaceous-Early Tertiary and Metallogenic Scheme of Northern Chile, Between 20° and 26° S
T. Bogdanic C. / S. Espinoza R. 213

Tectonic and Sedimentary Evolution of the Cretaceous-Eocene Salta Group Basin, Argentina
J.A. Salfity / R.A. Marquillas 266

Cretaceous Evolution of the Magallanes Basin
V.A. Ramos / M.B. Aguirre-Urreta 316

Editorial

In 1822, J.J. d'Omalius d'Halloy used the word Cretaceous to define the terrane known as "creta", "craie", "chalk" and "Kreide" in Latin, French, English and German, respectively.

In the geologic map which accompanied a new edition of his work, published in 1823, he considered the Cretaceous terranes to be those lying between the ammonitic sediments -including Jurassic limestones- below and the Tertiary terranes above.

The history of scientific knowledge of the Cretaceous in the Andes began shortly afterwards, with the celebrated journeys of two learned investigators, Alcide d'Orbigny (1826-1833, published in 1842) and Charles Darwin (1832-1836, published in 1846).

In his work d'Orbigny gave his attention to the Cretaceous terranes of the Andes, which "are very extensive in the New World, as they are found mainly along the Cordillera, from Colombia down to the Straits of Magellan" (1842, p. 238). He also mentioned a series of fossils, some of which -those of Colombia- he considered Neocomian. The outstanding capability of the French savant enabled him to announce in 1842 that five of the fossil species from Colombia were identical to those known in the Paris basin. He concluded from this that "the seas of Europe and America must have been interconnected, and that the Atlantic Ocean must already have existed at that time as a single basin, stretching from Europe to America" (1842, p. 244).

From the time in which the term Cretaceous was put forward (1822) until the time d'Orbigny (1826-1833) and Darwin (1832-1836) made their journeys, not many years went by. Therefore, the widespread use of this term in Europe and the characterization of the outcrops discovered by these travellers in the "New World" as Cretaceous came about almost simultaneously.

From that time on, studies of the Cretaceous System in the Andes have been carried out in almost uninterrupted fashion.

In this century, prior to 1970, the evolution of the South American Cretaceous System, especially in the Andes, was researched by several outstanding authors; among whom were Steinmann (1929), Weaver (1942), Weeks (1947), Groeber (1953) and Harrington (1962).

The information published during the last thirty years or so has been very abundant, and covers almost all the Cretaceous sedimentary basins and Cretaceous plutonic and volcanic processes in the Andes. This circumstance led us to prepare the present volume, with the idea of publishing a work which would summarize all the available information and interpretations concerning Cretaceous tectonic evolution in the Andean

Cordillera. The Sub-Andean System is also included in the widest sense; that is, from the Eastern Venezuela Basin to the Magallanes Basin of Patagonia.

The theme seems to us doubly relevant. In the first place, because this is a fascinating subject from the academic point of view, given the vast array of tectonic processes which governed the evolution of the Andes toward the end of the Mesozoic. Secondly, because of the economic importance conferred on this portion of the Earth's crust by the hydrocarbon, uranium and other metallic and non-metallic ore resources contained in its Cretaceous sedimentary and eruptive units.

Andean Cretaceous geology drew the attention of a great number or researchers, due to the wealth of geologic testimony it contains. This includes: the extensive development of Cretaceous sedimentary basins; the structural dynamics recorded along the Pacific margin of South America and at both ends of the Andes, as evidenced by subduction, collision and rifting processes which controlled subsidence, eruptivity, sedimentary facies and the presence of unconformities; an abundance of volcanic episodes, both marine and continental; the massive batholiths emplaced in several stretches of the Cordillera; the notable duration of the Andean Cretaceous sedimentary cycle, which commenced in the Malm and finished in the Paleogene; the complexity of the pre-Cretaceous basement; the thick Cenozoic deposits covering the Cretaceous basins; finally, the abundant marine and continental faunas, as well as the flora. All this makes the study of the Andean Cretaceous System singularly attractive.

As we are all aware, the abundant information available now allows us to offer the geologic community -through the medium of the Earth Evolution Sciences Series- a work of both practical and scientific interest, condensed into each one of the chapters of the present volume. The copious bibliography cited in each of the papers making up the volume is proof of this.

The first work, on the passive Jurassic-Cretaceous margin of Northern South America, mainly covering Venezuela and Trinidad, was prepared by James L. Pindell and Johan P. Erikson.

The second paper, by Jean François Toussaint and Jorge Julián Restrepo, refers to the Cretaceous in the Colombian Andes, and includes its relationship with the Cretaceous in Ecuador.

The paper which follows, covering the tectonic evolution of the Cretaceous in Peru, was written by Etienne Jaillard.

The fourth paper refers to the evolution of the Cretaceous basin in Bolivia, and is the work of Thierry Sempere.

The next paper, by Tomislav Bogdanic and Sergio Espinoza refers to the Cretaceous geology and metallogenesis in the north of Chile.

This is followed by a paper on Cretaceous evolution in the north of Argentina prepared by the undersigned and Rosa A. Marquillas.

Finally, the chapter on the Cretaceous Magallanes basin in Patagonia was written by Víctor A. Ramos and María B. Aguirre Urreta.

Unfortunately, we were unable to include a chapter on the Cretaceous evolution of central Argentina and Chile in the present volume, due to force of circumstances having prevented the authors concerned from completing their work.

The editor of this volume wishes to thank Mr. Pedro J. Edmunds for his great help in translating two of the papers in this volume and revising the English versions of the remainder. He also wishes to thank Mr. Sr. Henry R. Estrada for his care and efficiency in the work of preparing this volume for publication.

J.A. Salfity

References

Darwin, Ch. (1846): Geological observations of South America, being the third part of the Geology of the voyage of the Beagle during 1832-1836. London, Smith, Elder, vii + 279 pp.

Groeber, P. (1953): Andico *in* Geografía de la República Argentina. Mesozoico, (Groeber, P., Stipanicic, P.N., and Mingramm, A.R.G.), Buenos Aires, Sociedad Argentina de Estudios Geográficos, GAEA, 2, Part 1 (1952), 349-541.

Halloy, J.J. d'Omalius d' (1823): Observations sur un essai de carte géologique de la France, des Pays-Bas et des contrées voisines. Annales des Mines (1822), v. 7, Paris.

Harrington, H.J. (1962): Paleogeographic development of South America. American Association of Petroleum Geologists Bulletin, v. 46, 1773-1814.

Orbigny, A. d' (1842): Voyage dans l'Amérique Méridionale, 1826-1833. Paris, P. Bertrand, v. III, 3rd Part, Géologie, 290 pp.

Steinmann, G. (1929): Geologie von Peru. Heidelberg, Karl Winter Ed., 448 pp.

Weaver, Ch.E. (1942): A general summary of the Mesozoic of South America and Central America. American Scientific Congress, 8th, Washington, Proceedings, 4, 149-193.

Weeks, L.G. (1947): Paleogeography of South America. American Association of Petroleum Geologists Bulletin, v. 31, 1194-1241.

The Mesozoic Passive Margin of Northern South America

J. L. Pindell, J. P. Erikson
Department of Earth Sciences, Dartmouth College, Hanover, New Hampshire 03755 USA

Abstract

A paleogeographic model is presented for the Mesozoic geological history of northern South America from Colombia to Trinidad, concentrating on the tectonic controls on sedimentation and on the origin of allochthonous terranes and the time at which they were obducted onto the autochthonous margin of northern South America. The effects of latest Cretaceous-Cenozoic tectonic deformation and terrane accretion are assessed and accordingly removed to produce a more accurate palinspastic Cretaceous geometry of autochthonous portions of northern South America. By means of continental reconstruction of this geometry with Yucatán, the Bahamas, and southern North America, the Jurassic rifting history of northern South America is outlined. It is seen that an as-yet unknown Jurassic marine shelf section probably exists beneath northern Maturín Basin/Serranía del Interior/Trinidad which is correlative and probably similar to the western Cuban Jurassic stratigraphic section (initial conjugate rift margin). From the Jurassic rift configuration, the Cretaceous stratigraphic development of the northern South American passive margin is outlined in a framework of accurate relative plate motions, and models are presented for the creation, history, and origins of the allochthonous terranes in the light of Caribbean Plate migration history. Northeastern Venezuela's passive margin sedimentary section is analysed and reveals a stable, two-phase history of shallow marine, mixed siliciclastic and carbonate accumulation in the Early Cretaceous and dominantly pelagic limestone and shale deposition in the Late Cretaceous; this stable history reflects the continuation of passive margin conditions through the Cretaceous, in contrast to the tectonically active western part of the margin in Late Cretaceous time. Finally, a brief description of Cenozoic Caribbean-South American plate interaction shows how and when the allochthons were finally emplaced upon the margin of Venezuela and Trinidad.

Introduction

The Cretaceous development of northern South America, comprised of northern Colombia, Venezuela, and Trinidad, differed greatly from the Cretaceous evolution of the strictly Andean nations to the south. Unlike the Andean nations in which subduction, arc magmatism, backarc extension, or foreland thrusting were often common, the northern margin was a tectonically quiescent, Atlantic-type passive margin, which was created during the Triassic-Jurassic breakup of Pangea. This quiescent margin formed the autochthon during subsequent, Cenozoic orogenic phases of development. Late Jurassic and Cretaceous times along this margin were characterized mainly by sediment accumulation on a thermally subsiding platform. Differences in Cretaceous sediment accumulation rates and the development of regional facies patterns were most likely residual effects of differential amounts of rifting and crustal thinning during the Jurassic. Not until the Cenozoic did orogenesis begin along the northern margin, although the effects of a latest Cretaceous Andean deformation in Colombia were felt in western Venezuela. Cenozoic orogenesis was entirely due to South America's history of westward drift across the mantle and motion relative to the Caribbean Plate. Portions of, or accreted terranes within, the southeastern Caribbean Plate were obducted onto the autochthonous passive margin, and are collectively referred to here as "allochthons" within the "allochthonous belt". Many of the rocks of the allochthonous belt, although not accreted to the autochthon until the Cenozoic, are of Cretaceous age, and are briefly described herein in addition to the primary consideration of the autochthon.

An accurate assessment of the Cretaceous history and paleogeography of autochthonous northern South America cannot be made without first determining and palinspastically restoring the Cenozoic deformations, which were severe in northern South America. This includes assessing and removing the geometric effects of accretion of the allochthonous terranes to the original margin, bulk shortenings and extensions within the margin due to plate convergence or divergence, and strike-slip migrations of crustal elements along or within the margin. These concerns are considered quantitatively in this paper to allow a more accurate depiction of the Mesozoic configuration of the autochthon. On the larger scale, this Mesozoic configuration of northern South America formed an important element of western Pangea, and, therefore, its plate tectonic history can be documented in the light of plate dispersal during and after the breakup of the supercontinent. Thus, positioning northern South America in an accurate Early Mesozoic portrayal of Pangea, and then tracing the subsequent plate motions and plate boundary geometries during the opening of the Atlantic, place important quantitative constraints in the margin's tectonic history, in terms both of timing of events such as rifting and also of the magnitude and rate of plate drift and paleo-oceanic conditions along the margin. Such precise plate kinematic constraints cannot be achieved in the other Andean nations to the south, making assessments of their regional evolution relatively more interpretive.

In this paper, consideration of the above elements places autochthonous northern South America in a quantitatively constrained Mesozoic geometric and plate kinematic configuration and framework. This framework will then provide the foundations for subsequent summaries of Jurassic rift history and Cretaceous sedimentation patterns. From those summaries, sequential paleogeographic maps are then developed for Jurassic and Cretaceous times which show the main elements of northern Colombia's, Venezuela's, and Trinidad's Mesozoic geologic development. The Cenozoic orogenic tectonic history is then briefly outlined, in order to show how the Mesozoic passive margin has been destroyed and transformed into an active plate boundary zone between the allochthonous Caribbean and South American plates.

Regional Geologic Elements of Northern South America

Regional geologic elements of northern South America are shown in Figure 1, including the main active fault zones of the South American-Caribbean plate boundary zone, the boundary of the autochthonous and allochthonous rock sequences, thrust belts/uplifts of mainly Paleogene and mainly Neogene age, and sedimentary basins of both the autochthon and allochthon.

Active Plate Boundary Zone

The main causes of active deformation in northern South America and the southern Caribbean are (1) the relative motion between the Caribbean and South American plates, which is largely a function of the westward component of drift of South America across the mantle, and (2) the relative motions of various blocks within the plate boundary zone (Figure 2). Where the Caribbean and South American crusts are adjacent, relative motion is approximately E-W and accommodated at long, straight, dextral strike slip fault zones such as the Morón, Coche, El Pilar, and North Coast Fault Zones of eastern Venezuela and Trinidad (Figure 1). Motions among the blocks of the plate boundary zone (PBZ) over the last 15 Ma or so have been so significant that the relative motions between the blocks and the Caribbean or South America plates have been entirely different from Caribbean-South American relative motion. For example, in the Mérida Andes, separating the Maracaibo Block and Guyana Shield, relative motion is NE-SW. Likewise, at the South Caribbean Foldbelt separating the Guajira Block and the Caribbean Plate, relative motion is SE-NW. Many of these displacement zones have offsets of regional significance, and must be palinspastically restored to establish Cretaceous sedimentation patterns (see next section).

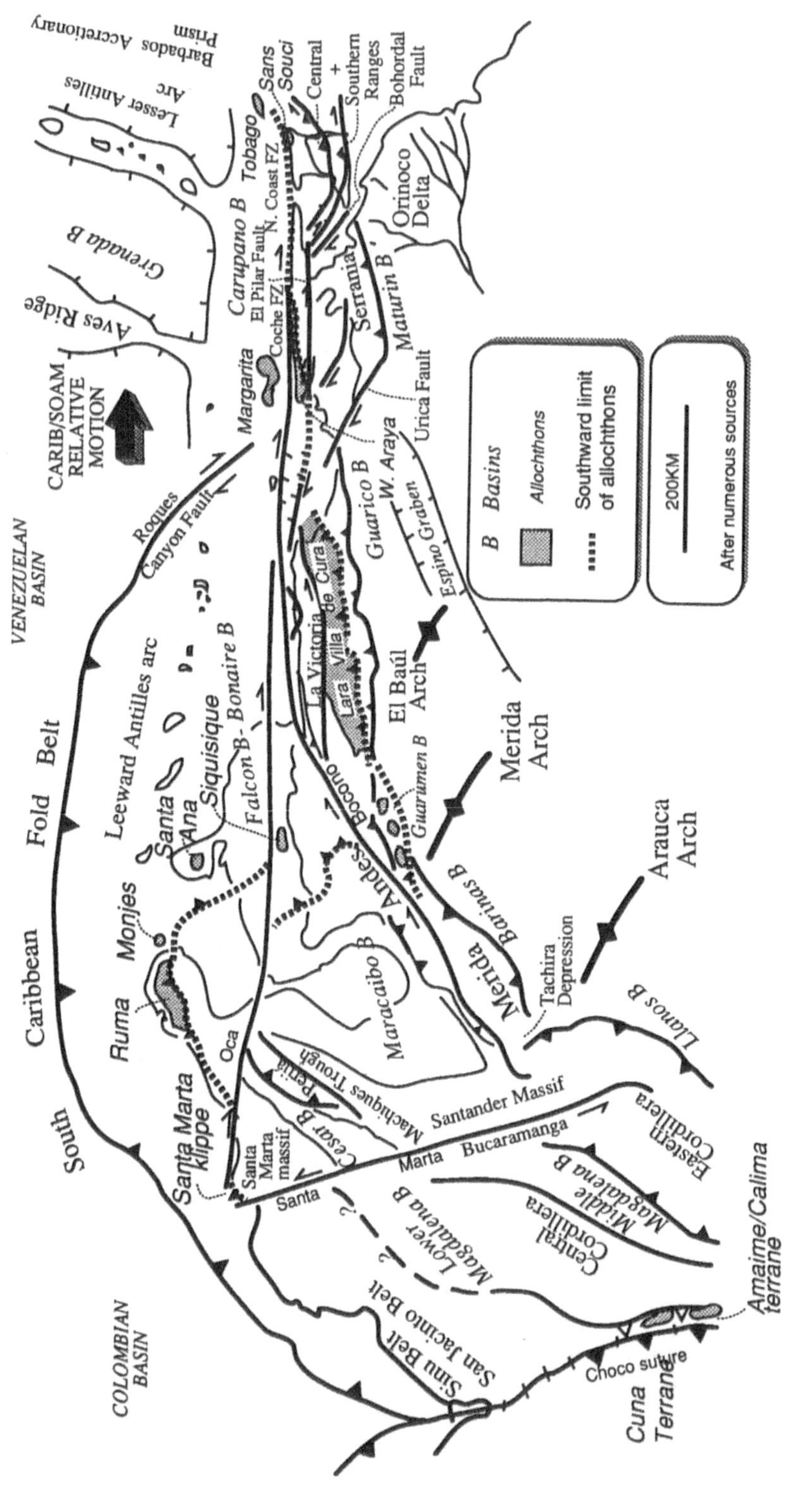

Fig. 1
Simplified geological and locality map of northern South America, with items mentioned in text, after numerous sources.

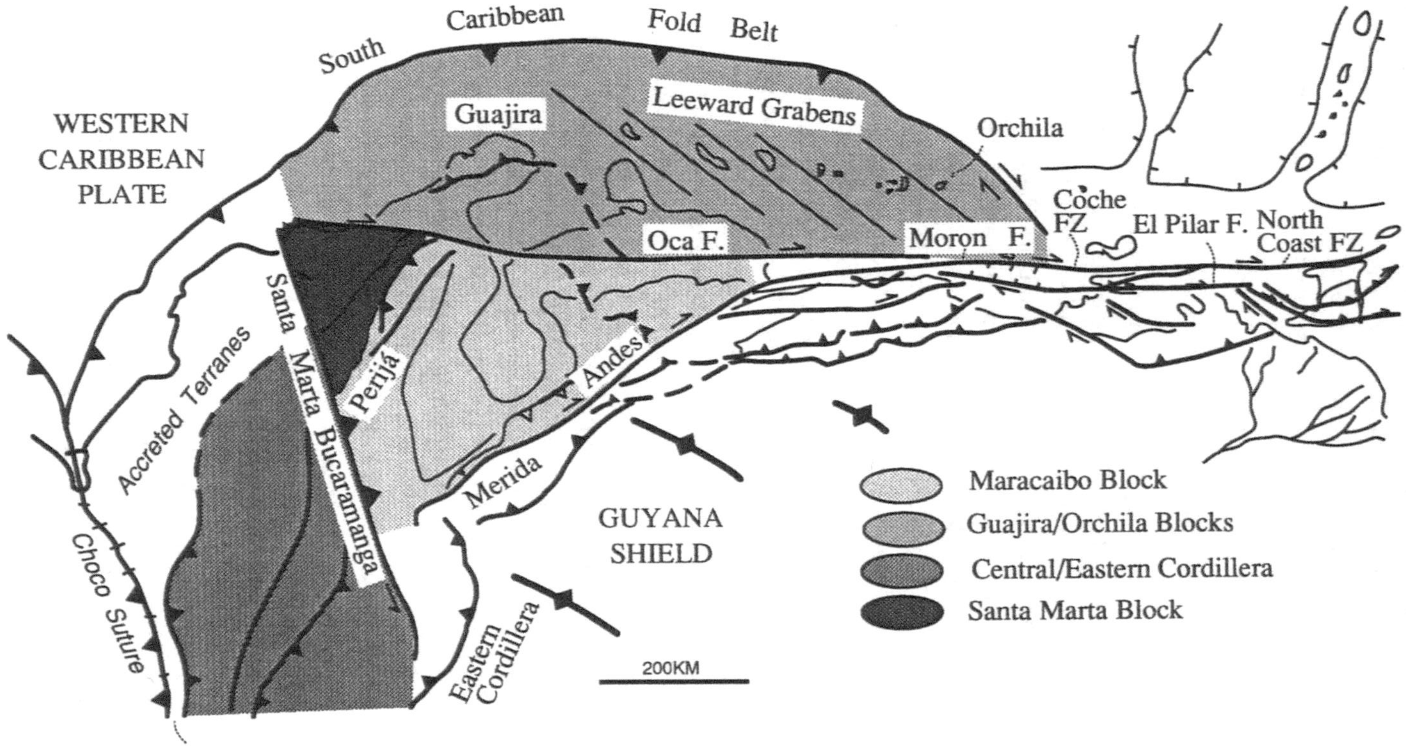

Fig. 2

Main blocks of the northern Andes undergoing relative displacements during Neogene orogenesis. Restoration of Andean displacements is required before pre-Cenozoic paleogeography can be acccurately protrayed.

The active deformations within the PBZ are responsible for most of the present day topography and for exposure of Cretaceous and other rock units. This uplift has exposed certain rock units and formerly active fault zones which allow interpretation of an earlier phase of northern South America's tectonic evolution, active from the Late Paleocene to the Middle Miocene.

Thrust Belts/Uplifts of Northern South America

Areas of Neogene topographic uplift largely coincide with structures of the active plate boundary zone. These include the Perijá Range, the Eastern Cordillera of Colombia, the Mérida Andes, the Falcón Anticlinorium, the Caribbean Mountains, the Serranía del Interior Oriental, the Araya-Paria Peninsula and Northern Range of Trinidad, and the Central and Southern Ranges of Trinidad. Many of these are the sites of active thrusting in association with differing degrees of strike slip offset, but the Caribbean Mountains, the Serranía del Interior Oriental, and the Araya-Paria-Northern Range of Trinidad, all of which had been thrust belts in the Early and Middle Miocene, are presently being uplifted because they probably are now (since Late Miocene) the footwalls of extensional fault zones. The uplift is due to isostatic rebound as a result of the removal of mass in the hanging walls (Bonaire and Carupano Basins) by the normal component of motion along northward dipping, obliquely-slipping fault planes.

Exposed in these various uplifts, and known from subsurface mapping as well, is an earlier Cenozoic thrust belt which extends from the Santa Marta Massif of Colombia to the north coast of the Northern Range of Trinidad, and presumably out into the Barbados accretionary prism (Figure 1). Thrusting in this belt was diachronous from west to east, and occurred in Santa Marta/Guajira Peninsula in the latest Cretaceous? and Paleocene, northern Lake Maracaibo in the Late Paleocene to Middle Eocene, the Guarumen sub-basin of the northeastern Barinas Basin in the Middle to Late Eocene, the western Guárico Basin in the Oligocene, the Maturín Basin in the latest Oligocene to Middle Miocene, and in Trinidad in Early to Middle Miocene. This thrust history is dated by the age of sediments involved in thrusts, by the age of the foredeep basinal section overthrust by the thrusts in various segments of the thrust belt, and by the onset of rapid subsidence in the foredeep basin ahead of the thrusting (Pindell et al., 1991). Of prime importance is that this Paleogene-Middle Miocene thrust belt has been significantly offset in places by structures of the active plate boundary zone, such as at the crossing of the belt by the active Mérida Andes. There, the dextral strike slip component of Mérida deformation is at least 80 km, and probably greater. This magnitude of neotectonic deformation demonstrates the need for palinspastic restoration of Neogene-Recent deformation in order to establish a more accurate depiction of paleogeography during earlier times.

Autochthonous vs. Allochthonous Suites of Rock in Northern South America

The uppermost nappes of the diachronous Paleogene-Middle Miocene thrust belt are represented by a semi-continuous belt of mostly metamorphosed rocks of oceanic and/or island arc affinity. This belt of oceanic/arc rock comprises the Ruma, Santa Ana, Siquisique, Villa de Cura (but not Caracas Group), El Copey Formation and equivalents of western Araya, Juan Griego Group of Margarita, Sans Souci Group of Northern Range of Trinidad, and Tobago metamorphic complexes (Case and MacDonald, 1973; Martín-Bellizzia and Arozena, 1972; Bartok et al., 1985; Maresch, 1974; Stephan et al., 1980; Avé-Lallemant, 1990; Snoke et al., 1990; Algar and Pindell, 1991a), and the metamorphism in most of these complexes is Aptian/Albian to early Late Cretaceous in age. Bellizzia (1972; 1986) recognized that these units are entirely allochthonous with respect to the interior sedimentary basins. It has long been thought, based largely on the metamorphic ages and on traditional Late Cretaceous age assignments on flysch units (Garrapata, Paracotos, etc.) that are now known to be Cenozoic, that the allochthons first collided with northern South America in the Late Cretaceous. However, as developed below (Concept of the Cretaceous Passive Margin), the arguments for Cretaceous orogenesis involving northern South American autochthonous rocks are now inadequate, and it is now clear that the allochthons and parautochthonous assemblages of the South American outer margin were thrust southward into entirely Cenozoic flysch basins which formed above the Mesozoic shelf sediments and basement (e.g., Pindell et al., 1988). In some areas, the autochthonous and parautochthonous sections have been exposed by erosion or tectonism within or north of the allochthons. For example, in Venezuela's Coast Range, basement and the Mesozoic Caracas Group shelf sediments have been uplifted between La Victoria and San Sebastián Faults since being overthrust by the Villa de Cura Klippe (Figure 1; Schubert, 1988), thereby isolating the Villa de Cura Klippe from its original parental terrane. Initial studies on the age of cooling after metamorphism on such rocks show Oligocene-Miocene (Kohn et al., 1984; Foland et al., 1992) rather than Cretaceous ages of metamorphism, which is consistent with Cenozoic obduction.

Sedimentary Basins and Basement Arches

At and to the south of the line delimiting the allochthonous bodies, parautochthonous Cenozoic flysch sequences are thrust southward onto their fully autochthonous counterparts in downflexed, asymmetric foredeep basins. The northern/central Maracaibo, the northern Barinas, the Guárico, the Maturín, and the Southern Basin of Trinidad are such downflexed foredeep basins with at least several km of Cenozoic clastic sediments above the Cretaceous sections. At the time of thrusting, the Cretaceous section shallowed toward the south in each basin, as in the Maturín Basin of today (e.g., Hedberg, 1950). However, in the Maracaibo Basin, subsequent basin inversion and enormous sedimentation rates in the southern part of the basin have imparted a southerly dip on the Cretaceous section across much of the basin (Bockmeulen et al., 1983). The Cretaceous sections of all basins, and sections that are now incorporated in the thrust belts as well (e.g., Eastern Cordillera of Colombia), show evidence of shallowing and becoming more proximal to the south or southeast, toward the Guyana Shield. It is widely accepted that the collective Early and Middle Cretaceous sections from the various basins formed a continuous marine platform (e.g., Krause and James, 1989), interrupted to varying degrees only by the El Baúl, Mérida, and Arauca basement arches which separated areas of greater basinal subsidence rates and sediment accumulation. However, the concept of Late Cretaceous tectonism involving the autochthon persists (e.g., James, 1990; Beets et al., 1984; Chevalier et al., 1988). In the following section, we review recent developments which diminish the probability that Cretaceous tectonism occurred along the autochthonous margin, and which, therefore, strengthen the case for continued passive margin conditions well into the Cenozoic. Only in the west, in the Maracaibo Basin, did tectonic events control sedimentation toward the end of the Cretaceous; there, foredeep subsidence occurred which was directly related to loading of the Central Cordillera of Colombia.

The Cretaceous passive margin conditions were initiated by Jurassic rifting between Yucatán and northern South America. In addition to the development of the continental margin itself (i.e., edge of continental crust and the morphological definition of the continental slope and rise to the north), several interior rift basins were developed in northern South America during the breakup of Pangea. The Espino, Táchira, and Machiques Troughs are examples (Figure 1), and these underlie, respectively, the southern Maturín Basin, the Mérida Andes, and the Perijá Range. The latter two have been inverted during Andean orogenesis, whereas the former has remained relatively stable probably because Caribbean-South American plate interactions are so much less severe in the east. The Espino Graben continues WSW to at least the Colombian border, and may also trend to the northeast beneath the Serranía del Interior, Gulf of Paria, and/or Orinoco Delta. The Táchira and Machiques Troughs, in particular, were active Jurassic rift zones of lithospheric thinning, as indicated by intense, basaltic volcanism and the thicker and more rapid accumulation of Cretaceous sediments above them during thermal subsidence, as compared to the surrounding platformal areas. Although basalts are known from the Espino Graben (Feo-Codecido et al., 1984), it, and especially its continuation to the west, seems more likely to be the surface expression of an upper crustal detachment than a full scale lithospheric rift, with little or no subsequent phase of significant thermal subsidence. The areas of Jurassic lithospheric attenuation became the main Cretaceous depocenters of the passive margin, whereas the gaps between these rifts have been preserved as structural and depositional highs, as denoted commonly by stratigraphic onlap patterns reflecting their relatively raised paleotopographies. Given the distribution and spacing of the rifts (Figure 1), the intervening arches were probably maintained or even enhanced by flexure of the lithosphere in response to Jurassic and Cretaceous sedimentary loading in the rifts. The basement arches have helped to control later sedimentation patterns and hydrocarbon migration pathways, and, therefore, are extremely important features.

Concept of a Cretaceous Passive Margin

Numerous arguments indicate that northern South America was a passive, Atlantic-type margin for Late Jurassic and Cretaceous times. The main ones are briefly reviewed below.

First, subsidence curves (Figure 3) for various positions along the autochthonous portion of northern South America (south of the allochthonous overthrust belt) show a smooth, slow sediment accumulation history typical of thermal subsidence for most of the Cretaceous and, in the east, well into the Tertiary. No important uplift events are indicated, and no periods are known of subsidence indicative of tectonically induced foredeep basin loading, except for, as noted above, the Andean effect on the Maracaibo Basin in the Campanian-Maastrichtian.

Second, orogenic flysch units of central Venezuela once thought to be of Late Cretaceous age and therefore representative of Cretaceous orogeny, such as the Paracotos and Garrapata Formations (González de Juana et al., 1980), have recently yielded Paleocene and Eocene faunal ages from the lutitic matrices (work in progress). The older ages were apparently derived from the larger clasts in the flysch. The new, younger ages from the matrix support the model of Tertiary emplacement of the allochthons onto the autochthonous passive margin, as well as reduce the validity of the previous interpretation of Late Cretaceous orogenesis (e.g., Maresch, 1974). To our knowledge, there is no unequivocal structural evidence in autochthonous portions of Venezuela for Cretaceous deformation.

Third, as in Venezuela, structural and stratigraphic field relationships in the Northern and Central Ranges of Trinidad, once thought to preserve evidence of Cretaceous deformations (e.g., Kugler, 1953), have been recently remapped and dated and support the view that all deformation and metamorphism in these Ranges is of Tertiary age, probably Early Oligocene and younger and occurred in conjunction with the arrival of the migrating Caribbean Plate (Algar and Pindell, 1991a; in review). For example, the Maastrichtian (to ?Paleocene) Galera Formation in the northeastern Northern Range has been shown to possess all deformation fabrics that are recorded in the Jurassic and Cretaceous formations, such that all deformation there must be Cenozoic in age. Furthermore, a brecciated zone in the Northern Range, which was previously mapped as a transgressive lag deposit (Barr, 1963) above older Cretaceous metamorphic rocks and beneath the less metamorphosed Galera Formation, is now recognized to be a cata-

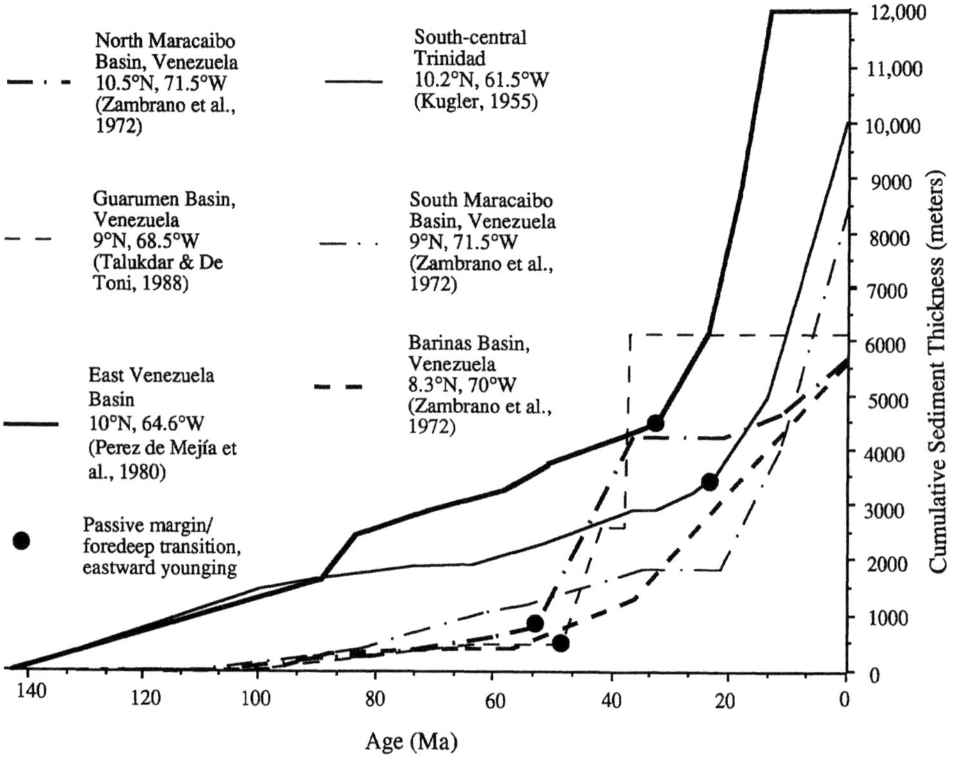

Fig. 3

Sediment accumulation histories for several localities along northern South America. Note that Cretaceous portions of the curves show smooth, progressive accumulation, denoting thermal subsidence only, and that accumulation rates involving tectonic subsidence of basement typical of foredeep loading do not occur until the Cenozoic. Note also the clear eastward progression of the age of onset of foredeep subsidence rates, from northern Maracaibo, to Guarumen, to Eastern Venezuela, to Trinidad. This progression matches the relative eastward migration of the Caribbean Plate, which is directly responsible for foredeep basin development. Modified after Pindell et al. (1991).

clastic zone along a strike-slip fault which possesses rock fragments as young as Oligocene, thereby dating the faulting as Oligocene or younger (Algar and Pindell, 1991a; in review).

Fourth, it is now clear that the Caribbean Plate is composed mostly of Late Jurassic and ?Early Cretaceous lithosphere which was generated in the Pacific, to the west of the Americas, and which has subsequently migrated eastwards to its present position between the Americas (Pindell, 1990; Pindell et al., 1988). Relative eastward migration began in the Albian, and the leading edge of the Caribbean crust, the Greater Antilles arc, reached the Colombian Andes and the southern Yucatán Block by the Maastrichtian (Pindell and Barrett, 1990). Continued migration progressively destroyed, from west to east, the pre-existing margin of northern South America during the Cenozoic. All of the allochthonous terranes of northern South America possessing Cretaceous metamorphic and/or arc volcanic rock, such as the Villa de Cura Klippe, have originated from either the Caribbean Plate or its accretionary prism and were emplaced onto the South American margin as a result of Caribbean migration and interactions. During the Late Jurassic and Cretaceous, there were no such Caribbean Plate interactions along northern South America. To the contrary, the northern South American margin during those times was that which was formed by rifting and plate separation from elements of the North American Plate, namely the Yucatán Block (see below).

To summarize, subsidence histories, syn-orogenic flysch units, stratigraphic and structural studies, and plate motions histories all indicate the existence of a Jurassic to Cretaceous passive margin in northern South America, which was destroyed by Caribbean plate motions from west to east mainly during Cenozoic times. We have yet to discern any specific, well documented evidence suggesting Cretaceous plate tectonic deformations involving the autochthonous northern South American margin. However, Caribbean-South America plate interactions have caused allochthons of Jurassic-Cretaceous oceanic/arc rocks, many of which were deformed and metamorphosed in the Cretaceous, to be obducted onto the passive margin in Cenozoic times (Figure 1). Therefore, the line separating these rocks from the overthrust autochthonous passive margin rocks to the south separates two unrelated stratigraphic realms, the allochthonous Caribbean Suite and the autochthonous Proto-Caribbean Suite, the latter of which represents the passive margin of the Jurassic-Cretaceous Proto-Caribbean Seaway (Pindell, 1991).

Palinspastic Reconstruction of Cenozoic Orogenesis

The Cretaceous configuration of northern South America can be reconstructed by restoring bulk displacements during Andean orogenesis, and by removing the Caribbean Plate and its obducted allochthons, and associated syn-emplacement shortening, from the autochthonous margin.

Restoration of Andean Strain

The net palinspastic restoration of displacements arising from Andean orogenesis in northwest South America requires the construction of vector diagrams denoting relative offsets among various semi-rigid blocks. This methodology (Dewey and Pindell, 1985; 1986) allows the prediction of strains in zones where direct measurements are lacking, by completing vector triangles whose other legs are zones of strain with known or estimated offsets. The most important elements of information for bulk palinspastic restoration are the total offset for each zone of strain and the time period over which offset took place. With this information, a "vector nest" of numerous vector triangles can be made which, in turn, defines a reconstruction which shows the configuration of basement at the onset of the strains.

Blocks in northwest South America which have behaved independently of one another during Andean orogenesis are shown in Figure 2. Primary zones of crustal strain between these more stable blocks include the Mérida Andes, the Eastern Andes of Colombia, the Perijá Andes, the Oca Fault, the faults between the offshore islands, the Roques Canyon fault zone, the Morón Fault/Falcón-Bonaire Basin along the central Venezuelan coastline, and the South Caribbean Foldbelt (Figure 1; Case and Holcombe, 1980). All terranes are entrained to varying degrees in the ongoing, generally east-west shear between the Caribbean Plate and northern South America. Most of the deformation is known to be mainly Neogene in age and presently active.

In northwest South America, an age of 15 Ma (Middle Miocene) is a reasonable estimate for the onset of bulk strain, which accelerated into the Pliocene. This age is indicated by the onset of rapid foredeep subsidence and sediment accumulation in basins adjacent to the Andes (Guayabo, El Fausto Groups, etc.), and by the Middle Miocene age of low-altitude sediments which have been uplifted to high altitudes in various localities in the Andes (Dengo and Covey, in press, 1991).

Figure 4 shows our estimate of Neogene northern Andean displacements and shortenings using the vector triangle nest method, measured at 10° N, 70.5° W. Note that the bulk of the deformation has occurred since the Middle Miocene (~15 Ma), but that pre-Andean, Oligocene-Early Miocene Oca Fault/Falcón Basin displacement (Tschanz et al., 1974; Muessig, 1984) has also been included in Figure 4 to provide a more accurate depiction of Cretaceous relationships. The vector nest of Figure 4 is constructed using published criteria for estimating the offset of certain fault zones. In some cases, we have slightly revised the cited values of offset by making our own measurements with the criteria noted by the original authors and additional information.

Removal of Caribbean Allochthonous Terranes

The line bordering the southern extent of allochthonous terranes in northern South America (Figure 1) is the present erosional limit of the thin-skinned thrusts along which the terranes were emplaced. It is not known by how much the continental passive margin of South America has been overthrust by these terranes, but values of 100 to 200 km are reasonable given that deep water sediments have been thrust onto shallow water marine deposits (shelf) of the Proto-Caribbean Suite (e.g., Las Brisas Formation of the Caracas Group, González de Juana et al., 1980). Despite this imprecision, the geometry of the Tertiary thrust belt in which the allochthons rest is indicative of an overthrust margin which was characterized

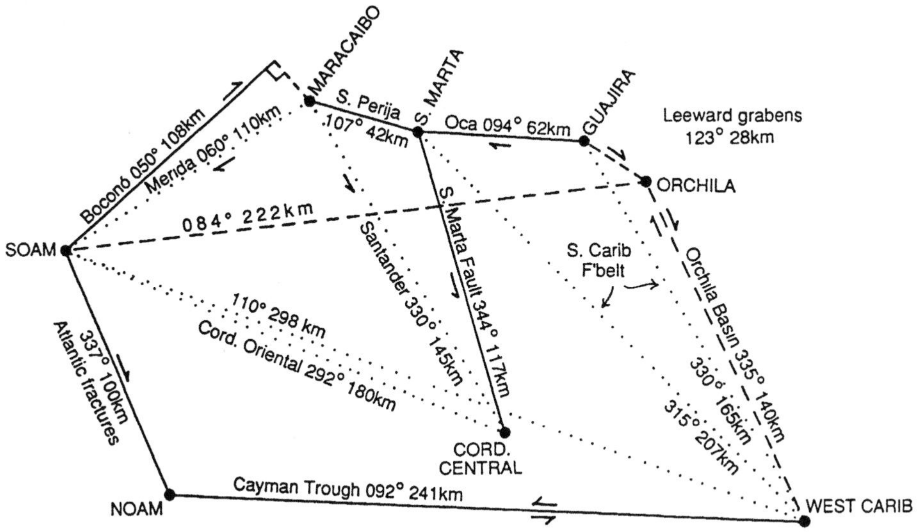

Fig. 4

Vector triangle construction of Andean displacements and Falcon Basin extension since the Oligocene, modified after Dewey and Pindell (1985). Dots are blocks under relative motion (see Fig. 2), and tielines are fault zones separating the blocks. Solid lines are fault zones of measurable offset, dashed lines are offsets of known trend only, and dotted lines are offsets inferred by the triangle construction process. Sources used are: Pindell et al. (1988); Kellogg (1984); Tschanz et al. (1974); Case et al. (1984); Ladd et al. (1984); Hilst (1990); Silver et al. (1975); Biju-Duval et al. (1983); Case and Holcombe (1980); Rosencrantz et al. (1988).

by a number of salients and re-entrants. Figure 1 shows at least three southeastward steps in the thrust belt, at the eastern Maracaibo Basin, the Gulf of Barcelona (Urica Fault Zone), and the Gulf of Paria (Bohordal Fault). Given that these SE-trending offsets in the thrust front are parallel to the separation direction between Yucatán and northern South America (see below), they probably define marginal offsets, or rift transfer zones, in the original Jurassic rifted margin. The marginal offsets would have been characterized in Jurassic times by normal faulting with a component of sinistral shear in basement blocks, stepping down to the east across each fault. Algar and Erikson (1992) have shown that the slightly different paleogeographic characteristics of the Cretaceous stratigraphies of the Serranía del Interior Oriental and Trinidad are highly compatible with their respective depocenters being located on different rift elements of the passive margin (separated by a marginal offset), with Trinidad's stratigraphy always being of slightly deeper water origin.

Figure 5 shows the paleogeographic effect of (1) restoring Andean orogenic and Falcón Basin offsets, and (2) removing the allochthonous terranes from northern South America.

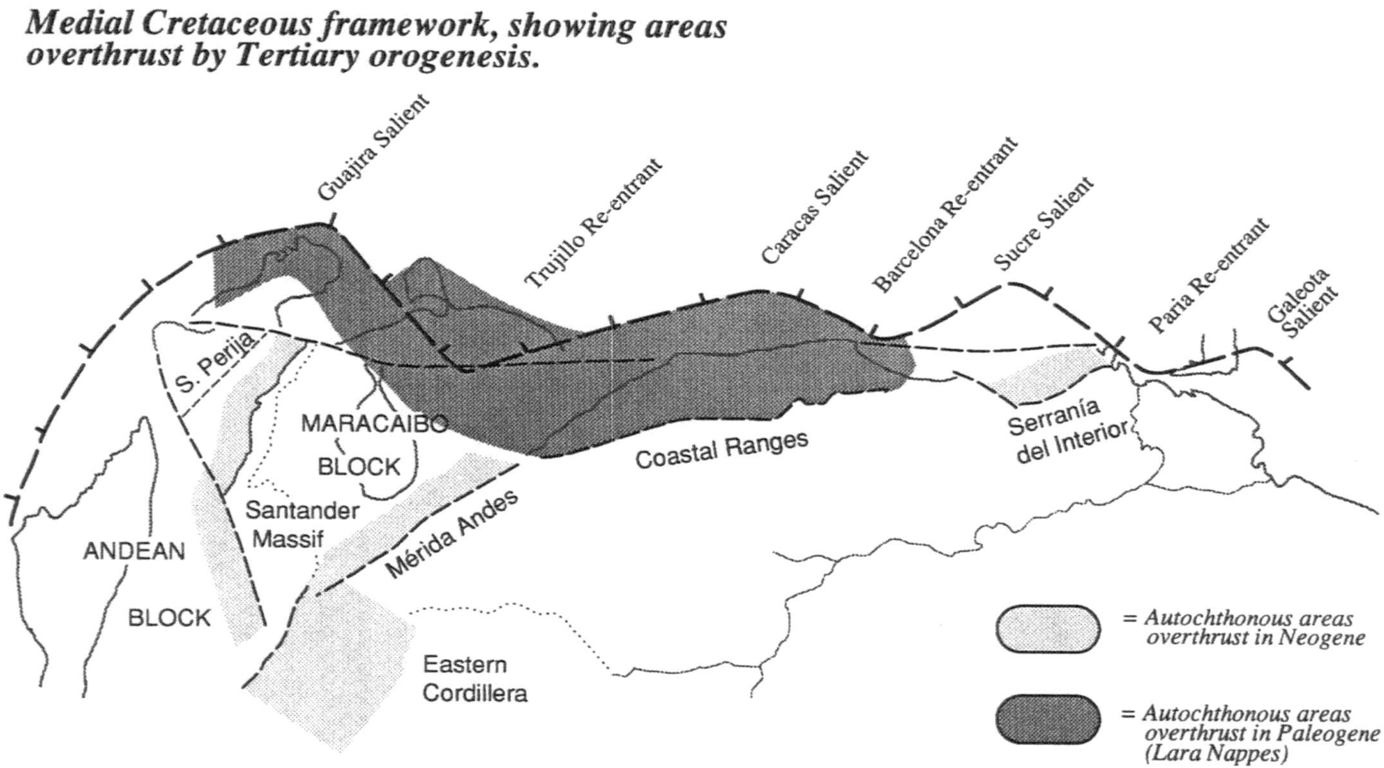

Fig. 5

Pre-Cenozoic reconstruction of northern South America, restoring offsets of Fig. 4 and removing the Caribbean Plate and its allochthons from the passive margin. Note the marginal offsets (salients and re-entrants) in the margin, which formed during plate separation from the Yucatan Block during Jurassic rifting. This basement configuration (after Pindell et al., in press, a) is used in this paper to portray Jurassic and Cretaceous paleogeographic development.

This reconstruction approximately represents Campanian times, during or just prior to the accretion of the Amaime-Chaucha Terrane along Colombia's Central Cordillera, and the continued relative eastward migration of the equivalent Greater Antilles arc and Caribbean crust into the Proto-Caribbean seaway. The marginal offsets, noted above, along the northern margin are shown as well, but it should be kept in mind that the positions of the NNW-trending marginal offsets are better constrained than the actual northward extent of the shelf and the position of the shelf edge along the ENE-trending rifted segments. This basement configuration (Figure 5) provides the regional geometry for discussions of Jurassic and Cretaceous paleogeographic development in the remainder of this paper.

The Yucatán Block and Northern South America: Conjugate Margins During Middle Jurassic Rifting of Western Pangea

Most Permo-Triassic, western Pangean reconstructions and early rift geometries (e.g., Pindell, 1985a) show the eastern margin of the Yucatán Block adjacent to northern South America. Figure 6 shows the pre-rift configuration favored here, using the reconstructed shape for northern South America defined above. Note the good match between the marginal morphologies of the two blocks. Both margins were formed by active rifting in the Middle Jurassic, followed by thermal subsidence into the Cretaceous. Mesozoic passive margin sequences of northeastern Yucatán are now exposed in western Cuba as Jurassic and Cretaceous parautochthonous shelf and deeper water rocks of Sierra Guaniguanico (Pindell, 1985a). There, nonmarine-transitional deposition of San Cayetano sands/shales in the rift environment (Pardo, 1975) gave way to deeper marine sedimentation of the Jagua and other formation by the Late Jurassic. Fairly stable marginal sedimentation in Cuba continued through at least latest Cretaceous.

Parautochthonous Jurassic shallow-marine shelf rocks along northern South America are known from the Cocinas and Cojoro Groups in Guajira Peninsula, Las Brisas Formation of the Caracas Group in the Venezuelan Coast Ranges, and possibly the Uquire Formation and "Paria evaporites" along Araya-Paria Peninsula and the Northern Range of Trinidad (González de Juana et al., 1980; Bray and Eva, 1983). Deeper water Late Jurassic and Cretaceous rocks in Trinidad's Northern Range and Venezuela's Araya-Paria Peninsula most likely represent passive margin slope and rise sediments of the original northern South American margin (Algar and Pindell, 1991b). The Jurassic facies in the Northern Range comprise siliciclastic and calciclastic turbidites and suggest, as is known from central and western Venezuela, the existence in the eastern basin of an adjacent continental shelf to the south. Subsidence models of the Cretaceous section in the Serranía del Interior Oriental (Erikson and Pindell, 1992) are compatible with a Middle Jurassic age for the onset of rifting, allowing sufficient time for the development of an extensive Jurassic shelfal area which is as yet unknown in the Eastern Venezuelan Basin.

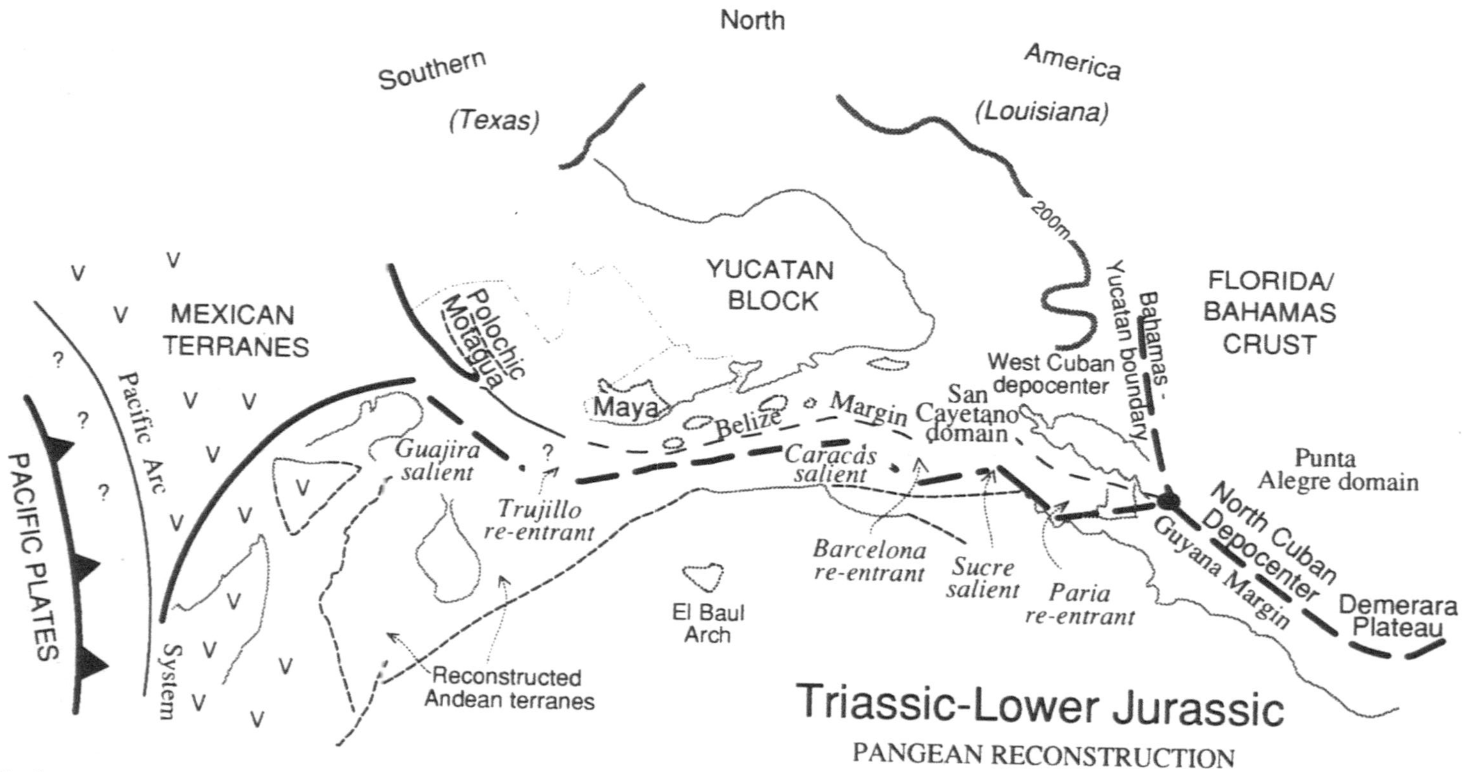

Fig. 6

Reconstruction of northern South America, the Yucatan Block, the Bahamas Block, and unspecified Mexican terranes, modified after and following arguments in Pindell (1985, b) and Rowley and Pindell (1989). Note alignment of probable marginal offsets, and continuity of Jurassic magmatic arc from Mexico to Colombia. Positon of Chortis block is unknown. Note also that two primary depocenters of Cuba, now partially overthrust by the Cuban arc, are located in northeastern Yucatan (terrigenous San Cayetano domain) and the northern Bahamas Block (evaporite/carbonate Punta Alegre domain). This reconstruction defines the pre-rift geometry of basement at the onset of the formation of northern South America's passive margin.

The Late Jurassic rocks of western Cuba are potential petroleum source rocks (Echevarría-Rodríguez et al., 1991). If similar, source-prone shelf sections exist at depth in the northern Eastern Venezuela Basin or beneath the Barranquín Formation in the Serranía del Interior, such unknown equivalent rocks may have contributed to that basin's hydrocarbon potential. In eastern Venezuela and Trinidad, most oils have been chemically tied to Cretaceous marine hemipelagites (e.g., Querecual, San Antonio, Naparima Hill Formations). However, if Jurassic shelf rocks of this margin are sufficiently similar to the Cretaceous shelf sediments, Jurassic contributions to Venezuela's and Trinidad's petroleum reserves may be as yet undetected. Cross-sections of Venezuela's Serranía del Interior Oriental (Talukdar et al., 1988) suggest that a thick, undrilled, undated sedimentary section is folded and thrusted with the overlying Cretaceous-Cenozoic section. Although this unknown section has been been considered Paleozoic in age, known Paleozoic rocks in the region (e.g., Dragón gneiss) were metamorphosed in the Paleozoic prior to rifting, and probably would not be folded with the overlying sediments.

Mesozoic-Cenozoic Relative Motions of the Americas, Yucatán, and Africa

Based on plate rotation parameters (Pindell et al., 1988), Figure 7 shows the former positions of South America relative to North America since the Triassic. Figure 7 shows that after initial breakup, fast separation occurred in the Late Jurassic until about the Jurassic- Cretaceous boundary, when separation slowed and changed direction slightly. By the Middle Neocomian, spreading had accelerated again. At some time between the Late Albian and the Santonian, seafloor spreading between the Americas ceased. For the remainder of the Cretaceous and into the Eocene, the two plates behaved essentially as one. An earlier rather than later time within this Albian-Santonian span is suggested for the termination of spreading by the fact that a significant area of crust with Central Atlantic fabric occurs west of anomaly 34 (Campanian) just east of the Lesser Antilles (Speed and Westbrook, 1984). We favor an Albian age for the termination of seafloor spreading because the loss of the ridge-push force within the Proto-Caribbean at that time may explain the well-known, rapid Albian subsidence of the Proto-Caribbean margins and the associated southward transgression in northern South America (reduction in the South American Plate's in-plane stress).

Early Tertiary relative motion between North and South America was negligible until the Eocene, at which time very slow convergence began in a NNW-SSE direction. This period of relative motion is partially responsible for Cenozoic compressional deformations around the Caribbean, in conjunction with the relative eastward migration of the Caribbean Plate. Thus, it has no direct effect on Cretaceous developments in northern South America. However, because North and South America were several hundred km further apart in Late Cretaceous to Eocene times than they are presently, latest

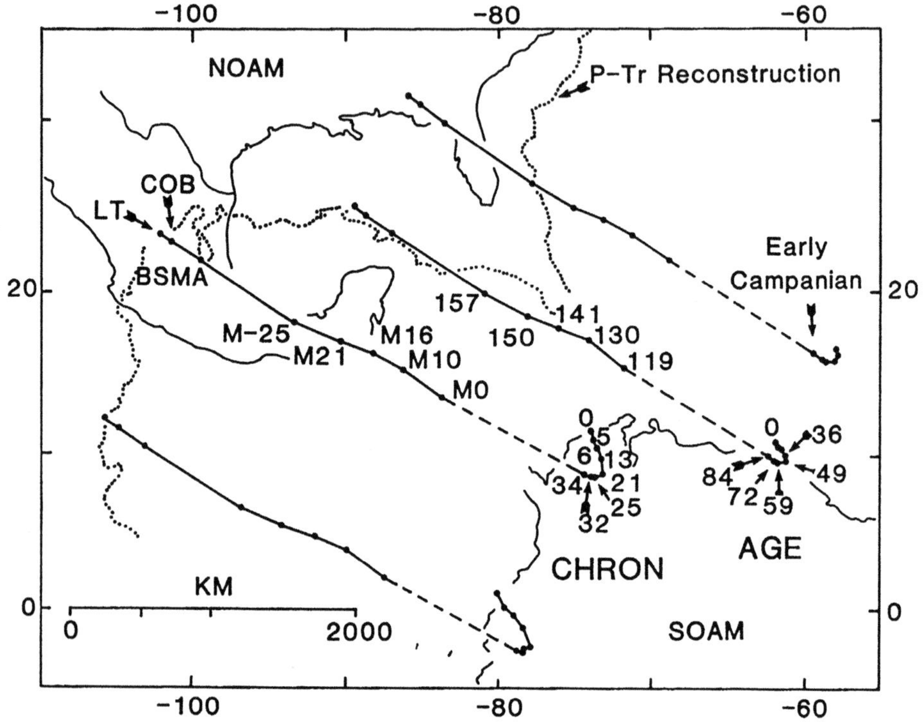

Fig. 7

Relative motion history between North and South American continents since the Permo-Triassic Pangean reconstruction, after Pindell et al. (1988). Lines denote flow lines of points on South America during the development of the intervening oceanic seaway. See text for discussion.

Cretaceous entrance and progressive relative eastward migration of Caribbean lithosphere between the Americas was facilitated.

The pre-rift position of Yucatán between Texas-Louisiana and northern South America (Figure 6) requires a rotational opening (counter-clockwise relative to North America) of the Gulf of Mexico, which is supported by the westward-widening fan-shaped area of oceanic crust in the deep Gulf of Mexico (Buffler and Sawyer, 1985), and also a corresponding rotational opening between Yucatán and South America (counter-clockwise rotation of Yucatán relative to South America) to create the early Proto-Caribbean Seaway (Pindell, 1985a). It is probable, therefore, that geological manifestations of rifting and passive margin development propagated westwards along northern South America, and that rifting along eastern Yucatán propagated southwards (present coordinates), toward a pole of rotation positioned somewhere within or slightly to the west of Colombia. The idea of southward rift propagation along eastern Yucatán is consistent with the age of known stratigraphic sequences there: the nearshore San Cayetano Formation and overlying marine Jagua Formation of western Cuba (original northeastern corner of Yucatán Block) are Middle and Late Jurassic, whereas the coeval sections in southern Yucatán remained fully non-marine (e.g., Todos Santos Formation; López-Ramos, 1975). Although predicted by the model, the E-W diachroneity of rifting and marine incursion in northern South America cannot be proved at present because the Serranía del Interior and Eastern Venezuelan and Trinidadian thrustbelts probably hide a potentially thick Middle to Late Jurassic shelf section, as suggested earlier.

Finally, it should be noted that the NW-trending margin along Surinam and the Guyanas, between Trinidad and the Demerara Rise, is a transform margin, rather than a rifted margin, which was formed by the left-lateral passage of the basement of the Bahamas Platform during Jurassic breakup. As with any long transform margin, imperfect shear could have produced significant tectonism in the area during the Jurassic, including uplift or subsidence of the lithosphere for distances up to several hundred km from the transform. Such tectonic control may have influenced the predicted Jurassic sedimentation in the eastern Venezuela Basin and Trinidad, but in ways that are presently unknown. One possibility is that regional uplift, possibly caused by convergence along the transform, may have interrupted the predicted initial Jurassic sedimentary cycle, after which renewed transgression and sedimentation produced a second cycle comprising the well-known Neocomian Barranquín to Eocene Caratas Formations and their equivalents in Trinidad.

Triassic-Jurassic Rifting and Creation of the Passive Margin

The Triassic-Middle Jurassic period marks the onset of continental rifting and passive margin development in northern South America from Guajira Peninsula eastward, and the establishment of the margin as a tectonic element which can be addressed in terms of Mesozoic and Cenozoic global plate kinematics as derived from the world's oceans. South of the Guajira Peninsula, Triassic-Jurassic lithospheric extension within the magmatic arc system (of the Santa Marta Massif, Sierra Perijá, Central Cordillera, and northern Eastern Cordillera) created intra-arc rift complexes within the Andean basement (based on radiometric ages from arc plutonic rocks; Irving, 1975).

Figures 8, 9, and 10 show the regional Jurassic paleogeography, and are pertinent to the following discussion of the rifting history.

Jurassic redbeds and other sediments are exposed in the Mérida Andes, the Sierra Perijá, the Central and Eastern Cordillera, the Santa Marta Massif, and the Guajira Peninsula, and are usually several kilometers thick (Maze, 1984; Irving, 1975). When Neogene displacements are accounted for (e.g., Figure 5), it is seen that all of the above occurrences represent only four main intracontinental rift features, the Táchira (or Uribante) Trough

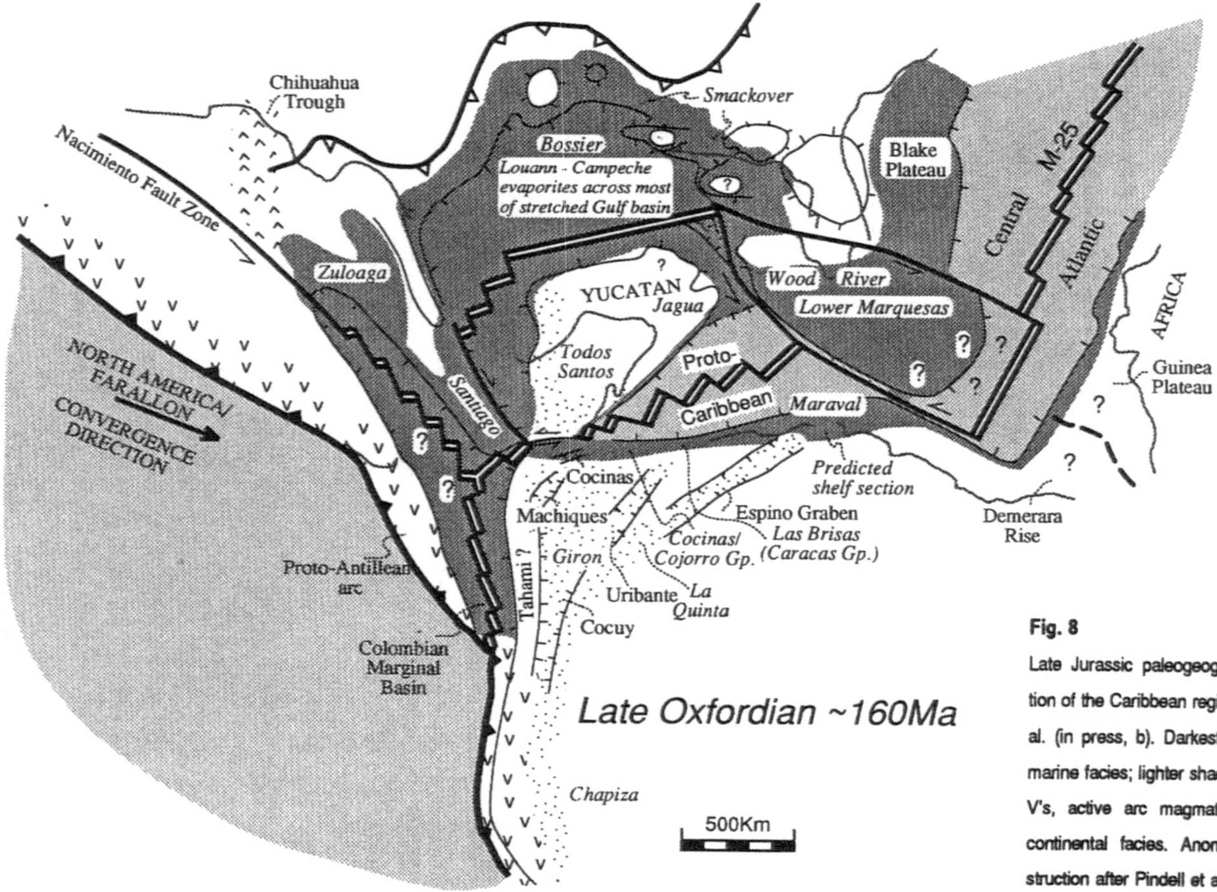

Fig. 8

Late Jurassic paleogeographic reconstruction of the Caribbean region, after Pindell et al. (in press, b). Darkest shading, shallow marine facies; lighter shading, deep marine; V's, active arc magmatism; dots, sandy continental facies. Anomaly M-25 reconstruction after Pindell et al. (1988). See text for discussion.

of the Mérida Andes, the Bogotá Basin of the Eastern Cordillera, the Machiques Trough of Sierra Perijá, and the Cocinas Trough of the Guajira Peninsula (Figures 8, 9). In each of these areas, rapid Early Cretaceous sedimentation relative to surrounding areas (Zambrano et al., 1972) is interpreted here as being due to thermal subsidence, and indicate significant lithospheric extension during the Jurassic rifting.

In the southern Maturín subsurface of eastern Venezuela, the Espino Graben (Figures 1, 8, 9) and smaller associated rifts have several kilometers of Jurassic redbeds, but a Early Cretaceous section is not apparent (Feo-Codecido et al., 1984; Daal et al., 1989). The Late Cretaceous section, which is present, probably pertains to deposition during the Cretaceous eustatic sea level high. The absence of a Early Cretaceous section indicates minimal thermal subsidence after Jurassic redbed deposition, suggesting that the graben effectively formed as the surface expression of a thin-skinned detachment, and not as a full scale lithospheric rift. The dip of the detachment was probably NNW, i.e., the basement of the El Baúl Arch area shifted NNW during Jurassic rifting. Basalts migrated into the Jurassic Espino section at around 162 Ma, but these could have been far-traveled.

As for the Jurassic rifted margin sequence along the actual continental margin of northern South America (i.e., true continental slope province), little can be ascertained because of the obscuring effect of Cenozoic overthrusting by the Caribbean allochthonous terranes. However, there must be a thick prism of rift deposits belonging to the true rifted margin to the Proto-Caribbean Seaway. This rifted margin probably had several re-entrants in it, one at northeastern Lake Maracaibo (Trujillo Trough), another at the Serranía del Interior Oriental (Gulf of Barcelona), and a third at Trinidad (Gulf of Paria) (Figures 5, 8, 9). These three sites are where the Tertiary thrust belt is truncated or offset by tear fault systems presumably related to deep basement structure. Because rifting from Yucatán occurred in a NNW-SSE separation direction, ENE segments of the margin are true rifts, whereas NNW segments are transfer zones dominated by transcurrent faulting. Therefore, the east side of each marginal offset is predicted to have greater basement depth and total syn- and post-rift stratigraphic section, and therefore be less resistant to Tertiary overthrusting. Depending on clastic influx rates, these marginal re-entrants may have influenced source rock (restricted?) or reservoir rock (deltaic sections in SW corners of re-entrants) sedimentation early on in the margin's development. Reservoir rocks in this setting would be unimportant presently, as such rocks would be metamorphosed by now. However, if source rocks were deposited at that setting, they may have contributed to Venezuela's hydrocarbon accumulations by updip (southward) migration from areas matured during either initial overthrusting (Tertiary) or during shelf edge sediment accumulation. We note that in the eastern re-entrant (Gulf of Paria), evaporite deposition is suggested by the presence of anhydrite in drill cores and occasional mines along the southern flank of the Northern Range. However, spores from tectonized evaporites in the Couva Marine-1 well in the Gulf of Paria (Bray and Eva,

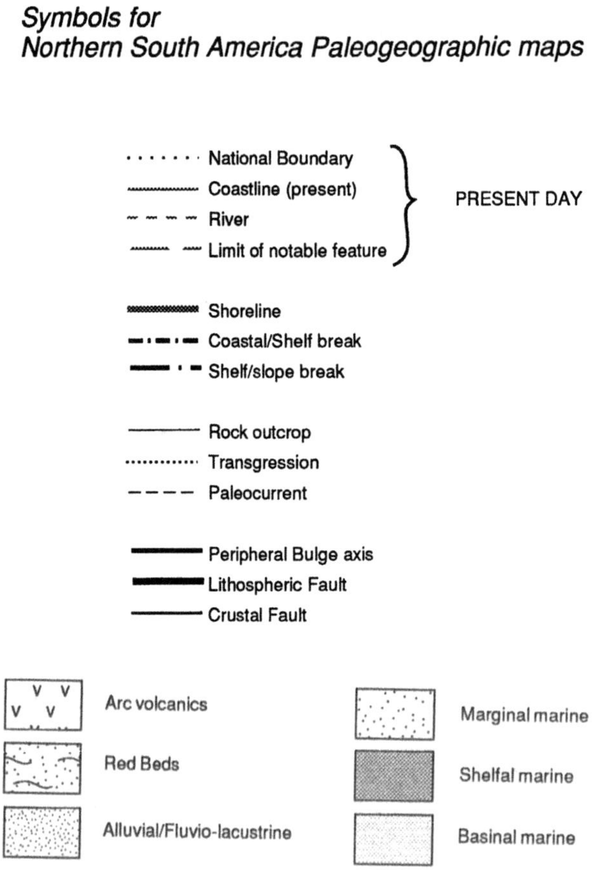

Figs. 9-15

Sequential paleogeographic maps of northern South America for Middle Jurassic to Paleocene times. See key for significance of symbology and patterns. All after Pindell et al. (in press, a). Southward limit of allochthonous bodies is shown in Figure 13 (Medial Cretaceous) to show approximately the full northward extent of Medial Cretaceous source rocks which existed prior to Tertiary thrusting and emplacement of the allochthons. ➤

1983) suggested a Berriasian rather than Jurassic age, but the spores did not necessarily originate from the evaporitic portion of the tectonized core (K. Burke, personal communication, 1984).

As mentioned earlier, the rotational opening of the Proto-Caribbean seaway (Figure 8) predicts a westward younging of passive margin development in northern South America. In the Guajira Peninsula of Colombia, redbed deposition was followed and accompanied by the mainly fluvial to deltaic deposits of the Cojoro Group, which had limited marine influence (e.g., Rancho Grande Formation; Early? or Middle? Jurassic). Presumably, the seas were transgressing southwards onto the margin, but chronologic control on early marine sedimentation from western and eastern (Eastern Venezuelan Basin) portions of this margin is insufficient to document a westward younging in the onset of first transgression. Nevertheless, shallow marine conditions were clearly present along the northern margin by the Late Jurassic (Figure 10), as evidenced by the Cocinas Group

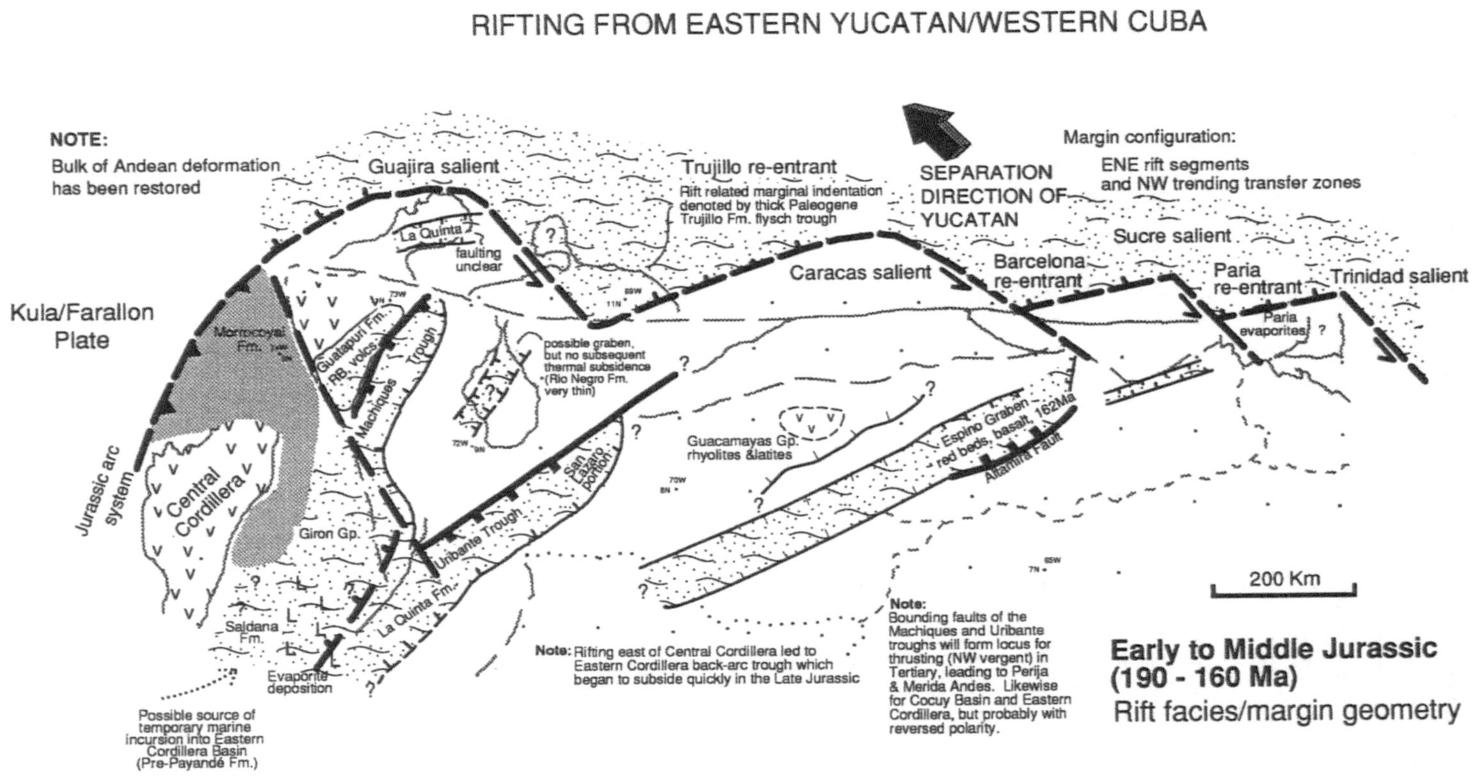

Fig. 9

See explanation on page 22.

sediments (MacDonald, 1965), and the Las Brisas Formation's shallow-marine faunal forms in the metamorphosed Caracas Group of the Venezuelan Coast Ranges (González de Juana et al., 1980). For unknown tectonic? reasons, the Cocinas Group was partially eroded prior to subsequent deposition of the Early Cretaceous transgressive Río Negro sandstones. Interactions with Yucatán, at the apex of the Proto-Caribbean seaway prior to plate separation (Figure 8), might have produced an uplift and an erosional unconformity.

In the "Llanos" basins, magnetic surveys (Corpoven, 1989) show that the Espino Graben continues WSW to at least the Colombian border (Apure-Mantecal Graben), and that possibly another depocenter occurs parallel and to the north of this one. In non-graben areas of the plains region, the Middle and Late Jurassic is mainly absent or very thin. As for the subsurface of the Colombian Llanos, rocks of this age are little known and one must bear in mind that the Eastern Cordillera has migrated eastwards by at least 150 km from its initial rifted depocenter, such that rift deposits might not be expected as far to the east as the present Llanos basin. There are, however, normal-fault bounded occurrences of Ordovician sediments beneath the Cretaceous in the southern Llanos (e.g., Paporis, Thampa, and Caguan grabens).

In the northern Andes of Colombia, Triassic and Early Jurassic basin formation, probably intra-arc rifting, is suggested by marine and non-marine sediments and by extensive mafic, intermediate, and silicic volcanism along the Magdalena Valley. Examples of sediments include the Los Indios, Corual, Guatapuri, Boca, Montebel, Payandé, Morrocoyal, and Arcabuco Formations (Irving, 1975) (Figure 9). In Sierra Perijá, the La Ge, Tinacoa and Macoíta redbeds and volcanics (Triassic-Early Jurassic) are deformed and unconformably underlie additional red beds of La Quinta Formation (variably Jurassic). The Triassic and Early Jurassic occurrences of marine sediments demonstrate an early marine incursion into central Colombia above the deformed and eroded Late Paleozoic orogen. Tectonic extension is suggested as the cause for subsidence because Triassic eustatic sea level was low and marine transgression at this time suggests a tectonic subsidence mechanism. Extension appears related to intra-arc rifting because of the coeval and adjacent Jurassic arc plutonism of the following ages (McCourt et al., 1984; Irving, 1975): in the Central Cordillera (174-142 Ma), the Santander Massif (198-160 Ma), and the Santa Marta Massif (202 to 160 Ma); a 195 Ma age on a pegmatite associated with a granite in Guajira might relate either to rifting or to the Andean arc. This plutonism began in the Early Jurassic and continued into the Medial and possibly the Late Jurassic. Possibly due to very rapid sedimentation rates, deposition for most of the remainder of the Jurassic in the northern Andes basins of Colombia was continental (e.g., Girón Formation and equivalents).

The reconstruction of Figure 5 (and shown in Figures 8, 9) produces a basement configuration in which the Magdalena and César Basins (i.e., Bogotá and Machiques Troughs) are flanked and probably underlain by Early to Middle Jurassic plutons. The radiometric ages (see above) suggest that magmatism in the eastern belt (Sierra Perijá, Santander Massif, northern Eastern Cordillera) terminated earlier than in the western belt (Central Cordillera, Santa Marta Massif, Figure 1). The intervening basinal area (Magdalena/César, and especially the basal Eastern Cordilleran section, or Bogotá Basin) possesses thick Girón Group redbed deposits of Jurassic age, in addition to the Triassic-Early Jurassic marine and non-marine sections outlined above. We suggest that this configuration represents syn-plutonic intra-arc extension during subduction from the west (Figure 8). A single Late Triassic-Early Jurassic arc may have undergone extension such that the western half migrated relatively westward by a considerable distance, thereby producing the Bogotá and Machiques Basins. The Machiques extension was accompanied in the Maracaibo region by coeval extension at the Táchira rift in the same cross-section, such that the combined extension on these might approximate that on the Bogotá rift(s). The western arc remained active until the end of the Jurassic, perhaps because it was able to maintain its position over the original Benioff Zone longer, which was probably undergoing rollback as the cause for intra-arc extension (Figure 8). Termination of plutonism in the eastern half (~160 Ma, Oxfordian; Irving, 1975) may record the approximate timing of the extensional event, which is supported by age estimates on the Girón redbeds (Irving, 1975).

In the above, rift-produced, explanation for the double Jurassic arc, heat flow in the Bogotá/Machiques/Táchira rifts would have been quite high in the Jurassic, providing a logical setting for marine inundation in the Late Jurassic and Cretaceous as thermal subsidence set in. For example, the Tithonian in the Bogotá Basin is represented by the Sáname Formation of the lower Cáqueza Group (Figure 10), a dark, pyritic marine shale with interbedded quartz sandstones above a transgressive basal conglomerate; the Sáname Formation is at least 300 and 800 meters thick near Quetame and Río Batá, respectively (Campbell and Burgl, 1965). This marine basin expanded in all directions during the Early Cretaceous and persisted until near the end of the Cretaceous as an epicontinental seaway without any subaerial exposure or unconformity. As no arc volcanism is known in the region until at least the latter Late Cretaceous, and probably the Miocene, it appears as though that portion of Colombia from the eastern flank of Central Cordillera toward the Llanos was part of the northern South American Cretaceous passive margin province. More westerly parts of Colombia, such as the Amaime Terrane of Aspden and McCourt (1986), the oceanic Calima and Cuna Terranes of Toussaint and Restrepo (this volume), and possibly the continental Tahamí terrane, were allochthonous relative to the eastern interior platform until at least the latter Cretaceous. Debate continues on the whereabouts of the Tahamí Terrane during Cretaceous time (see Toussaint and Restrepo, this volume). In this paper, the Tahamí Terrane is simplistically considered as the western flank of Central Cordillera, which represents the passive continental margin of the Colombian marginal seaway after termination of arc magmatism in the eastern part of the Central Cordillera (Figure 8). This interpretation places the Tahamí Terrane, which has a non-volcanogenic Aptian-Albian autochthonous sedimentary cover (Abejorral-La Soledad-San Luis-Berlín), well away from Cretaceous arc magmatism, but places it in a position where its epicontinental sedimentation should have persisted until the Santonian rather than the Albian. We suggest that Campanian deformation and subsequent erosion may have removed or obscured sediments of Cenomanian-Santonian age, which, if they were Villeta-like, would have predictably formed only a thin, easily metamorphosed and erodable portion of the sedimentary cover in this setting.

Cretaceous Development of Northern South America

In the following discussion, we develop separately the Cretaceous paleogeography of the western and eastern portion of the passive margin, and the allochthonous Caribbean terranes. However, regional Cretaceous to Paleocene sedimentation patterns are shown for the entire margin in Figures 11-16. We begin this discussion with a brief assessment of the age of basement, timing of arc magmatism, and the history of metamorphism in the allochthonous terranes which have been accreted to the passive margin.

The Allochthonous Belt: Intra-Oceanic Arcs of the Caribbean Domain

After Late Jurassic termination of plutonism and volcanism in Colombia, it is often assumed that the Central Cordillera became an arc again at about Aptian time (Aspden and McCourt, 1986; McCourt et al., 1984), and remained active for the remainder of the Cretaceous and into the Tertiary. It is also assumed that a similar accretion and magmatism took place in Venezuela at that time due to the age of metamorphic rocks of the Villa de Cura klippe (Beets et al., 1984). However, examination of the occurrence, relations with other rock units, and methods of dating of the various arc related complexes, as well as facies patterns in the Magdalena Basin, platformal cover of the Tahamí Terrane, and the central Venezuelan shelf, shows that post-Jurassic arc magmatism occurred in the allochthonous complexes of Colombia and Venezuela, far from the autochthon. South American continental crust did not become an arc again until the latest Cretaceous in the northern Central Cordillera (e.g., Antioquia Batholith area), and the Tertiary farther south. In particular, we question Aspden and McCourt's (1986) and Bourgois et al.'s (1987) claim that the ~Aptian radiometric ages on the blueschists of the eastern Cauca Valley date the collision between the Amaime Terrane and Central Cordillera, and consider instead that the Aptian blueschists were formed elsewhere and then emplaced against the Central Cordillera much later, probably in the Campanian. Such a post-metamorphic emplacement of the Villa de Cura Klippe is very clear in

Venezuela, except that the emplacement there was of Late Eocene to Oligocene age. This has significant implications for the Cretaceous paleogeography of northwestern South America, namely, that (1) the Cretaceous passive margin conditions of autochthonous South America extended all the way to Ecuador, (2) the stratigraphic sections of the Magdalena, Bogotá, Llanos, Machiques/César, Táchira and Maracaibo Basins were not affected by subduction-related tectonic convergent and vertical motions of the lithosphere for the bulk of the Cretaceous, and (3) there may have been long intervals such as the Cenomanian to Campanian during which there was no western clastic provenance area for the Magdalena Basin (Central Cordillera, including? Tahamí Terrane, was not subaerial). Furthermore, it points to an important regional orogenic event which must have taken place offshore in active, intra-oceanic magmatic arcs.

The better constrained ages on primary metamorphism from the allochthonous belt of northern South America point overwhelmingly to the Aptian-Albian (Snoke, 1991). This is true of rocks from Tobago (Snoke et al., 1990); Margarita/western Araya (Avé-Lallemant, 1990; Chevalier et al., 1988); the Villa de Cura Klippe (Beets et al., 1984); the Concha Schist of Santa Marta Massif (MacDonald et al., 1971) and amphibolites on Los Monjes, northeast of Guajira (Santamaría and Schubert, 1974) (both of the Ruma Metamorphic Zone), and along the eastern Cauca Valley in Colombia near Pasto (Los Azules Complex), east of Cali (Amaime Formation/Ginebra Massif), and SW of Medellín (Cauca ophiolitic complex) (Toussaint and Restrepo, this volume). In addition, Mattson (1984) argued that similar Aptian-Albian metamorphism characterizes the basement rocks of the Greater Antilles intra-oceanic arc in the northern Caribbean. These complexes collectively define an intra-oceanic arc system(s) in which Aptian-Albian orogenesis produced variable grades of metamorphism, including greenschist, amphibolite, and blueschist. In addition, the same areas define the roots of the bulk of Late Cretaceous-Early Tertiary magmatism throughout the region.

Figure 16 shows a hypothetical Albian plate reconstruction in which northern South America and the Yucatán Block are passive margin elements, and an offshore intra-oceanic arc floored by Jurassic and/or Early Cretaceous crust connects the Cretaceous ensialic arcs of Mexico/Chortis and Peru/southern Ecuador. All magmatism is restricted to the arcs as shown, forming a continuous belt from North to South America, with the Caribbean lithosphere west of the arc. We suggest that the Aptian-Albian orogenesis in the allochthonous terranes is directly correlative to the Sierra Madre (Sevier) and Peruvian orogenic pulses, and was caused by the onset of rapid migration of the Americas across the mantle, forcing the arcs to become compressional (in the sense of Dewey, 1980), during an acceleration of seafloor spreading in the Central Atlantic and the onset of spreading in the Equatorial Atlantic (Pindell et al., 1988). The backarc thrusting in Peru and Mexico was manifested in the intra-oceanic arc as a flip in the polarity of subduction (Figure 17a). This flip appears tied to the metamorphism in the terranes,

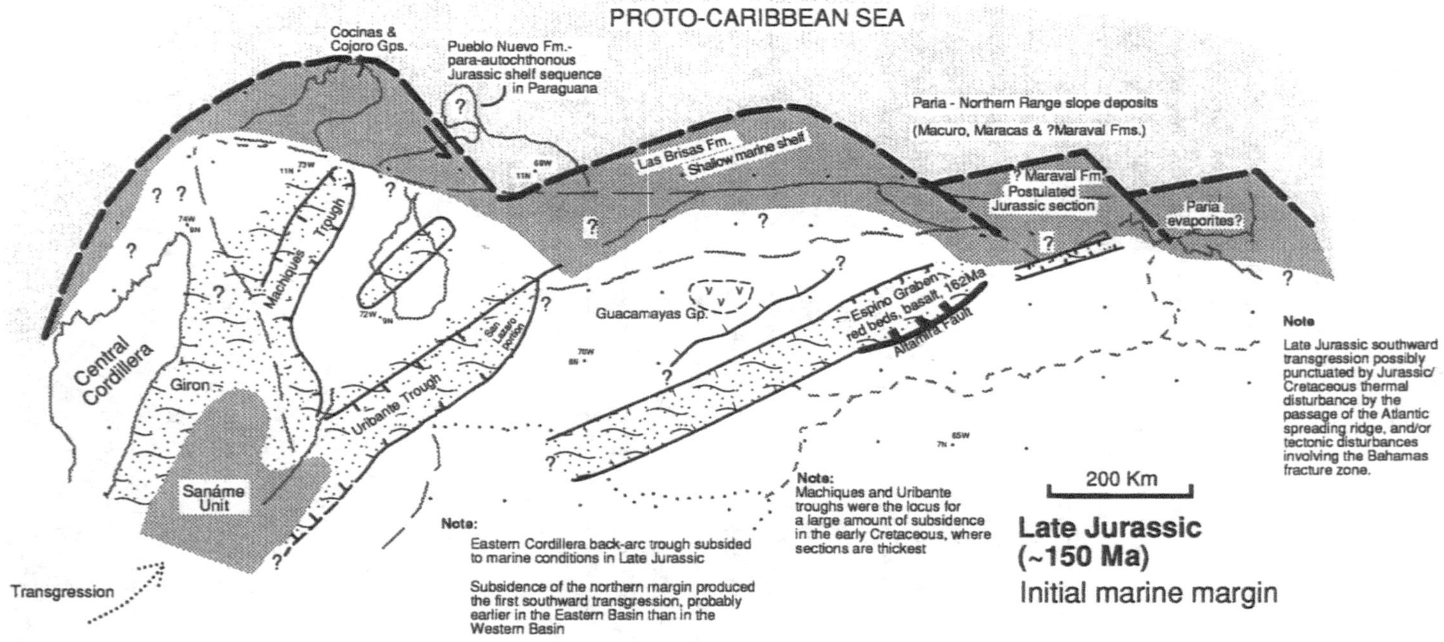

Fig. 10

See explanation on page 22.

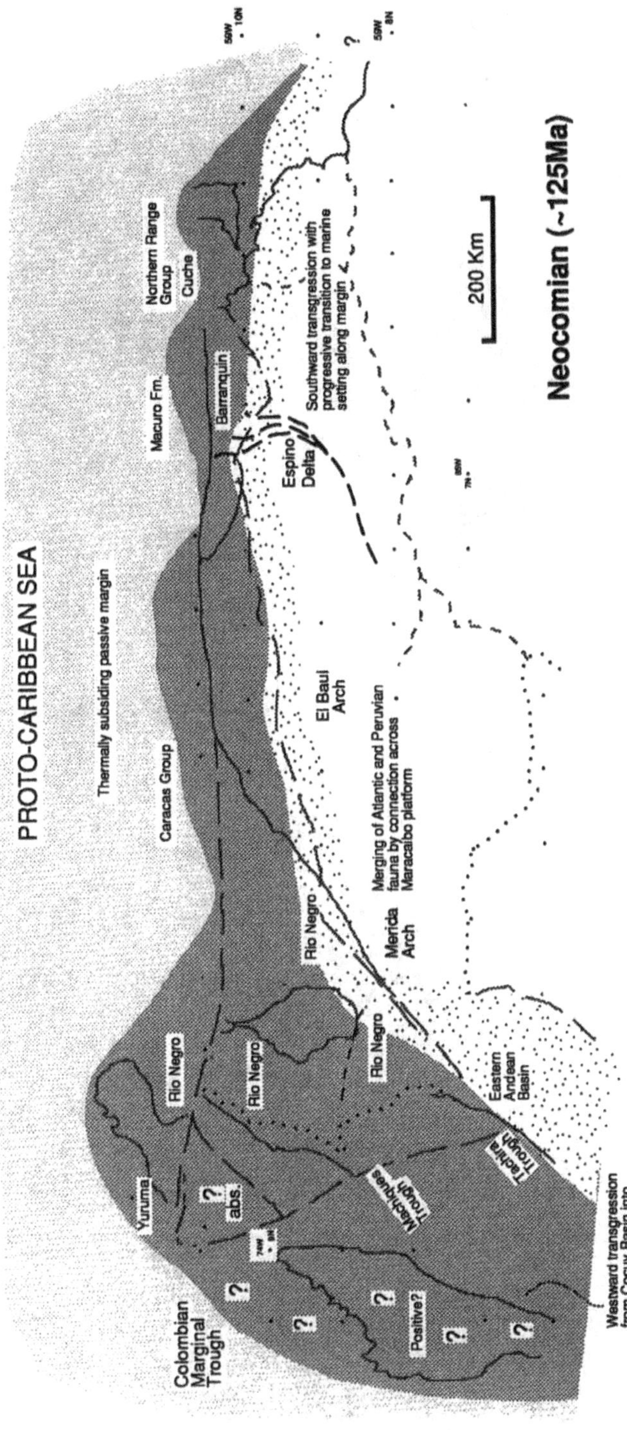

Fig. 11
See explanation on page 22.

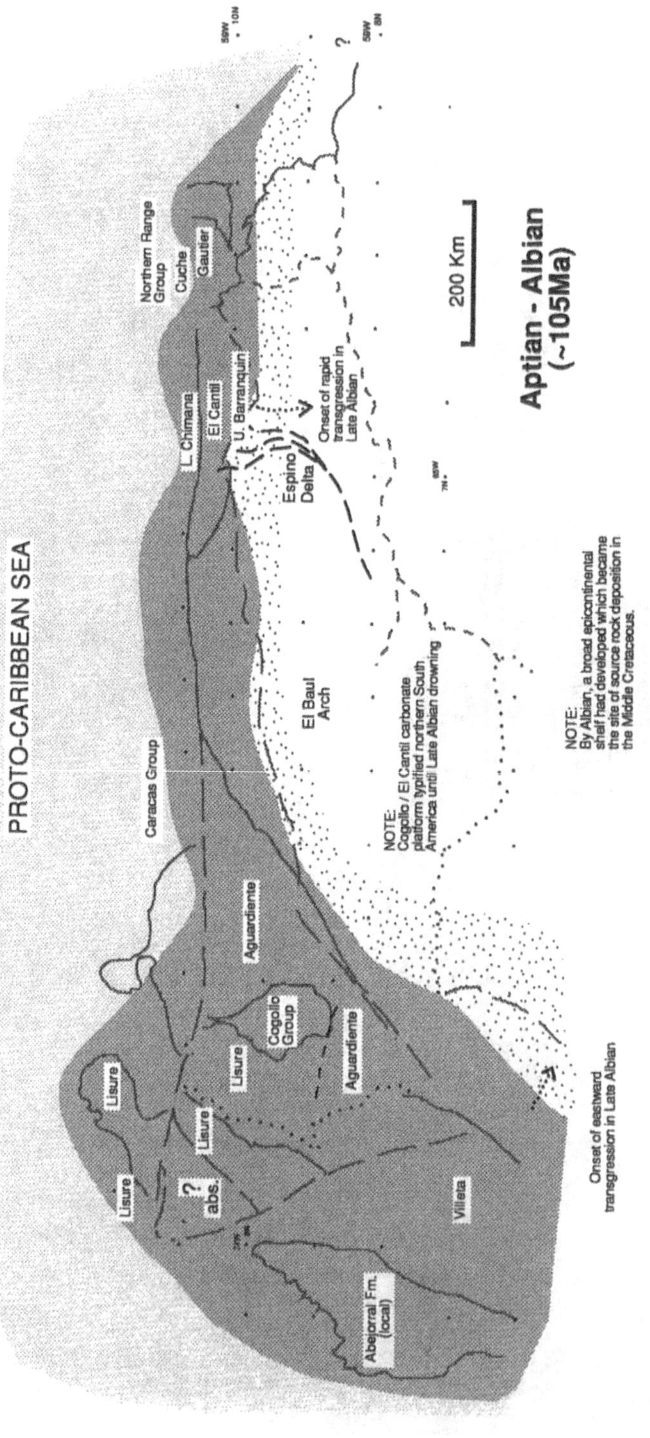

Fig. 12
See explanation on page 22.

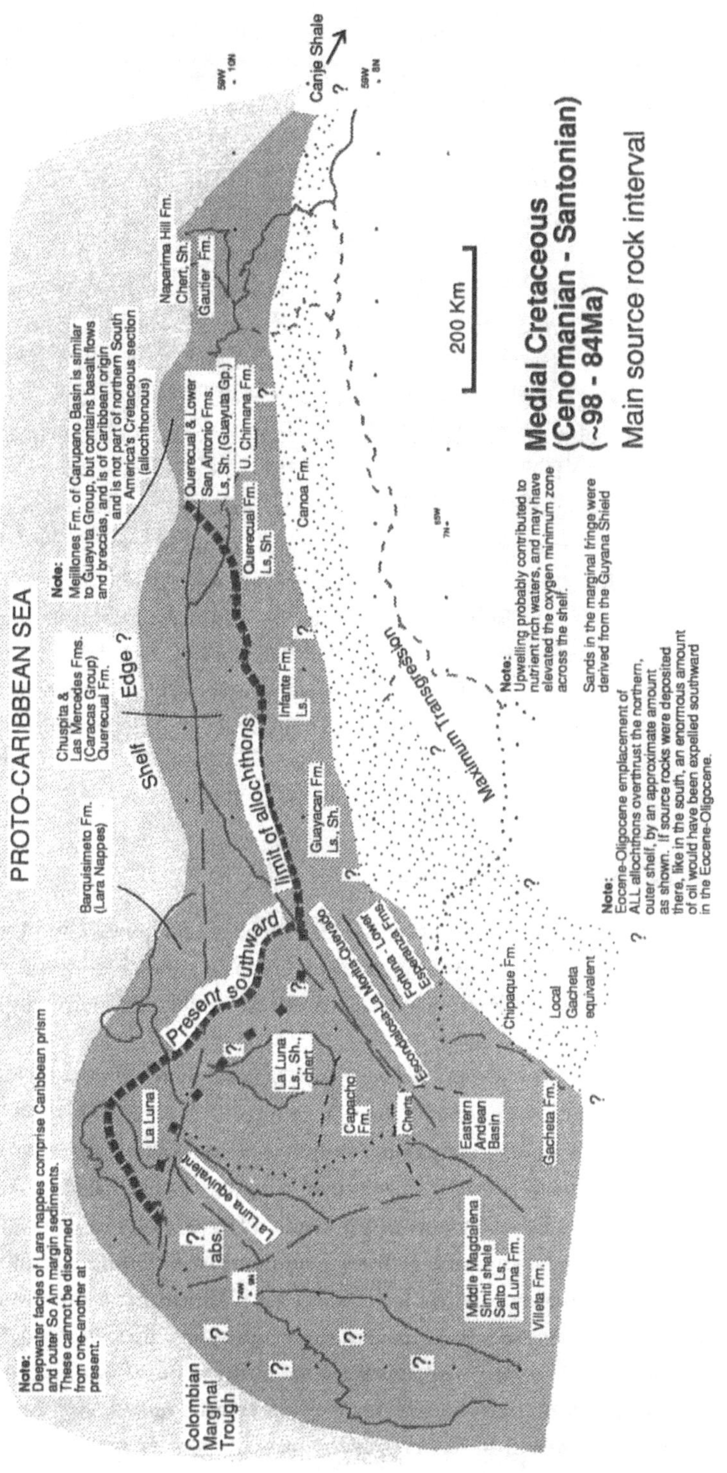

Fig. 13

See explanation on page 22.

either by rapid uplift and erosion to deeper levels, or by the attempted subduction of previously backarc elements in the new, west-dipping Benioff Zone (Pindell and Barrett, 1990; Snoke, 1991; Mattson, 1984). From the Albian on, the Caribbean Plate was able to migrate ENE relative to the Americas, with the metamorphosed Early Cretaceous arc serving as the roots of the Late Cretaceous to Eocene arc. The northern end of the arc, and not the Chortis Block, collided with southern Yucatán in the Late Campanian (Pindell and Dewey, 1982; Rosenfeld, 1990). To the south, the accretion of the arc against the Colombian-Ecuadorian margin probably progressed northwards, culminating in Colombia during the Campanian. In regard to Colombia's Cretaceous evolution (Toussaint and Restrepo, this volume), the Aptian-Albian polarity reversal and Campanian arc accretion correspond to the Early and Late Cretaceous "tectono-metamorphic events". In the interpretation here, the accretion of the Calima and Tahamí Terranes occurred in the Campanian rather than the Albian: this is why we predict that Cenomanian to Santonian sequences should have been deposited on the Tahamí Terrane but have been obscured by the later deformation. In regard to Venezuela's and Trinidad's evolution, the composite arc terrane of Figure 16 in the eastern Pacific is the origin of all the allochthonous terranes obducted onto the autochthonous margin in the Cenozoic. During the migration of the arc, a developing accretionary prism probably comprised sediments and basaltic blocks of the Proto-Caribbean Seaway. Such rocks may be represented now by the Sans Souci Group of Trinidad (Algar and Pindell, in review), and by various sections of what has been regarded as thrusted "Querecual Formation" (deep marine shales and pelagic carbonates) in the Lara Nappes (González de Juana et al., 1980).

The Cretaceous Western Passive Margin Platform

In this section, we avoid detailed discussion of stratigraphy, and concentrate instead on tectonic controls of sedimentation in the western platform. Figure 18 is a representative composite stratigraphic section for the western margin of Venezuela, with emphasis on the Maracaibo Basin.

As mentioned earlier, the areas where strong Jurassic lithospheric extension had taken place (Bogotá Basin, Táchira and Machiques Trough, and theoretically the northern rifted margin) were the sites of most rapid marine sedimentation, driven by thermal subsidence, in the Late Jurassic to Early Cretaceous (Figures 10, 11). Earliest Cretaceous sedimentation in the Bogotá Basin possessed a Peruvian fauna, derived from the south. Northward onlap out of the basin and into the Táchira and Machiques Troughs merged with southward marine advance across the Venezuelan Platform from the Proto-Caribbean seaway during the Hauterivian-Aptian interval, bringing European and Peruvian faunas to Colombia and Venezuela, respectively, for the first time. The platformal areas received a blanket of fluvio-deltaic transgressive quartz sandstones and conglomerates (Río Negro Formation), while the deeper basinal areas in the Bogotá Basin received black marine shales and quartz sandstones of the Cáqueza Group. Sedimentary slumping and turbidity current deposition is common in the Berriasian section of the Bogotá Basin (Campbell and Burgl, 1965), possibly suggesting syn-depositional tectonism or, perhaps, differential subsidence of rift blocks. However, Dengo and Covey (in press, 1991) suggest that Early Cretaceous sequences change thickness across basement faults indicate an Early Cretaceous tectonic extension. Tithonian through Valanginian deposition seems to onlap progressively more distal areas, as would be expected during thermal subsidence after rifting.

In the Hauterivian, the marine depocenter was greatly enlarged. The Middle Magdalena Basin was inundated by this time (Tambor sandstone). Deposition of sandstone and black shale continued in the Bogotá Basin. The accelerated transgression could have been caused by a eustatic rise, or perhaps by another, or continued, extensional faulting.

In the Aptian-Albian (Figure 12), a marine shelf developed after the southward Neocomian marine transgression. The formations of the shallow-water Cogollo Group record thermal subsidence of the margin, with carbonate and clastic facies being controlled by relative sea level changes and/or lateral migrations of channels across the shelf, as in the equivalent El Cantil carbonate to the east (see below). The Cogollo becomes sandier to the southeast toward the craton, where the Barinas Basin was invaded by marine waters for the first time in the Aptian-Albian (Zambrano et al., 1972).

By the Albian, platformal sedimentation continued in the various members of the Villeta Group in the Bogotá and Magdalena Basins, with marine shales and limestones and some sandstones (Fabre, 1985). Southward and eastward transgression occurred in the Early? Albian, leading in the Upper Magdalena to deposition of the paralic to transitional Caballos Formation followed by the lower Villeta Formation (Kroonenberg and Diedrix, 1984), and in the Llanos Basin to deposition of transitional marine to deltaic sandstones (McCollough, 1990). In general, the Aptian to Albian of northwestern South America comprised a broad marine passive margin shelf with shales, carbonates and local sandstones, with a sandy fringe influencing deposition toward the craton, extending from the Demerara Plateau to northern Ecuador. We note that Aptian-Albian isopachs in western Venezuela continue to show more rapid accumulation in the Táchira and Machiques Troughs, the sites of the greatest Jurassic lithospheric extension and subsequent thermal subsidence (Zambrano et al., 1972).

Northern South America's best and repeatedly described source rock interval of medial Cretaceous age (Figure 13) was deposited from Ecuador to the Guyanas (including Napo, Simití, Villeta, La Luna, Capacho, Fortuna, Querecual, San Antonio, Naparima Hill, Canje, and equivalent Formations). To the southeast, toward the shield, equivalents of these units become sandier, and source potential becomes greatly reduced. Intra-formational unconformities are rare, indicating stable relative sea level behavior.

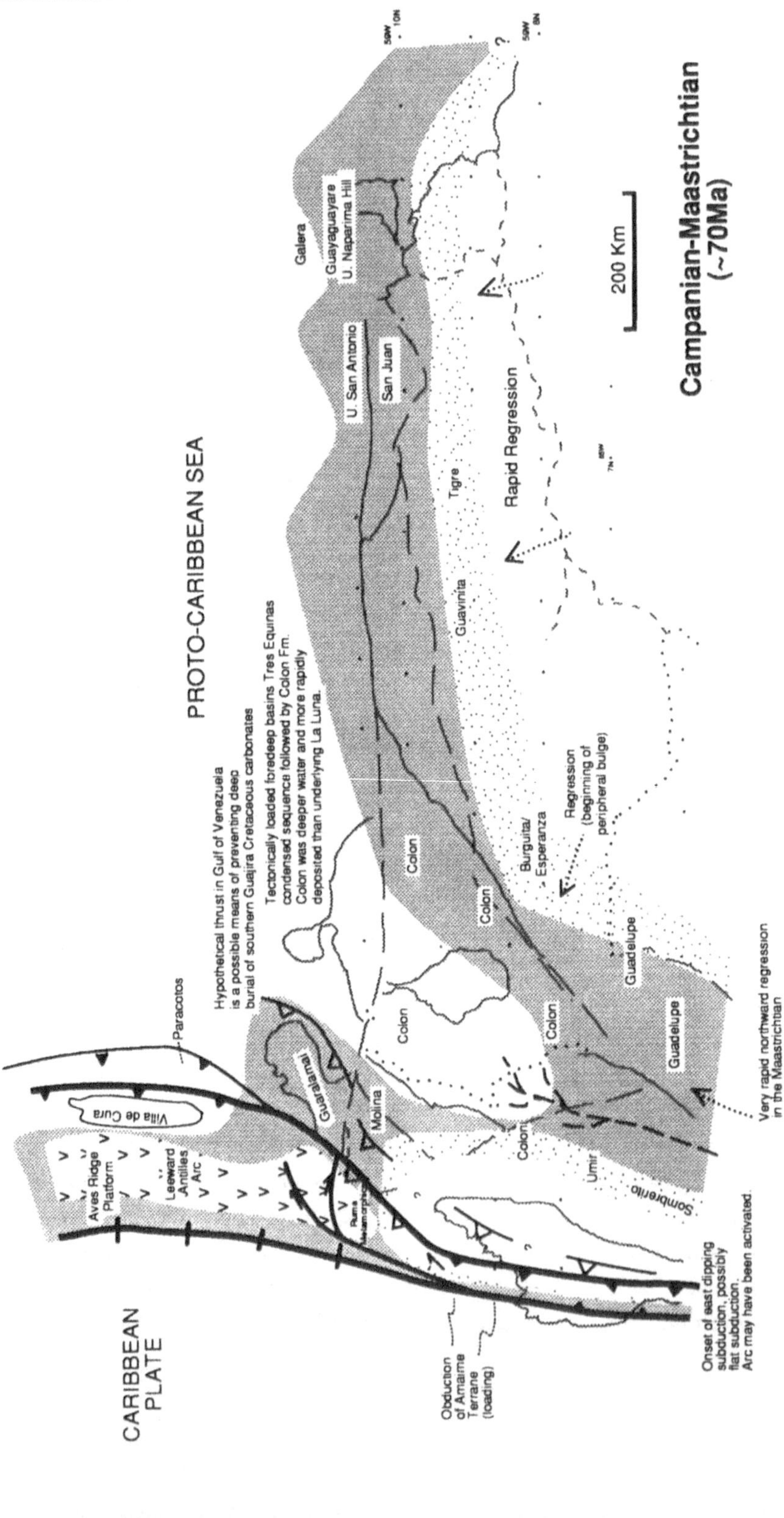

Fig. 14

See explanation on page 22.

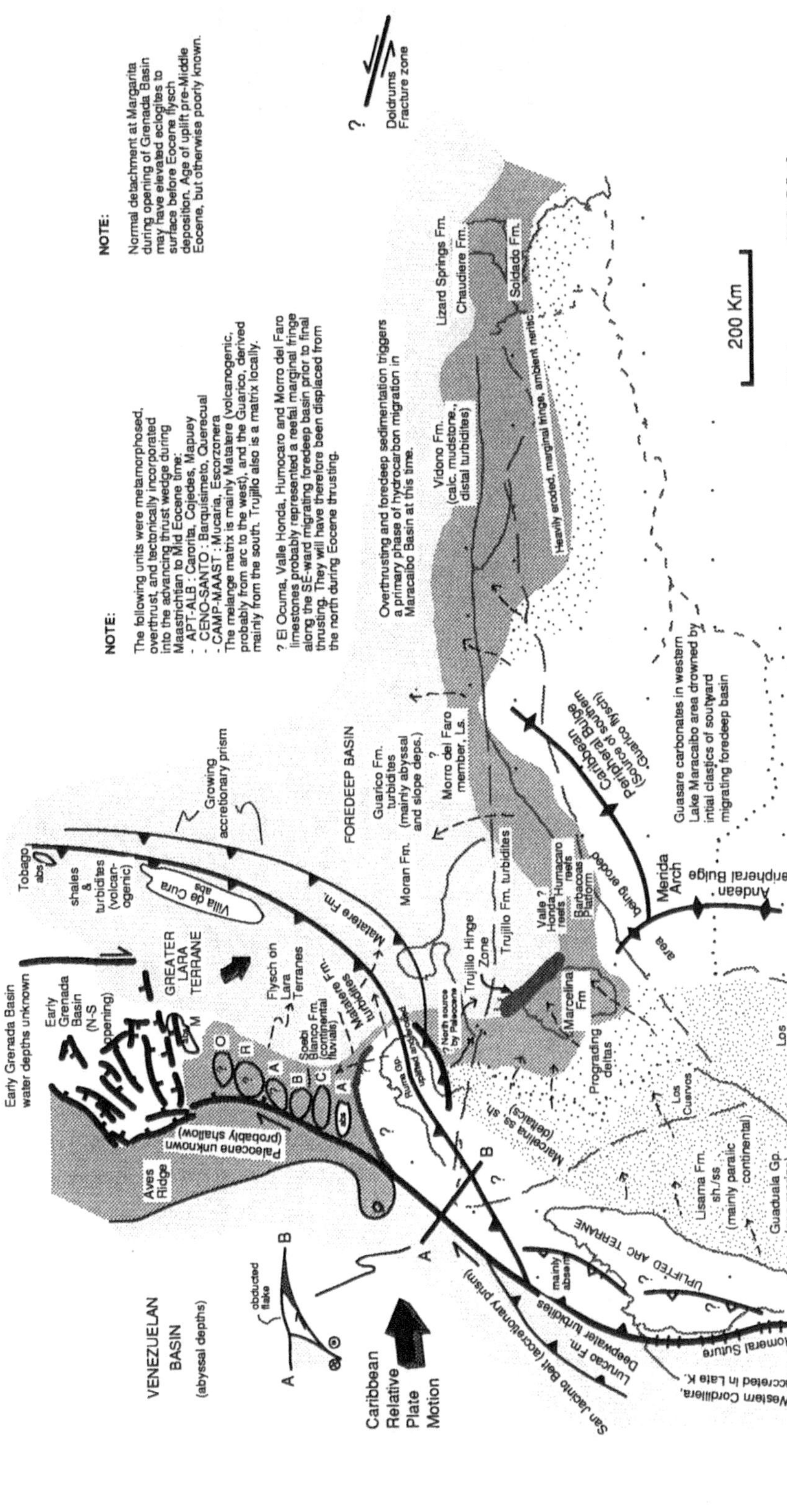

Late Paleocene (59 Ma)

Fig. 15

See explanation on page 22.

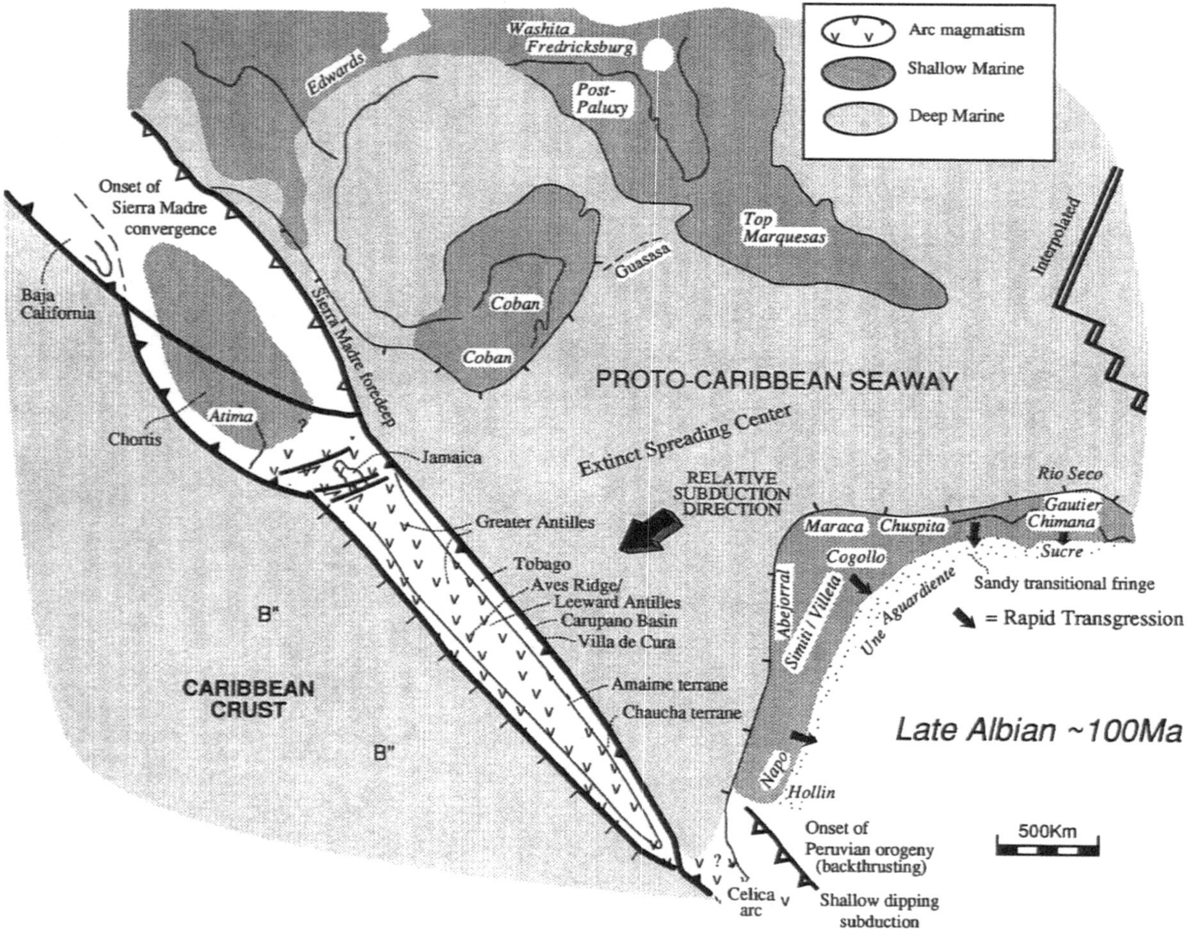

Fig. 16

Albian paleogeographic reconstruction of the Caribbean region, after Pindell et al. (in press, b). Seafloor spreading had ceased in the Proto-Caribbean Seaway, and the Caribbean-Amaime intraoceanic magmatic arc stretched from Mexico/Chortis to Ecuador/Peru. Northern South America was an entirely passive margin during Medial Cretaceous source rock deposition, with no tectonic barrier to the north.

Along the Proto-Caribbean Seaway, the margin was probably the site of upwelling which could have influenced marine sedimentation southward onto the platform (Macellari and De Vries, 1987). The margin's Cretaceous latitude was ~5°-10°N and located in the easterly trade wind belt, which is a setting for upwelling. Presumably environmental or ecological factors (hypernutrification?) inhibited the continued growth of Cogollo/El Cantil reefs along the margin, letting it subside beneath the wave base. The resulting platform was dysoxic and characterized by quiet sedimentation, although water depths in the west were probably not great. It is important to note that the La Luna and Querecual and other formations to the south were deposited on the continental platform, well above the actual rifted margin slope of northern South America.

In the west, sedimentation in the Táchira and Machiques Troughs and Bogotá Basin may have continued at higher rates than in inter-rift regions. The rifted areas have the shaliest sections of the platform, with sands only rarely entering the section. On the Central Cordillera of Colombia, later erosion obscures the Central Cordillera's Middle Cretaceous paleogeography. We know of no significant proximal clastic contribution to the western Magdalena from the Cordillera at this time, suggesting that it was not present or submarine.

It has been a widely held belief that the Albian deepening of marine conditions along the northern margin was related to the approach of island arcs which loaded the shelf platform (e.g., Maresch, 1974; James, 1990). However, we know of no sedimentological evidence of an orogenic terrane near the autochthonous portions of the margin in the Late Cretaceous. Furthermore, subsidence histories of autochthonous sections show no rapid acceleration in Late Cretaceous sedimentation that would imply the arrival of a large tectonically driven load. In addition, the relative motion history of known plates in the region is not consistent with compression along the margin until the Cenozoic arrival of the Caribbean Plate. Deepening of marine conditions appears to be eustatic, but as pointed out earlier, the Late Albian? termination of seafloor spreading in the Proto-Caribbean Seaway could have reduced in-plane stress and led to an overall marginal subsidence around the entire Proto-Caribbean.

In Campanian to Maastrichtian time (Figure 14), tectonism in the Colombian Andes is indicated by (1) a generally eastward vergent thrust system which brought allochthonous oceanic materials eastward onto the continental crust of the Central Cordillera; (2) numerous radiometric ages reflecting a deep, thermal heating event; by evidence of marine deepening within the Tres Esquinas condensed sequence (Ghosh, 1984) and the overlying Colón Formation in the Maracaibo/northern Andes region; (3) rapid accumulation rates in the Colón Formation, the Umir Formation in the Middle Magdalena Basin, and the Monserrate and Guaduas Formations in the Upper Magdalena Basin; and (4) by the regressive Guadalupe Formation in the Llanos and Bogotá Basins. The chronological and structural problems of sorting out the details of the event in the

Andes proper are outlined by Toussaint and Restrepo (this volume) and not repeated here. In general, however, the Late Cretaceous formations of the platform denote (1) a deepening of water conditions (e.g., Tres Esquinas, Colón), (2) a coeval acceleration of sedimentation rates, relative to the Villeta/La Luna Group, within the flexural wavelength of the Andes (e.g., Colón and Umir), and (3) a rapid regression (e.g, Guadalupe) at the strand line in areas proximal to the Shield. Assuming the mentioned formations are generally coeval, as is believed, these developments can be interpreted as the development of moderate foredeep basin in which subsidence rates in the basin increase, and uplift of the peripheral bulge causes regression near the Shield. Such a basin requires the onset of tectonic loading in the northern Andes, which was undoubtedly related to the eastward vergent "Late Cretaceous tectono-genesis" of Toussaint and Restrepo (this volume). Figure 17b shows a hypothetical cross-section of these proposed events. By the Maastrichtian, sediment infilling and falling long-term eustasy caused emergence of most of the platformal areas. In the Middle Magdalena, the Campanian-Maastrichtian and possibly Danian? Umir Formation, 1000 to 1400 meters of coaly mudstone with sand stringers deposited in paralic conditions, records accelerated basement subsidence and then regression as well.

Northwards, the Colón Formation remained marine into the Paleocene, demonstrating northward regression during foredeep basin infilling.

If the eastward vergent thrusting of the Amaime (or Calima) Terrane in the northern Andes represented the leading edge of the Caribbean crust, as we believe, convergence could only have proceeded a short distance as the arc climbed up the passive margin (perhaps 100-150 km) before a new Benioff Zone would be formed because of trench choking by the Central Cordillera's continental crust. We suggest that a new east-dipping subduction zone formed west of the accreted Amaime-Chaucha Terrane by the Mid-Campanian, thereby making the Central Cordillera the overriding plate for the first time since the Jurassic (Figure 17c). We note that the Campanian is also the time of initiation of the Panama-Costa Rica magmatic arc (Pindell and Barrett, 1990), which probably spanned from Ecuador to Chortis, much like the earlier Antillean arc, thereby bounding and defining the Caribbean Plate for the first time (Figure 19). Thus, convergence rates in Colombia from Campanian onward would have matched Caribbean-

Fig. 17

Schematic cross-sections of tectonic events in the northern Andes. (A) Hypothetical model of Aptian-Albian intra-oceanic arc polarity reversal as the Caribbean Plate began its relative eastward migration between the Americas, by the subduction of Proto-Caribbean crust. (B) Model showing Campanian emplacement of the Amaime Terrane onto the Tahamí Terrane as the Caribbean-Amaime arc converged upon Colombia, as discussed in text. (C) Late Campanian-Maastrichtian model for the onset of renewed subduction beneath northern South America, which produced Antioquia and other magmatic rocks in the north, as discussed in text. After Pindell et al. (in review). ➤

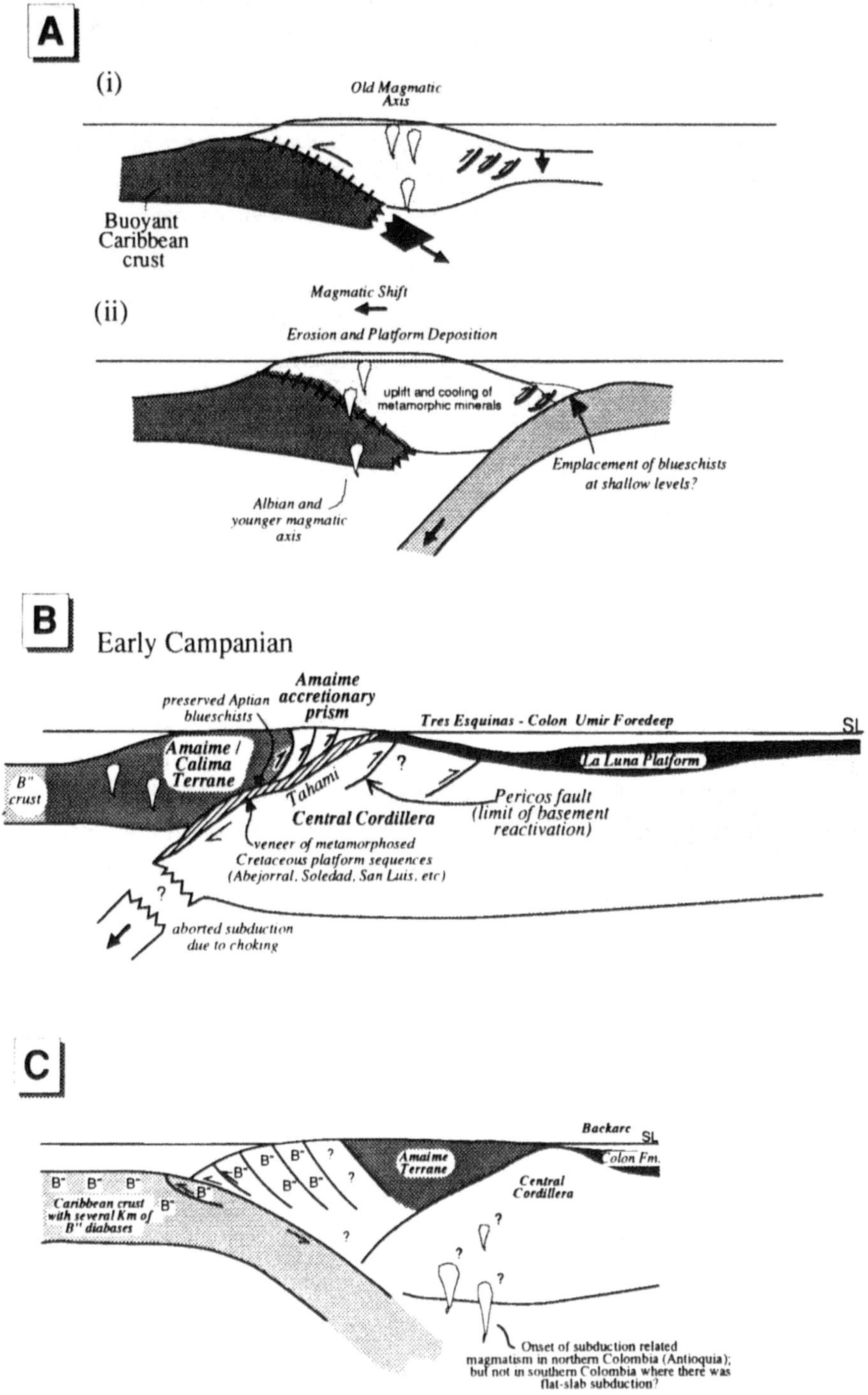
A
(i)
Old Magmatic Axis
Buoyant Caribbean crust
(ii)
Magmatic Shift
Erosion and Platform Deposition
uplift and cooling of metamorphic minerals
Emplacement of blueschists at shallow levels?
Albian and younger magmatic axis
B
Early Campanian
Amaime accretionary prism
preserved Aptian blueschists
Tres Esquinas - Colon Umir Foredeep
SL
Amaime / Calima Terrane
B" crust
La Luna Platform
Tahami
Central Cordillera
Pericos fault (limit of basement reactivation)
veneer of metamorphosed Cretaceous platform sequences (Abejorral, Soledad, San Luis, etc)
aborted subduction due to choking
C
Backarc
SL
Amaime Terrane
Colon Fm.
Central Cordillera
Caribbean crust with several Km of B" diabases
Onset of subduction related magmatism in northern Colombia (Antioquia); but not in southern Colombia where there was flat-slab subduction?

South American relative motion rates, which probably were only about 30 mm/yr, rather than Farallon relative motion rates, which were about 150 mm/yr. The difference in rate was taken up at the Panama-Costa Rica arc.

In view of the tectonic driving factor for deposition in the Late Cretaceous foredeep basin, we suggest that any attempts to correlate sequence boundaries in the Magdalena, César, Maracaibo, Llanos, or Putumayo Basins with global curves must be held in suspicion. In Eastern Venezuela and Trinidad, this effect is not seen, as there was no foredeep basin there at the time.

The part of the Caribbean crust that avoided collision with South America continued to consume Proto-Caribbean crust during westward drift of the Americas (Figure 19). A transpressional transform, the extension of the Campanian Amaime thrust front, must have progressively lengthened between the Greater Antilles trench, where west-dipping subduction occurred, and the western Colombian trench, where east-dipping subduction occurred. The zone at which this transform lay was along the Lower Magdalena Basin margin, the NW edge of the Santa Marta block, and the northwestern Guajira margin. This transform zone was transpressional, as thin-skinned thrusts of Caribbean-derived metamorphic rocks and pre-existing plutonic bodies were emplaced up onto the Santa Marta Massif and the Guajira and Paraguana blocks (Ruma metamorphic zone). Paleocene to Eocene cooling ages in these areas probably record the actual emplacement by overthrusting, but the protoliths were mostly Cretaceous so the present mixture of reported ages (e.g., Snoke, 1991) is to be expected. The transpressional emplacement of the Ruma Zone may have caused accelerated subsidence in the César Basin, in which the Maastrichtian Molina sands and shales (Colón equivalent) reach 3000 m in thickness. As relative plate motion continued, the Caribbean crust began to round the corner of Guajira, and by the Paleocene the north-south opening of the Grenada Basin contributed to SE-ward convergence of the Lara Nappes in the northern Maracaibo region (Figure 15; see section on Tertiary, below).

West of the accreted Calima or Amaime Terrane, sedimentary offscraping during subduction of Caribbean crust began to produce the San Jacinto accretionary prism, which was thickened to the point of becoming subaerial by the Eocene (Duque-Caro, 1984). The onset of east-dipping subduction in the Late Cretaceous after accretion of the Calima Terrane would have produced uplift in the Central Cordillera, which probably contributed to cooling and the setting of radiometric clocks in the northern Andes. Finally, we note that the intrusion of the Antioquia Batholith and smaller stocks in the northern Andes relates to this phase of east-dipping subduction. In the region of the Upper Mag-

Fig. 18

Composite stratigraphic column for the western portion of the passive margin platform, based on Lake Maracaibo region (after numerous sources). ➤

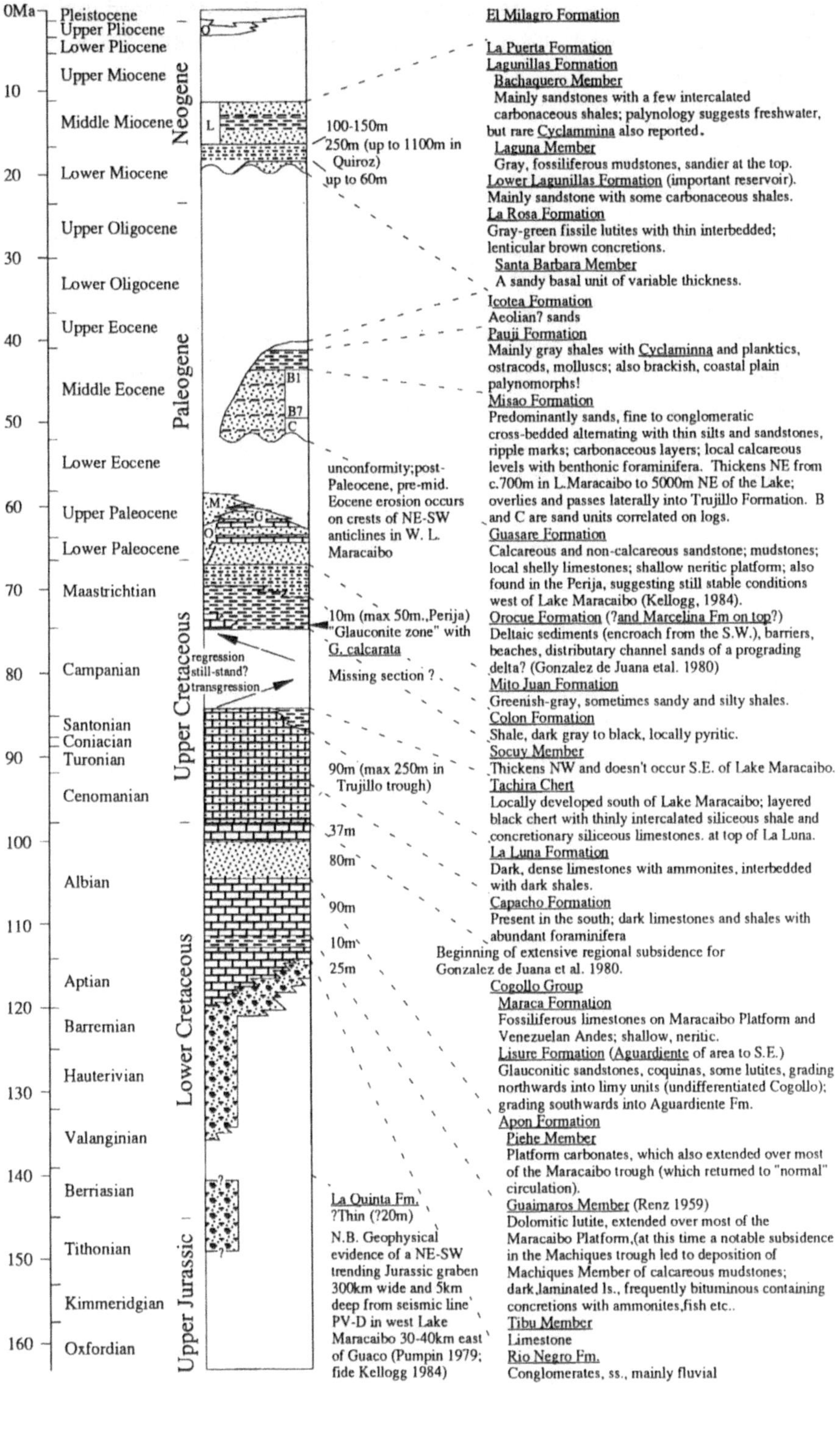
0Ma
10
20
30
40
50
60
70
80
90
100
110
120
130
140
150
160
Pleistocene
Upper Pliocene
Lower Pliocene
Upper Miocene
Middle Miocene
Lower Miocene
Upper Oligocene
Lower Oligocene
Upper Eocene
Middle Eocene
Lower Eocene
Upper Paleocene
Lower Paleocene
Maastrichtian
Campanian
Santonian
Coniacian
Turonian
Cenomanian
Albian
Aptian
Barremian
Hauterivian
Valanginian
Berriasian
Tithonian
Kimmeridgian
Oxfordian
Neogene
Paleogene
Upper Cretaceous
Lower Cretaceous
Upper Jurassic
regression
still-stand?
transgression
L
B1
B7
C
M
G
O
100-150m
250m (up to 1100m in Quiroz)
up to 60m
unconformity;post-Paleocene, pre-mid. Eocene erosion occurs on crests of NE-SW anticlines in W. L. Maracaibo
10m (max 50m.,Perija) "Glauconite zone" with G. calcarata
Missing section ?
90m (max 250m in Trujillo trough)
37m
80m
90m
10m
25m
La Quinta Fm.
?Thin (?20m)
N.B. Geophysical evidence of a NE-SW trending Jurassic graben 300km wide and 5km deep from seismic line PV-D in west Lake Maracaibo 30-40km east of Guaco (Pumpin 1979; fide Kellogg 1984)
El Milagro Formation
La Puerta Formation
Lagunillas Formation
Bachaquero Member
Mainly sandstones with a few intercalated carbonaceous shales; palynology suggests freshwater, but rare Cyclammina also reported.
Laguna Member
Gray, fossiliferous mudstones, sandier at the top.
Lower Lagunillas Formation (important reservoir).
Mainly sandstone with some carbonaceous shales.
La Rosa Formation
Gray-green fissile lutites with thin interbedded; lenticular brown concretions.
Santa Barbara Member
A sandy basal unit of variable thickness.
Icotea Formation
Aeolian? sands
Pauji Formation
Mainly gray shales with Cyclaminna and planktics, ostracods, molluscs; also brackish, coastal plain palynomorphs!
Misao Formation
Predominantly sands, fine to conglomeratic cross-bedded alternating with thin silts and sandstones, ripple marks; carbonaceous layers; local calcareous levels with benthonic foraminifera. Thickens NE from c.700m in L.Maracaibo to 5000m NE of the Lake; overlies and passes laterally into Trujillo Formation. B and C are sand units correlated on logs.
Guasare Formation
Calcareous and non-calcareous sandstone; mudstones; local shelly limestones; shallow neritic platform; also found in the Perija, suggesting still stable conditions west of Lake Maracaibo (Kellogg, 1984).
Orocue Formation (?and Marcelina Fm on top?)
Deltaic sediments (encroach from the S.W.), barriers, beaches, distributary channel sands of a prograding delta? (Gonzalez de Juana etal. 1980)
Mito Juan Formation
Greenish-gray, sometimes sandy and silty shales.
Colon Formation
Shale, dark gray to black, locally pyritic.
Socuy Member
Thickens NW and doesn't occur S.E. of Lake Maracaibo.
Tachira Chert
Locally developed south of Lake Maracaibo; layered black chert with thinly intercalated siliceous shale and concretionary siliceous limestones. at top of La Luna.
La Luna Formation
Dark, dense limestones with ammonites, interbedded with dark shales.
Capacho Formation
Present in the south; dark limestones and shales with abundant foraminifera
Beginning of extensive regional subsidence for Gonzalez de Juana et al. 1980.
Cogollo Group
Maraca Formation
Fossiliferous limestones on Maracaibo Platform and Venezuelan Andes; shallow, neritic.
Lisure Formation (Aguardiente of area to S.E.)
Glauconitic sandstones, coquinas, some lutites, grading northwards into limy units (undifferentiated Cogollo); grading southwards into Aguardiente Fm.
Apon Formation
Piche Member
Platform carbonates, which also extended over most of the Maracaibo trough (which returned to "normal" circulation).
Guaimaros Member (Renz 1959)
Dolomitic lutite, extended over most of the Maracaibo Platform,(at this time a notable subsidence in the Machiques trough led to deposition of Machiques Member of calcareous mudstones; dark,laminated ls., frequently bituminous containing concretions with ammonites,fish etc..
Tibu Member
Limestone
Rio Negro Fm.
Conglomerates, ss., mainly fluvial

dalena Basin, plutons of this age are lacking, and backthrusting developed in the Paleogene (Butler and Schammel, 1988). These relationships may indicate a tear in the downgoing slab beneath Colombia at that time, dipping steeply in the north and shallowly in the south (Figure 17c).

The Eastern Venezuelan Cretaceous Passive Margin Platform

In contrast to the western portion of the passive margin platform which is known mainly from the subsurface and from limited, highly deformed outcrops in the Andes, the eastern Cretaceous platform can be studied in great detail in the excellent exposures of the Serranía del Interior Oriental. In fact, the Serranía section and its correlative units in Trinidad (Figure 20), which are generally of a deeper water nature, probably provide the western hemisphere's best exposed passive margin section, nearly ideal for the study of Cretaceous passive margin processes such as relative sea level behavior during non-glacial times in the absence of tectonism (Pindell and Drake, in review). In the section below, we discuss the stratigraphy and paleogeography of the Serranía del Interior's section (Erikson, 1992; Erikson and Pindell, in press) (Figure 20) with the hope that the analysis highlights processes important to the entire northern South American margin and to Cretaceous passive margin development worldwide.

The Valanginian? to Aptian Barranquín Formation is the oldest exposed formation of the Serranía del Interior, and its base is seen nowhere. The Barranquín Formation is as much as 1500-2500 m thick in the Serranía del Interior. Most of the Barranquín Formation was deposited in the marine realm, with fluvial deposits primarily concentrated in the southwestern Serranía del Interior.

In the northwestern Serranía del Interior and nearby islands, the lowest member's ~400 m of shallow marine sandstone and shale were deposited on a shelf exposed to marine currents and are only exposed in the northernmost Serranía del Interior (Von der Osten, 1957). A coastline with rivers disgorging large quantities of coarse, quartzitic to arkosic sands lay relatively close to the south. The bioclastic wackestone limestone beds of the overlying Valanginian?-Hauterivian member (i.e., Morro Blanco member) were formed as biostromes and remobilized bioclastic flows on a ramp or steep shelf. Water depths increased upwards through the ~300 m thickness of the limestone member, indicating a southward or southeastward migration of facies. A possible depositional hiatus or submarine erosion surface in the upper and paleobathymetrically deeper part of the limestone member is probably due to increasing distance from a shallow marine source. The upper part of the Barranquín Formation (upper Hauterivian?-Aptian Picuda and Taguarumo members) is about 1 km thick and consists of an aggradational to progradational succession of shelfal sandstone and shale that has limestone locally present in its uppermost part. The upward increase in limestone content may reflect incipient development of a carbonate platform.

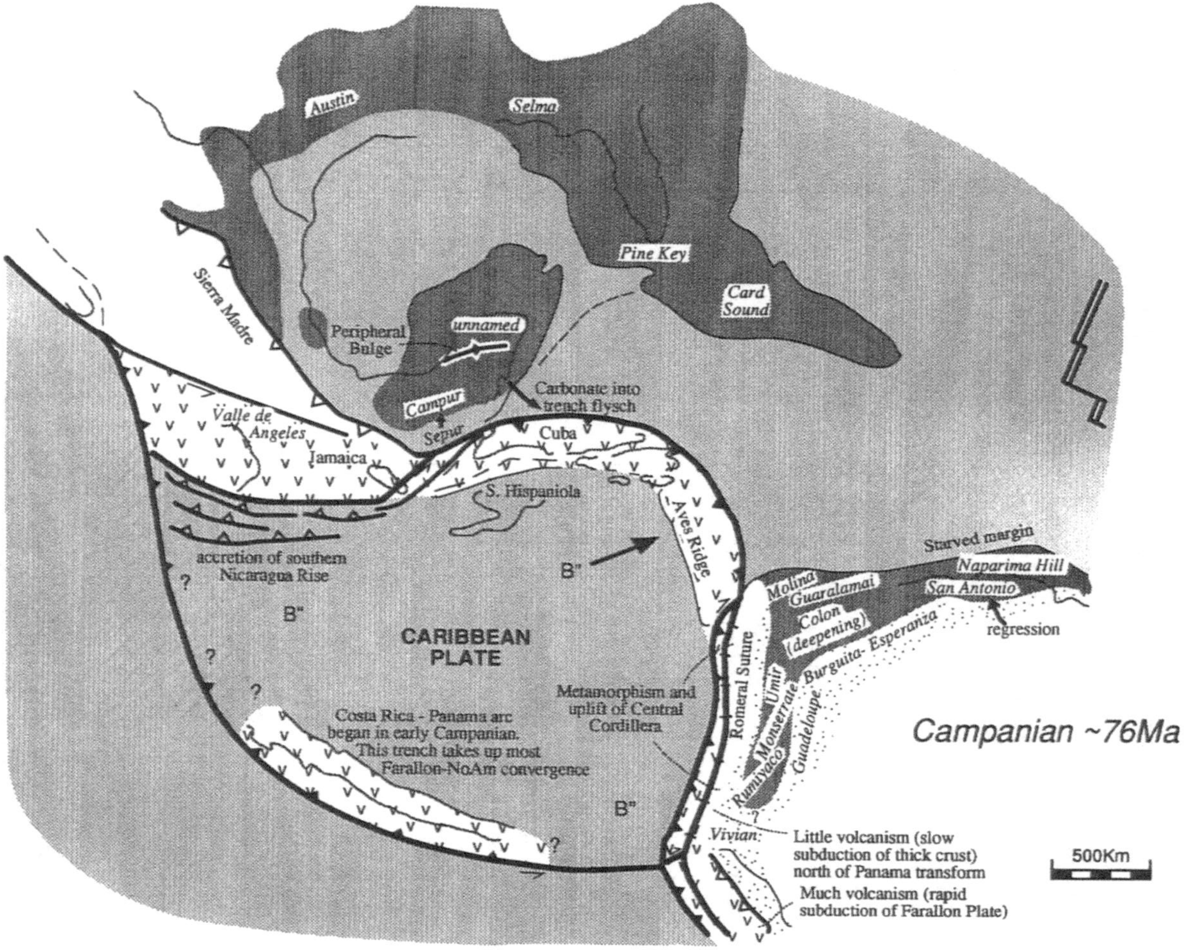

Fig. 19

Campanian paleogeographic reconstruction, after Pindell et al. (in press, b). Caribbean-Amaime arc had collided with Colombia and Yucatan, and, in Colombia, subduction of Caribbean crust beneath northern South America produced late Cretaceous magmatism there. Onset of subduction at this time at the Panama-Costa Rica arc greatly reduced the Farallon Plate's convergence velocity at the Colombian terench, and defined the Caribbean Plate for the first time.

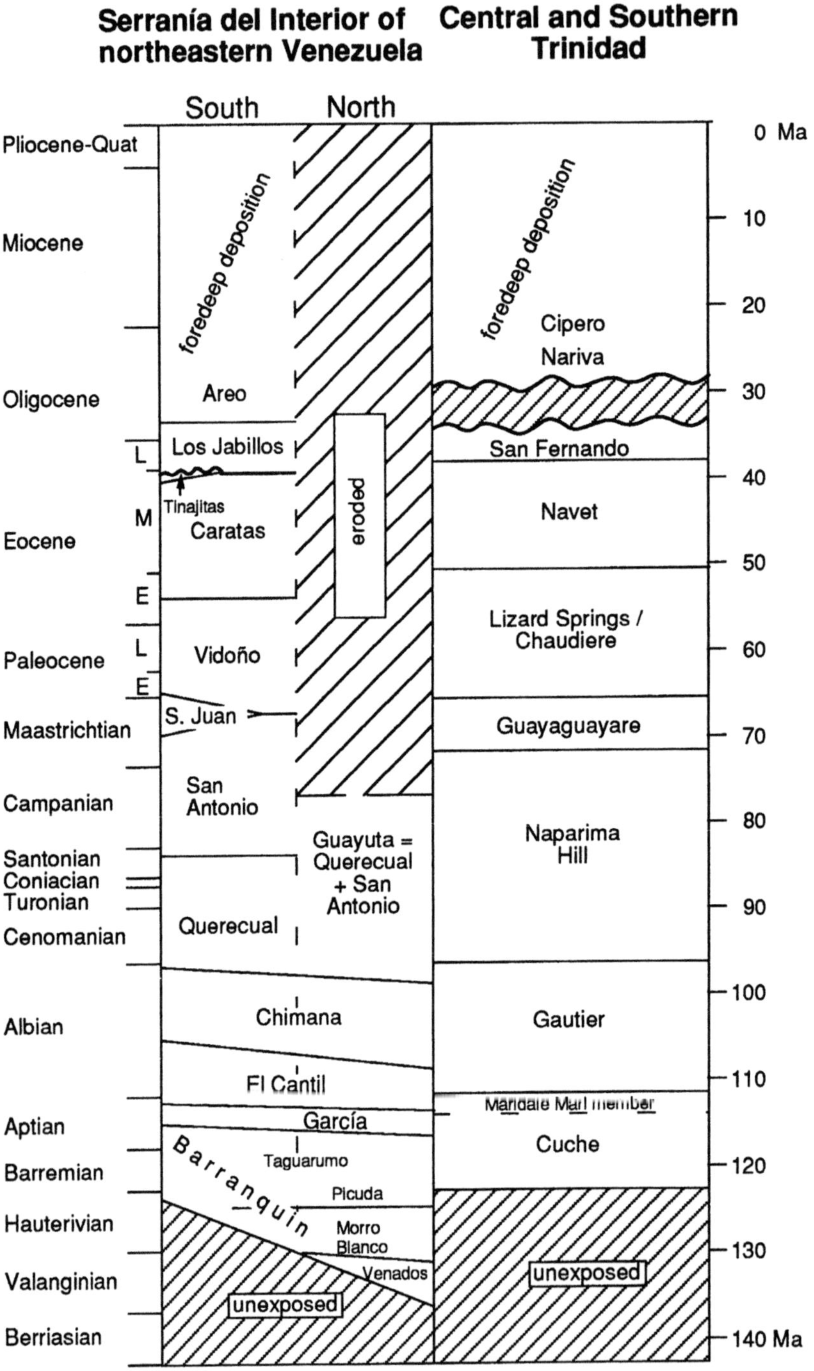

Fig. 20

Composite Cretaceous-Early Tertiary stratigraphic column for Eastern Venezuelan Basin, from the Serranía del Interior and Central and Southern Trinidad, after Algar, 1993; Erikson, 1992; Kugler, 1953.

The Barranquín Formation of the majority of the Serranía del Interior is typified by more proximal facies than it is in the islands and northwestern Serranía del Interior. A limestone horizon equivalent to the Morro Blanco member is not seen in most of the Serranía del Interior. The depositional environments of these regions were variable but generally very shallow marine to deltaic. In the southeastern Serranía del Interior, the ~2000 m thick Barranquín exposure preserves at a gross scale a single upward-fining and -thinning sequence within the sandstone and shale beds (Falcón, 1988), due to a protracted marine transgression. Some but not all of the Barranquín Formation sections become increasingly limestone-rich near the top. We interpret these limestone occurrences as due to a protracted regional transgression that reduced siliciclastic sediment transport to these exposed sites. Large relative falls or rapid relative rises of sea level could not have occurred during deposition of the Barranquín Formation.

A broad, very shallow shelf had prograded northward by the end of Barranquín Formation deposition. Localized carbonate build-ups increase in abundance at the top of the Barranquín Formation, particularly in the far north of the Serranía del Interior, indicating a reduction in siliciclastic progradation and conditions more amenable to carbonate growth. Water depths increased within a shelf/platform environment from the uppermost Barranquín Formation to the lower García Formation. In general, the base of the García Formation is a condensed section that was deposited under increasing water depths, whereas the majority of the formation preserves generally fine-grained facies in a shallowing sequence. The mixed but dominantly calcareous and glauconitic lithologies of the islands near the northwestern Serranía del Interior were deposited on an upward-shallowing, deep platform. Low-energy marine conditions were present during hemipelagic accumulation of shale on the upward-shallowing shelf in the southeastern Serranía del Interior. During García Formation deposition, the platform/slope break was always north of all exposed outcrops.

During lower García Formation deposition, postulated carbonate build-ups north of the Serranía del Interior probably aggraded faster than the platform to their south, whereas on the platform, the shallow marine limestone of the uppermost Barranquín Formation were succeeded by the deeper marine, lower García Formation. Siliciclastic progradation from the south, and calciclastic debris from the north filled in the deep, broad lagoon now preserved as the upper García Formation in the archipelago region.

The Aptian García Formation (Figure 20) is 60-150 m thick in most of the Serranía del Interior, but typical, fine-grained García Formation lithology is not present between the Barranquín and El Cantil formations in the southwestern Serranía del Interior. Much of the upper Barranquín Formation in the western and southeastern Serranía del Interior has a pronounced deltaic influence, generally subaqueous delta plain or delta front. Major river(s) emanating from the Guyana Shield delivered large quantities of coarse sediment to the northeastern Venezuela coast, as evidenced by the Barranquín Formation. These river(s) formed a delta(s). Wave and current energy were sufficiently high to redistribute this material from the delta and onto the shelf. The distribution of the fine-grained García Formation suggests that a large source of coarse sediments, presumably a major delta or several coalesced smaller deltas, was situated southwest of the Serranía del Interior. A major re-entrant of the continental crust is located beneath northeastern Venezuela (Figure 5), and probably is responsible for having localized the major Early Cretaceous delta southwest of the Serranía del Interior.

Transgression and/or retrogradation is most pronounced at the base of the García Formation, whereas the upper part of the García Formation preserves a shallowing sequence as the sandy delta readvanced northward onto the shelf. In distal portions of the deltaic system, primarily dark, variably glauconitic, organic-rich silts were deposited and minor sandstone is found locally at the top of the García Formation. Contemporaneously, in the more proximal deltaic settings of the southwestern Serranía del Interior, the transgression resulted in decreased, but sand-dominated, sedimentation and the continuation of Barranquín-type accumulation.

The upper Aptian-lower Albian El Cantil Formation preserves some of the most varied lithologies and depositional paleoenvironments of the Cretaceous strata in northeastern Venezuela. The thickness of the El Cantil Formation ranges from approximately 40 m in the southwestern Serranía del Interior, through 230 m in the archipelago region northwest of the Serranía del Interior, to 600-800 m in the southeastern Serranía del Interior. The lithology of the El Cantil Formation similarly varies, but is dominated by limestone: in the archipelago it is a generally clean, bedded, biostromal limestone whereas siliciclastic sediments comprise a major portion of the El Cantil Formation in the southern and eastern Serranía del Interior. Siliciclastic sediments of the southern and eastern Serranía del Interior are interbedded with limestone in platform facies. The thin, foraminiferal wackestone/packstone limestone of the El Cantil Formation in the southwestern Serranía del Interior was deposited in an inner platform environment above a condensed section on the Barranquín Formation.

In the archipelago region, the upward-shallowing transition from the García Formation to the El Cantil Formation occurs on the landward side of the platform break, as indicated by the low energy, rudist biostrome facies preserved in the lower carbonate unit of the El Cantil Formation. This transition indicates that the limestone depositional environment migrated southward and shoreward onto the platform facies of the García Formation (Erikson, 1992).

The eastern Serranía del Interior regions were part of a broad platform when El Cantil limestone growth began in the north. Water depths may have been deeper in the eastern part of the platform than they were in the archipelago region, and they were certainly deeper than in the southwestern Serranía del Interior. Accumulation of the prodelta and distal delta front sedimentary rocks of the García and lower El Cantil formations in the eastern Serranía del Interior caused shallowing of the depositional surface on a shallow, current-modified shelf. Shallow marine limestone deposition became possible in the eastern Serranía del Interior as the depositional surface entered the euphotic zone.

A possibly correlative, 10-25 m thick, lithostratigraphic marker interval within the El Cantil Formation may extend over much of the Serranía del Interior (Rod and Maync, 1954; González de Juana et al., 1980; Vivas, 1987; Yoris, 1988; Erikson, 1992; Erikson and Pindell, in press). The siliciclast-rich unit locally consists of ooid beds, conformable dolomite, and crossbedded, quartz sandstone, and may represent a small but significant shallowing event near the time of the Aptian-Albian boundary (Erikson, 1992; Erikson and Pindell, in press).

The end of El Cantil Formation deposition in the archipelago region occurred as the lower Albian shelfbreak ceased aggrading and migrated southward. The uppermost 25-30 m of the El Cantil Formation in the archipelago region is an outer platform rudist-dominated bank that passes upwards across a sharp contact to glauconite-dominated greensand of the lower Chimana Formation. Early Albian accumulation of the El Cantil Formation continued in the rest of the Serranía del Interior after Chimana Formation glauconitic shale deposition had begun in the archipelago.

The Chimana Formation is the shaley transitional formation between the platform limestone of the El Cantil Formation and the deeper water, organic-rich calcareous shale and limestone of the Querecual and San Antonio formations. The Chimana Formation has a relatively uniform 100-200 m thickness across much of the Serranía del Interior, with a probable increase in the far southeast. Its lithology is characterized by the abundance of glauconite in dark shale, limestone, and sandstone; dark sandstone and limestone are more abundant in the southeastern Serranía del Interior. The base of the Chimana Formation is a condensed section that was deposited in probably increasingly deep water, whereas the middle and upper parts may preserve progradational sequences and possibly shallowing paleowater depths. The depositional environment of the Chimana Formation is predominantly shelfal, but with notable variations of energy levels and grain provenance.

The El Cantil-Chimana formational contact probably is a southwardly younging time-transgressive surface. The 600-800 m thickness of the El Cantil Formation in the southeastern Serranía del Interior (Yoris, 1986) is composed of intermixed carbonate and siliciclastic units within protected platform facies, suggesting that the upper El Cantil- age shelfbreak was north of southeastern Serranía del Interior after deposition of the Chimana Formation had begun in the archipelago region. The age relationship and facies geometry during deposition of the El Cantil and Chimana formations describes a retrograding shelf break (Erikson, 1992; Erikson and Pindell, in press). The shelf break in northeastern Venezuela was "reefal" (dominated by rudist biostromes), and the rapid transition from a shelf break in the archipelago region to a younger shelf break in the upper El Cantil limestone in the east-central Serranía del Interior and then in the southeastern Serranía del Interior conform closely to the model of a retrograding carbonate shelfbreak (James and Mountjoy, 1983; also Kendall and Schlager, 1981; Erlich et al., 1990) (Figure 21). The observed facies geometry is what would be expected for a retrograded carbonate shelfbreak at a passive margin undergoing transgression and carbonate platform drowning (Schlager, 1981, 1989, 1991; Erlich et al., 1990). The primary Aptian-Albian continental shelfbreak was north of all exposed Aptian-Albian exposures. However, by the time of deposition of the upper El Cantil Formation, a secondary shelfbreak in the eastern Serranía del Interior had retrograded in successive steps to the far southeastern Serranía del Interior. As relative sea level rose during El Cantil Formation deposition, the shelfbreak retrograded much farther in the eastern Serranía del Interior than in the western Serranía del Interior.

The organic-rich shale and micritic limestone of the Late Cretaceous Guayuta Group, consisting of the Querecual and San Antonio formations (Figure 20), have generated considerable interest, as they are the petroleum source for much of the petroleum-prolific Eastern Venezuela Basin. The base of the Querecual Formation extends into the upper Albian in much of the central and northern Serranía del Interior, but probably becomes younger with its base in the Cenomanian in the southern Serranía del Interior and in the subsurface of the Eastern Venezuela Basin. The contact of the Querecual and San Antonio formations is imprecisely dated and has been placed by various authors within the range of Late Coniacian to very Early Campanian. The top of the San Antonio Formation lies within the Middle or Late Maastrichtian.

The thicknesses of the Querecual and San Antonio formations are more regionally uniform than other Cretaceous formations. The Querecual Formation is approximately 650-750 m thick over much of the Serranía del Interior, and thins slightly to the south. The San Antonio Formation's thickness is less precisely known, but is within the range of 275-500 m.

The Guayuta Group's lithology is much more uniform throughout the Serranía del Interior than are the lithologies of the underlying formations, reflecting more regionally uniform paleoenvironmental conditions during Guayuta Group deposition. Calcareous shale and carbonaceous, micritic limestone are typical of the Querecual and San Antonio formations throughout the Serranía del Interior. These lithologies are organic-rich as noted above, and are commonly very finely laminated in the Querecual Formation, but are more massive in the San Antonio Formation due to more extensive bioturbation. Micrite concretions are abundant and occur throughout the Querecual and San Antonio formations, though are much more common in intervals within the Querecual Formation. Chert increases in abundance upwards from the middle part of the Querecual Formation and is most prominent in the San Antonio Formation, particularly in the northeastern Serranía del Interior. The San Antonio Formation preserves more lithologic var-

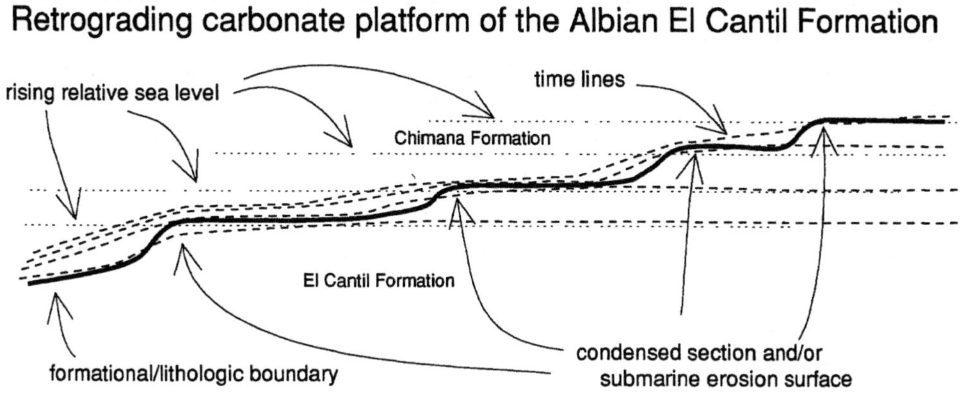

Fig. 21

Schematic interpretation of the El Cantil-Chimana formational boundary, and the retrograding carbonate platform edge, after Erikson (1992).

iation than does the Querecual Formation: quartz sandstone is prominent but still volumetrically a minor component in the lower part of the San Antonio Formation in the western and southern Serranía del Interior, but diminishes in abundance northwards; the principal regional distinction within the San Antonio Formation is the transition to massive white cherty beds with rare siliciclastic contamination and without preserved organic matter in the north, but the continuation of black, organic-rich sedimentation in the southern Serranía del Interior.

The depositional environment of the pervasively very thinly laminated Querecual Formation was generally anoxic or suboxic, and sediment accumulation was dominated by hemipelagic or pelagic deposition and at least some silt and mud turbidite deposition. Negligible siliciclastic sediment coarser than very fine sand occurs within the Querecual Formation, reflecting the great distance to the Cenomanian-Santonian littoral zone.

The Late Cretaceous littoral zone transgressed well onto the craton, trapping continental epiclastic sediments far to the south of the Serranía del Interior outcrops. Shallow marine sedimentary rocks are found within 60 km of the Orinoco River, and mark the southernmost known extent of Cretaceous marine deposition (van Erve, 1986). This transgression that culminated in Turonian time trapped much of the siliciclastic sediment derived from the craton. Marine deposition spanned a shallow to moderately deep shelf with a north-south extent of at least 250 km (present) and probably more than 350 km (predeformation). Regression of the littoral zone in the region of the Orinoco oil belt began in Coniacian-Santonian time (van Erve, 1986). Marine rocks of Campanian through Early Tertiary age are not found in the southern Eastern Venezuelan Basin (González de Juana et al., 1980).

A low gradient ramp was the setting of Querecual and San Antonio deposition. The pelagic limestone and the hemipelagic and fine turbiditic shale accumulated on a depositional surface with a sufficiently low gradient that slumps and sediment gravity flows were not locally initiated, yet in the San Antonio Formation of the southern and western Serranía del Interior there are numerous sandstone sediment gravity flow deposits. The implication is that the gradient of the depositional surface was low and that sand was derived from a region with relatively steeper gradients where sandy sediment gravity flows were initiated. Based on the distribution of sandy facies in the San Antonio Formation and the thickness of the overlying San Juan Formation, relatively shallow regions were located west, south, and southeast of the Serranía del Interior.

The San Juan Formation (Figure 20) is very distinctive as usually easily identified exposures of predominantly clean, quartzose sandstone and minor shale. The Maastrichtian San Juan Formation is the prominently sandy lens between the calcareous, black shales of the San Antonio Formation and Vidoño Formation. The sedimentary structures of the San Juan Formation are indicative of energetic transport and deposition of its sandstone and shale. All structures and lithologies are consistent with and conjunctively suggest a submarine fan setting of deposition (Di Croce, 1989). The relatively coarse grain sizes and energetic sedimentary structures of the San Juan Formation, relative to the underlying San Antonio Formation and the overlying Vidoño Formation, are indicative both of a significant change in sediment source and of the shelfbreak location and configuration during Maastrichtian time. The San Juan Formation forms a lens elongated east-west, reflecting the presence of a shelfedge south of the thickest San Juan accumulations and presently in the subsurface between the Serranía del Interior and the Pirital Fault (Erikson, 1992).

The presented history of Cretaceous deposition in northeastern Venezuela can be viewed as two distinct phases: an Early Cretaceous phase of shallow shelf, fine- to coarse-grained, mixed siliciclastic and carbonate sedimentation, and a Late Cretaceous phase of deeper marine, dominantly pelagic accumulation of limestone, marl, and shale with a Maastrichtian, sand-rich cap. All Cretaceous sedimentation occurred on a stable passive continental margin. The primary paleogeographic and relative sea level event of the Cretaceous, the significant water depth increase that started in the Late Albian, reflects global events. The stability of the northeastern Venezuelan margin through the Cretaceous contrasts sharply with the tectonic activity seen in the western part of the northern South American margin. Based partly on this contrast of tectonic activity, conclusions can be made of the tectonic interactions of the Caribbean and South American plates.

Cenozoic Interactions with the Caribbean Plate, and the Eastwardly Progressive Destruction of the Passive Margin

At the start of the Cenozoic, deposition along the shelf was entirely derived from the Guyana shield or Central Andes of Colombia to the south or southwest, except possibly in the northwestern Maracaibo and César basins. In those areas, NW-derived sediments from the Santa Marta Massif, Guajira Peninsula, and/or Caribbean Plate/Ruma complex began to accumulate in Paleocene or, possibly, Maastrichtian times. The Guajira basement and its Cretaceous section were probably imbricated by thrusts well beneath the thrusts carrying the allochthonous Ruma metamorphic rocks, which might explain why the Guajira Peninsula was uplifted and eroded in the Paleocene rather than depressed and buried like the basement of the northern Maracaibo Basin.

The southward obduction of allochthonous rocks at various points along northern South America can be dated between (1) the age of accelerated (load-induced) subsidence of the autochthonous platform (Figure 3), and (2) the age of the youngest overthrust foredeep deposits and/or the oldest overlap assemblage across the thrusts responsible for the emplacement of the allochthon. In the Siquisique area, obduction is dated as Early to Medial Eocene by the age of overthrust flysch (Paleocene-Early Eocene Trujillo Formation) and by the onset of rapid subsidence in the northern Maracaibo Basin. At Guarumen Medial-Late Eocene obduction is indicated both by the age of overthrust flysch and accelerated subsidence. At Barcelona Late Oligocene-Early Miocene obduction is indicated by the overthrust Roblecito Formation and by accelerated subsidence. In south-central Trinidad Early to Middle Miocene obduction is indicated by the overthrust Cipero Formation and by the onset of accelerated subsidence. There is clearly an eastward migration in the position of tectonically driven peak sediment accumulation and basement subsidence (curves with dots in Figure 3). This migration is directly associated with Caribbean Plate migration. Given the north-south mechanism for opening of the Grenada Basin, rates of migration appear to be about 20% slower than those measured at the Cayman Trough, consistent with motion around the Minster and Jordan (1978) pole near southern South America.

Although not the focus of this paper, Figure 22 shows stages of Cenozoic development of northern South America, after Pindell et al. (in press, a). Some of the main points to make are as follows.

1. The southward component of Paleogene compression in the Maracaibo-Lara area is best explained by a dextral, north-south opening model for the Grenada Basin (Figure 22a). Opening of the Grenada Basin appears to have occurred from the Paleocene to the Early Oligocene, and was probably driven by transform drag and subduction rollback of the Proto-Caribbean margin and adjacent Jurassic oceanic crust. Opening continued until a sufficient thickness of buoyant continental crust of the margin entered the subduction zone, culminating in NW-SE collision of the Lara Nappes.

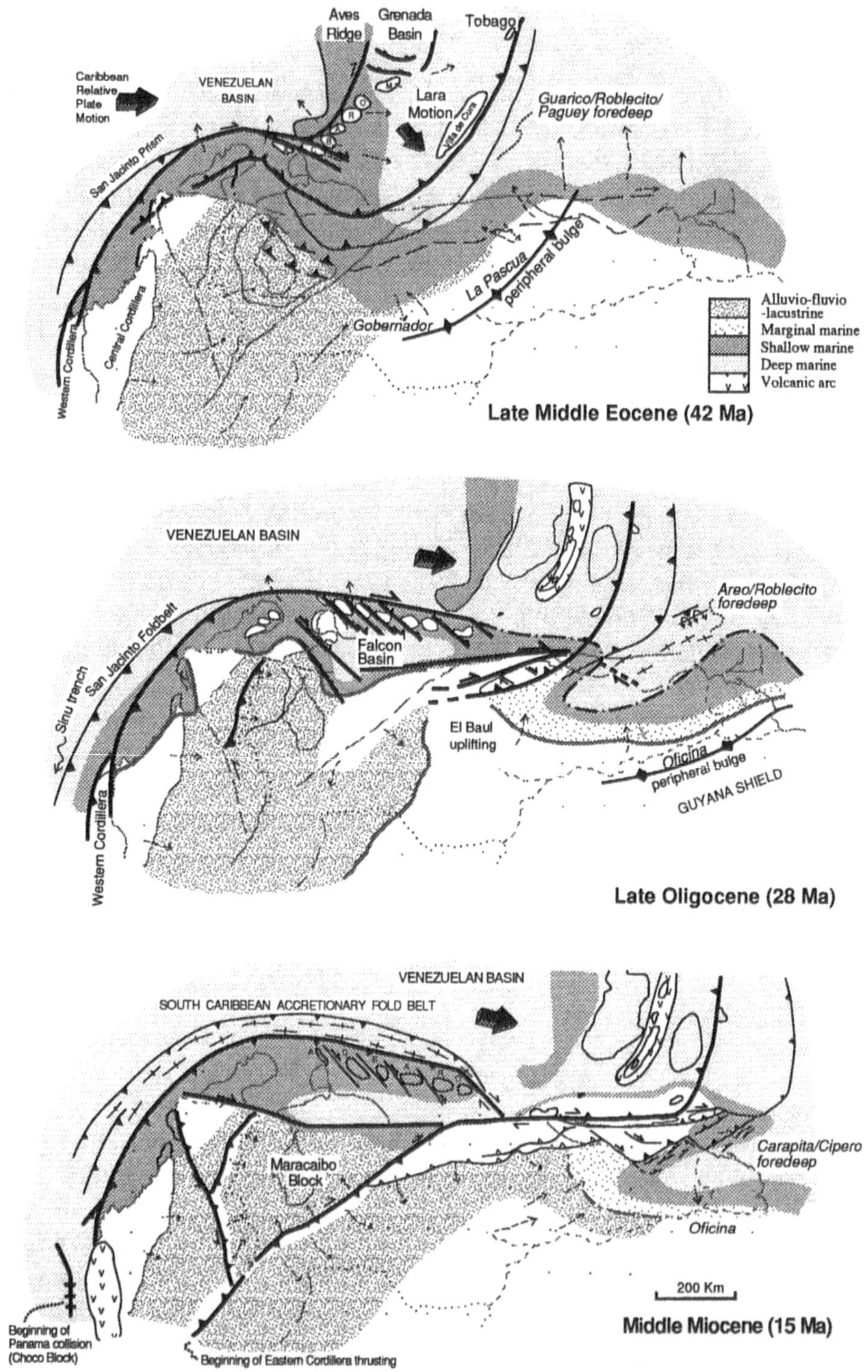
Aves Ridge
Grenada Basin
Tobago
Caribbean Relative Plate Motion
VENEZUELAN BASIN
Lara Motion
Villa de Cura
Guarico/Roblecito/ Paguey foredeep
San Jacinto Prism
La Pascua peripheral bulge
Gobernador
Central Cordillera
Western Cordillera
Alluvio-fluvio -lacustrine
Marginal marine
Shallow marine
Deep marine
Volcanic arc
Late Middle Eocene (42 Ma)
VENEZUELAN BASIN
Areo/Roblecito foredeep
San Jacinto Foldbelt
Sinu trench
Falcon Basin
El Baul uplifting
Oficina peripheral bulge
GUYANA SHIELD
Western Cordillera
Late Oligocene (28 Ma)
VENEZUELAN BASIN
SOUTH CARIBBEAN ACCRETIONARY FOLD BELT
Maracaibo Block
Carapita/Cipero foredeep
Oficina
200 Km
Beginning of Panama collision (Choco Block)
Beginning of Eastern Cordillera thrusting
Middle Miocene (15 Ma)

2. After the Grenada Basin had ceased to open, relative motion between the Caribbean terranes and South America was much more oblique (Figure 22b). Important strike-slip fault systems (e.g., Falcón Basin) began to develop in the Oligocene, allowing continued eastward displacement of the Caribbean Plate relative to the already emplaced allochthons in the west.

3. Given a north-south opening mechanism for the Grenada Basin, rates of plate migration of 16 and 24 mm/yr before and after 26 Ma (scaled to the Cayman Trough, Rosencrantz et al., 1988) appear to explain the timing of foredeep and peripheral bulge migration along northern SouthAmerica.

4. Strike-slip offset between the Caribbean Plate and South America is distributed along the basal emplacement thrusts of the allochthons (primary site of displacements), and on offshore and onshore E-W trending faults, extension within the Falcón-Bonaire Basin, and the Oca and Mérida fault systems (secondary sites of offsets). Because the emplacement of the allochthons was coincident with over 1000 km of Cenozoic relative motion, rather than an earlier, independent Cretaceous event, there is absolutely no expectation of ~1000 km Cenozoic offsets superimposed upon a pre-existing orogenic terrane. The allochthons were emplaced coevally with, and as a direct function of, the strike-slip history, and not before.

5. Middle Miocene encroachment of the Cuna Terrane (Panama Arc) with Western Colombia produced strong compression in the northern Andes, driving the Maracaibo and Guajira Blocks northwards from the site of the Eastern Andes, which began to undergo uplift at that time. The independent northward component of motion of these blocks produced convergence between the Caribbean Plate and the Guajira Block, creating the South Caribbean Foldbelt (Figure 22c; Dewey and Pindell, 1985, 1986; Mann and Burke, 1984).

Conclusions

The northern margin of South America, comprising Colombia, Venezuela, and Trinidad, was formed during the Triassic-Jurassic breakup of Pangea. Eastern Yucatán was the opposing conjugate margin of Venezuela and Trinidad, whereas Mexican terranes were rifted from Colombia during arc magmatism pertaining to subduction of Pacific oceanic plates. Thermal subsidence from the Late Jurassic through most of the Cretaceous produced an entirely passive continental margin in the autochthonous portions of all three countries, but Late Cretaceous (Campanian) tectonism involving the autochthon in the northern Colombia, probably involving the accretion and loading of the Amaime and/or

Fig. 22

◄ Stages of Cenozoic development of northern South America, as discused in text, modified after Pindell et al. (in press, a).

Calima Terranes onto the Central Cordillera, initiated foredeep basinal conditions to the east of the Andes with accelerated sediment accumulation rates under deeper water conditions. Passive margin conditions persisted in eastern Venezuela and Trinidad until the Eocene, as reflected in the lithologies and paleogeographic history presented here.

The rocks of the Caribbean allochthons from Ecuador to Trinidad that were metamorphosed in the Aptian-Albian owe this metamorphism to an orogenic event which took place well offshore in the eastern Pacific, and not to collision with the passive margin. The orogenesis probably pertains to arc polarity reversal in the Aptian-Albian, and is correlative to the Peruvian and Sevier orogenic pulses onshore in both North and South America, in which backthrusting rather than polarity reversal occurred because the crust in those areas was continental rather than oceanic. A chief cause of the event was the acceleration of the Americas westward over the mantle during separation from Africa.

Late Cretaceous to Recent eastward relative motion of the Caribbean arc produced diachronous, cratonward-vergent, collision and obduction of terranes from Ecuador to Trinidad. The autochthonous sections of Colombia, western Venezuela, Eastern Venezuela, and Trinidad became involved in the Campanian- Maastrichtian, the Paleocene-Eocene, the Late Oligocene-Middle Miocene, and the Early to Late Miocene, respectively. From northern Colombia to Trinidad, this diachronous arc-collision progressively destroyed the Mesozoic passive margin, and also diachronously drove prolific hydrocarbon maturation with attendant southward migration from the site of overthrusting.

Acknowledgments

We are very grateful to Dr. Sam Algar, a close colleague on much of the research presented herein, for his assistance with the preparation of this manuscript. We also thank Dr. Francia Galea for discussions on Venezuelan Cretaceous stratigraphy, and Dr. Edward Robinson for compiling the stratigraphic column for Lake Maracaibo. We are indebted to the consortium of companies funding research on northern South America at Dartmouth College, namely Amoco, Arco, British Gas, British Petroleum, BHP Petroleum, Conoco, Lasmo, Mobil, Petrocanada, Phillips, Texaco, Total, and Unocal.

References

Algar, S.T. (1993): Structural, stratigraphic and thermo-chronologic evolution of Trinidad. Ph.D. Dissertation, Dartmouth College, Hanover, 430 pp.

Algar, S.T., and Erikson, J.P. (1992): Stratigraphic correlation of Trinidad and eastern Venezuela. Caribbean Geological Conference, 13th, Pinar del Río, Cuba, Abstracts.

Algar, S.T., and Pindell, J.L. (1991a): Structural development of the Northern Range of Trinidad, and implications for the tectonic evolution of the southeastern Caribbean, *in* Geological Conference of the Geological Society of Trinidad and Tobago, 2nd, (Ed.: Gillezeau, K.A.), Port of Spain (1990), Transactions, 6-22.

Algar, S.T., and Pindell, J.L. (1991b): Stratigraphy and sedimentology of the Toco region of the Northern Range, N.E. Trinidad, *in* Geological Conference of the Geological Society of Trinidad and Tobago, 2nd, (Ed.: Gillezeau, K.A.), Port of Spain (1990), Transactions, 56-69.

Algar, S.T., and Pindell, J.L. (in review): Structure and deformation history of the Northern Range of Trinidad and adjacent areas: implications for the tectonic history of northeastern South America. Tectonics.

Aspden, J.A., and McCourt, W.J. (1986): Mesozoic oceanic terrane in the central Andes of Columbia. Geology, v. 14, 415-418.

Avé-Lallemant, H.G. (1990): The Caribbean-South American Plate boundary, Araya Peninsula, Eastern Venezuela, *in* Caribbean Geological Conference, 12th, Christianstadt, Saint Croix, (Eds.: Larue, D.K., and Draper, G.), Transactions, 461-471.

Barr, K.W. (1963): The Geology of the Toco District, Trinidad W.I., Overseas Geological Surveys H.M.S.O.

Bartok, P.E., Renz, O., and Westerman, G.E.G. (1985): The Sisquisique ophiolites northern Lara State, Venezuela: A discussion on their Middle Jurassic ammonites and tectonic implications. Geological Society of America Bulletin, v.96, 1050-1055.

Beets, D.J., Maresch, W.V., Klaver, G.T., Mottana, A., Bocchio, R., Beunk, F.F., and Monem, H.P. (1984): Magmatic rock series and high-pressure metamorphism as constraints on the tectonic history of the southern Caribbean, *in* The Caribbean-South American Plate Boundary and Regional Tectonics, (Eds.: Bonini, W.E., Hargraves, R.B., and Shagam, R.), Geological Society of America Memoir, 162, 95-130.

Bellizzia G., A. (1972): Sistema Montañoso del Caribe, borde sur de la Placa Caribe ¿es una cordillera alóctona? Caribbean Geological Conference, 6th, Margarita, Venezuela (1971), Transactions, VI, 247-258.

Bellizzia G., A. (1986): Sistema Montañoso del Caribe -una cordillera alóctona en la parte norte de América del Sur. Congreso Geológico Venezolano, 6th, Caracas (1985), Sociedad Venezolana de Geólogos, Memoria, 6657-6836.

Biju-Duval, B., Mascle, A., Rosales, H., and Young, G. (1983): Episutural Oligo-Miocene basins along the north Venezuelan margin. American Association of Petroleum Geologists Memoir, 32, 347-358.

Bockmeulen, H., Barker, C., and Dickey, P.A. (1983): Geology and geochemistry of crude oils, Bolivar Coastal Fields, Venezuela. American Association of Petroleum Geologists Bulletin, v. 67, 242-270.

Bourgois, J., Toussaint, J., González, H., Azema, J., Calle, B., Desmet, A., Murcia, L.A., Acevedo, A.P., Parra, E., and Tournon, J. (1987): Geological history of the Cretaceous ophiolitic complexes of Northwestern South America (Colombian Andes). Tectonophysics, v. 143, 307-327.

Bray, R., and Eva, A. (1983): Age, depositional environment, and tectonic significance of the Couva marine evaporite, offshore Trinidad. Caribbean Geological Conference, 10th, Cartagena. Bogotá, Ingeominas, Memorias, p. 372.

Buffler, R.T., and Sawyer, D.S. (1985): Distribution of crust and early history, Gulf of Mexico Basin. Gulf Coast Association of Geological Societies, Transactions, 35, 333-334.

Butler, K., and Schamel, S. (1988): Structure along the eastern margin of the Central Cordillera, Upper Magdalena Valley, Columbia. Journal of South American Earth Sciences, v. 1, 109-120.

Campbell, C.J., and Burgl, H. (1965): Section through the Eastern Cordillera of Colombia, South America. Geological Society of America Bulletin, v. 76, 567-590.

Case, J.E., and Holcombe, T.L. (1980): Geologic-tectonic map of the Caribbean region. U. S. Geological Survey, Misc. Invest. Map I-1100, scale 1:2,500,000.

Case, J.E., and MacDonald, W.D. (1973): Regional gravity anomalies and crustal structure in northern Colombia. Geological Society of America Bulletin, v. 84, 2905-2916.

Case, J.E., Holcombe, T.L., and Martin, R.G. (1984): Map of geologic provinces in the Caribbean region, *in* The Caribbean-South American Plate Boundary and Regional Tectonics, (Eds.: Bonini, W.E., Hargraves, R.B., and Shagam, R.), Geological Society of America Memoir, 162, 1-29.

Chevalier, Y., Stephan, J.-F., Darboux, J.-R., Gravelle, M., Bellon, H., Bellizzia, A., and Blanche, R. (1988): Obduction et collision pre-Tertiaire dans les zones internes de la Chaine Caraïbe vénézuélienne, sur le transect Ile de Margarita-Péninsule d'Araya. Comptes Rendus à l'Académie des Sciences de Paris, série II, 307, 1925-1932.

Corpoven (1989): Mapa de Anomalías Magnéticas de Venezuela, Escala 1:500.000. Caracas, Corpoven.

Daal, A., González, A., Hernández, V., Uzcategui, M., Rodríguez, H., Pizon, J., and Choppin, H. (1989): Evolución geológica de la región occidental de la Cuenca Oriental de Venezuela. Congreso Geológico Venezolano, 7th, Barquisimeto, Sociedad Venezolana de Geólogos, Memoria, 388-402.

Dengo, C.A., and Covey, M.C. (1991, in press): Structure of the Eastern Cordillera of Colombia: a tectonic model of the Colombian Andes. Simposio Bolivariano, Exploración Petrolera en las Cuencas Subandinas, 4th, Bogotá.

Dewey, J.F. (1980): Episodicity, sequence and style at convergent plate boundaries, *in* The Continental Crust and Its Mineral Deposits, (Ed.: Strangeway, D.W.), Geological Association of Canada Special Paper, 20, 553-573.

Dewey, J.F., and Pindell, J.L. (1985): Neogene block tectonics of eastern Turkey and northern South America: Continental applications of the finite difference method. Tectonics, v. 4, 71-83.

Dewey, J.F., and Pindell, J.L. (1986): Reply to comment on "Neogene block tectonics of eastern Turkey and northern South America: continental applications of the finite difference method". Tectonics, v. 5, 703-705.

Di Croce, J. (1989): Análisis sedimentológico de la Formación San Juan en la Cuenca Oriental de Venezuela (Estados Anzoátegui y Monagas). M.Sc. Thesis, Universidad Central de Venezuela, Caracas, 129 pp.

Duque-Caro, H. (1984): Structural style, diapirism, and accretionary episodes of the Sinú-San Jacinto terrane, southwestern Caribbean borderland, *in* The Caribbean-South American Plate Boundary and Regional Tectonics, (Eds.: Bonini, W.E., Hargraves, R.B., and Shagam, R.), Geological Society of America Memoir, 162, 303-316.

Echevarría-Rodríquez, G., Hernández-Pérez, G., López-Quintero, J.O., López-Rivera, J.G., Rodríguez-Hernández, R., Sánchez-Arango, J.R., Socorro-Trujillo, R., Tenreyro-Pérez, R., and Yparraguirre-Pena, J.R. (1991): Oil and gas exploration in Cuba. Journal of Petroleum Geology, v.14, 259-274.

Erikson, J.P. (1992): Stratigraphic and paleogeographic evolution, and their tectonic and possible eustatic controls, of the Cretaceous of northeastern Venezuela and the Neogene Cibao basin of the Dominican Republic. Ph.D. Thesis, Dartmouth College, Hanover, 472 pp.

Erikson, J.P., and Pindell, J.L. (1992): Subsidence analysis of northeastern Venezuela's exposed Mesozoic passive margin. EOS, Transactions, v. 73, p. 564.

Erikson, J.P., and Pindell, J.L. (in press): Cretaceous-Eocene passive margin stratigraphy and paleogeography of northeastern Venezuela, *in* Mesozoic-Cenozoic stratigraphy and tectonic evolution of the Caribbean region/northern South America: Implications for eustasy from exposed sections of a Cretaceous-Eocene passive margin setting, (Eds.: Pindell, J.L., and Drake, C.L.), Geological Society of America Special Paper.

Erlich, R.N., Barrett, S.F., and Guo Bai Ju (1990): Seismic and geologic characteristics of drowning events on carbonate platforms. American Association of Petroleum Geologists Bulletin, v. 74, 1523-1537.

Fabre, A. (1985): Dinámica de la Sedimentación Cretácica en la región de la Sierra Nevada de Cocuy (Cordillera Oriental de Colombia), *in* Proyecto Cretácico, (Dir.: Etayo, F.), Bogotá, Ingeominas, Contribuciones, Publicaciones Geológicas Especiales, 16, XIX/1-XXIV/47.

Falcón L., R.A. (1988): Estratigrafía de las Formaciones Barranquín y García, en el flanco sur de la Serranía del Interior Oriental de Venezuela. M.Sc. Thesis, Universidad Central de Venezuela, Caracas, 186 pp.

Feo-Codecido, G., Smith, F.D., Jr., Aboud, N., and de Di Giacomo, E. (1984): Basement and Paleozoic rocks of the Venezuelan Llanos basins, *in* The Caribbean-South American Plate Boundary and Regional Tectonics, (Eds.: Bonini, W.E., Hargraves, R.B., and Shagam, R.), Geological Society of America Memoir, 162, 175-187.

Foland, K.A., Speed, R., and Weber, J. (1992): Geochronologic studies of the hinterland of the Caribbean Mountains orogen of Venezuela and Trinidad. Geological Society of America Abstracts with Programs, v. 24, no. 7, A148.

Ghosh, S.K. (1984): Late Cretaceous condensed sequence, Venezuelan Andes, *in* The Caribbean-South American Plate Boundary and Regional Tectonics, (Eds.: Bonini, W.E., Hargraves, R.B., and Shagam, R.), Geological Society of America Memoir, 162, 317-324.

González de Juana, C., Iturralde de Arozena, J., and Picard, X. (1980): Geología de Venezuela y de sus cuencas petrolíferas. Caracas, Ediciones FONINVES, 1031 pp.

Hedberg, H.D. (1950): Geology of the Eastern Venezuela basin (Anzoátegui-Monagas-Sucre-eastern Guárico portion). Geological Society of America Bulletin, v. 61, 1173-1216.

Hilst, R.D. van der (1990): Tomography with P,PP and pP delay-time data and the 3-dimensional mantle structure below the Caribbean region. Geologica Ultraiectina, Mededelingenvan de Faculteit Aardwetenschappen der Rijksuniversiteit te Utrecht, 67.

Irving, E.M. (1975): Structural evolution of the northernmost Andes, Colombia. U. S. Geological Survey Professional Paper, 846, 1-47.

James, K.H. (1990): The Venezuelan hydrocarbon habitat, *in* Classic Petroleum Provinces, (Ed.: Brooks, J.), Geological Society Special Publication, 50, 9-35.

James, K.H., and Mountjoy, E.W. (1983): Shelf-slope break in fossil carbonate platforms: an overview, *in* The shelfbreak: critical interface on continental margins, (Eds.: Stanley, D.J., and Moore, G.T.), Society of Economic Paleontologists and Mineralogists Special Publication, 33, 189-206.

Kellogg, J.N. (1984): Cenozoic tectonic history of the Sierra de Perijá and adjacent basins, *in* The Caribbean-South American Plate Boundary and Regional Tectonics, (Eds.: Bonini, W.E., Hargraves, R.B., and Shagam, R.), Geological Society of America Memoir, 162, 239-262.

Kendall, C.G.St.C., and Schlager, W. (1981): Carbonates and relative changes in sea level. Marine Geology, v. 44, 181-212.

Kohn, B.P., Shagam, R., and Subieta, T. (1984): Results and preliminary implications of 16 fission-track ages from rocks of western Caribbean Mountains, Venezuela, *in* The Caribbean-South American Plate Boundary and Regional Tectonics, (Eds.: Bonini, W.E., Hargraves, R.B., and Shagam, R.), Geological Society of America Memoir, 162, 415-421.

Krause, H., and James, K.H. (1989): Hydrocarbon resources of Venezuela -their source rocks and structural habitat, *in* Geology of the Andes and its relations to energy and mineral resources, Santiago (1985), (Eds.: Ericksen, G.E., Cañas Pinochet, M.T., and Reinemund, J.A.), Circum-Pacific Council for Energy and Mineral Resources, Earth Science Series, v. 11, 405-414.

Kroonenberg, S.B., and Diedrix, H. (1984): Geology of South-Central Huila, Uppermost Magdalena Valley, Colombia. CIF, Unidad de Geología, Field Trip Guide, 1-39, Bogotá.

Kugler, H.G. (1953): Jurassic to recent sedimentary environments in Trinidad. Bulletin Association Suisse des Géologues et Ingénieurs du Pétrole, v. 20, 27-60.

Kugler, H.G. (1955): Trinidad, *in* Lexique Stratigraphique International, v. V: Amérique Latine (Dir.: Hoffstetter, R.), Fasc. 2 b, Antilles, Paris, Centre National de la Recherche Scientifique, 39-116.

Ladd, J.W., and 7 others (1984): Seismic reflection profiles across the southern margin of the Caribbean, *in* The Caribbean-South American Plate Boundary and Regional Tectonics, (Eds.: Bonini, W.E., Hargraves, R.B., and Shagam, R.), Geological Society of America Memoir, 162, 153-159.

López-Ramos, E. (1975): Geological summary of the Yucatán Peninsula, *in* The Ocean Basins and Margins, (Eds.: Nairn, A.E.M., and Stehli, F.G.), v. 3, The Gulf of Mexico and The Caribbean, New York, Plenum, 257-282.

MacDonald, W.D. (1965): Geology of the Serania de Macuira area, Guajira Peninsula, northeast Colombia. Caribbean Geological Conference, 4th, Port-of-Spain, Trinidad, Transactions, 267-274.

MacDonald, W.D., Doolan, B.L., and Cordani, U.G. (1971): Cretaceous-Early Tertiary metamorphic K/Ar age values from the South Caribbean. Geological Society of America Bulletin, v. 82, 1381-1388.

Macellari, C.E., and De Vries, T.J. (1987): Late Cretaceous upwelling and anoxic sedimentation in northwestern South America. Paleogeography, Paleoclimatology, Paleoecology, v. 59, 279-292.

Mann, P., and K. Burke (1984): Neotectonics of the Caribbean. Reviews of Geophysics and Space Physics, v. 22, 309-362.

Maresch, W.V. (1974): Plate tectonics origin of the Caribbean Mountain System of northern South America; discussion and proposal. Geological Society of America Bulletin, v. 85, 669-682.

Martín-Bellizzia, C., and de Arozena, J.M.I. (1972): Complejo ultramáfico zonado de Tausabana-El Rodeo, gabro zonado de Siraba-Capuana y complejo subvolcánico estratificado de Santa Ana, Paraguaná, Estado Falcón. Caribbean Geological Conference, 6th, Margarita, Venezuela (1971), Transactions, 337-356.

Mattson, P. (1984): Caribbean structural breaks and plate movements, *in* The Caribbean-South American Plate Boundary and Regional Tectonics, (Eds.: Bonini, W.E., Hargraves, R.B., and Shagam, R.), Geological Society of America Memoir, 162, 131-152.

Maze, W.B. (1984): Jurassic La Quinta Formation in the Sierra de Perijá, northwestern Venezuela: geology and tectonic environment of redbeds and volcanic rocks, *in* The Caribbean-South American Plate Boundary and Regional Tectonics, (Eds.: Bonini, W.E., Hargraves, R.B., and Shagam, R.), Geological Society of America Memoir, 162, 263-282.

McCollough, C.N. (1990): Geology of the Super Giant Cano Limon Field and the Llanos Basin, Colombia, *in* Circum-Pacific Hydrocarbon and Mineral Resources Conference, 4th, (1986), Transactions, 299-316.

McCourt, W.J., Aspden, J.A., and Brook, M.J. (1984): New geological and geochronological data from the Colombian Andes: Continental growth by multiple accretion. Geological Society of London Journal, v. 141, 831-845.

Minster, J.B., and Jordan, T.H. (1978): Present-day plate motions. Journal of Geophysical Research, v. 83, 5331-5354.

Muessig, K.W. (1984): Structure and Cenozoic tectonics of the Falcón basin, Venezuela, and adjacent areas, *in* The Caribbean-South American Plate Boundary and Regional Tectonics, (Eds.: Bonini, W.E., Hargraves, R.B., and Shagam, R.), Geological Society of America Memoir, 162, 217-230.

Pardo, G. (1975): Geology of Cuba, *in* The Ocean Basins and Margins, (Eds.: Nairn, A.E.M., and Stehli, F.G.), v. 3, The Gulf of Mexico and The Caribbean, New York, Plenum, 553-615.

Pérez de Mejía, D., Kiser, G.D., Maximowitsch, B., and Young, G. A. (1980): Geología de Venezuela, *in* Evaluación de Formaciones en Venezuela, (Ed.: Felder, B.), Caracas, Schlumberger Surenco, I/1-I/23.

Pindell, J.L. (1985a): Alleghenian reconstruction and subsequent evolution of the Gulf of Mexico, Bahamas, and Proto-Caribbean. Tectonics, v. 4, 1-39.

Pindell, J.L. (1985b): Plate tectonic evolution of the Gulf of Mexico and Caribbean region. Ph.D. Thesis, Durham University, Durham, England, 227 pp.

Pindell, J.L. (1990): Arguments suggesting a Pacific origin for the Caribbean Plate, *in* Caribbean Geological Conference, 12th, Christianstadt, Saint Croix, (Eds.: Larue, D.K., and Draper, G.), Transactions, 1-4.

Pindell, J.L. (1991): Geologic rationale for hydrocarbon exploration in the Caribbean and adjacent regions. Journal of Petroleum Geology, v. 14, 237-257.

Pindell, J.L., and Barrett, S.F. (1990): Geological evolution of the Caribbean region; a plate tectonic perspective, *in* The Caribbean Region, (Eds.: Dengo, G., and Case, J.E.), Geological Society of America, The Geology of North America, v. H, 405-432.

Pindell, J.L., and Dewey, J.F. (1982): Permo-Triassic reconstruction of western Pangea and the evolution of the Gulf of Mexico/Caribbean region. Tectonics, v. 1, 179-212.

Pindell, J.L., and Drake, C.L. (in review): Mesozoic-Cenozoic stratigraphy and tectonic evolution of the Caribbean region/northern South America: Implications for eustasy from exposed sections of a Cretaceous-Eocene passive margin setting. Geological Society of America Special Paper.

Pindell, J.L., Cande, S.C., Pitman III, W.C., Rowley, D.B., Dewey, J.F., LaBrecque, J., and Haxby, W. (1988): A plate-kinematic framework for models of Caribbean evolution. Tectonophysics, v. 155, 121-138.

Pindell, J.L., Erikson, J.P., and Algar, S.T. (1991): The relationship between plate motions and sedimentary basin development in northern South America: from a Mesozoic passive margin to a Cenozoic eastwardly-progressive transpressional orogen, *in* Geological Conference of the Geological Society of Trinidad and Tobago, 2nd, (Ed.: Gillezeau, K.A.), Port of Spain (1990), Transactions, 191-202.

Pindell, J.L., Algar, S.T., Erikson, J.P., Dewey, J.F. (in press, a): Mesozoic-Cenozoic tectonic and stratigraphic evolutionary synthesis of northern South America, *in* Mesozoic-Cenozoic stratigraphy and tectonic evolution of the Caribbean region/northern South America: Implications for eustasy from exposed sections of a Cretaceous-Eocene passive margin setting, (Eds.: Pindell, J.L., and Drake, C.L.), Geological Society of America Special Paper.

Pindell, J.L., Robinson, E.G., and Dewey, J.F. (in press, b): Paleogeographic evolution of the Gulf of Mexico/Caribbean Region, *in* Mesozoic-Cenozoic stratigraphy and tectonic evolution of the Caribbean region/northern South America: Implications for eustasy from exposed sections of a Cretaceous-Eocene passive margin setting, (Eds.: Pindell, J.L., and Drake, C.L.), Geological Society of America Special Paper.

Pindell, J.L., Tabbutt, K., Villamil, T., and Erikson, J.P. (in review): Neogene palinspastic restoration and models for the Mesozoic-Cenozoic evolution of the northern Andes region: implications for hydrocarbon maturation history in the Andean basins of Ecuador, Colombia, and western Venezuela, *in* Mesozoic-Cenozoic stratigraphy and tectonic evolution of the Caribbean region/northern South America: Implications for eustasy from exposed sections of a Cretaceous-Eocene passive margin setting, (Eds.: Pindell, J.L., and Drake, C.L.), Geological Society of America Special Paper.

Pumpin, V. (1979): El marco estructural de Venezuela noroccidental. Caracas, EPC Report, Maraven.

Renz, O. (1959): Estratigrafía del Cretáceo en Venezuela occidental. Dirección de Geología, Boletín de Geología, v. 5, 3-48, Caracas.

Rod, E., and Maync, W. (1954): Revision of Lower Cretaceous stratigraphy of Venezuela. American Association of Petroleum Geologists Bulletin, v. 38, 193-283.

Rosencrantz, E., Ross, M.I., and Sclater, J.G. (1988): Age and spreading history of the Cayman Trough as determined from depth, heat flow, and magnetic anomalies. Journal of Geophysical Research, v. 93, 2141-2157.

Rosenfeld, J.H. (1990): Sedimentary rocks of the Santa Cruz Ophiolite, Guatemala -A Proto-Caribbean history, *in* Caribbean Geological Conference, 12th, Christianstadt, Saint Croix, (Eds.: Larue, D.K., and Draper, G.), Transactions, 513-519.

Rowley, D.B., and Pindell, J.L. (1989): End Paleozoic-Early Mesozoic western Pangean reconstruction and its implications for the distribution of Precambrian and Paleozoic rocks around Meso-America. Precambrian Research, v. 42, 411-444.

Santamaria, F., and Schubert, C. (1974): Geochemistry and geochronology of the Southern Caribbean-Northern Venezuela plate boundary. Geological Society of America Bulletin, v. 85, 1085-1098.

Schlager, W. (1981): The paradox of drowned reefs and carbonate platforms. Geological Society of America Bulletin, v. 92, 197-211.

Schlager, W. (1989): Drowning unconformities on carbonate platforms, in Controls on carbonate platform and basin development (Eds.: Crevello, P.D., Wilson, J.L., Sarg, J.F., and Read, J.F.), Society of Economic Paleontologists and Mineralogists Special Publication, 44, 15-25.

Schlager, W. (1991): Depositional bias and environmental change -important factors in sequence stratigraphy. Sedimentary Geology, v. 70, 109-130.

Schubert, C. (1988): Neotectonics of La Victoria Fault Zone, north-central Venezuela. Annales Tectonicae, v. II, 58-66.

Silver, E.A., Case, J.E., and MacGillavry, H.J. (1975): Geophysical study of the Venezuelan Borderland. Geological Society of America Bullettin, v. 86, 213-226.

Snoke, A.W. (1991): An evaluation of the petrogenesis of the Accreted Mesozoic Island Arc of the Southern Caribbean, in Geological Conference of the Geological Society of Trinidad and Tobago, 2nd, (Ed.: Gillezeau, K.A.), Port of Spain (1990), Transactions, 222-232.

Snoke, A.W., Yule, J.D., Rowe, D.W., Wadge, G., and Sharp, W. D. (1990): Stratigraphic and structural relationships on Tobago, West Indies, and some tectonic implications, in Caribbean Geological Conference, 12th, Christianstadt, Saint Croix, (Eds.: Larue, D.K., and Draper, G.), Transactions, 389-402.

Speed, R.C., and Westbrook, G.K. (Eds.) (1984): Atlas 10, Ocean margin drilling program, Lesser Antilles and adjacent ocean floor. Woods Hole, Massachusetts, Marine Science Institute Publications.

Stephan, J.-F., Beck, C., Bellizzia, A., Blanchet, R. (1980): La chaîne Caraïbe du Pacifique à l'Atlantique. International Geological Congress, 26th, Paris, v. C-5, in Bureau de Recherches Géologiques et Minières Memoire, 115, 38-59.

Talukdar, S., and De Toni, B. (1988): Modelaje geoquímico en la cuenca inexplorada de Guarumen, Venezuela. Simposio Bolivariano, Exploración Petrolera de las Cuencas Subandinas, 3rd, Caracas, Memorias, 755-781.

Talukdar, S., Gallango, O., and Ruggiero, A. (1988): Generation and migration of oil in the Maturín subbasin, Eastern Venezuela Basin. Organic Geochemistry, v. 13, 537-547.

Tschanz, C.M., Marvin, R.F., Cruz, B., Mehnert, H.H., and Cebula, G.T. (1974): Geologic evolution of the Sierra Nevada de Santa Marta, northeastern Colombia. Geological Society of America Bulletin, v. 85, 273-284.

van Erve, A. (1986): Palynostratigraphy and palaeogeography of the Cretaceous Temblador Group in the area of Zuata -Faja Petrolifera del Orinoco, eastern Venezuela. Congreso Geológico Venezolano, 6th, Caracas (1985), Sociedad Venezolana de Geólogos, Memoria, 1210-1242.

Vivas, V. (1987): Bioestratigrafía del Cretáceo de la región de Bergantín-Santa Inés, Estado Anzoátegui, Venezuela nororiental. Dirección de Geología, Boletín de Geología, v. 16, 29, 3-128, Caracas.

Von der Osten, E. (1957): Lower Cretaceous Barranquín Formation of northeastern Venezuela. American Association of Petroleum Geologists Bulletin, v. 41, 679-708.

Yoris, F.G. (1986): Revisión de la estratigrafía del Cretáceo inferior al sur y este de la Serranía del Interior, Venezuela nororiental. Congreso Geológico Venezolano, 6th, Caracas (1985), Sociedad Venezolana de Geólogos, Memoria, 1343-1393.

Yoris, F.G. (1988): Localidades tipo y secciones de referencia para los miembros de la Formación El Cantil en la Serranía del Interior, Venezuela nororiental. Sociedad Venezolana de Geólogos, Boletín, v. 34, 52-69, Caracas.

Zambrano, E., Vásquez, E., Duval, B., Latreille, M., and Coffinieres, B. (1972): Paleogeographic and petroleum synthesis of western Venezuela. Paris, Editions Technip, 59 pp.

The Colombian Andes During Cretaceous Times

J. F. Toussaint, J. J. Restrepo
Universidad Nacional, A. A. 3840, Medellín, Colombia.

Abstract

During Cretaceous times, the Colombian Andes underwent major geological changes.
In the Colombian East, which is located east of the Otú-Pericos fault, distension and subsidence brought about the formation of a large basin in which thick sequences of epicontinental sediments were deposited.
In the Colombian West, an oceanic domain with ophiolites was generated in an allochthonous position and later thrust over the continental margin. This first tectonic event is associated with a mid to high pressure metamorphic event. Later an island-arc magmatism developed, characterized by the formation of volcanic-sedimentary assemblages and an intermediate plutonic belt.
A second important tectono-metamorphic event affected the entire Colombian West during Late Cretaceous times, and a later magmatism of intermediate composition marks the termination of the tectono-genesis.
The assemblage between the Colombian East and the Colombian West occurred during the end of the Cretaceous or beginning of the Cenozoic, with large-scale lateral displacement along the Otú-Pericos fault. This movement continued later along the Cauca-Romeral system that crosses the tectono-metamorphic structures formerly created.
A comparison between the Colombian and Ecuadorian Andes shows similarities with regard to the lithostratigraphy of several terranes, both oceanic and continental, and with regard to several accretion ages. However, the distension and subsidence that affected the Colombian East seem to be features exclusive to the northern region.

Introduction

The Colombian Andes are formed by several ranges of approximately N-S strike, separated by inter-Andean valleys. They are -from east to west- the Eastern Cordillera, the Rio Magdalena Valley, the Central Cordillera, the Rio Cauca Valley, the Western Cordillera, the Rio Atrato-Rio San Juan valleys and the Serrania de Baudó. On the eastern side of the Andes there is a region of vast plains called the "Llanos Orientales" (Figure 1).

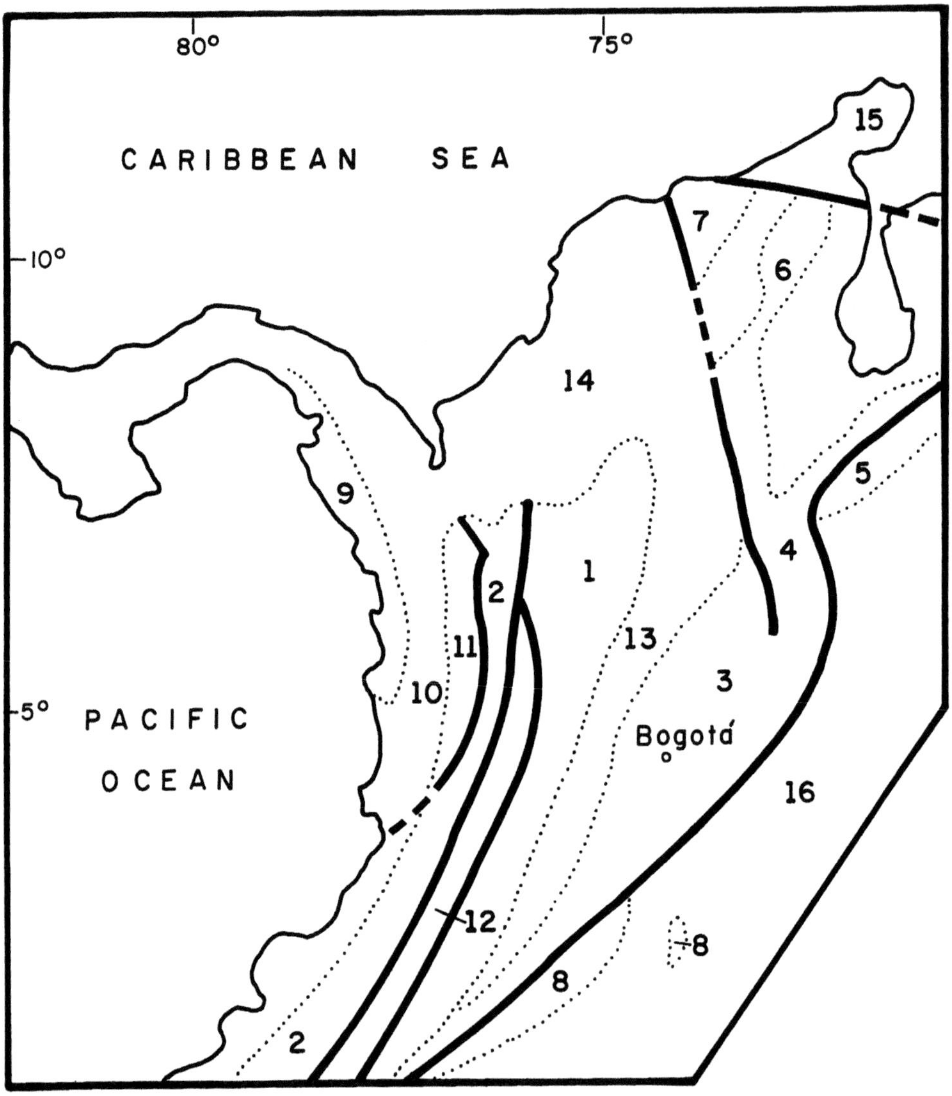

Fig. 1

Physiographic map of Colombia. 1- Central Cordillera, 2- Western Cordillera, 3- Santander Massif, 5- Andes de Mérida, 6- Sierra de Perijá, 7- Sierra Nevada de Santa Marta, 8- Sierra de la Macarena and Garzón Massif, 9- Serranía de Baudó, 10- Atrato Valley, 11- Mandé Arc, 12- Cauca Valley, 13- Magdalena and César Valleys, 14- Caribbean Coastal plains, 15- La Guajira Peninsula, 16- Llanos Orientales.

The structural position of the Colombian Andes (Figure 18) has been examined in the general context of the Andean chain, in particular by Gansser (1973), Aubouin (1977), and Zeil (1979), who pointed out the fundamental differences between the Northern and Central Andes, such as, for example, the presence or absence of ophiolites and the tectonic style. However, few papers have been published with specific reference to the geotectonic evolution of the Colombian Andes, and until the beginning of the 70's they mainly focused on the autochthonist view, based on traditional geosynclinal notions (e.g. Radelli, 1967; Irving, 1971). The hypotheses regarding the allochthony of the western part of the Colombian Andes began to be outlined in papers by Case et al. (1971), Estrada (1972), and Restrepo and Toussaint (1973), while in recent years the idea has been proposed that the Eastern region of the country is also allochthonous (e.g. Case et al., 1984; Etayo et al., 1986; Restrepo and Toussaint, 1988).

Thus, in the past few years a succession of papers concentrating on the notion of allochthonous terranes have proposed that the NW corner of South America is composed of a mosaic of terranes accreted to the Guyana Shield during different geological periods, and in particular during the Cretaceous (Etayo et al., 1986; Restrepo and Toussaint, 1988; Toussaint and Restrepo, 1988; Duque-Caro, 1989).

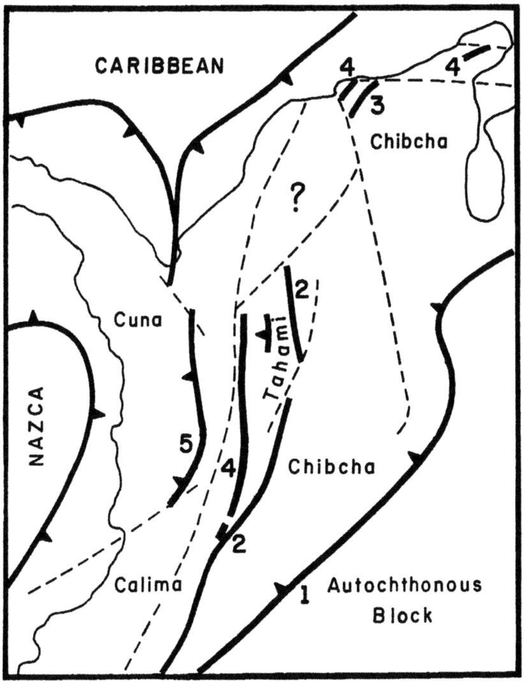

Fig. 2

Map of Colombian terranes (after Toussaint and Restrepo, 1989). 1- Guaicaramo Fault, 2- Otú-Pericos Fault, 3- Sevilla Alignment, 4- Tahamí and Calima boundary, 5- Dabeiba- Pueblo Rico Fault.

A brief description of the chief characteristics of the Colombian terranes is presented below using the terminology recently employed by Toussaint and Restrepo (1989).
From east to west, the following are found (Figure 2):
An Autochthonous Block (BA) joined to the Guyana Shield since the end of the Precambrian, at least, that coincides with the "Llanos Orientales", Caquetá and Amazon regions. It has a continental crust with a thickness of some 35 km. Radiometric Precambrian datings suggest an important Transamazonic tectono-metamorphic event followed by a Nickerian rejuvenation. The Early Paleozoic is represented by Cambrian-Ordovician sediments, while the Late Paleozoic is missing.
The Chibcha terrane (Ch) -with a Precambrian continental basement- comprises the Eastern Cordillera, The Santander Massif, the eastern flank of the Central Cordillera and the southeastern part of the Sierra Nevada de Santa Marta (SNSM). The last tectonic event that affected it was Caledonian. Late Laleozoic marine sedimentary rocks unconformably cover the metamorphic units of the Early Paleozoic. The Chibcha terrane was accreted to the Autochthonous Block at the end of the Paleozoic by means of the Guaicaramo fault, and later strong magmatism affected its eastern border during the Jurassic. An extensive Meso-Cenozoic sedimentary cover was deposited on it, reaching a thickness of 12.000 m of marine Cretaceous sediments in the Bogotá area.
The Tahamí terrane (Ta) with a continental crust about 45 Km. thick, essentially embraces the northern part of the Central Cordillera. It has experienced several tectono-metamorphic events of Precambrian(?), Acadian, Hercynian and Cretaceous age. The thin marine sedimentary cover is limited to the Early Cretaceous.
The Calima terrane (Ca) with an oceanic basement some 35 Km. thick, encompasses a great part of the Western Cordillera but is also displayed on the western flank of the Central Cordillera. It is formed by units of exclusively Cretaceous age representing oceanic crust and island arc environments. It is probable that the Calima terrane is a megaterrane formed by several terranes which have not yet been fully recognized.
Important Cretaceous tectono-metamorphic events mark the amalgamation between the Calima and Tahamí Terranes. These were followed by the post-tectonic intrusion of the huge Antioquian Batholith. Later, at the end of the Cretaceous, the composite Calima-Tahamí terrane was accreted, through the Otú-Pericos wrench megafault, to the continent composed of the Autochthonous Block and the Chibcha terrane.
The Cuna terrane (Cu), with a Cretaceous oceanic basement, includes the Serranía del Baudó, the Atrato Basin and the northwestern border of the Western Cordillera. It is formed by oceanic crust and island arc materials, with ages ranging from Late Cretaceous to Early Miocene. The great Mandé Batholith, with porphyry copper deposits is representative of the Paleo-Eocene island-arc magmatism.
During Miocene times, the Cuna terrane was accreted to the Andean Block, formed by a mosaic of formerly accreted terranes. The Dabeiba-Pueblo Rico suture coincides with a

series of eastward thrusts. During this final collision, the Colombian Andes underwent a notable reduction in width, marked in particular by the formation of the Eastern Cordillera Megahorst, with its eastern border thrust eastward over the eastern plains of the "Llanos Orientales", and its western border thrust westward over the Magdalena Valley. Consequently, the morphotectonism of the Colombian Andes depicted in the cross section located at 6°N (Figure 3) seems to be in great part a consequence of the Miocene collision of the Cuna terrane against the Andean Block.

General Comments on the Cretaceous

The Cretaceous is, without a doubt, the most complicated period in the geologic evolution of the Colombian Andes, since at this time many sedimentary, magmatic, metamorphic and tectonic phenomena occured, the interrelations and exact datings of which are not fully clear.

On a regional scale, the feature of Cretaceous evolution that draws most attention is the clear difference between the Colombian East, which includes the Autochthonous Block and the Chibcha terrane, on the one hand, and the Colombian West, comprising the Tahamí, Calima and Cuna terranes, on the other. The boundary between both domains, presently located on the eastern flank of the Cordillera Central, is represented by the Otú-Pericos fault, whose extension in the SNSM is the Sevilla alignment (see Figure 2).

The fundamental differences between the Colombian East and Colombian West showed up very strongly during the Cretaceous, and are mainly related to the contrasts in tectonic environment governing both domains.

Furthermore, the differences between East and West are not completely related to their basement type, since continental basements exist both in the East and in the Tahamí terrane in the West, while oceanic basements are found exclusively in the Calima and Cuna terranes.

During the Cretaceous, the Colombian East was affected by extensional tectonism and subsidence which caused the deposition of an epicontinental marine sedimentation more than 12000 m thick in the Bogota Basin, in a very calm environment. No unconformity is present in this sequence, and magmatism is limited to a few gabbro stocks.

In contrast, significant mainly tectono-metamorphic and magmatic events took place in the Colombian West, imprinting its essential features on the chain, especially the accretion of an oceanic terrane to the Northern Andes. The formation of allochthonous oceanic crust and island arcs, followed by their tectonic emplacement, impart a clearly Caribbean feature to this part of the Andes. These characteristics were noted in 1955 by Gerth, who gave the Colombian-Ecuadorian Andes the name of Caribbean Andes.

Furthermore, it was at the end of the Cretaceous that the composite Tahamí-Calima terrane and the Chibcha terrane were joined together by a large wrench displacement, forming the Andean Block.

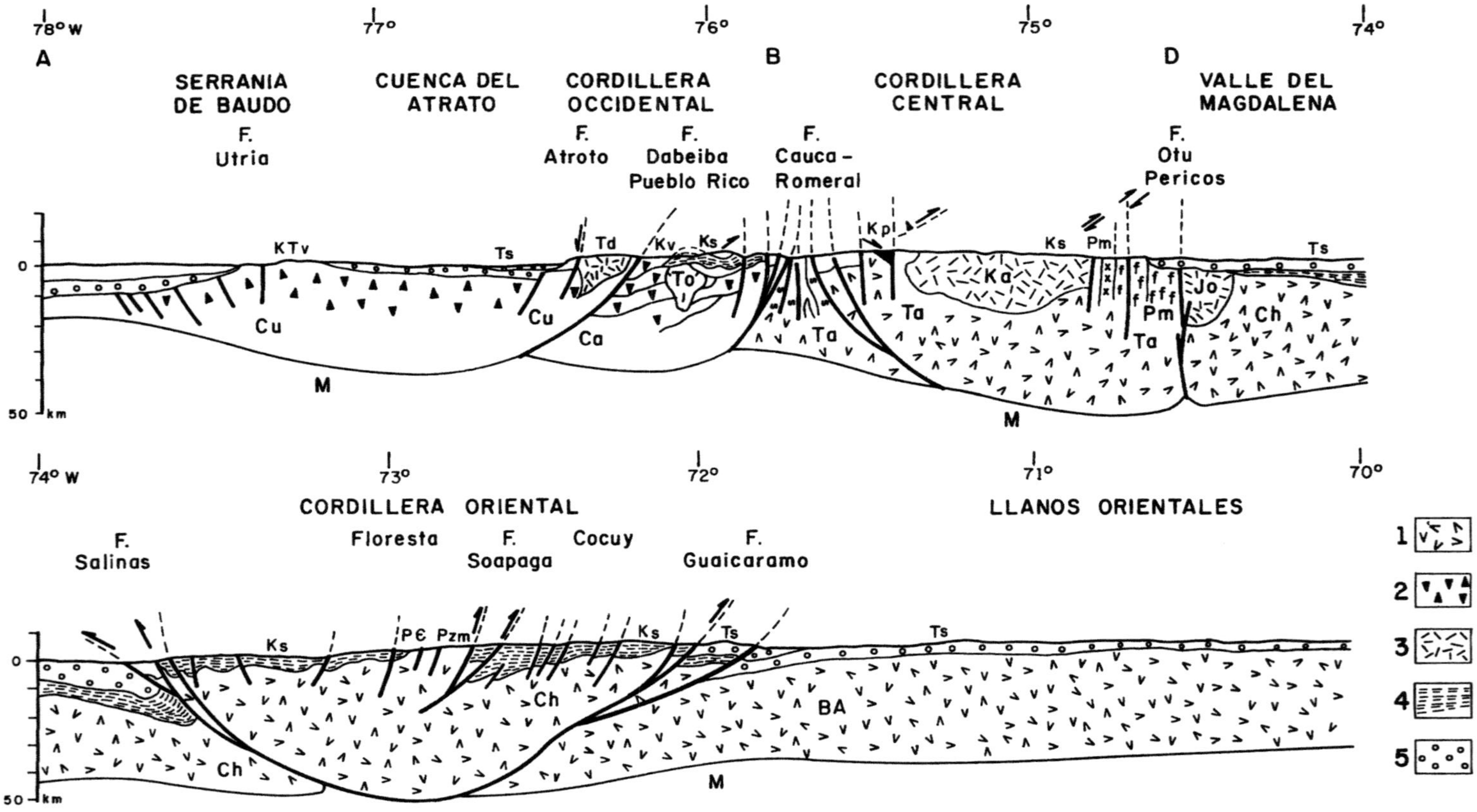

Fig. 3

6°N Transection. BA- Autochthonous Block, Ch- Chibcha Terrane, Ta- Tahamí Terrane, Ca- Calima Terrane, Cu- Cuna Terrane, M- Moho, 1- Continental basement, 2- Oceanic basement, 3- Intermediate plutonism, 4- Mesozoic sediments, 5- Cenozoic sediments.

Many problems remain unsolved in the Colombian West of Cretaceous times. The origin of the basic and ultrabasic assemblages, the region's tectonic emplacement and the igneous and metamorphic events associated with the polyphasic tectono-genesis that occurred in the Cretaceous are strongly debated issues.

In this study we first present the geologic features of the Colombian East and then analyze diverse aspects and characteristics of the western region, while endeavoring to outline the various working hypotheses currently being discussed.

Evolution of the Colombian East

During Cretaceous times one of the most noteworthy phenomena in the evolution of the Colombian Andes was the formation of an extensional basin in the Colombian East (Figure 18), in which a thick sequence of epicontinental marine sediments was deposited with no unconformities whatsoever (see Etayo et al., 1985).

Extension During Tithonian-Albian Times

The extension that produced an important marine transgression in the Colombian East during the Cretaceous is probably the continuation of what occurred during the Early Mesozoic in the same region. Thus, during the Jurassic the Colombian East was affected by extensional events that gave place to the formation of grabens, located in an area limited to the west by a N-S magmatic arc which developed on the western margin of the Chibcha terrane and to the east by the continental platform in the emerged eastern plains of the "Llanos Orientales".

The Jurassic extensional basin has been interpreted as a back-arc basin located eastward of the magmatic belt and characterized by the intrusion of numerous batholiths. In this depression, affected by normal faults that produced more or less depressed blocks, a continental sedimentation characterized by the red sandstone of the Giron Group was deposited. However, the sea was relatively close, since it sometimes invaded the edge of the basin.

At the end of the Jurassic, movements in a NS, NW-SE and NE-SW fault net allowed block readjustments and rotation. According to Radelli (1967) the inclination of these blocks was generally eastward during the Jurassic sedimentation, but this situation seems to have changed during Nevadian movements, with an inverse inclination; the depth of the grabens was therefore greater in the west during Cretaceous times. In any case, when the Cretaceous marine transgression began the region was already affected by extensive fracturing. Many faults that had reverse type displacements during the Cenozoic Andean phases were really older faults having their origin, at least, in the Early Mesozoic extension. This is the case with the Salinas, Suarez, Boyacá, Sóapaga and Honda faults, that had normal-type movements both during the Jurassic and the Early

Cretaceous. In relation to the Guaicaramo fault that separates the Andes from the eastern plains, it is an even older fault that probably acted as a limit between terranes from the Precambrian onward (Figure 4).

According to Fabre (1984), the Jurassic extension continued during a good part of the Early Cretaceous, bringing about a rapid subsidence of some grabens separated by raised blocks having a slower rate of subsidence. The extension phase may also explain why, during the Early Cretaceous, formation facies and thickness varied sharply from one region to another. It is estimated that subsidence in the troughs located in the Cocuy region and in the Middle Magdalena region lay within the 150-250 m per m.y. range, while the intermediate Santander Massif remained in a stable position. The extension produced a thinning of the whole crust and probably, at least in the Middle Magdalena region, of the whole lithosphere (Fabre, 1984). According to this author, the thinning of the crust favored an increased heat flow, which may have reached 2 HFU with a geothermal gradient of approximately 40°C/Km (Figure 5).

Thinning of the lithosphere may also have brought about a partial melting of the mantle and further emplacement of alkaline basic intrusives in the regions with the greatest subsidence.

It is in this environment of general expansion in which the Cretaceous marine transgression must be situated. At the end of Jurassic the continental and fluvial-lacustrine facies, represented by the Arcabuco Formation in the Villa de Leyva region and the Los Santos Formation near Bucaramanga, were still being deposited, although the Caqueza Group in the Bogota basin was already clearly marine. The Berriasian and Valanginian are present in shallow marine levels, such as the limestones of the Rosablanca Formation in Santander and Boyacá, but sedimentation was then incresed by euxinic facies, particularly in the Paja Formation of the Middle Magdalena Basin and in the Villeta Group of the Bogotá Basin. The euxinic character of this sedimentation became even more accentuated during the Late Cretaceous.

Marine sedimentation began at the end of Jurassic time in the Middle Magdalena Basin, and extended progressively northward up to the southern part of the Cesar Valley during the Barremian (Figure 6). This transgression came from the south, and was joined at the end of the Aptian by that which affected the Guajira Region to the north. However, both the SNSM and the Serrania de Perijá remained emerged during a great part of the Early Cretaceous.

The Santander Massif was also an uplifted area until the end of the Hauterivian. For this reason, the thickness of marine formations deposited in the Magdalena Valley decreases progressively and their facies change as they approach the massif. It was only during the Aptian that the Santander Massif was submerged, remaining covered by the sea during much of the Late Cretaceous. This opened a means of communication between

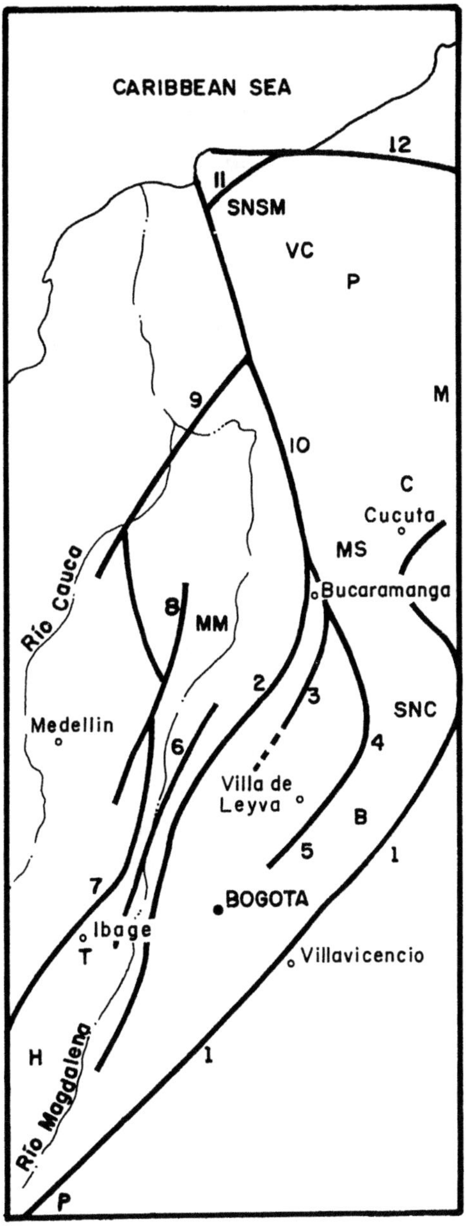

Fig. 4

Main structural units and faults of the Colombian East. 1- Guaicaramo Fault, 2- Salinas Fault, 3- Suárez fault, 4- Soapaga Fault, 5- Boyacá Fault, 6- Honda Fault, 7- Otu-Pericos Fault, 8- Palestina Fault, 9- San Marcos Fault, 10- Bucaramanga- Santa Marta Fault, 11- Sevilla Alignment, 12- Oca Fault, SNSM- Sierra Nevada de Santa Marta, VC- Valle del Cesar, SP- Serranía de Perijá, M- Maracaibo Basin, C- Catatumbo Basin, B- Boyacá, P- Putumayo, H- Huila, T- Tolima, MM- Middle Magdalena Basin.

the Middle Magdalena and the Catatumbo and Maracaibo Basins. In the south the sea reached Tolima and Huila and later the Putumayo region during Aptian time.

The Early Cretaceous depression seems to have been a somewhat enclosed basin, protected from the actively open ocean located to the W and NW. This feature may be

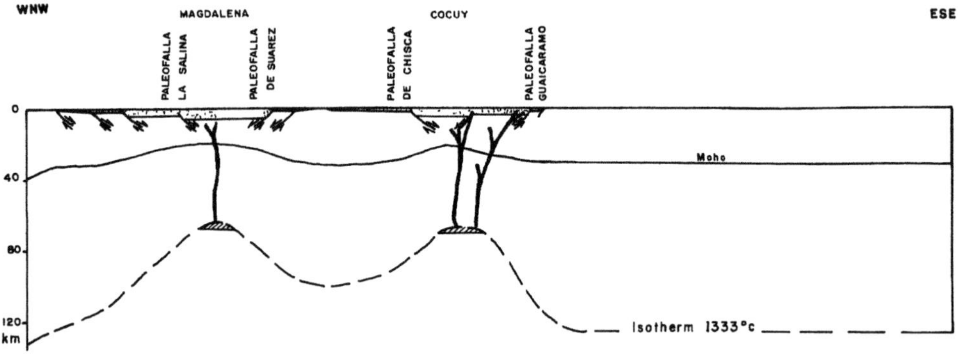

Fig. 5

Depth section of the Colombian East along 6° 30'N during Aptian times (after Fabre, 1984).

indicative of the presence of a barrier between the East Andean epicontinental basin and the open sea.

Many authors consider that this barrier is represented by the Central Cordillera as a whole. However this hypothesis is not completely convincing, due to the great paleogeografic and tectonic differences that exist between the regions located to the West and East of the Otú-Pericos megafault during both the Early Cretaceous and the Late Cretaceous. These differences between the East and West suggest that the Tahamí and Calima terranes, now located west of the Autochthonous Block-Chibcha terrane, were not in this position during the Cretaceous. Taking the foregoing into account, it seems more feasible to think that the barrier that isolated the East Andean basin was the magmatic belt formed during Jurassic time on the western border of the Autochthonous Block-Chibcha terrane block, stretching from the SNSM to the south of the country. This belt, formed by great batholiths emplaced between 190 and 140 Ma ago, was probably associated with a volcanic arc that could have constituted a barrier to the west of the eastern epicontinental basin. The erosion of this barrier probably took a very long time, since the transgression only reached Ibagué at the end of the Aptian (Núñez, 1987).

Subsidence between Albian and Santonian Times

From the beginning of Albian, time onward, practically all the Andean zone of the Colombian East was covered by an epicontinental sea, which progressively invaded the western part of the "Llanos Orientales". Maximum extension was attained during the Santonian, when a fairly rapid general regression began.

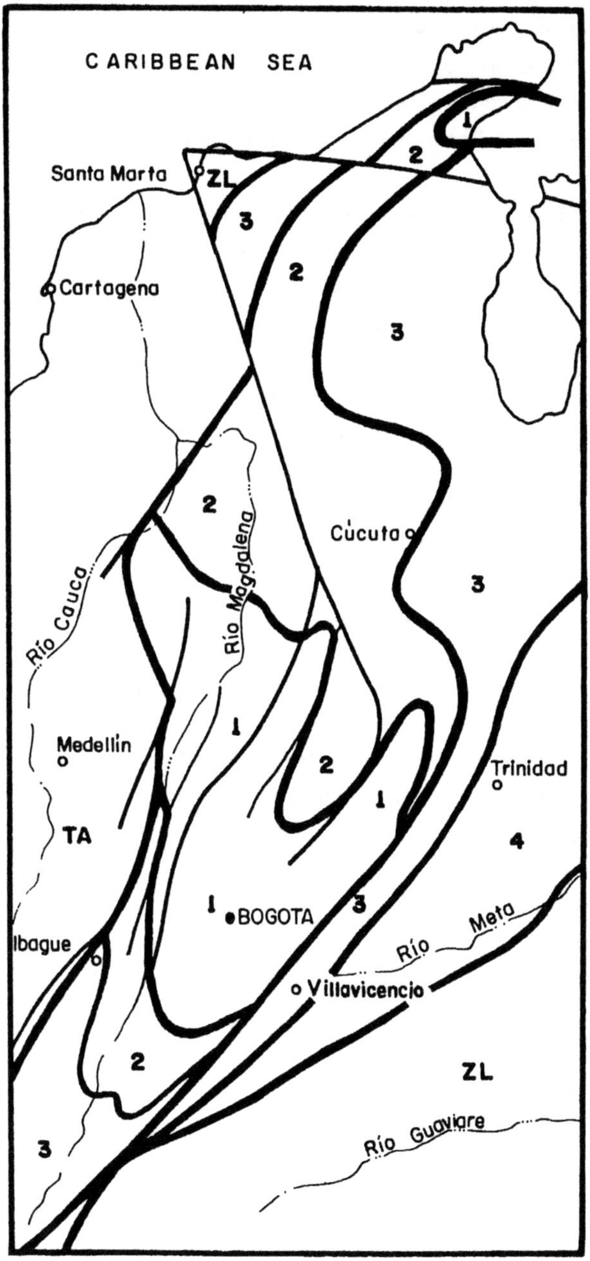

Fig. 6

Map of the Cretaceous marine transgression in the Colombian East. 1- Berriasian, 2- Aptian, 3- Albian, 4- Campanian, Ta- Tahamí Terrane, ZL- Uplift area.

According to Fabre (1984), the extensional period that brought about the formation of grabens bounded by normal faults was followed by a period of thermal subsidence. The faults that had been, active during the Early Cretaceous momentarily ceased their activity, and the subsidence rate diminished. This author also believes that the lithosphere began a progressive recovery of its rigidity by slowly cooling,during which time subsidence also progressively began to affect an ever wider area.

The rapid facies in changes that characterized sedimentation from Tithonian to Aptian times, caused by differences in subsidence between the various blocks, now came to an end. Sedimentation became much more uniform, both in the facies laid down and in the thickness of each formation (Figure 7).

In the Eastern Andes, the epicontinental sea became deeper at the beginning of the Late Cretaceous, and euxinic facies favorable to mother rock genesis were deposited. This was the case with the La Luna Formation in the Middle Magdalena, Santander Massif and Cocuy regions, and with the Churivita, San Rafael and Conejo Formations in the Villa de Leyva area. Phosphate levels indicating a hot climate appear as from the Turonian, becoming quite significant on both sides of the Santander Massif during Coniacian time. In the Bogotá Basin the upper part of the Villeta Group (containing the dark shales of the Chipaque Formation associated with evaporite levels) was laid down. Toward the "Llanos Orientales", shale and dark carbonate sedimentation progress to more detritic facies with the Gacheta Formation sandstones. This would tend to show that those sediments came from the Guyana Shield.

The Albian-Santonian transgression was quite rapid, probably due to a rise in the eustatic sea level. However this level fell again during Campanian times (Fabre, 1984).

There then began the emergence that extended over all the Colombian East. The prevailing facies were sandstones, frequently associated with coal deposits.

The sea moved back from the "Llanos Orientales" during the Campanian and from the Bogotá Basin during the Maastrichtian, with the deposition of the still marine Guadalupe Group sandstones, followed conformably by the continental sediments of the Guaduas Formation.

In the Middle Magdalena Basin sedimentation continued with a predominance of Campanian-Maastrichtian marine shales, followed by lagoon and deltaic deposits of the Paleocene Lisama Formation. In the region between Catatumbo Basin and the Sierra Nevada del Cocuy the Colon, Mito-Juan and Catatumbo Formations of coastal sandstones and shales were deposited.

Therefore, the marine sedimentation which commenced at the end of the Jurassic came to a close at the end of the Cretaceous. In the Bogotá Basin more than 12000 m of sediment were deposited, and in the Magdalena Basin more than 6.000 m. During the 75 m.y. of continuous sedimentation no compressional tectonic movement was produced. The normal faults that operated during the Early Cretaceous became inactive in the Albian, but subsidence continued throoghout the Late Cretaceous. The marine regression during the Campanian and the Maastrichtian was quite rapid, and at the beginning of the Cenozoic practically the entire East Andean region had emerged.

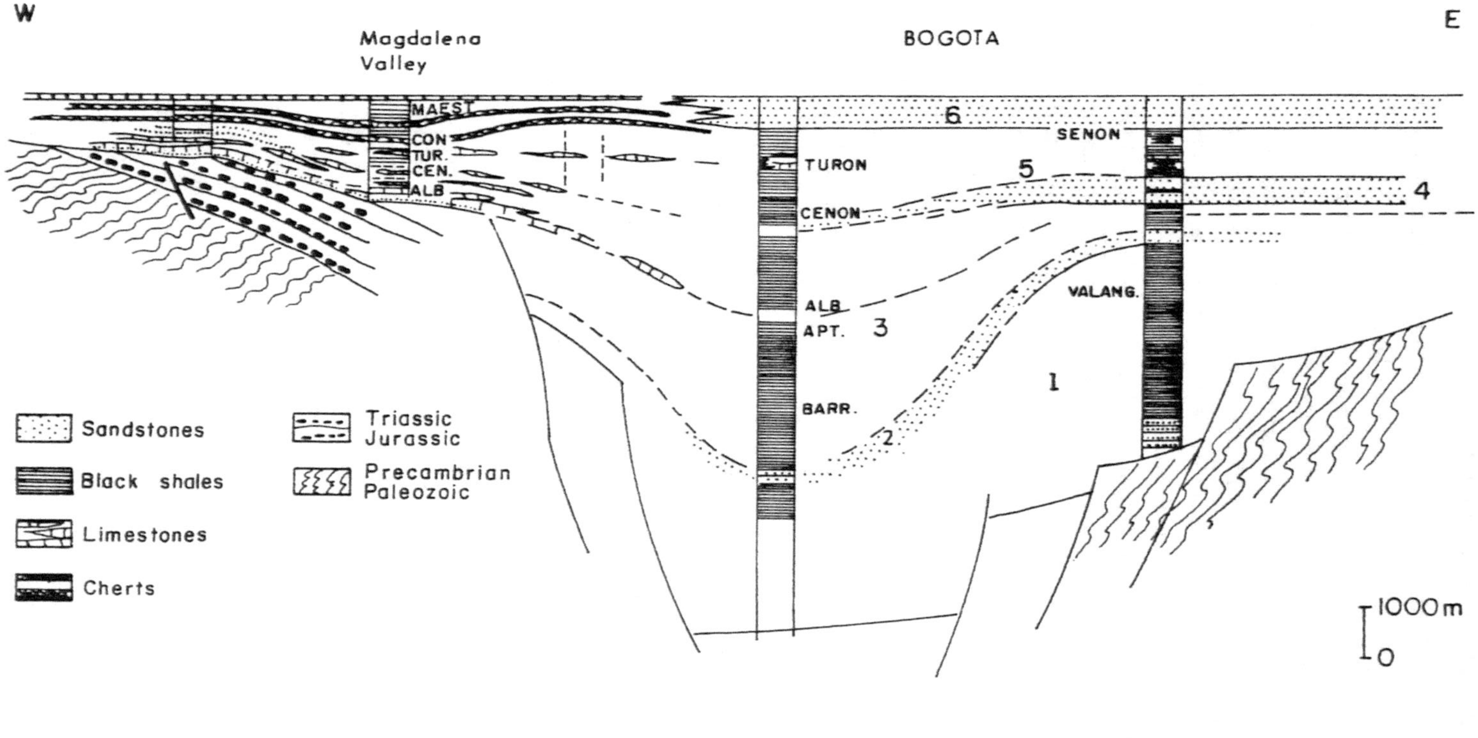

Fig. 7

Cross-section of the Bogotá Basin at the end ofthe Cretaceous (modified from Julivert, 1973). 1- Caqueza Group, 2- Caqueza top sandstones, 3- Villeta Group, Fomeque Formation, 4- Villeta Group, Une Formation, 5- Villeta Group, Chipaque Formation, 6- Guadalupe Group.

Conclusion with regard to the Colombian East

The extension and subsidence that were active during the Cretaceous in the region which coincides approximately with the old Chibcha terrane were in contrast to the compressive tectonic environment that prevailed in the Colombian West.

As noted above, the Cretaceous extension took place in a region affected by a previously established fault net, which no doubt favored the formation of the basins. Also remarkable was the long duration of the subsidence, which gave place to a continuous conformable sedimentation, in other words, there were no abrupt changes either in the direction or in the intensity of regional tectonic strains.

In a speculative fashion, it can be considered that the extension of the Colombian East was a consequence of northeastward movements of the oceanic domain located to the West. Thus, regionally, the Tethys sea that may have been formed at the beginning of the Mesozoic by the division between North and South America was later substituted by the Caribbean Plate, which came from the west and was progressively inserted as a wedge between both continents. This phenomenon implies a dextral movement of the oceanic domain in relation to the South American continent. Such a displacement continued throughout Cretaceous times, and the old Chibcha terrane could have reacted to this general dextral movement with the opening of rhomboid and pull-apart basins, thus permiting deposition of thick sedimentary sequences in a perfectly calm environment.

As will be discussed later, this hypothesis -although highly speculative- is quite in accordance with the phenomena that developed in the Colombian West; phenomena that will be analyzed below.

Evolution of the Colombian West

Introduction

The geologic evolution of the Colombian West (Figures 16 and 18) during the Cretaceous is highly complex. The first apparent event to be detected was the formation of an ophiolite complex, representing an oceanic crust, followed by a first tectono-metamorphic event, characterized by the emplacement of these ophiolites on the continental margin and by the formation of a high- to mid-pressure metamorphic belt.

At the end of the Early Cretaceous, several volcanic-sedimentary units were formed in the Calima terrane. They seem to represent arc-island environments, where the basic rocks are associated with marine sediments, intruded by a magmatic belt of intermediate composition. Contemporary with the former, an epicontinental sedimentation was deposited on the Tahamí terrane.

A second tectono-metamorphic event took place during the Late Cretaceous, marked by tangential tectonism in the Western Cordillera that also affected the Central Cordillera. This second event was followed by post-tectonic magmatism in the Tahamí terrane.

Jurassic(?) and Early Cretaceous Ophiolitic Complexes

The presence of basic and ultrabasic rocks in the western flank of the Central Cordillera and along the Valle del Cauca depression (Figure 8) suggests the existence of ophiolitic complexes representative of old oceanic crusts emplaced on the continental margin (Restrepo and Toussaint, 1974). However, although the spatial association between ultrabasic and basic rocks is frequent, there is no clear evidence that they form complete ophiolite sets, or that the rocks that constitute them are co-magmatic.

In general, the ultrabasic rocks correspond to dunites, wherlites and harzburgites, with different degree of serpentinization, and the basic rocks are represented by spilites, banded and massive gabbros, diabases, pillow basalts and tuffs, associated with some

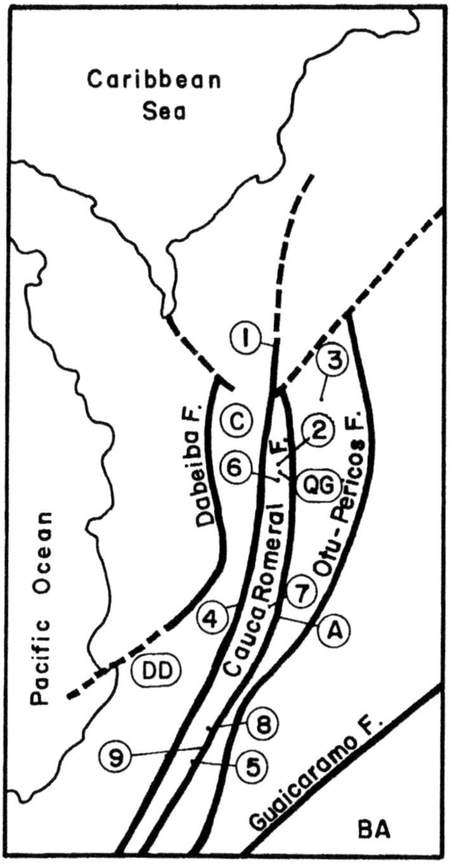

Fig. 8

Main geologic units of the Colombian West duringthe Cretaceous. 1- Cerro Matoso Ultrabasics, 2- Cauca Op' olithic Complex, 3- Río Nechi Ophiolithic Complex, 4- Bolivar Massif, 5- Los Azules Complex, 6- Arquia Group, 7- Pijao Region, 8- Barragán Region, 9- Jambaló Region, C- Cañasgordas Group, QG- Quebradagrande Formation, A- Amaime Formation, DD- Dagua and Diabasic Groups.

marine sediments. Plagiogranites, diorites and norites have been also documented. The ophiolite complexes crop out currently in two parallel belts located on both sides of the Cauca depression, continuing on in Ecuador as far as the Gulf of Guayaquil.
Of the several outcrops of ultrabasic rocks, the most important are those on the Cerro Matoso containing important Ni ore deposits, the Ophiolite Complexes of the Cauca and Nechi rivers in the Central Cordillera, the Bolivar massif on the eastern flank of the Western Cordillera, and the Azules Complex located in the Patia Valley (Figure 8). Radiometric datings obtained from basic rocks supposedly related to ultrabasic rocks have shown ages ranging from 160 Ma to 100 Ma.

Tectono-Metamorphic Event During the Early Cretaceous

The amalgamation between the Calima and Tahamí terranes is one of the fundamental phenomena in the northern Andes, inasmuch as for the first time in the Meso-Cenozoic history of the region oceanic material was accreted to a continental terrane. This tectonism, which strongly influenced the region, is characterized by a series of interrelated phenomena of wich the main one was the emplacement of mafic and ultramafic sets in the Central Cordillera during the Early Cretaceous, associated with the formation of a mid to high pressure metamorphic belt in the Cauca region.
On the western flank of the Central Cordillera -that is, on the eastern border of the Calima terrane- a mid to high pressure metamorphic belt, which is always associated with ophiolite sets, was formed during the Early Cretaceous. This belt extends in discontinuous fashion from 7°30′ N to 4° S.
At 6°N the belt is dominated by the Arquia Group, which includes greenschists, muscovite schists, quartzites and garnet amphibolites, for which an age of 113 Ma was obtained (Restrepo and Toussaint, 1975).
The Medellin Amphibolites and their associated paragneisses, which have traditionally been included in the Paleozoic metamorphic basement of the Tahamí terrane, have also given K/Ar and Rb/Sr isochrone datings of Cretaceous age.
At 4°N latitude the most western metamorphic units of the Central Cordillera were originally included in the Cajamarca Group of supposedly Paleozoic age (Nelson, 1957). However ages of between 120 Ma and 107 Ma were obtained from several of the amphibolites and metagabbros there (Toussaint and Restrepo, 1976, 1978; McCourt et al., 1984). High-pressure rocks, such as glaucophane schists and eclogites, have also been recognized south of 4°N in the regions of Pijao, Tacueyó, Barragán and Jambaló. The blueschists are generally composed of glaucofane, paragonite, phengite, epidote and albite, but lawsonite has only been recognized in Barragán. In Jambaló, the sequence was dated between 125 Ma and 104 Ma (Orrego et al., 1980b; Souza et al., 1984).
To the south of the Gulf of Guayaquil, in Ecuador, blueschists also crop out, as well as the Raspas eclogites in El Oro province, giving ages of 132 Ma (Feininger, 1980).

Several origins have been proposed for the mid to high pressure metamorphic belt. In the opinion of Feininger (1980), the eclogites and blueschists formed in the subduction zone may have been uplifted by serpentinite diapirs. Orrego et al. (1980a) consider that these sequences are part of a melange that may mark the Jurassic-Early Cretaceous subduction zone which was tectonically intercalated with Paleozoic metamorphic units.

Restrepo and Toussaint (1975) hold the view that the mid to high metamorphic belt associated with an ophiolite set is the consequence of a first continental-oceanic collision, which may have taken place during the Early Cretaceous. The metamorphic belt could have been formed in a subduction zone and subsequently raised by obduction, but it could also have been formed directly during the overthrusting of ophiolites along the continental margin.

The description of several cross sections located on the boundary between the Tahamí and Calima terranes may facilitate understanding of this situation.

To the east of Medellin, a dunite body forms a belt about 35 Km. long. Restrepo and Toussaint (1973) point out that this dunite lies in sub-horizontal contact on an amphibolite unit, and that the contact zone is marked by the presence of strongly microfolded talc, chlorite and actinolite schists, derived from the amphibolite by retrograde metamorphism (Figure 9). These schists have been K/Ar dated at 105 Ma, which may indicate the age of tectonic activity. A Paleozoic age for the Medellin Amphibolites has generally been presumed up to the present, but due to recent whole rock isochrone dating of a paragneiss intercalated with the amphibolites and several K/Ar datings, all indicative of the Cretaceous, it is now considered that the Amphibolites belong to the Arquia Complex, a mid to high pressure metamorphic body formed during the Cretaceous tectonogenesis. In this new interpretation, the Medellin Amphibolites are part of the basic top of an ophiolite that may have metamorphosed during or just before its tectonic emplacement.

It is noticeable that the amphibolite foliations, like those of the paragneisses, are enterely sub-parallel to the tectonic contact between dunite and amphibolite. For this reason Tamayo (1984) interprets the regional structure as a large fold overturned eastward.

Toussaint and Restrepo (1986) consider that the amphibolites and dunites may have been been emplaced tectonically over the Tahamí terrane, which crops out to the SW and is represented there by migmatites, orthogneisses and Paleozoic schists.

In the Jambaló region, Bourgois et al. (1985) show that schists with glaucofane are present below the San Antonio Amphibolites, formed by a set of amphibolites, metagabbros and metadiabases. Both the blueschists and the amphibolite show a remarkable westward dipping cleavage, and its tectonic contact is marked by serpentinized peridotite flakes (Figure 10).

According to Bourgois et al. (1985) the amphibolites and peridotites represent an ophiolitic complex, metamorphosed by obduction in HP-LT conditions during their emplace-

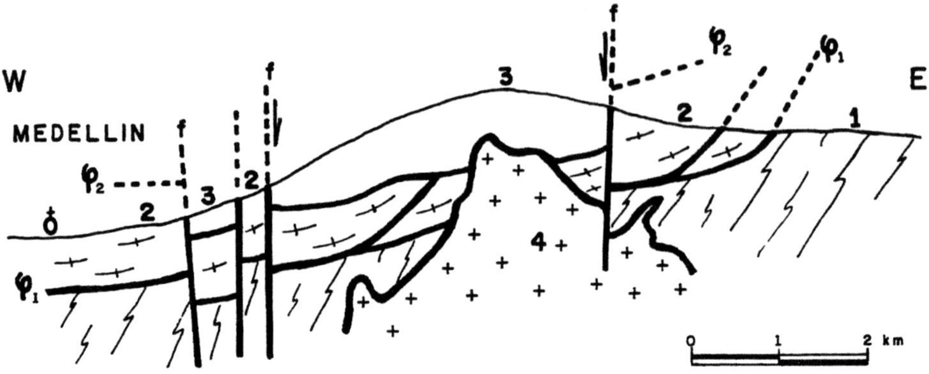

Fig. 9

Structural position of the Cauca Ophiolithic Complex near Medellín (modified from Restrepo and Toussaint, 1973). 1- Paleozoic basement, 2- Medellín amphibolites (Cretaceous), 3- Medellín dunites (Cretaceous), 4- Antioqueño batholith (Cretaceous).

ment in Early Cretaceous time. These authors consider that the blueschists were formed from the Paleozoic basement, with chloritic schists, amphibolites, quartzites and graphitic schists. However, Orrego et al. (1980a) and Restrepo and Toussaint (1975) consider that both the blueschists and the amphibolites belong to the oceanic rocks that constitute the Calima terrane formed during the Early Cretaceous.

To the south of Popayán, near the locality of Paispamba (Figure 10), Bourgois et al. (1985) observed both non metamorphosed ophiolites and metaophiolites, separated by peridotite flakes. According to these authors the metamorphosed ophiolites, represented by amphibolites, metagabbros and metabasalts, are associated with Early Cretaceous tectonogenesis, while the non-metamorphosed ophiolites, which include gabbros, basalts and pillow lavas similar to those of the Western Cordillera, may have been emplaced during the Late Cretaceous tectono-metamorphic event.

Considering that mid to high pressure metamorphic rocks are always intimately related to ultrabasic rocks, it is reasonable to suppose that an important tectono-metamorphic event, tied in with the emplacement of oceanic material, on or alongside the Tahamí terrane, took place during the Early Cretaceous. However, after this first event, a second one -recorded mainly in the Western Cordillera- took place, and also affected the Central Cordillera.

Early Cretaceous Epicontinental Sedimentation in the Tahamí Terrane

In various locations of the Tahamí terrane, epicontinental sedimentary sequences deposited unconformably over the Paleozoic metamorphic basement of this terrane have

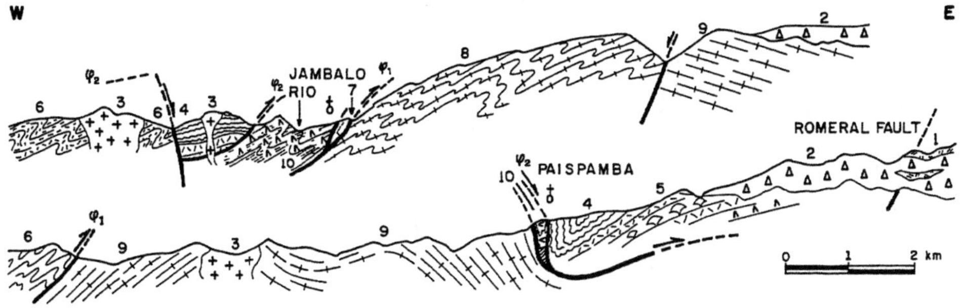

Fig. 10

Geologic cross-sections in the Jambaló and Paisbamba Regions (modified from Bourgois et al., 1985). 1- Present vulcanism, 2- Recent volcanic ash, 3- Meso- Cenozoic intrusives, 4- Río Calima Unit, siliceous limestones, 5- Río Calima Unit, basalts, 6- Río

been recognized (see Figure 11). They are the Abejorral and La Soledad Formations and the San Luis and Berlin sediments. The sedimentation was initially of a coastal type, which then changed to a submarine-platform environment. Locally euxinic conditions in very closed basins were developed, perhaps bounded by high submarine bottoms. Fossils reveal a middle Albian age (Burgl and Radelli, 1962), but this can extend as far as the Late Jurassic (Restrepo, 1986).

According to Toussaint and Restrepo (1974), these sediments were deposited over a vast continental platform, during a short marine transgression which covered the Tahamí terrane at the end of the Early Cretaceous. These sediments are contemporary with the Fomeque and Une Formations in the Bogotá basin, and the Tablazo and Simiti Formations in the Middle Magdalena basin. However, because the Tahamí terrane was not yet accreted to the Colombian East, it cannot be argued that both sequences belong to the same basin. It is also necessary to note that while in the Colombian East this sedimentation continued during the whole of the Late Cretaceous, on the Tahamí terrane it stopped suddenly at the end of the Albian, when strong tectonogenesis affected both the basement and the covering of this terrane. On the other hand, the Cretaceous sedimentary sequences of the Tahamí terrane, with no volcanic contributions, contrast with the volcanic-sedimentary environment that prevailed in the Calima terrane during this time. According to Restrepo and Toussaint (1989), these two different environments were generated in separate locations but were juxtaposed afterwards during the Late Cretaceous tectonic event.

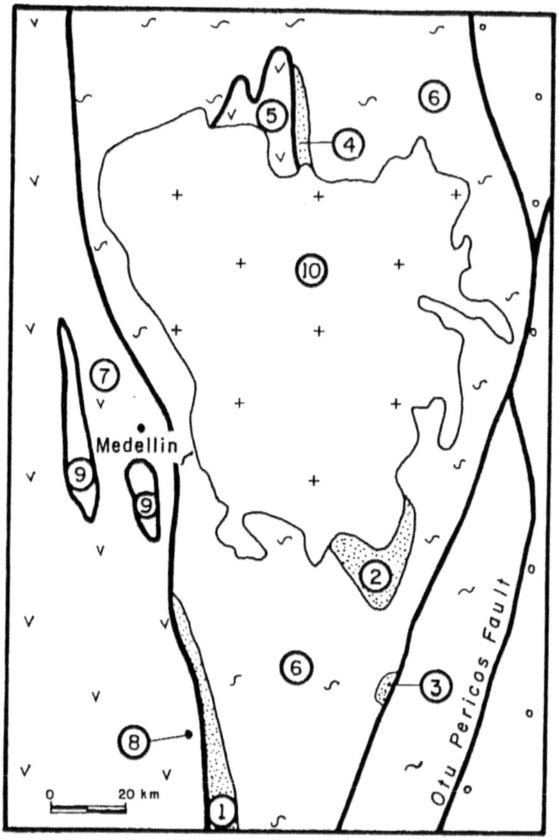

Fig. 11

Location of some geologic units in the northern end of the Central Cordillera. 1- Abejorral Formation, Cretaceous, autochthonous, 2- San Luis sediments, Cretaceous, autochthonous, 3- Berlín sediments, Cretaceous, autochthonous, 4- La Soledad Formation, Cretaceous, autochthonous, 5- San Pablo Formation, Cretaceous, allochthonous, 6- Paleozoic basement, 7- Calima Terrane, 8- Quebradagrande Formation in the Calima Terrane, 9- Tahamí Terrane surrounded by Calima Terrane.

Volcanic-Sedimentary Complexes and a Magmatism of Intermediate Composition

Volcanic-sedimentary assemblages crop out on the western flank of the Central Cordillera and on a great part of the Western Cordillera; they are sequences of basic volcanic rocks associated with marine sediments, sometimes partly metamorphosed. In the Central Cordillera these sets mainly correspond to the Quebradagrande and Amaime Formations; in the Western Cordillera they correspond mainly to the Dagua and Diabasic groups in the central region, and to the Cañasgordas Group in the northern region (Figure 8).

The Quebradagrande Formation (Botero, 1963) is formed of basic volcanic rocks interbedded with marine sediments and is located between 5°N and 7°N.

Fossils of Late Aptian-Early Albian age, are found in the sediments interbedded with basic rocks, making the Quebradagrande Formation apparently younger than the Cauca Ophiolite Complex. Furthermore, the presence of abundant pyroclastic material in the Quebradagrande Formation would seem to equate it more to an arc island than to an oceanic crustal environment.

The Amaime Formation (Mc Court et al., 1984) is located on the western flank of the Central Cordillera between 3°N and 4°30'N. It is formed mainly of massive or pillow basalt lavas, which Nivia (1989), through detailed geochemical analyses, found to correspond to a thickened oceanic-crust environment of Icelandic type.

In the Western Cordillera, the Diabasic Group was chosen by Nelson (1962) to designate a sequence of basic submarine volcanic rocks with associated sediments. They are in general spilitized diabases, massive and pillowed potassium-poor tholeiitic basalts and agglomerated tuffs, with minor intercalations of cherts, graywacke or limestones. The unit is metamorphosed in prehnite-pumpellyte and greenschist facies (Rodríguez, 1981). Paleontologically, the sediments interbedded in the basic rocks indicate an age in the Albian-Campanian range. Also, numerous radiometric datings are available, whith one group between 115 Ma and 90 Ma, that may correspond to the igneous age, while another group of age determinations between 78 Ma and 60 Ma could mark the tectono-metamorphic event that affected the region during the Late Cretaceous (Souza et al., 1984; Restrepo et al., 1985).

The Dagua Group (Nelson, 1957) is subdivided into the Espinal and Cisneros Formations. The Espinal Formation is formed of cherts and dark siliceous shales, often with pyrite, and of mudstones and graywackes with some fine limestones. Barrero (1977) considers that it is predominently a turbiditic set, with some pelagic sediments mainly with radiolarites of Late Cretaceous age. The Cisneros Formation is composed of green and purple siliceous phyllites, carbonaceous slates with intercalations of metacherts, metagraywackes and metalimestones, that come from the low grade metamorphism of a pelagic sedimentary sequence, particularly from offshore turbidites interbedded with pyroclastic rocks.

According to Etayo et al. (1982) the Dagua Group corresponds to contemporary heterotypic facies covering the foot of the continental slope, the oceanic trench and the foot of the volcanic relief, and that may have been superposed by an accumulation of tectonic origin.

Fossil traces such as chondrites and zoophycos are often found in the phyllites, indicating an abyssal environment; they have been dated as Cretaceous (Rodríguez, 1981). In addition, Barrero (1977) obtained two radiometric K/Ar ages of 81.8 Ma and 61.9 Ma in the phyllites, which would seem to indicate their probable metamorphic age.

In the northern part of the Western Cordillera, the basic and sedimentary rocks have been grouped in the Cañasgordas Group. This is divided into the Barroso Formation, corresponding to the basic rocks with some cherts, and the Penderisco Formation, containing flysch sediments associated with pelagic deposits. The Barroso Formation is composed of diabases, basalts, spilites with hyaloclastites, tuffs and agglomerates, sometimes interbedded with thin layers of black chert dated as middle Albian (Etayo et al., 1982). This age agrees with a K/Ar dating of 105 Ma obtained from a basalt (Toussaint and Restrepo, 1978), and with datings from gabbros and intrusive tonalites in the basic rocks in the 97 Ma and 92 Ma range (Restrepo and Toussaint, 1975; González et al., 1978). The abundance of agglomerates and tuffs, as well as the association of basic rocks of the Barroso Formation with gabbro stocks, such as the Altamira gabbro, and with tonalite plutons (mainly the Buriticá stock and Sabanalarga Batholith), seem to indicate an island arc environment, which may have developed over an oceanic crust.

The Penderisco Formation is formed by an alternation of graywackes and lithic arkoses, as well as shales and claystones, in beds of variable thickness, occasionally associated with some conglomerates. This unit has a turbiditic origin.

The hypothesis that a large part of the basic rocks can be assigned to an island-arc environment is supported by the presence on the eastern margin of the Calima terrane of a magmatic belt of intermediate composition that intrudes into the Cretaceous basic rocks (Figure 8).

The oldest pluton seems to be the Tamesis Stock dated at 124 Ma (Calle et al., 1980). This body crops out on the eastern flank of the Western Cordillera at 5°40′N. At latitude 4°N, on the western border of the Central Cordillera, the Buga Batholith, clearly intrusive in the basic rocks of the Amaime Formation (Schwinn, 1969), has given K/Ar ages of 113 Ma (Toussaint et al., 1978) and of 71 Ma (McCourt and Aspden, 1984), as well as an Rb/Sr age of 99 Ma (McCourt and Aspden, 1984).

At the beginning date of the Late Cretaceous, some other bodies intruded the basic rocks on the eastern border of the Western Cordillera. In the northern part they are the Buriticá Stock and the Altamira Gabbro. In the central part, near Cali, several stocks, such as the Zabaletas, El Tambor, El Palmar, El 18 and Vijes felsites, also crop out.

In the northern part of the Cauca depression the Sabanalarga Batholith, dated at 97 Ma (González et al., 1978), is intrusive both in the basic rocks of the Barroso Formation and in the sequence of metamorphic rocks of uncertain age (since they could either belong to the Paleozoic basement of the Tahamí terrane or could be part of the Arquía Complex of Cretaceous age and assigned to the Calima terrane).

At present, it is not clear if the subduction zone that brought about the formation of the elongate belt on the Eastern border of the Calima terrane during the Aptian-Cenomanian lapse remained active, and was also responsible for the younger belt located to the east and characterized by the Antioquian Batholith, or if on the contrary the younger plutons were formed by a new subduction zone.

Late Cretaceous Tectono-metamorphic Event

Calima Terrane deformations. In Nelson's paper (1957), the Western Cordillera was considered to be made up of two lithostratigraphic units: the Diabasic Group and the Dagua Group; with the second overlaying the first in normal fashion. Also, Irving (1971) shows a succession of wide anticlines and synclines in the Buga-Buenaventura cross-section, and Gansser (1973) suggests an isoclinal folding style with a subvertical axial plane. In turn, Barrero (1977) gives an interpretation based mainly on the current eastern dip position of the main schistosity evident in the Dagua Group phyllite, but he does not take into account the fact that the present position of the first schistosity is not only the result of the first tectonic phase but also of subsequent refolding movements.

Barrero (1977), is of the opinion that the tectonic event which affected the Western Cordillera, known as the Calima Orogeny, could have been produced by a change in position of the subduction zones, which may have moved from the Cauca to the Atrato-San Juan basin. This change may have allowed the formation of an accretion prism, whose internal structures would dip eastward, below the continent.

Recent research (Bourgois et al., 1982a, 1982b; Bourgois et al., 1985) has made it possible to formulate the hypothesis that the central part of the Western Cordillera may consist of a tectonic pile-up of nappes. Evidence of thrusting may be represented by low-angle reverse faults, that separate structural units of different tectonic styles. Thus, the lower unit -called the Rio Dagua Unit, mostly made up of phyllites, cherts and basic rocks- is affected by two deformational phases. The first is characterized by the formation of a flow schistosity, S1, associated with decametric to hectometric isoclinal folding. A greenschist metamorphism developed during this phase, forming the Dagua phyllites, which were dated between 61 Ma and 82 Ma by Barrero (1977). The second phase is characterized by a second schistosity, S2, from 10 to 100 cm. in width and in general parallel to the axial planes of overturned asymmetric folds of hectometric size. The S2 schistosity has an overall dip of less than 30 degrees, slightly folded, in a broad open regional anticline that affected the whole Cordillera during the Late Tertiary.

Thus, Bourgois et al. (1982) believe that the main foliation of the phyllites was developed during the first phase of isoclinal folding, and was then deformed during the second phase, characterized by a fold overturned toward the SE.

On the Buga-Buenaventura highway the intermediate Loboguerrero unit reveals a wide fold overturned toward the SE, associated with an incipient fracture cleavage, parallel to the axial plane.

The upper unit, known as the Rio Calima Unit, is mainly represented by calcareous cherts, affected by a fairly strong similar folding, a thick flysch sequence and a large basic volcanic rocks unit. When tectonism was active, this unit was always on a high structural level. Pillow lava polarity seems to indicate that it was affected by E-SE overturned folding.

In addition, deformation analysis shows that tectonic unit emplacement took place from the NW toward the SE, and that the Rio Calima Unit, the highest in the structural sequence, has a more distal origin than that of the lower unit (see Figure 12).

With this interpretation of the Late Cretaceous tectonogenesis in the Western Cordillera, we consider that this Cordillera has clearly "Alpine" characteristics that make it similar

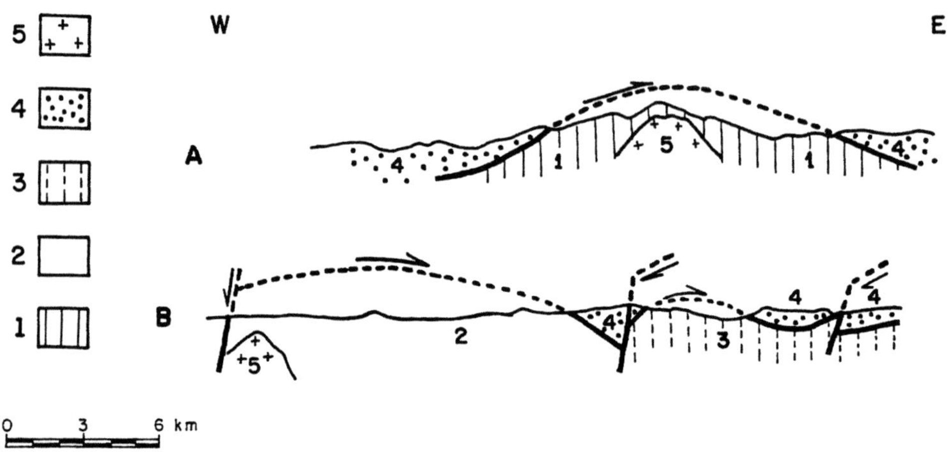

Fig. 12

Simplified crosscuts of the Calima Terrane (after Bourgois et al., 1987). A- Section at 7° N, B- Section at 4° N, 1- Dabeiba-Uramita Unit, 2- Río Dagua Unit, 3- Ventana de Loboguerrero Unit, 4- Río Calima nappe, 5- Cenozoic plutonism.

to the Caribbean Cordillera of Venezuela and departs strongly from the Central Andean style.

Recently, a new cross-section between Buga and Buenaventura was presented by McCourt et al. (1984), in which these authors lay special emphasis on the large NE trending faults, of normal, reverse and wrench type, that would seem to cut the Cordillera and bound a sequence of rhomboid blocks. This faulting is assigned to the Calima Orogeny, and would be the consequence of an oblique collision between an oceanic plate and the continent. However, in our opinion the wrench fault system that affects the Western Cordillera and the Cauca region seems to be younger than the Late Cretaceous tectono-metamorphic event, characterized by the accretion of an oceanic domain to the continental domain and by the development of the observed metamorphism in the Dagua Group. The change from collision tectonics, marked by thrusting, to a horizontal tectonic shear, marked by wrench faulting, is due to progressive changes in the relative direction of movements between continental and oceanic blocks at the end of the Cretaceous or the beginning of the Cenozoic. This former tectonic shearing is probably also responsible for the final assemblage between the Colombian East and the composite Tahamí-Calima terrane, by means of the Otú-Pericos fault and the large wrench movements characteristic of the Cauca-Romeral system in the Rio Cauca Valley (Figure 13).

There are also several hypotheses with regard to the evolution of the northern part of the Western Cordillera. The first such hypothesis originally presented by Hoyos and Zuluaga (1978) and later extended by Restrepo et al. (1985), considers that prior to Cretaceous tectonogenesis, the Cañasgordas Group was formed by an oceanic basement representative of an oceanic crust and/or island arcs, called the Barroso Formation, over which a mainly turbidite and siliceous limestone sequence called the Penderisco Formation was deposited. According to these authors, during the Cretaceous tectonic event the

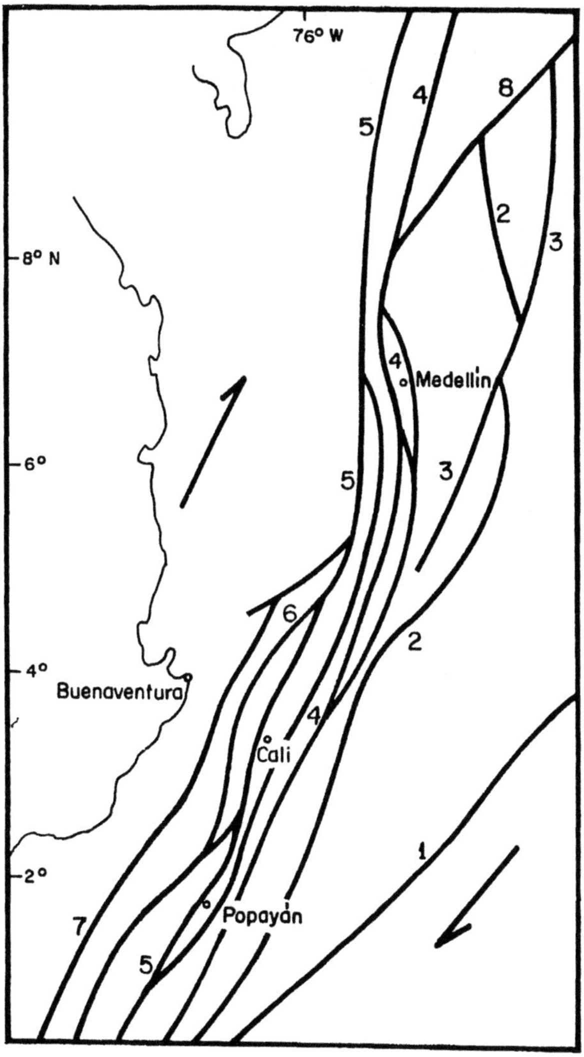

Fig. 13

Main dextral-striking faults of Cretaceous age in the Colombian Andes. 1- Guaicaramo Fault, 2- Otu-Pericos Fault, 3- Palestina Fault, 4- Romeral Fault, 5- Cauca Fault, 6- Dagua-Calima Fault, 7- El Tambor Fault, 8- Espíritu Santo Fault.

sedimentary cover was in great part detached from its basic volcanic basement, and was then deformed by straight to eastward-overturned folding. The thick sandstone and graywacke layers would have been affected by isopac folds of kilometric wavelength, while in the thin layers of shales and sandstones there appear closed metric folds with thinning of the flanks. The basement would have been affected by large reverse faults.

In the opinion of Bourgois et al. (1982, 1985) the setting is different. In fact, according to these authors, the upper tectonic unit that includes basic volcanic rocks, cherts and thick flysch was thrust over a lower tectonic unit composed of fine turbidites affected by two schistosities (Figure 12).

This thrusting mainly affected the Western Cordillera, but its effects seem to have extended up to the axis of the Central Cordillera, as for example in the Rio Nechi Ophiolite Complex (Figure 14). In this way, near Yarumal, on the axis of the Central Cordillera, the Rio Nechi Complex was tectonically pushed over the Paleozoic basement of the Tahamí terrane (Restrepo and Toussaint, 1973; Bourgois et al., 1985). The basic and marine sedimentary rocks are there in subhorizontal tectonic contact with both the underlying Valdivia Group Paleozoic schists (Restrepo and Toussaint, 1973) and the underlying sedimentary autochthonous Tahamí terrane cover (Bourgois et al., 1985), here represented by the Soledad Formation dated paleontologically as Early Albian (Hall et al., 1972). Also, to the east, a large gabbro body with dunite flakes in its base and numerous isolated ultrabasic and basic small klippes over the Paleozoic basement is located between the Central Cordillera axis and the Rio Cauca Valley.

Restrepo and Toussaint (1973) believe that the Rio Nechi ophiolites were thrust from the Valle del Cauca in an easterly direction, over both the basement and the Tahamí terrane cover, so that this obduction would have been about 70 Km wide. The thrusted layer has now been removed by erosion, leaving only some elongate belts which were pro-

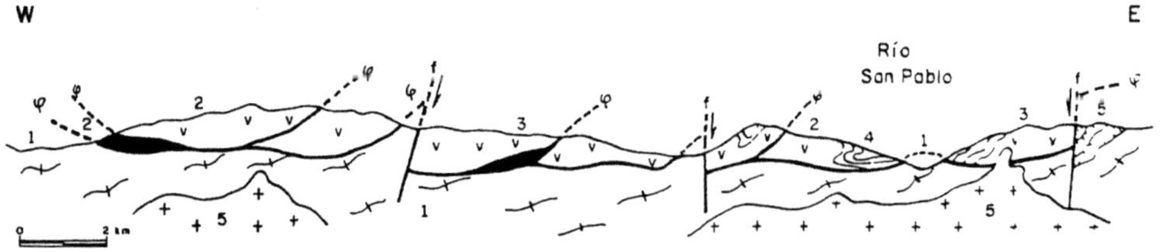

Fig. 14

Structural position of the Río Nechí Ophiolithic Complex (modified from Restrepo and Toussaint, 1973) 1- Paleozoic basement, 2- Dunites, 3- Basic volcanic rocks, 4- Allochthonous marine sediments, 5- La Soledad Formation, autochthonous

tected by their position in grabens. In the opinion of Bourgois et al. (1985), the tectonic emplacement of the Rio Nechi Complex may imply that the obduction extension could be larger than is supported by Restrepo and Toussaint (1973), since its origin would be to the West of the Western Cordillera.

Tahamí terrane deformations. During the Late Cretaceous tectono- metamorphic event, both the sedimentary cover and the Tahamí terrane basement were strongly deformed.

The Abejorral Formation, for exemple, deposited unconformably over the metamorphic basement of the Tahamí terrane, was strongly tectonized just before the Middle Albian. Rodríguez and Rojas (1985) analyzed the facial sequences of the formation and recognized several environments, such as fluvial fan, foreshore, offshore and turbiditic currents. The authors consider that these sequences were superimposed over one another by a thrusting process. According to Restrepo and Toussaint (1974) the deformation affected the whole sedimentary sequence. Consequently, the conglomerates are strongly deformed and their fragments have been realigned.

The shales show rounded to angular folds of anisopac type with flank thinning, and in some levels of slaty shales fracture cleavage is present. The few fossils found also show evident deformation with well defined lengthening and squeezing.

According to Restrepo and Toussaint (1974) the deformation was due to E-W compression produced by the tectonic emplacement of oceanic units on the margin and also, in part, over the continental materials of the Tahamí Terrane.

Rodríguez and Rojas (1985) consider that the tectonic piling-up of the facial sequences can also be interpreted as due to the collision of an oceanic island arc with the continental plate (Figure 15). Furthermore, this tectonic event is probably the cause of the definite retreat of the sea from the Tahamí terrane in the Late Albian. According to Rodríguez and Rojas (1985), the thrusting affected not only the sedimentary cover but also a great part of the continental crust, which in turn may have been affected by several shear zones with a small eastward dip.

The basement was also affected radiometrically, since samples from clearly Paleozoic rocks have given Cretaceous K/Ar dates. This phenomenon seems to indicate that there was a significant isotopic readjustment during the Late Cretaceous tectonogenesis.

In any case, an important tectono-metamorphic event affected all the Colombian West at the end of the Cretaceous. This event is mainly recorded in the Western Cordillera, but also affected the Central Cordillera, which was superimposed on the eastern border of the Cauca depression, during the Early Cretaceous event.

Post-Tectonic Magmatism at the End of the Cretaceous

After the amalgamation between the Calima and Tahamí terranes, the northeastern part of this unit was intruded by several plutons, and in particular by the Antioquian Batholith (Figures 8 and 16).

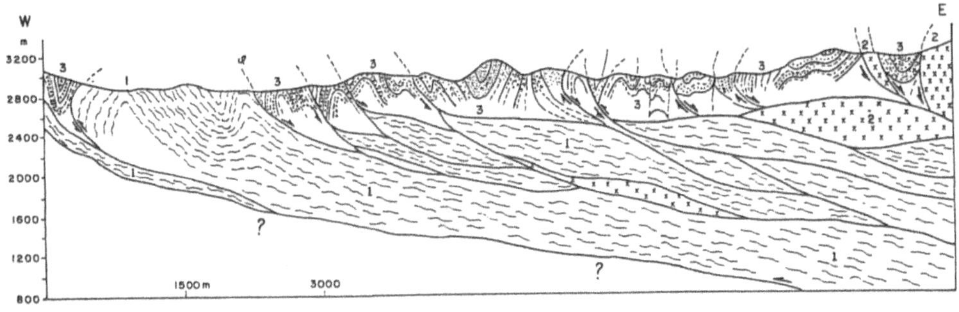

Fig. 15

Deformation of the Tahamí Terrane basement and cover in the Abejorral region during late Cretaceous tectono-genesis (modified from Rodríguez and Rojas, 1985). 1- Paleozoic basement, 2- Jurassic plutons, 3- Aptian- Albian Abejorral Formation.

According to Toussaint (1978), this batholith was intruded during an extensional phase that followed the strong tectono-metamorphic event that affected all the Colombian West. This extension is marked by normal faults younger than the emplacement of the ophiolites of the Rio Nechi Complex to the north of the batholith, but older than the intrusion. Also, the batholith is clearly later than the Medellin dunites and amphibolites
The age of the Antioquian Batholith is known from numerous datings, which range from 90 Ma to 56 Ma. However, the more recent dates are representative of the age of a dynamic deformation after the intrusion.
Nevertheless, the relation in time between the magmatic activity of intermediate composition and the tectono-metamorphic events of the Colombian West is not always clear. The radiometric ages of the plutons seem to indicate that they are youger than the tectono-metamorphic event of the Early Cretaceous (marked by the tectonic emplacement of some ophiolites and the development of middle- to high-pressure metamorphism), but it is not clear which of these plutons are older and which are younger than the Late Cretaceous metamorphic event (which affected the Western Cordillera and, to a lesser degree, the Central Cordillera).

Assemblage Between the Colombian East and West

The differences between the Colombian East and West were well defined during Paleozoic and Early Mesozoic times, and it was only from the beginning of the Early Cretaceous that the Chibcha y Tahamí terranes seemed to bear similar litho-stratigraphic sequences, since a marine epicontinental sedimentation was deposited on both terranes. However, important differences appear again during the Late Cretaceous, when marine sedimentation continued over the Chibcha terrane, whereas the sea definitely retreated

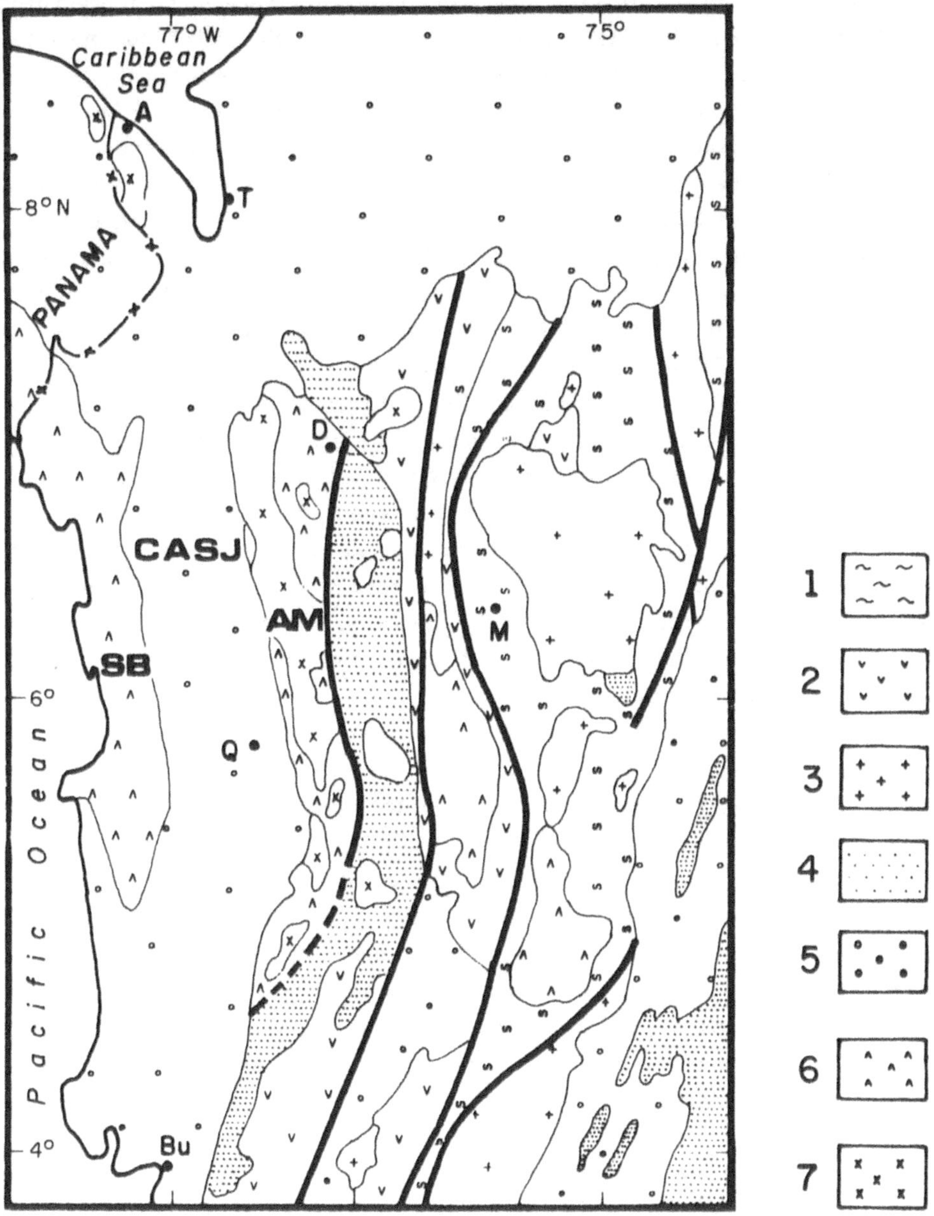

Fig. 16

Geologic map of Northwest Colombia. SB- Serranía de Baudó, CASJ- Atrato River and San Juan River Basin, AM- Mandé Magmatic Arc, A- Acandí, T- Turbo, D- Dabeiba, M- Medellín, Q- Quibdó, Bu- Buenaventura, 1- Polimetamorphic complex, 2- Basic Mesozoic volcanic rocks, 3- Mesozoic plutonic rocks, 4- Pre-cenozoic sediments, 5- Cenozoic sediments, 6- Cenozoic volcanic rocks, 7- Cenozoic plutonic rocks.

from the Tahamí terrane during the Albian. In addition, intermediate type magmatism was significant in the Tahamí terrane, while it was absent in the Chibcha terrane. However, the fundamental differences are mainly of tectonic and metamorphic type. Thus, while important tectono-metamorphic events were taking place in both the Tahami and the Calima terranes on the western side of the Otu-Pericos fault -which marks the current boundary between the East and the West- extension and subsidence prevailed in the Colombian East, without any unconformity in the marine sedimentary sequence.

In an autochthonous framework, that may imply East-West continuity before or at the beginning of Cretaceous times, it is difficult to imagine how the whole of the Colombian west experienced some of the most important tectono-metamorphics events in the entire geologic history of Colombia, while to the east of the Otú-Pericos megafault only an extensional environment was in evidence. The transition between both domains is so sharp that in San Luis and Berlin, a few kilometers to the west of the fault, the Early Cretaceous sedimentary rocks were strongly tectonized before the Campanian, while a few kilometers east of the fault the whole Cretaceous sequence is concordant and without any tectonic effect except that of an extensional type. Again, near latitude 4°30'N, an important geochronologic event is recorded in numerous K/Ar radiometric datings of the metamorphic rocks located to the west of the Otú-Pericos megafault, while no radiometric rejuvenation was detected in samples of the Ibague Batholith, located just east of the megafault.

Taking these differences into consideration it must be reasoned that during the Cretaceous the Colombian West was not adjacent to the Colombian East, and that assemblage between both domains took place after the tectono-metamorphic phenomena formerly described.

Since there are no paleomagnetic studies of the area from which to measure the displacement of the terranes, nor microtectonic studies of the Otú-Pericos megafault, it is difficult to trace the movements which caused the union of both domains.

A more regional view, however, makes it possible to relate this phenomenon to general movements of the Colombian West and the Caribbean during Cretaceous times. It is a fact that all recent models of the Caribbean propose that this domain was formed to the west and was then inserted as a wedge between the North American and South American continents.

The Caribbean dextral movements in relation to South America started at the end of the Early Cretaceous, which may imply a general dextral movement of the Colombian West in relation to the Colombian East and the Guyana Shield in the Late Cretaceous. If this hypothesis is correct, the origin of the Tahamí and Calima terranes may be located south of their current position.

Large wrench faults, that extended form the Guayaquil Gulf in Ecuador up to the Northern Andes, slice the structures that were formed originally during the Tahamí and Calima amalgamation.

The Otú-Pericos suture may have functioned only during the Cretaceous, but later the general movement between the Caribbean and South American domains took place, partly along the Palestina Fault that cuts the Otú-Pericos fault, but mainly along the dextral wrench Cauca-Romeral fault system, that slices the weld zone amalgamating the Calima and Tahamí terranes. It is probable that by this time the ophiolitic massifs were arraigned in two parallel belts, located on either side of the Cauca-Romeral System.

Thus, both the junction between the Colombian East and West along the Otú-Pericos fault and the large-scale shearing that affected the whole of the Colombian West at the close of the Cretaceous, immediately after the Calima tectono-genesis ended, would be directly related to the dextral movements of the Caribbean domain in relation to the NW South American margin.

Cretaceous Basement of Terranes Accreted During Cenozoic Times

The Cuna terrane is located west of the Calima terrane (see Figure 2), and is separated from the former by the Dabeiba-Pueblorico suture formed by a Miocene collision. (Toussaint and Restrepo, 1986).

The Cuna terrane is formed from east to west by the Mandé magmatic arc, the Rio Atrato basin and the Serranía de Baudó.

The Mandé magmatic arc, of Paleocene to Eocene age, was probably developed over a Cretaceous oceanic basement, but this has not been established for certain. The Cuna terrane basement has been recorded only in the Serranía de Baudó and in the Gorgona and Gorgonilla islands, which may represent the southern prolongation of the Serrania de Baudó, although McGeary and Ben Avraham (1985) consider that these islands belong to a different terrane. The Serranía de Baudó basement is composed of gabbros, diabases, and massive and pillow basalts (Gansser, 1950; Galvis, 1980) with a 70 $\pm$ 3.5 Ma dating (Bourgois et al., 1982a, 1982b).

On the Gorgona and Gorgonilla islands, Gansser (1950) gave the name Gorgona Igneous Complex to a set of dunites and wehrlites, poikilitic gabbros and massive or pillow basalt flows, covered by a basic tuff that contains Inoceramus. Echeverria (1980) recorded Cretaceous komatiites with spinifex texture which may represent a magma produced by partial melting of the mantle with an eruption temperature somewhere between 1400 °C and 1500 °C.

Duque-Caro (1985) consider that the Cuna terrane was accreted to the Andean Block during the Mid Miocene, and McGeary and Ben Avraham (1985) believe the accretion of the Gorgona terrane came at the beginning of the Miocene.

Relation Between the Colombian and Ecuadorian Andes

As in Colombia, the Ecuadorian Andes located to the north of the Gulf of Guayaquil show a western part that corresponds to an oceanic domain formed by oceanic crust topped by island arcs (Figure 17). This oceanic domain, considered as allochthonous, was named the Piñon terrane by Feininger (1986). Later it was divided into several smaller terranes, such as the Nono - Shobol and Piñon - Macuchi terranes by Mégard and Lebrat (1986). According to these authors, the Piñon - Macuchi terrane is located to the west of the Pallatanga and Milagro faults, and has an ophiolitic basement that corresponds to the Piñon Formation of Late Jurassic(?) - Early Cretaceous age, over which an essentially basic incipient island arc -the Macuchi Arc- was developed. It presents flysch - type sequences, partly interbedded with the volcanic arc units, with Albian to Maastrichtian fossils covered by deep marine sediments such as the Maastrichtian to Paleocene cherts of Guayaquil. Eocene coral - reef limestones followed by a flysch sequence of the same age, cover the former set.

The Nono - Shobol terrane, bounded to the west by the Milagro fault and to the east by the Pallatanga fault, corresponds to an island arc environment partially interbedded and covered by a flysch sequence of Late Cretaceous to Paleocene age. The Nono - Shobol terrane was separated from the Piñon - Macuchi terrane by Mégard and Lebrat (1986), who supposed a different accretion age for both terranes.

Accretion of the Nono - Shobol terrane would therefore have taken place at the end of the Cretaceous, whereas that of the Piñon - Macuchi terrane would have occured in Eocene time. In both cases, the accretions were due to occlusion of the subduction zone, when the island arc reached the trench. This clogging permitted the formation of a new subduction, located farther west.

The Nono - Shobol terrane shows some similarity to the Colombian Calima terrane, both in lithology and in age, although the metamorphism that affected the Calima terrane in Colombia during the Late Cretaceous does not seem to be present in Ecuador.

The terranes with oceanic features are separated from the continental domain by the Calacali - Pallatanga suture, marked by several flakes of ophiolite complexes, such as Calacali, Saloya and Pallatanga. These ophiolites are the prolongation of the ophiolite belt that crops out along the Cauca - Patia fault on the eastern border of the Colombian Cordillera Central.

To the east of the Calacali - Pallantanga suture lies the Inter - Andean Valley, which corresponds to the southern part of the Cauca - Patia Valley in Colombia. This depression, filled by a large volume of Late Tertiary and Quaternary volcanic - sedimentary deposits, shows its basement only in the Chaucha region, at latitude 1°N. This consists of quartzites and schists, whose ages have not yet been determined. Feininger (1986) considers that the metamorphic sequence of Chaucha belongs to a small continental terrane located between latitude 2°S and 3°S, but Aspden et al. (1988) suggest that this

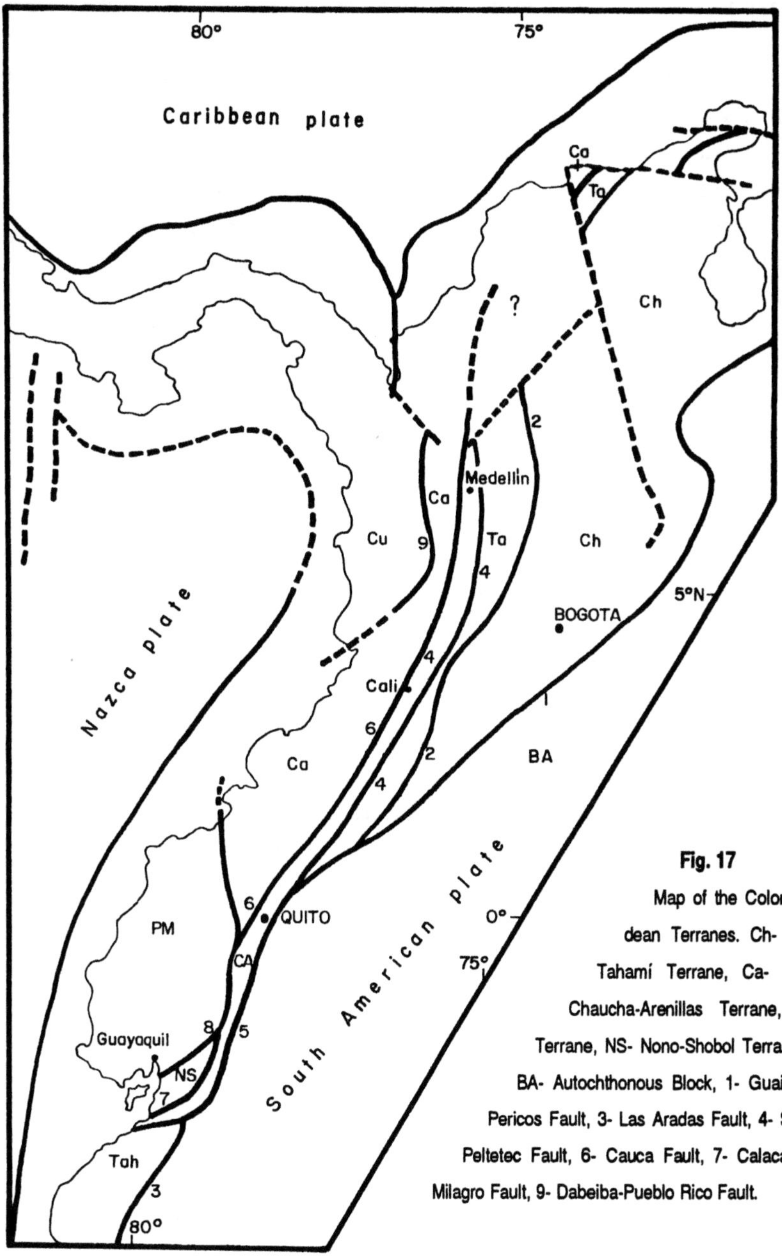

Fig. 17

Map of the Colombian- Ecuadorian Andean Terranes. Ch- Chibcha Terrane, Ta- Tahamí Terrane, Ca- Calima Terrane, CA- Chaucha-Arenillas Terrane, PM- Piñon-Macuchi Terrane, NS- Nono-Shobol Terrane, Cu- Cuna Terrane, BA- Autochthonous Block, 1- Guaicaramo Fault, 2- Otu-Pericos Fault, 3- Las Aradas Fault, 4- San Jerónimo Fault, 5- Peltetec Fault, 6- Cauca Fault, 7- Calacali-Pallatanga Fault, 8- Milagro Fault, 9- Dabeiba-Pueblo Rico Fault.

terrane, know as the Chaucha - Arenilla, may be present under the volcanic - sedimentary deposits of the Inter - Andean Valley from the Gulf of Guayaquil up to the Medellin region of Colombia. However, it should be noted that the basement of the Valle del

Cauca in Colombia was initially considered to be of continental type (Meissner et al., 1976), but new hypotheses now suppose it to be an oceanic basement (INGEOMINAS in preparation).

The Chaucha - Arenillas terrane is bounded on its eastern side by the Peltetec megafault that separates it from the South American continental craton. This megafault is marked by a discontinuous belt of ophiolite complexes, such as those of Rio Blanco and Peltetec, which are aligned with the Los Azules and Jambaló complexes in Colombia; in other words, it coincides with the most westerly faults of the Cauca - Romeral system.

South of the Gulf of Guayaquil, both Feininger (1986) and Mégard and Lebrat (1986) supposed the most western part of the continental basement to be allochthonous in relation to the South American Craton, the Las Aradas fault being the suture between them. Feininger (1986) named the terrane located to the west of Las Aradas fault the Tahuin terrane, although later Aspden et al. (1988) named it the Bocana terrane.

The Tahuin terrane shows a polymetamorphic basement with Precambrian(?) and Paleozoic events, covered by Cretaceous epicontinental sedimentation. On the NW border of the Tahuin terrane there is a high - pressure metamorphic belt with eclogites and glaucophane schists, associated with ophiolites that correspond to the Raspas Formation dated at 132 Ma by K/Ar. This situation shows some similarity to the situation manifest in the Tahamí terrane in Colombia, although Restrepo and Toussaint (1988) conclude that the middle - pressure (MP) to high - pressure (HP) metamorphism affected the east side of the oceanic Calima terrane, rather than the basement of the continental Tahamí terrane.

It is difficult to give an exact age for the accretion of the Tahuin terrane to the South American craton, in part due to the fact that the suture is hidden by youger sediments. In the opinion of Mégard and Lebrat (1986) the suture took place before the andesitic vulcanism of the Celica Arc, dated between Albian and Maastrichtian times, but Feininger (1986) considers its age may be Eocene.

In conclusion, it seems that during Cretaceous and Early Tertiary times there were considerable similarities between the Colombian and Ecuadorian sectors of the Northern Andes, although there were also important differences.

In both regions, terranes of oceanic origin, representing island arcs developed over an ophiolitic material, were generated in an allochthonous position and then accreted to the South American continent: This collage process was achieved either by obduction, as in the Northern part of Colombia, or by accretion due to westerly migration of the subduction zones, as may have happened in Ecuador. This is not a simple case of a juncture between an oceanic domain and a continental one; rather, several successive accretions took place, both during the Cretaceous and the Cenozoic. Other similarities are shown by the western borders of the continental basements, composed of sialic terranes which are allochthonous in relation to the South American Craton.

The chronology of the successive accretions is complex, as the sedimentary, magmatic, metamorphic and tectonic events took place in a very narrow lapse of time. Furthermore, horizontal shear phenomena, during the Late Cretaceous and the Early Tertiary, played an equally important role, giving place to large displacements of mainly dextral type along the megafaults superposed on the ophiolitic structures, as in the case of the Cauca-Romeral system.

However, phenomena as tectonically important as the extension and subsidence that affected the entire Colombian East during the Cretaceous, and brought about the continuous deposition of thick epicontinental sedimentary sequences, were exclusively limited to Colombia.

Furthermore, it is not clear whether there is a terrane in Colombia equivalent to the Ecuadorian Piñon-Macuchi; nor is it clear whether there are Ecuadorian terranes equivalent to the Chibcha and Cuna terranes of Colombia.

A general correlation between the diverse events that occurred in the Northern Andes is still over-hypothetical, due to lack of precise data concerning regional structures, geochronology and paleomagnetism. Therefore, the general evolution pattern presented below is highly speculative.

Hypothetical Chronology of the Northern Andean Accretions

An oceanic material, generated at the end of the Jurassic and the beginning of the Cretaceous, was tectonically emplaced next to and apparently also partly on top of the continental basements of the Tahamí and Tahuin terranes. During this emplacement an important mid to high pressure metamorphic event took place, represented by the Arquía, Pijao, Jambaló and Raspas complexes. This phenomenon probably occurred south of the current position of the Tahamí and Tahuin terranes, which were displaced northward by the Aradas fault in northern Perú and southern Ecuador, and the Otú-Pericos fault in Colombia. This dextral movement probably ceased in the south before the beginning of the Celica Arc volcanic activity during Aptian times, but continued in the north until the beginning of the Paleocene.

In this setting, the role of the presumed Chaucha-Arenillas terrane is not clear, and even its existence between 1°S and 7°N is questionable. Since Aptian time a magmatism of island-arc type developed on the eastern side of the Calima and Nono-Shobol terranes. Later, at the end of the Cretaceous, these two terranes were affected by an important tectonic event. This tectonogenesis is recorded in the Calima terrane by overthrusting and metamorphism, and it is probable that at that time the units emplaced during the Early Cretaceous tectono-metamorphic event were moved. In addition, the western side of the continental terranes also experienced the effects of this new tectonogenesis, as in the case of the Tahamí terrane. The Nono-Shobol terrane was also accreted at the end of

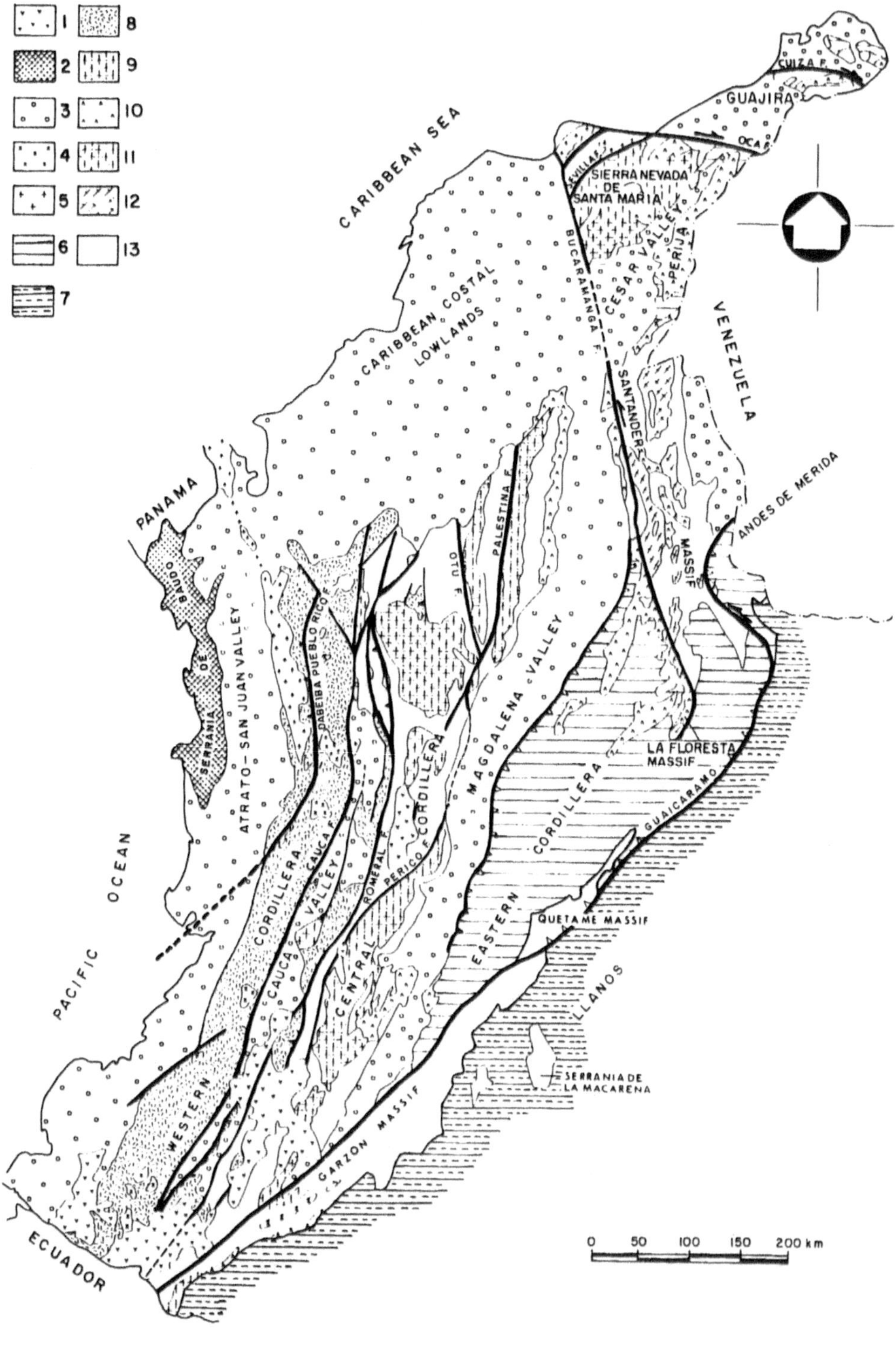

CARIBBEAN SEA
GUAJIRA
CUIZA F.
OCA F.
SEVILLA F.
SIERRA NEVADA DE SANTA MARIA
CESAR VALLEY
PERIJA
BUCARAMANGA F.
VENEZUELA
CARIBBEAN COSTAL LOWLANDS
SANTANDER MASSIF
ANDES DE MERIDA
PANAMA
SERRANIA DE BAUDO
ATRATO - SAN JUAN VALLEY
DABEIBA PUEBLO RICO F.
OTU F.
PALESTINA F.
MAGDALENA VALLEY
LA FLORESTA MASSIF
GUAICARAMO
PACIFIC OCEAN
WESTERN CORDILLERA
CAUCA F.
CAUCA VALLEY
ROMERAL F.
PERICO F.
CENTRAL CORDILLERA
EASTERN CORDILLERA
QUETAME MASSIF
LLANOS
SERRANIA DE LA MACARENA
GARZON MASSIF
ECUADOR
0 50 100 150 200 km
1 2 3 4 5 6 7 8 9 10 11 12 13

the Cretaceous. A post-tectonic magmatism, that mainly included the Antioquian Batholith, marked the end of this Late Cretaceous tectono-metamorphic event.
The Macuchi Arc was developed during the Late Cretaceous in the Piñon-Macuchi terrane, but this terrane was completely accreted to South America only during the Late Eocene by a clogging process of the subduction zone located between the Calima and Nono-Shobol terranes, already accreted. Furthermore, the basement of the Cuna terrane, in NW Colombia, which was only accreted during the Middle Miocene, was formed during the Cretaceous.
The Otú-Pericos wrench fault ceased its activity at the end of the Cretaceous, allowing a definite juncture between the Colombian West and East, but large-scale wrenching movements continued along the Guayaquil-Dolores and Cauca-Romeral fault systems, leading to the formation of tectonic features, such as the Ecuadorian Inter-Andean Valley and the Colombian Patia-Cauca Valley, that may be pull-apart basins.

Conclusions

In the Northern Andes, the Cretaceous is a period of deep changes that imprinted on this region its most essential features. Oceanic domains that include crust and island arcs were generated in allochthonous positions in relation to the continental nucleus of South America, and were then emplaced tectonically alongside and in part on top of the sialic crust. Ophiolite flakes and metamorphic units of mid to high pressure generally define the continental-oceanic limit, where as the limits between terranes are also characterized by dextral wrench megafaults, that show the accretions were not always frontal. Another important characteristic is the superposition of the large Cauca-Romeral wrench faults over the ophiolitic sutures.
These changes affected the continental margin, and small terranes were displaced from their original location and then emplaced in apparently more northern positions.
Two major tectono-metamorphic events seem to have taken place during Cretaceous time: one previous to the Albian and another at the close of the Cretaceous. However, the chronology of diverse sedimentary, magmatic, metamorphic and tectonic events still remains unclear. Also, mainly within Colombia, a magmatic belt of intermediate type and Aptian to Cenomanian age is observed on the eastern side of the Calima Terrane,

Fig. 18

◄ Geologic map of the Colombian Andes. 1- Pliocene-Quaternary vulcanism, 2- Serranía de Baudó, basic Cretaceous and Cenozoic rocks, 3- Cenozoic sediments, 4- Calima Terrane, Cenozoic magmatism, 5- Cuna Terrane, Cenozoic magmatism, 6- Mesozoic sediments, 7- Cenozoic sediments of the Llanos Orientales, 8- Basic Cretaceous rocks, 9- Late Cretaceous magmatism, 10- Triassic and Jurassic vulcanism, 11- Jurassic plutonism, 12- Paleozoic magmatism, 13- Paleozoic and Precambrian basement.

while a second belt, formed between the Campanian and Paleocene, was emplaced in a more eastely position, mainly on the Tahamí terrane.

At the end of the Cretaceous the general landscape of the Northern Andes had changed drastically, due to displacement of small continental terranes on the South American Craton margin, and in great part due to accretions of oceanic terranes on the northwestern side of the continent. These collisions continued during the Cenozoic with the emplacement of the Piñon-Macuchi terrane during the Eocene and the Cuna terrane during the Miocene, thus giving this region a style quite different from that of the Central Andes, where oceanic domains were subducted in direct convergence with the continental margin.

This work is a contribution to the IGCP Nº 279 Project, "Terranes in Latin America".

References

Aspden, J.A., Litherland, M., and Salazar, E. (1988): Una interpretación preliminar de la historia colisional del Centro y Sur del Ecuador y posibles controles para la geología cenozoica y de mineralización polimetálica. Politécnica, Monografía de Geología, 6, v. 13, 49-75, Quito.

Aubouin, J. (1977): Reflexión sobre la organización de las cordilleras sud-americanas. Congreso Latinoamericano de Geología, 2nd, Caracas (1973), Memoria, III, Caracas, Dirección de Geología, Boletín de Geología, Publicación Especial, 7, 1999-2013.

Barrero, D. (1977): Geology of the Central Western Cordillera west of Buga and Roldanillo, Colombia. Ph.D. Thesis, Colorado School of Mines, Golden, 154 pp.

Botero, G. (1963): Contribución al conocimiento de la geología de la zona central de Antioquia. Universidad Nacional, Facultad de Minas, Anales, v. 57, 1-101, Medellín.

Bourgois, J., Calle, B., Tournon, J., and Toussaint, J.F. (1982a): The ophiolitic Andes megastructures on the Buga-Buenaventura transverse (Western Cordillera-Valle, Colombia). Tectonophysics, v. 82, 207-229.

Bourgois, J., Azema, J., Tournon, J., Bellon, H., Calle, B., Parra, E., Toussaint, J.F., Glacon, G., Feinberg, H., De Wever, P. and Origlia, I. (1982b): Age et structure des complexes basiques et ultrabasiques de la facade pacifique entre 3°N et 12°N (Colombie, Panamá et Costa Rica). Bulletin de la Société géologique de France, v. 3, 545-554.

Bourgois, J., Toussaint, J.F., González, H., Orrego, A., Azema, J., Calle, B., Desmet, A., Murcia, A., Pablo, A., Parra, E., and Tournon, J. (1985): Les Ophiolites des Andes de Colombie: évolution structurale et signification géodynamique. Symposium "Géodynamique des Caraïbes", Paris, Ed. Technic, 475-493.

Burgl, H. and Radelli, L. (1962): Nuevas localidades fosilíferas de la Cordillera de Colombia. Universidad Nacional, Geología Colombiana, 3, 133-133, Bogotá.

Calle, B., Toussaint, J.F., Restrepo, J.J., and Linares, E. (1980): Edades K/Ar de dos plutones de la parte septentrional de la Cordillera Occidental de Colombia. Geología Norandina, 2, 17-20, Bogotá.

Case, J.E., Durán, L.G., López, A., and Moore, W.R. (1971): Tectonic investigations in Western Colombia and Eastern Panamá. Geological Society of America Bulletin, v. 82, 2685-2712.

Case, J.E., Helcombe, T.L., and Martin, R.G. (1984): Map of geologic provinces in the Caribbean region, in The Caribbean-South America Plate Boundary and Regional Tectonics, (Eds.: Bonini, W.E., Hargraves, R.B., and Shagan, R.), Geological Society of America Memoir, 162, 1-29.

Duque-Caro, H. (1985): La cuenca del Atrato en el bloque del Chocó (Suramérica Noroccidental) y sus implicacciones estratigráficas y estructurales. Congreso Latinoamericano de Geología, 6th, Bogotá, Memoria, 1, p. 61.

Duque-Caro, H. (1989): El arco de Dabeiba: nuevas aportaciones al conocimiento del NW de la Cordillera Occidental. Congreso Colombiano de Geología, 5th, Bucaramanga, Memoria, 1, 108-126.

Echeverría, L.M. (1980): Tertiary or Mesozoic Komatiites from Gorgona Island Colombia. Field realtions and geochemistry. Contributions to Mineralogy and Petrology, v. 73, 253-266.

Estrada, A. (1972): Geology and Plate tectonic History of the Colombia Andes. M.Sc. Thesis, Stanford University, Stanford, 115 pp.

Etayo, F. (Dir.) (1985): Proyecto Cretácico. Bogotá, Ingeominas, Contribuciones, Publicaciones Geológicas Especiales, 16, 352 pp.

Etayo, F., Parra, E., and Rodriguez, G. (1982): Análisis facial del "Grupo Dagua" con base en secciones aflorantes al oeste de Toro (Valle del Cauca). Geología Norandina, 5, 3-12, Bogotá.

Etayo, F., and 20 others (1986): Mapa de terrenos geológicos de Colombia. Bogotá, Ingeominas, Publicaciones Geológicas Especiales, 14, 235 pp.

Fabre, A. (1984): La subsidencia de la cuenca del Cocuy (Cordillera Oriental de Colombia) durante el Cretáceo y el Terciario. Geología Norandina, 8, 14-21, Bogotá.

Feininger, T. (1980): Eclogite and related high-pressure regional metamorphic rocks from the Andes of Ecuador. Journal of Petrology, v. 21, 107-140.

Feininger, T. (1986): Allocthonous terranes in the Andes of Ecuador and northwestern Perú. Geological Survey of Canada, Cont. 22686, p. 266-278.

Galvis, J. (1980): Un arco de islas terciario en el occidente Colombiano. Universidad Nacional, Geología Colombiana, 11, 7-43, Bogotá.

Gansser, A. (1950): Geological and petrographical notes on Gorgona Island in relation to North Western South América. Schweizerische Mineralogische und Petrographische Mitteilungen, v. 30, 219-237.

Gansser, A. (1973): Facts and Theories on the Andes. Journal of the Geological Society, v. 129, 93-131.

Gerth, H. (1955): Der geologishe Bau der südamerikanischen Kordillere. Berlin, Gebruder Borntraeger, 264 pp.

González, H., Restrepo, J.J., and Toussaint, J.F. (1978): Edad radiométrica K/Ar del batolito de Sabanalarga. Universidad Nacional, Publicación Especial, Geología, 8, 1-5, Medellín.

Hall, R.B., Alvarez, J., and Rico, H. (1972): Geología de parte de los Departamentos de Antioquia y Caldas (Subzona II-A). Ingeominas, Boletín Geológico., v. 20, 1, 1-85, Bogotá.

Hoyos, P., and Zuluaga, J. (1978): Estudio geológico del grupo Cañasgordas. Thesis, Universidad Nacional, Medellín, 147 pp.

Irving, E.M. (1971): Evolución estructural de los Andes más septentrionales de Colombia. Ingeominas, Boletín Geológico, v. 19, 2, 1-90, Bogotá.

Julivert, M. (1973): Les traits structuraux et l'évolution des Andes colombiennes. Revue de Géographie Physique et de Géologie Dynamique, v 15, 143-155.

McCourt, W., and Aspden, J.A. (1984): A Middle Mesozoic oceanic terrane in the Central Cordillera of Western Colombia. Geología Norandina, 9, 19-26, Bogotá.

McCourt, W., Aspden, J., and Brook, M. (1984): New geological and geochronological data from the Colombian Andes: continental growth by multiple accretion. Journal of the Geological Society, v. 141, 831-845.

McGeary, S., and Ben Abraham, Z. (1985): The accretion of Gorgona Island, Colombia: Multichannel Seismic Evidence, *in* Tectonostratigraphic Terranes of the Circum-Pacific Region, (Ed.: Howell, D.G.), Circum-Pacific Council for Energy and Mineral Resources, Earth Science Series, v. 1, 543-554.

Mégard, F., and Lebrat, M. (1986): Los terrenos exóticos del Occidente Ecuatoriano y sus relaciones con Suramérica. Coloquio sobre la Geología de Ecuador, Quito, 10 pp.

Meissner, R., Flueh, E., Stibane, F., and Berg, E. (1976): Dynamics of the active plate boundary in Southwest Colombia according to recent geophysical measurements. Tectonophysics, v. 35, 115-136.

Nelson, H. (1957): Contribution to the geology of the Central and Western Cordilleras of Colombia. Leidse Geologische Mededelingen, v. 22, 1-76.

Nelson, H. (1962): Contribución al conocimiento de la Cordillera Central de Colombia, Sección entre Ibagué y Armenia. Ingeominas, Boletín Geológico, v. 10, 161-202, Bogotá.

Nivia, A. (1989): El terreno Amaime-volcánica: una provincia acrecionada de basaltos de meseta oceánica. Congreso Colombiano de Geología, 5th, Bucaramanga, Memoria, I, 1-30.

Núñez, A. (1987): El Cretáceo sedimentario de Rioblanco-Planadas, SW del Dpto. del Tolima. Memoria Seminario, G. Botero, 109-120, Medellín.

Orrego, A., Cepeda, H., and Rodríguez, G.I. (1980a): Esquistos glaucofánicos en el área de Jambaló. Geología Norandina, 1, 5-10, Bogotá.

Orrego, A., Restrepo, J.J., Toussaint, J.F., and Linares, E. (1980b): Datación de un esquisto sericítico de Jambaló, Cauca. Medellin. Universidad Nacional, Publicación Especial, Geología, 25, 1-2, Medellín.

Radelli, L. (1967): Geologie des Andes Colombiennes. Thesis, University of Grenoble, Grenoble, 457 pp.

Restrepo, J.J. (1986): Metamorfismo en el sector norte de la Cordillera Central de Colombia. Medellín, Universidad Nacional, Informe, 276 pp.

Restrepo, J.J., and Toussaint, J.F. (1973): Obducción Cretácea en el Occidente Colombiano. Universidad Nacional, Publicación Especial, Geología, 3, 1-26, Medellín.

Restrepo, J.J., and Toussaint, J.F. (1974): Algunas consideraciones sobre la evolución estructural de los Andes Colombianos. Universidad Nacional, Publicación Especial, Geología, 4, 1-14, Medellín.
Restrepo, J.J., and Toussaint, J.F. (1975): Edades radiométricas de algunas rocas de Antioquia, Colombia. Universidad Nacional, Publicación Especial, Geología, 6, 1-24, Medellín.
Restrepo, J.J., and Toussaint, J.F. (1978): Datación de una metadiabasa del grupo Cajamarca. Universidad Nacional, Publicación Especial, Geología, 16, 1-4, Medellín.
Restrepo, J.J., and Toussaint, J.F. (1988): Terranes and Continental Accretion in the Colombian Andes. Episodes, v.11, 189-193.
Restrepo, J.J., and Toussaint, J.F. (1989): Terrenos alóctonos en los Andes Colombianos: explicación de algunas paradojas. Congreso Colombiano de Geología, 5th, Bucaramanga, Memoria, I, p. 92.
Restrepo, J.J., Toussaint, J.F., Hall, M., Pimentel, N., Cordani, U., Kawashita, K., and Linares, E. (1985): Evolución magmática de los Andes septentrionales. Universidad de Chile, Departamento de Geología y Geofísica, Comunicaciones, 35, 201-208, Santiago.
Rodríguez, G.I. (1981): Facies prehnita-pumpellita en rocas de los grupos Diabásico y Dagua en la parte central y sur de la Cordillera Occidental (Colombia). Geología Norandina, 3, 3-10, Bogotá.
Rodríguez, G.I., and Rojas, R. (1985): Estratigrafía y tectónica de la serie infracretácica en los alrededores de San Félix, Cordillera Central de Colombia, *in* Proyecto Cretácico, (Dir.: Etayo, F.), Bogotá, Ingeominas, Contribuciones, Publicaciones Geológicas Especiales, 16, XXI/1-XXI/21.
Schwinn, W.L. (1969): Guidebook to the geology of the Cali area. Colombian Society of Petroleum Geologists and Geophysicists, Field Trip, 10th, 29 pp., Bogotá.
Souza, H., Espinosa, A., and Delaloye, M. (1984): K/Ar ages of basic rocks in the Patia Valley, Southwest Colombia. Tectonophysics, v. 107, 135-145.
Tamayo, L.M. (1984): Análisis y mediciones de microestructuras en la anfibolita del Grupo Ayurá-Montebello. Thesis, Universidad Nacional, Bucaramanga, 75 pp.
Toussaint, J.F. (1978): Grandes rasgos geológicos de la parte septentrional del Noroccidente Colombiano. Universidad Nacional, Boletín de Ciencias de la Tierra, 3, 128 pp., Medellín.
Toussaint, J.F., and Restrepo, J.J. (1974): La Formación Abejorral y sus implicaciones sobre la evolución de la Cordillera Central durante el Cretáceo. Universidad Nacional, Facultad de Minas, Anales, v. 58, 13-29, Medellín.
Toussaint, J.F., and Restrepo, J.J. (1976): Modelos orogénicos de tectónica de placas en los Andes Colombianos. Universidad Nacional, Facultad de Ciencias, Boletín de Ciencias de la Tierra, 1, 1-47, Medellín.
Toussaint, J.F., and Restrepo, J.J. (1978): Edad K/Ar de dos rocas básicas del flanco noroccidental de la Cordillera Central. Universidad Nacional, Publicación Especial, Geología, 15, 1 pp., Medellín.
Toussaint, J.F., and Restrepo, J.J. (1986): Límites de placas y acortamientos recientes entre los paralelos 5°N y 8°N, Andes Colombianos. San Juan (Argentina), Symposium Megafaults of South America, 1-8.
Toussaint, J.F., and Restrepo, J.J. (1988): ¿Son alóctonos los Andes Colombianos?. Universidad Nacional, Revista ICNE, 1, 17-41, Medellín.
Toussaint, J.F., and Restrepo, J.J. (1989): Acreciones sucesivas en Colombia: un nuevo modelo de evolución geológica. Congreso Colombiano de Geología, 5th, Bucaramanga, Memoria, I, 92-107.
Toussaint, J.F., González, H., Restrepo, J.J., and Linares, E. (1978): Edad radiométrica K/Ar del neis de Puquí. Universidad Nacional, Publicación Especial, Geología, 7, 3 pp., Medellín.
Zeil, W. (1979): The Andes, a geological review. Berlin, Ed. Gebruder Borntraeger, 260 pp.

Kimmeridgian to Paleocene Tectonic and Geodynamic Evolution of the Peruvian (and Ecuadorian) Margin

E. Jaillard

ORSTOM, UR 1H, cs1 rue La Fayette, 75480 Paris Cedex 10, France, and Institut Dolomieu, 15 rue Maurice-Gignoux, 38031 Grenoble Cedex, France.

Presently at: ORSTOM, Apartado 17.11.06596 or 17.11.05149, Quito, Ecuador

Abstract

Three main tectonic periods are recognized between Kimmeridgian and Paleocene times in the Peruvian-Ecuadorian margin.

The "Virú" period comprises a Kimmeridgian event probably equivalent to the Araucan phase of Argentina and Chile, a Tithonian phase related to terrane accretions and collision tectonics along the Ecuadorian margin and to a sudden extension along the north Peruvian margin, and a Berriasian event most probably originated by the incipient South Atlantic rifting.

The "Mochica" period begins with tensional and volcanic precursor events (Late Aptian-earliest Albian). It continues with extensional effusions of coastal, back-arc or arc volcanic centres, which alternate with compressive crises (Early and Middle Albian). It ends with the accretion and deformation (thrusting?) of the Albian volcanic arc or back-arc volcanic system (Late Albian-Early Middle Cenomanian).

The "Peruvian" phase starts with a paleogeographic change probably triggered by the incipient coastal uplift (Turonian-Coniacian boundary), and continues in the Late Coniacian-Early Santonian with the initiation of northeastward overthrusts located at the southwestern boundary of the western Trough. It culminates in the latest Campanian, with the creation of intermontane basins which express the onset of the southwestern thrusts, and of foreland basins related to the onset of new overthrusts located at the northeastern boundary of the western Trough.

An extensional regime was probably dominant during latest Jurassic and Early Cretaceous times, leading to formation of the main sedimentary basins, with an apparent quiescence of the subduction-related volcanic systems. In Late Aptian times, the rapid convergence and rejuvenation of the subducted lithosphere induced a general compressional and/or wrenching regime, triggering the continentalward migration of the trench and the subsequent consumption, deformation and accretion of the volcanic

arc or aborted marginal basin ("Mochica period"). After Cenomanian-Turonian times, the northeastward spread of the compressive strain within the South American plate caused the latter's deformation ("Peruvian period").
The age and velocity of the subducted slab are decisive factors in the general strain regime within the subduction system; whereas variations of the convergence rate at the trench and inherited extensional structures determine the age and location of the short-lived compressional events. These generally correlate with plutonic gaps. Although poorly known, wrenching movements may have been important in the Cretaceous evolution of the Peruvian margin.

Introduction

Aim of this paper

The Andean chain was built up mostly during Tertiary times on a continental margin, as a response to compressional stress induced by the subduction of the Farallón (Nazca) oceanic plate. Nevertheless, during Mesozoic times subduction occurred beneath the central South American margin, but no important chain was formed. One can thus conclude that, though subduction processes are necessary to create an Andean-type chain, they are not sufficient in themselves. In order to recognize the other relevant parameters which act in such an orogeny, it is necessary to study the behaviour of a continental margin submitted to oceanic subduction during a non-orogenic period.
Between Kimmeridgian and Paleocene times the Peruvian-Ecuadorian margin was subjected to the subduction of the Farallón oceanic plate, but tectonic stress, magmatic manifestations, detrital sedimentation and mobility of the substratum were alternately absent, weak or important. The purpose of this synthesis is therefore to describe, to analyse and to interpret when possible the tectonic events recorded by the Peruvian-Ecuadorian margin. Their relationships with geodynamic processes are then examined, and a geodynamic model is proposed.

Methodology

For such a purpose various tools have been used:
Sedimentology studies make it possible to define the environment, evolution and geometry of both depositional sequences and sedimentary basins, to define the source areas and the flooding directions of the detrital supply, and to calculate subsidence curves. The synthesis of the above data, recorded on margin- or basin-scale paleogeographic or isopach maps, can then be used to define the internal shape and deformation of the margin, and to determine the tectonic setting of the sedimentary basins (see Miall, 1984 for review).

Structural analysis makes it posible to determine the direction of paleostress, using various tectonic features.

The analysis of striated synsedimentary faults makes it possible to determine precisely the strain pattern (e.g. Moulin, 1989).

Unstriated synsedimentary faults are assumed to have affected a homogeneous, recently deposited sediment. There, Anderson's principle (1951) has been applied. However, the data are sometimes difficult to interpret, since fracturing may be caused by gravity processes (slumpings and slidings). In this case their orientation is of purely local value, and independent of tectonic strains.

Planes of vertical clastic dykes, in-filled per descensum (opened upward) have been measured, and have been supposed to trend normally to the tensional stress. This can be due either to gravity processes with the dykes then roughly running perpendicular to the paleoslope (Parize et al., 1989), or to tectonic stress, the dykes then trending parallel to the compressional strain (Winslow, 1983). Their interpretation is therefore somewhat difficult.

The slump fold axis and vergences have been measured to try to determine the paleoslope direction.

All measurements have been rotated in order to restore the horizontal stratification, assuming that they were affected by a single folding event, the axis of which was parallel to the observed strata dip. However, in many cases insufficient data, uncertainties in structural identification or measurements, and errors in subsequent rotations may lead to approximate or ambiguous conclusions.

Finally, geochemical and petrological studies on magmatic rocks have led some workers to important tectonic conclusions, which are succinctly mentioned in this paper.

Field work and observations have been made mainly in Peru. However, the tectonic-sedimentary evolution of the Eastern Basin and Cordillera Real of Ecuador will be briefly described together with the northernmost areas of Peru, since they belong to the same paleogeographic units.

Geological Setting

Pre-Cretaceous History and resultant Paleogeography

During the Early Mesozoic, the Peruvian-Ecuadorian margin underwent a complex tectonic and sedimentary evolution (Jaillard et al., 1990).

During the Late Triassic and the Liassic, a thick carbonate shelf developed throughout the margin (Loughmann and Hallam, 1982). Then a Middle to Late Liassic extensional tectonic activity progressively destroyed the shelf. Its destruction was achieved in the Middle Jurassic. At this time, a NNE-trending continental volcanic arc developed in northern Peru and in Ecuador (Mourier et al., 1988a; Aspden et al., 1988), and poorly known subduction-related volcanic rocks overflowed onto the SW-trending coast of southern Peru (Romeuf, 1990). Meanwhile, continental sedimentation and erosions occurred on most of the Peruvian margin, except in southwestern Peru where a turbiditic trough was formed and filled (Vicente et al., 1982).

The Late Liassic and Mid-Jurassic tectonic events resulted in a contrasting paleogeographic pattern, which was later modified during latest Jurassic and earliest Cretaceous times. In this paper we shall describe the tectonic events according to the following paleogeographic zones, from West to East (Figure 1):

- The Coastal Zone included a coastal Cordillera and coastal Troughs. The Coastal Cordillera presently consists of mainly Paleozoic and older rocks, because Tertiary erosion removed most of the Mesozoic rocks. It now crops out in southern Peru. In northern Peru and southern Ecuador the Paleozoic Amotape massif is regarded as an allochthonous terrane accreted by latest Jurassic times (Mourier et al., 1988a). During the Cretaceous it constituted a morphological equivalent of the southern Coastal Cordillera. The coastal Troughs are infilled by very thick sequences of volcanic flows and volcaniclastic deposits of latest Jurassic and Mid-Cretaceous age. They are now exposed along the coast. In northwestern Peru, the Lancones basin constitutes an equivalent of the coastal Troughs which is prolonged into southwestern Ecuador (Celica Trough)(Figure 1). Farther north, coastal Ecuador is made up of allochthonous oceanic terranes accreted during Late Cretaceous to Tertiary times (Feininger and Bristow, 1980) and is not studied in this work.

- The Western Trough received a thick, mainly marine sedimentation during Mesozoic times. It presently crops out in the Western Cordillera. The Western Trough became an incipient Western Cordillera in Senonian times. Its northern end is a lesser subsident area, which constitutes the southward prolongation of part of the Cordillera Real and Subandean zone of Ecuador.

- The Axial Swell is a positive paleogeographic zone characterized by thin deposits during most of the Cretaceous. It now constitutes the Oriental Cordillera of northern Peru (Marañon geanticline), and the southwestern Altiplano of southern Peru (Cuzco-Puno axis).

- The Eastern Basins of Peru and Ecuador were moderately subsiding regions located on the western border of the Brazilian and Guyanese Precambrian shields respectively. They received a mixed marine and continental sedimentation during Mesozoic times. Some positive areas within the Peruvian Eastern Basin were long considered Mesozoic emergent areas (Eastern Cordillera of southern Peru), because they were uplifted and eroded during the Andean orogeny.

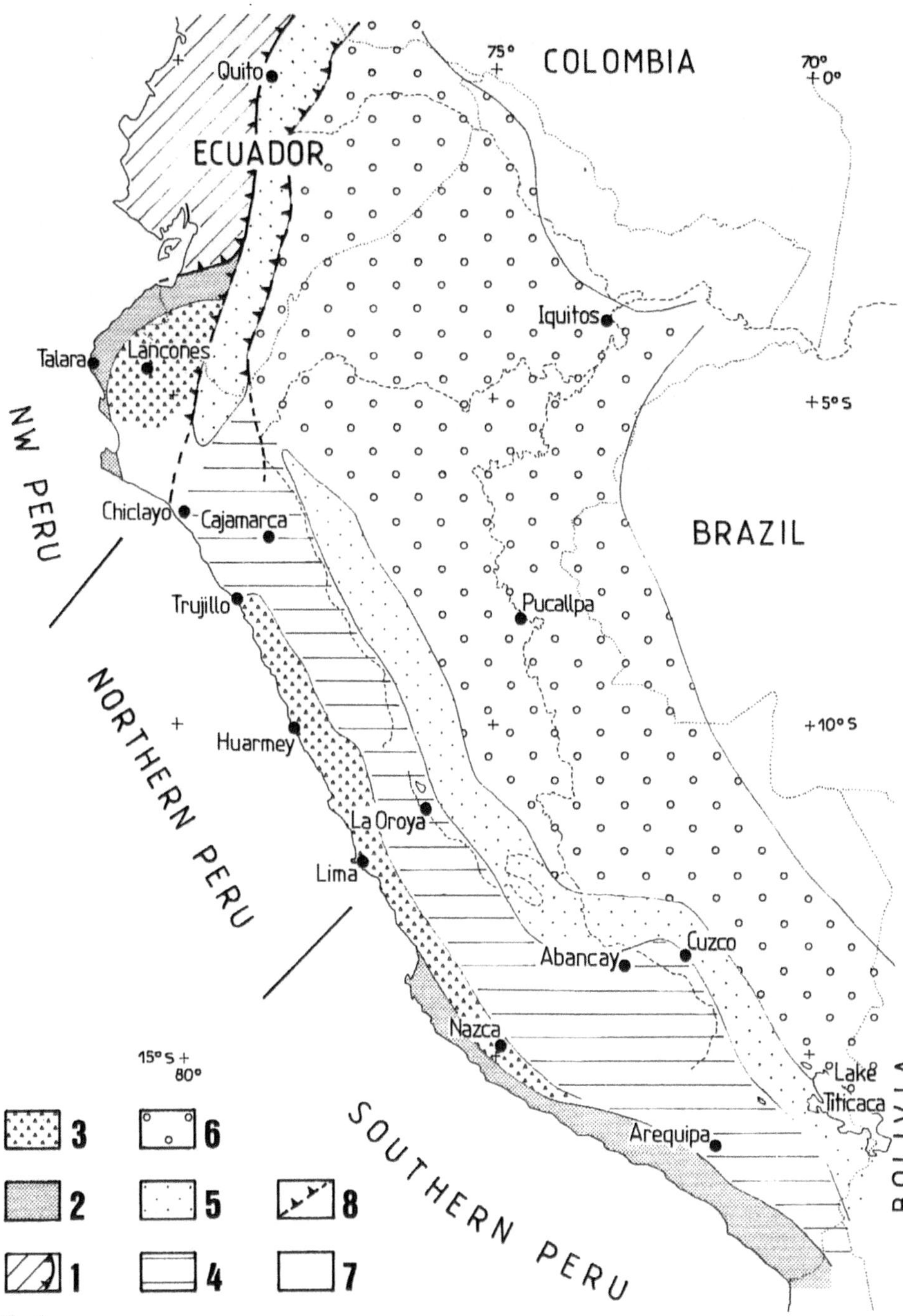

Fig. 1

Paleogeographic sketch of the Peruvian margin, and location of main areas cited. 1: Cretaceous allochthonous terranes, and suture. 2: Coastal Cordillera. 3: Coastal Troughs. 4: Western Trough. 5: Axial Swell. 6: Eastern Basin. 7: Brazilian and Guianese (Colombian) Precambrian shields. 8: Jurassic suture.

In this paper the Peruvian-Ecuadorian margin is divided into three NE trending zones (Figure 1).
Northwestern Peru includes the northernmost part of Peru, west of Chiclayo (Talara-Lancones area), and the southwesternmost part of Ecuador.
Northern Peru is located north of Lima and its eastern part will be described together with Eastern Ecuador. This zone is well-known in the Oriente of Ecuador (Tschopp, 1953; Bristow and Hoffstetter, 1977; Baldock, 1982), and, in Peru, along the Huarmey-Trujillo-Cajamarca-Iquitos line (Benavides, 1956; Soto, 1979; Myers, 1980; Jaillard, 1987; Mourier, 1988), and along a Lima-La Oroya-Pucallpa transect (Rodríguez and Chalco, 1975; Mégard, 1978; Dalmayrac, 1978; Moulin, 1989).
Southern Peru begins south of Lima. It is well-known along the Arequipa-Puno transect (Newell, 1949; Benavides, 1962; Laubacher, 1978; Vicente et al., 1982; Batty and Jaillard, 1989), and in the Cuzco area. The area located west and south of Abancay (Figure 1) is poorly known because of a widespread Tertiary volcanic cover, difficult accesses and recent political troubles.

Cretaceous Stratigraphic Chart and Main Tectonic Periods

Stratigraphic studies make it possible to date the sediments within which sedimentologic and tectonic information is contained. Therefore, revisions of published stratigraphic works, collections of new paleontologic and stratigraphic data and sequence stratigraphy analysis have been carried out and recently synthesized (Jaillard and Sempere, 1989). They make it possible to propose a stratigraphic correlation chart as a working hypothesis (Figure 2). The conversion from stratigraphic stages into absolute ages is made following the timescales of Haq et al. (1987) and Odin and Odin (1990).
Various periods can be recognized during the Cretaceous (Jaillard and Sempere, 1989, Figure 2). However, because the first one begins during latest Jurassic times and the last one ends during Early Tertiary times, we shall consider the Kimmeridgian-Paleocene interval. The Peruvian margin behaved successively: (1) as a mobile unstable region, locally affected by strong volcanic activity (Kimmeridgian-Berriasian); (2) as a stable anorogenic and non-volcanic extensional margin (Valanginian-Aptian); (3) as a stable region whose western part underwent extension-compression alternation, and was the site of outstanding magmatic activity (Late Aptian-Turonian); (4) as a mobile margin undergoing progressive compressional tectonic activity and affected by scarce plutonic intrusions (Coniacian-Campanian); finally, (5) as a relatively stable region which registered some noteworthy magmatic activity (Maastrichtian-Paleocene).

Tectonic Evolution of the Peruvian Margin Between Kimmeridgian and Paleocene Times

Kimmeridgian-Late Berriasian

This is tectonically a very unstable period, which resulted in a contrasting paleogeography (Batty et al., 1990). Several tectonic events can be recognized. However, the paucity of outcrops and the abundance of unfossiliferous continental deposits make accurate analyses and precise paleogeographic reconstructions difficult.

The Kimmeridgian (?) Event

The Kimmeridgian (?) tectonic event is poorly documented, except in southern Peru and in Bolivia (Sempere, this volume).

Northern Peru and Eastern Ecuador

In the Western domains, no Kimmeridgian deposits are known (Benavides, 1956; Rivera et al., 1975; Mégard, 1978; Moulin, 1989; Jaillard and Jacay, 1989). However, they might be represented in northern Peru by the last effusions of the Middle to Late Jurassic volcanic arc, partly dated as Callovian-Oxfordian (Colán Formation, Mourier, 1988) and by the coeval formations of Ecuador intruded by the Zamora and Abitagua plutons (Aspden et al., 1990).

In the Eastern Basins of Peru, undated coarse-grained red beds were deposited during Late Jurassic times (upper Sarayaquillo or Boquerón formations, Rodriguez and Chalco, 1975; Seminario and Guizado, 1976; Pardo and Zuñiga, 1976). They problably correlate with the upper Chapiza Formation of the Ecuadorian Oriente (Tschopp, 1953; Bristow and Hoffstetter, 1977; Baldock, 1982).

Southern Peru

Along the coast, Kimmeridgian deposits could be present in the upper part of the Jurassic detrital deposits of the Guaneros Formation (Vicente, 1981). Farther NE, marine to continental sands and shales may be equivalent to the Labra Formation of the Western Trough (Zuñimarca Formation, Vicente, 1981).

In the Western Trough, widespread silico-clastic shallow marine deposits overlie Callovian-Oxfordian black shales (Labra, Chuquibambilla and Chachacumane formations, Benavides, 1962; Pecho, 1981; Vicente, 1981, 1990; Vicente et al., 1982). In some areas, they disconformably overlie the Callovian shales (Yauca Formation of Olchauski, 1980).

On the Axial Swell and in the Eastern Basin, undated conglomerates (Chupa Formation, Klinck et al., 1986, Huambutío Formation, Carlotto, 1989) rest unconformably upon

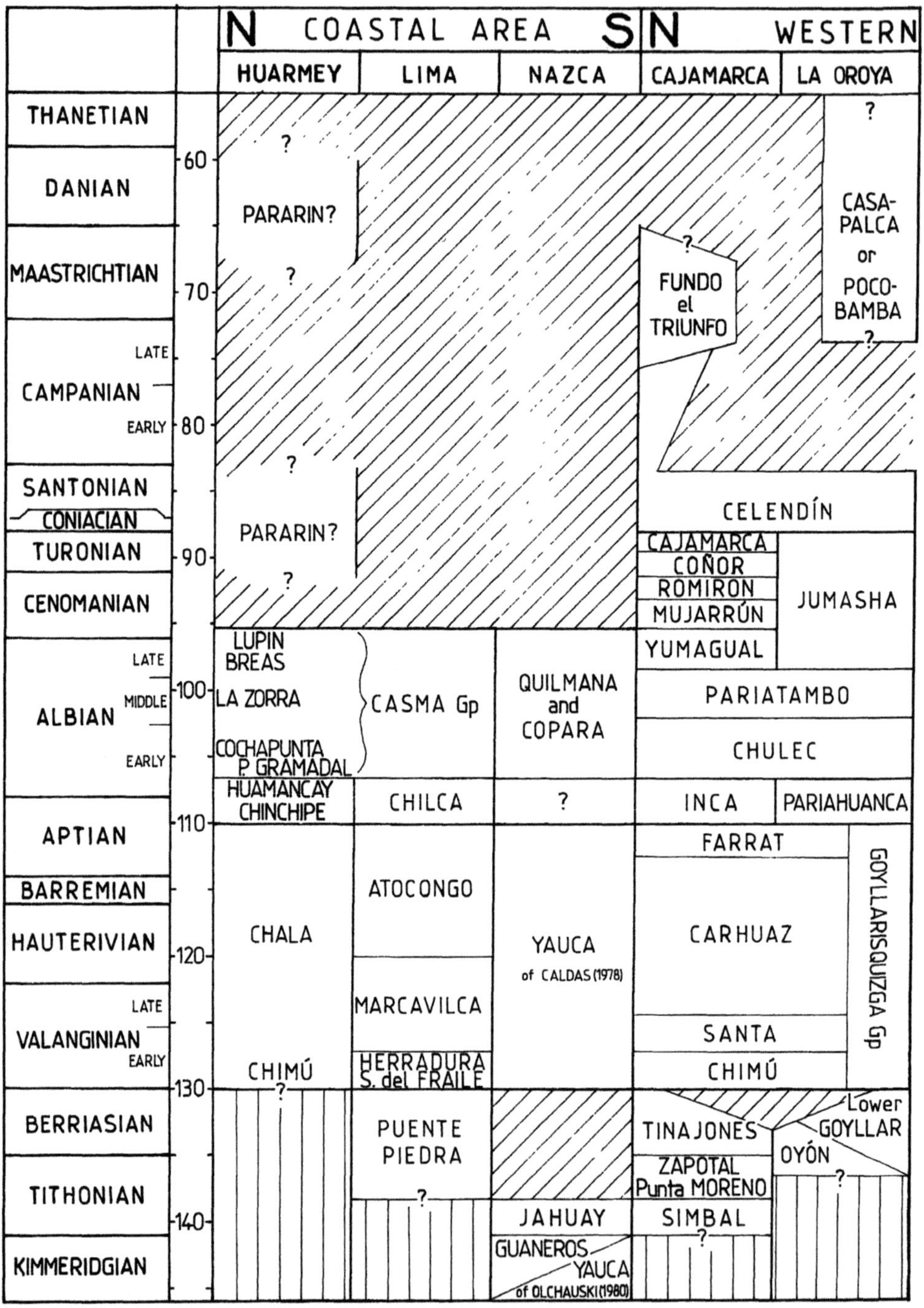

Fig. 2

Correlation chart of the main Kimmeridgian - Paleocene stratigraphic formations of the Peruvian margin. Location on Figure 1.

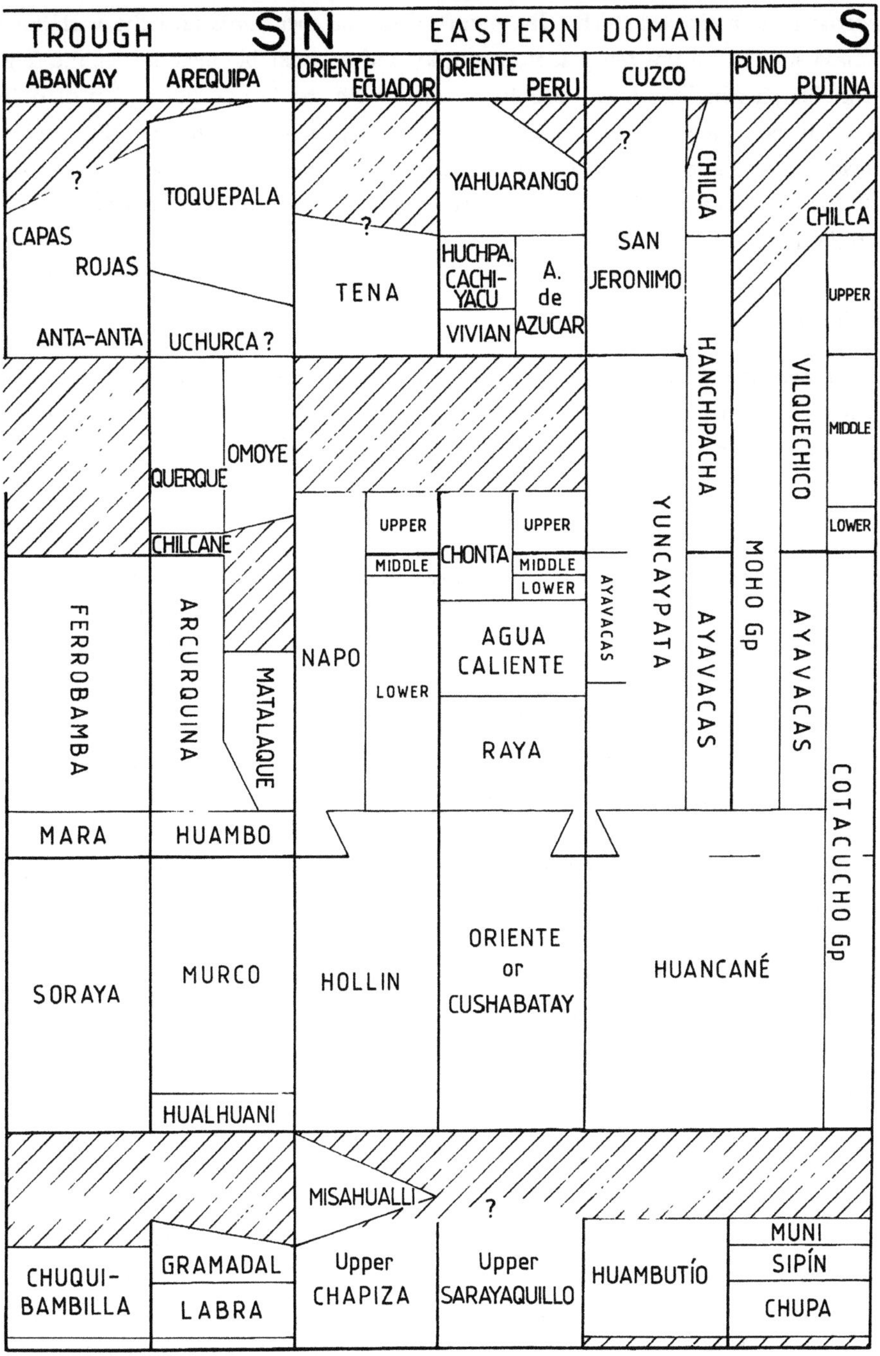

Fig. 2 (Continued)

Paleozoic to Triassic rocks. The dispersion of the paleocurrents and the variations in thickness suggest an uneven paleotopography and an unstable tectonic regime (Figure 3). The Chupa conglomerates roughly correlate with the Condo Formation of Bolivia, which has been interpreted (Sempere et al., 1989) as related to the Kimmeridgian Araucan tectonic event of Argentine and Chile (Stipanicic and Rodrigo, 1969; Riccardi, 1988). Farther northeast, these conglomerates probably correlate with part of the Sarayaquillo Formation.

The Middle Tithonian Phase

In all the western areas of Peru, the Early Tithonian is represented by partly calcareous deposits. These are abruptly overlain by detrital, continental or deep-marine deposits, thus evidencing an important tectonic/paleogeographic event.

Northern Peru and Eastern Ecuador

In the coastal area near Lima, a NW trending trough probably opened during Tithonian times (Figure 6), since marine hemipelagic black shales interbedded among volcanic and volcaniclastic deposits are dated as Late Tithonian-Late Berriasian (Rivera et al., 1975, Wiedmann, 1980). These high-K basalts and andesites are interpreted as the products of a volcanic arc deposited in a back-arc basin (Atherton et al., 1983, 1985).

In the western part of the Western Trough, the partly calcareous lagoonal Simbal Formation is sharply overlain by hemipelagic black shales and then by a 2500 m-thick turbiditic series of Late Tithonian age (Punta Moreno Formation, Jaillard and Jacay, 1989; Jacay, 1991. Numerous olistolites and internal disconformities express a syntectonic depositional setting. The creation of such a subsident trough constitutes a major extensional tectonic event, of Early Late Tithonian age (Jaillard and Jacay, 1989). In the eastern part of the Western Trough, Moulin (1989) describes estuarine red shales and sands of probable Tithonian-Berriasian age (Lower Goyllarisquizga Formation). The geometry of clastic dykes indicates a local NNE-SSW tensional strain (Moulin, 1989, Figure 4). Erosional features beneath the disconformable Valanginian sandstones suggest a hiatus of part of the Late Tithonian-Berriasian deposits (Moulin, 1989).

In the Eastern Basin of northern Peru and Ecuador, the lack of data about the unfossiliferous continental deposits precludes the identification of the Tithonian tectonic event. However, bimodal volcanic flows occurred in the Ecuadorian Oriente and in the northwestern Peruvian Oriente during Tithonian-Berriasian times (Misahualli Formation, Bristow and Hoffstetter, 1977; Hall and Calle, 1982). Farther south, Arthaud et al. (1977) mention NE- and NW-trending Tithonian synsedimentary normal faults.

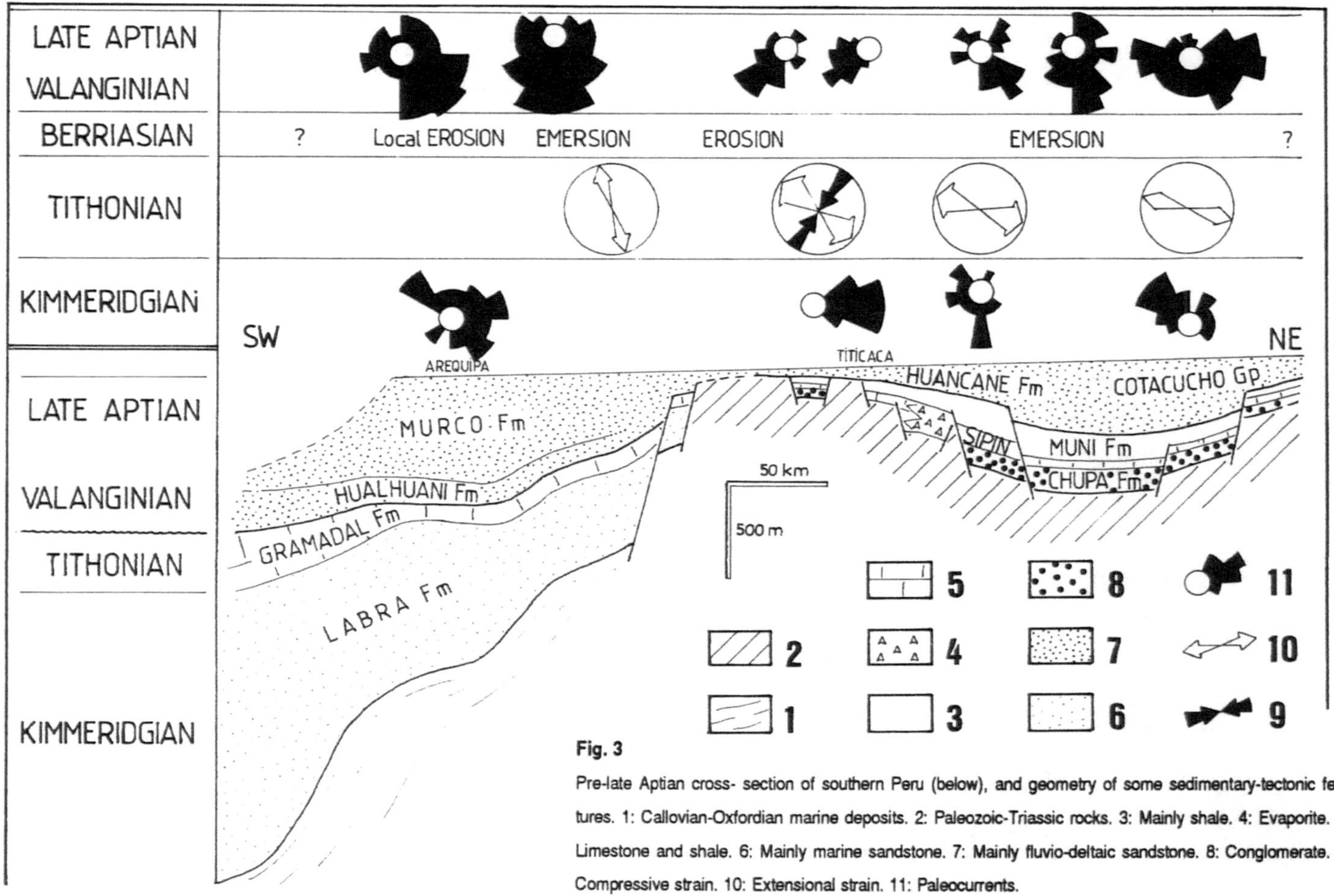

Fig. 3

Pre-late Aptian cross- section of southern Peru (below), and geometry of some sedimentary-tectonic features. 1: Callovian-Oxfordian marine deposits. 2: Paleozoic-Triassic rocks. 3: Mainly shale. 4: Evaporite. 5: Limestone and shale. 6: Mainly marine sandstone. 7: Mainly fluvio-deltaic sandstone. 8: Conglomerate. 9: Compressive strain. 10: Extensional strain. 11: Paleocurrents.

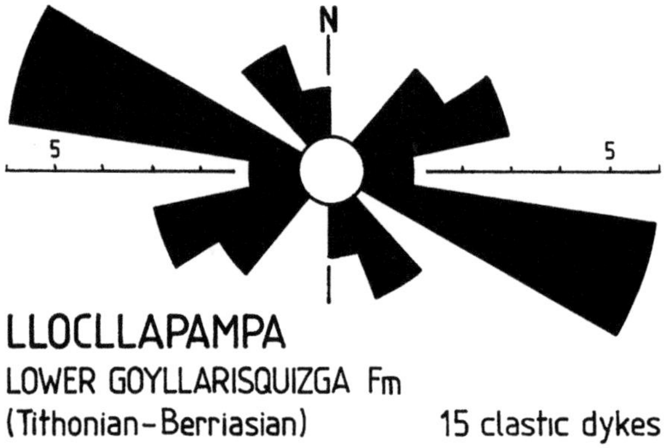

Fig. 4
Orientation of sedimentary clastic dykes in the Tithonian-Berriasian (?) Lower Goyllarisquizga Formation (after, Moulin, 1989).

Southern Peru

In the Western Trough near Nazca, the Early Tithonian partly calcareous Jahuay Formation (Rüegg, 1961) is directly overlain by the Valanginian sandstones (Yauca Formation of Caldas, 1978). There, a stratigraphic gap of Late Tithonian and Berriasian beds is thus probable. In the Arequipa area, Early Tithonian lagoonal sandstones and limestones (Gramadal Formation, Chávez, 1982; Batty and Jaillard, 1989) are overlain by a few tens of meters of red shales (Batty, in preparation, Figure 6). In other areas, the Valanginian sandstones seem to directly and conformably overlie the Kimmeridgian (?) sandstones (e.g. Olchauski, 1980; Pecho, 1981, 1983).

On the axial Swell and in the eastern region (Altiplano), Late Jurassic limestones and shales (Sipín Formation, Newell, 1949) probably correlate with the Gramadal Formation (Laubacher, 1978; Batty and Jaillard, 1989). They give way upward to tidal and continental red shales and thin-bedded sandstones (Muni Formation, Newell, 1949), which are in turn disconformably overlain by the Valanginian sandstones (Huancané Formation). In the Cuzco area, the undated Huambutío Formation exhibits a comparable sequence (Carlotto, 1989). On the Altiplano and in the northeastern part of the Western Trough, the Early Tithonian is marked by synsedimentary tectonic structures, such as normal faults, slumpings and clastic dykes (Figures 3 and 5). A geometrical study of these features indicates a NW- to WNW-trending tensional stress during Tithonian times (Jaillard and Batty, 1989), whereas scarce reverse synsedimentary faults suggest a normal NE-SW compressional component (Figure 5).

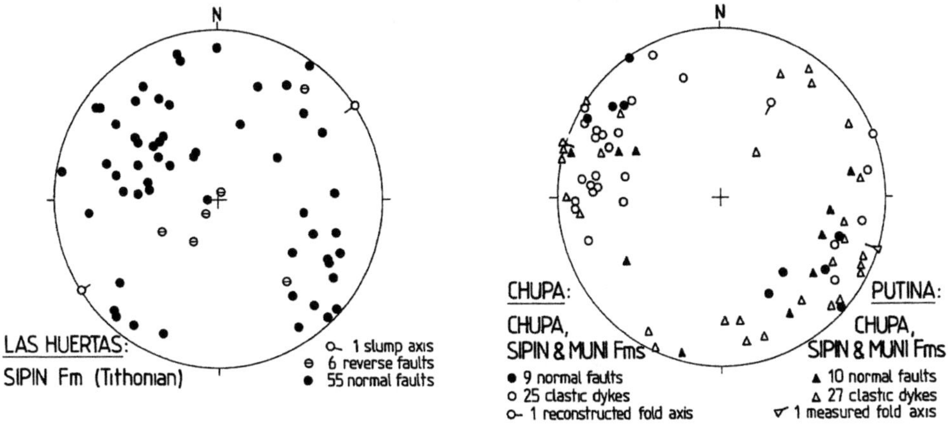

Fig. 5

Poles of synsedimentary faults and clastic dykes in the early Tithonian Sipín Formation of the Altiplano (eastern area, southern Peru).

The Berriasian Event

This is mainly expressed by a regional disconformity beneath the Valanginian sands (see Tschopp, 1953; Benavides, 1956; Pardo and Zúñiga, 1976; Laurent, 1985; Jaillard and Sempere, 1989; Figures 5 and 6).

Northern Peru and Eastern Ecuador

Along the coast, marine sedimentation and volcanic arc effusions went on until latest Berriasian times in the Lima area (Puente Piedra Formation, Wiedmann, 1980; Atherton et al., 1985, Figure 6).

In the Western Trough, near Trujillo, Late Tithonian turbidites are overlain by latest Tithonian black shales, and then by Berriasian (?) shallow marine to deltaic sandstones (lower part of Tinajones Formation, Cobbing et al., 1981; Jaillard and Jacay, 1989). This formation exhibits numerous synsedimentary normal faults, thus expressing an extensional instability (Jaillard and Jacay, 1989). It probably correlates southward with the deltaic Oyón Formation of the Lima area (Mégard, 1978; Cobbing et al., 1981). The overlying fluvial red beds (upper part of Tinajones Formation, Jaillard and Jacay, 1989) are only very locally preserved beneath the Valanginian sandstones, thus illustrating the pre-Valanginian erosional unconformity (Figure 6).

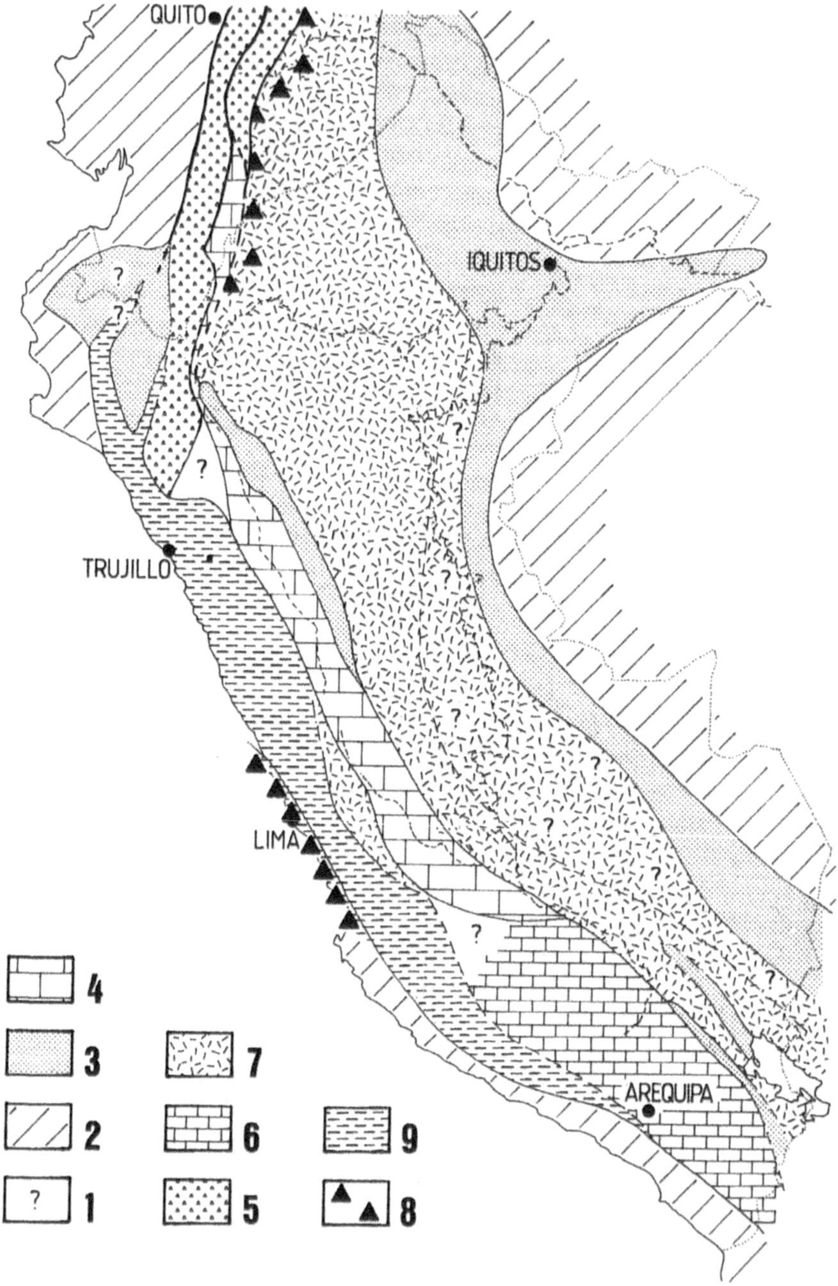

Fig. 6

Pre-Valanginian geological sketch map. 1: No pre-Valanginian outcrops. 2: No Valanginian deposits. 3: Pre-Norian rocks. 4: Norian to Middle Jurassic limestone. 5: Middle to Late Jurassic volcanic and volcaniclastic rocks (volcanic arc). 6: Early Tithonian limestone and shale. 7: Late Jurassic continental Red Beds. 8: Late Tithonian - Berriasian volcanic rocks. 9: Berriasian deltaic to marine deposits.

In the Oriente of Peru and Ecuador, the bimodal effusion of the Misahualli Formation could have continued during Berriasian times (Bristow and Hoffstetter, 1977).

Southern Peru

In the western part of the Western Trough, Late Tithonian-Berriasian marine to deltaic black shales and fine-grained sandstones probably correlate with the Oyón Formation (Bellido, 1956; Tiabaya outcrops, Geyer, 1983).
In the Western Trough and near the Axial Swell, no tectonic event has been clearly recognized, since the Valanginian sandstones disconformably overlie the often eroded post-Early Tithonian red beds, and Late Tithonian-Berriasian deposits are most probably lacking (see above, Figures 3 and 6). It suggests that these areas were emergent during Berriasian time (Batty and Jaillard, 1989; Batty et al., 1990, Figure 6).

Tectonic Interpretations

The Kimmeridgian tectonic event (ca 145-142 Ma): A distal response to the Argentina-Chilean Araucan phase?.
No information is available in northern Peru about the Kimmeridgian event. It is mostly expressed in southern Peru by the resumption of detrital sedimentation and by the instability of the uneven substratum (Figure 3). It seems to represent a distal response to the Argentina-Chilean Araucan tectonic phase of Kimmeridgian age (see Sempere, this volume).
The Middle Tithonian tectonic event (ca 138-137 Ma): Oblique collision and coastal extension.
The shape of the turbiditic trough of northwestern Peru, the high sedimentation rate and the strong initial tectonic subsidence strongly suggest this trough to be an extensional, probably pull-apart basin (Jaillard and Jacay, 1989), related to the oblique dextral collision of the Amotape allochthonous continental terrane (Mourier et al., 1988a; Mourier, 1988). In the Cordillera Real of Ecuador, a major collisional tectonic phase caused the folding, thrusting and metamorphism of Jurassic arc-related volcanic rocks, and is also regarded as related to the accretion of continental blocks (Aspden et al., 1988; Litherland and Aspden, 1990). It is probable that the latter phase correlates with the Tithonian tectonic event of northern Peru. Although no structural data is available about the pre-Cretaceous deformations of the Peruvian Jurassic volcanic arc, the contrasting paleogeography sealed by the Valanginian sandstones (Figure 6), the significant variations in thickness of the latter (Figure 7), and the abundance of volcanic clasts in the eastern upper Sarayaquillo Formation suggests that an important pre-Cretaceous tectonic event affected the Jurassic volcanic rocks.

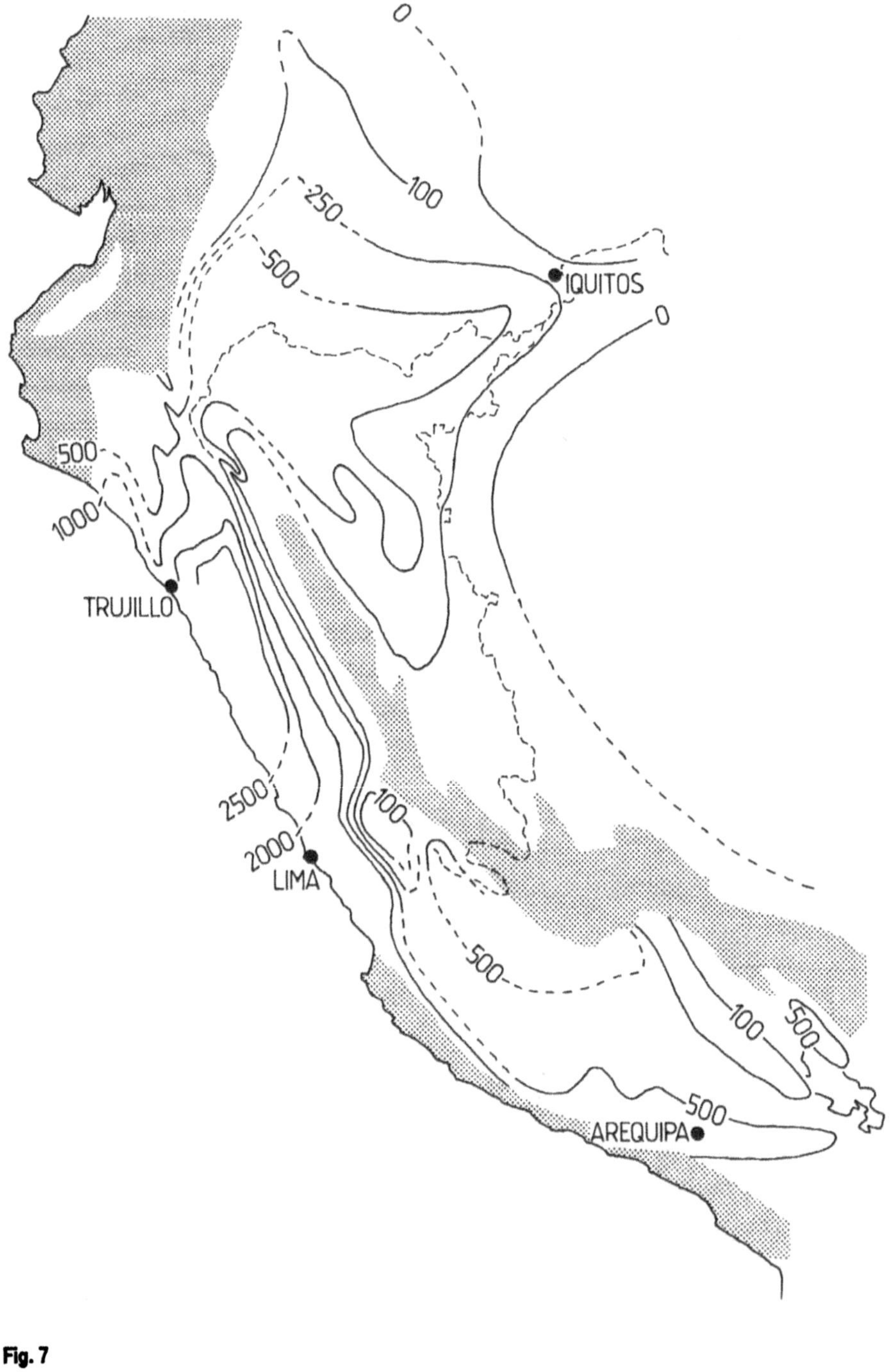

Fig. 7

Isopach map of Valanginian- late Aptian deposits. Shaded areas: no early Cretaceous deposits.

Nevertheless, farther south, extension seems to have been dominant. The opening of a Tithonian back-arc basin in the Lima area expresses an extensional tectonic regime, which could be associated with important strike-slip movements (Atherton et al., 1985). In eastern and southern Peru, a roughly NNW-SSE tensional strain, possibly associated with a NE-SW compression, and a subsequent regional emergence are recorded.

The Tithonian tectonic event can therefore be interpreted as the result of an oblique dextral collision of allochthonous terranes, generating oblique compression in northwestern Peru and Ecuador, whereas a roughly NNW-SSE extensional stress seems to have been dominant in the rest of Peru. However, the relationships between the northern compressional-wrenching regime and the southern extensional one are not well understood, and the NE-SW compression may have been more important than presently assumed.

The Berriasian tectonic event (ca 134-129 Ma): Further extensional phase and incipient northern South Atlantic rifting.

In northernmost Peru, the Berriasian event is characterized by an extensional tectonic instability and by bimodal volcanic effusions regarded as a Late stage of the continental accretion, probably represented by dextral wrenching movements (Jaillard and Jacay, 1989). In the rest of the Peruvian margin, the Berriasian event is marked by a stratigraphic gap and/or by erosions (Figures 3 and 6), and probably corresponds to a regional uplift. However, the Tithonian and Berriasian events are frequently confused and cannot be distinguished through the sedimentary record, because of the Late Tithonian gap.

The subsequent deposition of widespread, east-deriving sandstones suggests that an uplift of the eastern areas and a sinking of the western ones occurred during latest Berriasian times. Thus, the Berriasian event can be interpreted, according to the regions, either as a new manifestation of the extensional stress which dominated since Tithonian time, or as a genetically independant tectonic event related to the incipient rifting of the South-Atlantic ocean and resulting in the westward tilting of the entire South-American plate.

Valanginian-Early Late Aptian

This period is characterized by widespread silico-clastic deposits (Goyllarisquizga Group, Hollín Formation) and by a tectonic and magmatic quiescence. The fluvio-deltaic sedimentation is mainly controlled by eustatic sea-level fluctuations (Moulin, 1989; Moulin and Séguret, in press). In the coastal zone, marine partly calcareous deposits prevailed locally after Valanginian times (Rivera et al., 1975).

The isopach map for Early Cretaceous times (Figure 7) suggests that the shape of the sedimentary basins is controlled by and is the consequence of the Tithonian and Ber-

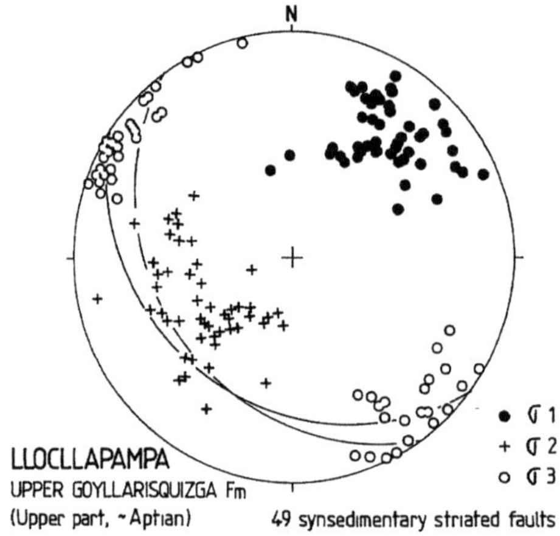

Fig. 8

Local Aptian extensional stress (top of Goyllarisquizga Formation) (after, Moulin, 1989).

riasian tectonic events. It reveals the three classical components of Cretaceous paleogeography (Benavides, 1956): a very subsident western region (West-Peruvian Trough), and an eastern less subsident one (East-Peruvian Trough, Ecuadorian Oriente), both regions being separated by a paleogeographic high with reduced sedimentation (Marañon Geanticline). Moreover, one can recognize a northern subsiding region and a southern much less subsident one, separated by a positive block located NE of Lima.

In northern Peru, the strong differential subsidence (Figure 7) suggests that synsedimentary normal faults parallel to the trench were active, and a tensional strain regime can be inferred. Paleocurrents indicate that the detrital material was supplied by the eastern shields, and that it was transported perpendicularly to the isopach curves, indicating a general trenchward gently-dipping paleoslope. In southern Peru, the extensional stress seems to have been much weaker.

The only reported synsedimentary tectonic features occurred in the Western Trough of central Peru. There, in the Goyllarisquizga Formation, Moulin (1989) mentions synsedimentary normal faults, indicating a progressive rotation of the local tensional stress from WNW-ESE in the lower part to NNW-SSE in the Middle part and to NE-SW in the upper part (Figure 8).

Except for the single 122 Ma K-Ar whole-rock age of Southeastern Peru (Crucero area, Clark et al., 1990), no volcanic manifestations have been reported so far within the deposits of the Western Trough. Along the coast, some Neocomian K-Ar ages within volcanic-sedimentary formations (124-128 Ma, Vidal et al., 1990) suggest that magmatic activity occurred locally. However, they could also represent minimum ages of older Tithonian-Berriasian rocks.

The eastern origin of the clastic supply suggests that it was generated by a large-scale westward tilting of the South American plate, which can be interpreted as the result of both the incipient rifting of the northern South Atlantic ocean (doming?), and the extensional subsidence of the western areas. However, the sudden development, and very large extension of the fluviodeltaic sedimentation suggests that the climate also changed drastically, becoming much more humid.

Late Aptian-Late Turonian

This time-span is characterized by a widespread marine transgression which permitted the deposition of successive carbonate shelves in the Western Trough, grading laterally to deltaic deposits in the Eastern Basin. Two tectonic periods can be recognized. The first (Late Aptian-Early Middle Cenomanian), corresponds to the evolution and accretion of a volcanic arc, or an extensional marginal basin ("Mochica" events, Mégard, 1984). The second (Middle Cenomanian-Late Turonian), is characterized by high depositional rates and by tectonic and magmatic quiescence.

The Late Aptian-Earliest Albian Event

Late Aptian-earliest Albian times are characterized by a widespread marine transgression (Jaillard, 1987; Moulin, 1989; Batty and Jaillard, 1989), by scattered volcanic manifestations and by local extensional tectonic events.

Northwestern Peru

In the Lancones Trough, possible Late Aptian-earliest Albian volcanic events could be represented by tuffaceous layers interbedded in the thick, poorly-dated sedimentary San Pedro Group (Reyes and Caldas, 1987).

Northern Peru and Eastern Ecuador

In the coastal area, thick conglomerates (Chinchipe Formation) are overlain by a Late Aptian-earliest Albian shaly sequence (Myers, 1974, 1980). In the Lima area, the presence of *Parahoplites* in the 2500 m-thick volcanic Chilca Formation (Rivera et al., 1975) shows that volcanic activity began locally as early as the Late Aptian-earliest Albian.
In the Cajamarca area of the Western Trough, the base of the Inca Formation locally exhibits erosional features, disconformities, or thin conglomerates or breccias (Jaillard, 1987). Scarce small-scale synsedimentary faults and slumps are also present, as well as local sulphide-bearing acidic volcanic flows and tuffs (Paredes, 1982, Figure 9). Farther south, the Pariahuanca Formation contains scarce basaltic flows, which exhibit an alkaline chemical trend, indicating an intracontinental extension (Soler, 1989, Figure 9).

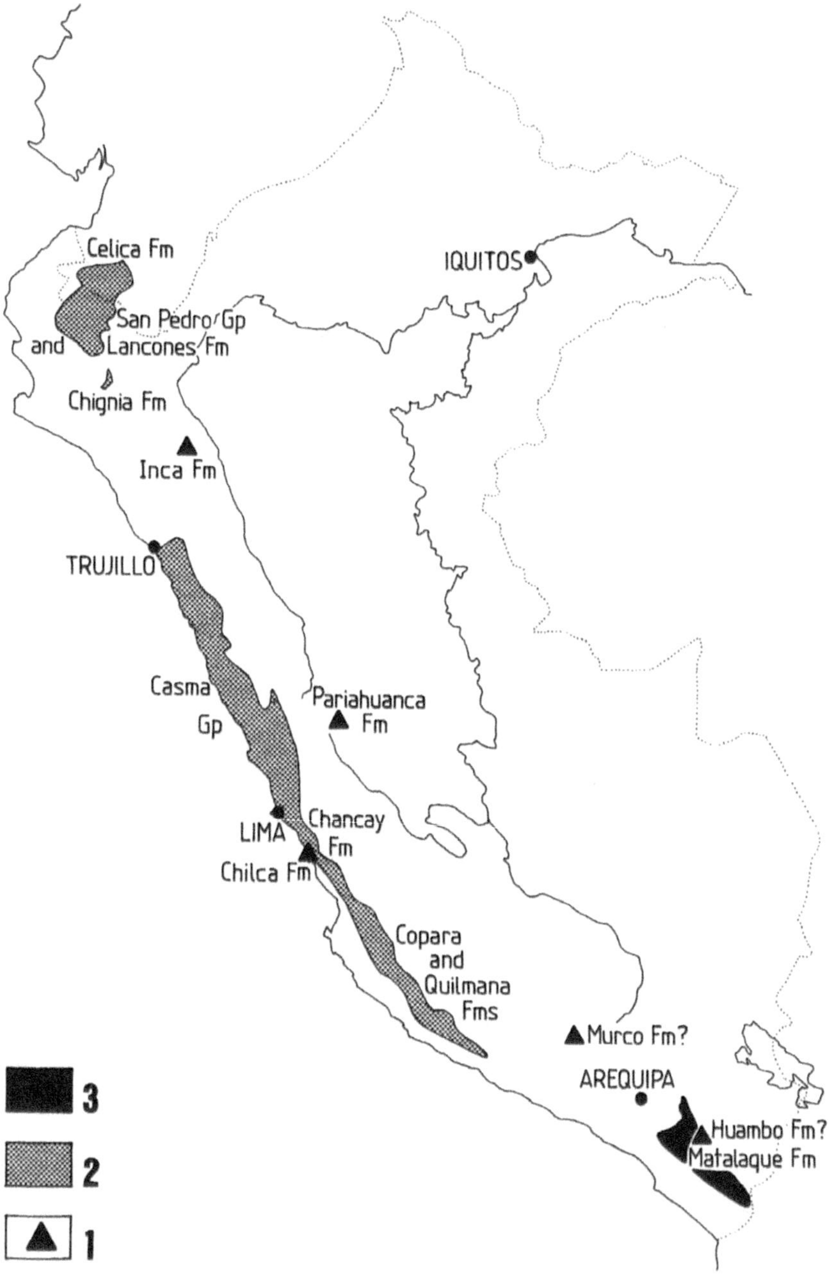

Fig. 9

Sketch map of the main late Aptian - Cenomanian volcanic events in Peru. 1: Local late Aptian-earliest Albian volcanic flows. 2: Marine, thick volcanic sequences of early Albian-Cenomanian age. 3: Continental, thick volcanic sequences of Albian-Cenomanian age.

the base of the same formation, Moulin (1989) mentions clastic dykes expressing a NE-SW extension.
In the Oriente of Peru and Ecuador, the marine transgression is only observed in the western areas. Farther east, sandstone deposition seems to have continued.

Southern Peru

Along the coast, near Ilo, plutons began to be emplaced (111 Ma, Beckinsale et al., 1985). In the Western Trough, dacite, rhyolite and andesite flows are interbedded in the Huambo Formation (Pérez, 1981; Loza, 1988; Batty, in preparation, Figure 9). However, the correlation of this undated formation with the Inca Formation is not verified throughout, and the associated volcanic flows might represent either earlier (Mid-Aptian?) or younger events (base of the Matalaque Formation, Mid-Albian?).

The Early and Middle Albian Events

During this time-span, the marine transgression reached its maximum extent, and a discontinuous belt of coastal volcanic centres was active (Figure 9).

Northwestern Peru and Southwestern Ecuador

The Lancones basin received a thick volcanic and volcaniclastic sequence of basaltic, andesitic, dacitic and rhyolitic rocks of Albian age (Ereo and La Bocana formations, Reyes and Caldas, 1987, San Lorenzo Group, Beaufils, 1977 in Mourier, 1988, Figures 9 and 13). It correlates with the Celica Formation of Ecuador (Feininger and Bristow, 1980; Baldock, 1982). The Celica Formation has been interpreted by most of the authors as a volcanic arc (Baldock, 1982; Lebrat et al., 1987; Wallrabe-Adams, 1990) but more recently Aguirre (1990) suggested that it could represent an aborted ensialic marginal basin. The Celica Formation is intruded by a pluton which yielded a 113 Ma K-Ar age (Baldock, 1982), thus suggesting that volcanic activity started before Aptian times.

Northern Peru

Along the coast, thick volcanic flows and volcaniclastic deposits are interbedded with marine shales, limestones and cherts of Early and Middle Albian age (Casma Group, Myers, 1974, 1980; Guevara, 1980) (Figure 9). Numerous coarse-grained turbidites and slumpings indicate the vicinity of an active volcanic arc and/or an unstable tectonic regime (Atherton and Webb, 1989). The thick basalt and andesite flows, pillow-lavas, tuffs and hyaloclastic rocks, with subordinate dacites and rhyolites of the Casma Group, exhibit a tholeiitic trend (Soler, 1991b). They were laid down in an extensional tectonic regime, which has been attributed (1) to a back-arc ensialic marginal basin setting (Atherton et al., 1983, 1985; Aguirre et al., 1989), (2) to a true ocean-floored marginal basin (Atherton, 1990), or (3) to the extensional tectonic subsidence of a volcanic arc (Soler, 1991a, 1991b)(see discussion in Mochica Period, p. 156-157). The great apparent thickness of the volcanic series (2000 to 8000 m, Figure 13) and the high geothermal gradient led Aguirre and Offler (1985) to admit to a considerable stretching of the underlying continental crust.

However, Mid-Albian plutons (102 Ma, Wilson, 1975) cross-cut folded Albian volcanic rocks of the Casma Group (Myers, 1975; Cobbing et al., 1981; Bussel and Pitcher, 1985), thus indicating that compression also occurred during the Early Albian. Moreover, pre -and syn-tectonic foliated early basic intrusions were emplaced (Patap super-unit, Beckinsale et al., 1985; Soler and Bonhomme, 1990).

In the Western Trough, numerous basalt and andesite sills and dykes in the Mid-Albian deposits (Pariatambo Formation) exhibit an alkaline chemistry, and indicate that an intracontinental extensional stress also affected the surrounding areas (Soler, 1989).

Southern Peru

Along the coast, between Lima and Nazca, thick accumulations of basalts and basaltic andesites (1000 to 2000 m, Copara and Quilmana formations, Figure 9) are interbedded with marine partly bituminous sediments, and unconformably overlie the Neocomian-Aptian sandstones (Caldas, 1978). Though an earlier age has been proposed (Caldas, 1978), they contain fauna of Albian affinity. The volcanic material has a clear calc-alkaline affinity, and is interpreted as having issued from a volcanic arc (Injoque, 1985; Soler, 1989). South of Arequipa, tonalite intrusions continued near Ilo (111 to 99 Ma, Beckinsale et al., 1985; Clark et al., 1990), thus indicating some early magmatic activity of Late Aptian to Mid-Albian age.

In the southwestern part of the Western Trough, up to 2500 m of terrestrial andesites, dacites and overlying agglomerates (Matalaque Formation, Marocco and Del Pino, 1966; Vicente, 1981, Figures 9 and 13) unconformably overlie either the Neocomian-Aptian Murco Formation or the Late Aptian-Early Albian (?) Huambo Formation, through an erosional surface (Batty and Jaillard, 1989). Geochemical study of the trace elements indicates a subduction-related volcanic arc origin for the Matalaque Formation (Carlier and Soler, personal communication, 1990). Although its age is poorly coordinated, this well-defined volcanic crisis seems to correlate roughly with the Albian to Mid-Cenomanian volcanic event.

Neither volcanic sills nor dykes have been observed in the western and eastern regions of southern Peru.

The Late Albian-Early Middle Cenomanian Phase

This period is characterized by a major marine regression that permitted deltaic deposits to extend southwestward (Figure 10). It is contemporaneous with the compressional deformation of the coastal volcanic centres, with precursory intrusions of the coastal Batholith, and with important synsedimentary deformations.

Northwestern Peru and Southwestern Ecuador

In the eastern part of the Lancones Trough, the thick Albian volcanic effusions are unconformably overlain (basal conglomerate) by Late Albian-Cenomanian breccias, pyroclastites and intermediate to acidic flows (Lancones Formation, Reyes and Caldas, 1987; Beaufils, 1977 in Mourier, 1988). These rocks merge laterally westward into Cenomanian to Senonian turbidites, interbedded with volcanic flows (Copa Sombrero Formation of Peru, Morris and Aleman, 1975; Reyes and Caldas, 1987, Alamor Group of Ecuador, Bristow and Hoffstetter, 1977; Baldock, 1982). Farther west, the turbidites overlie the Albian-Cenomanian limestones of the western border of the Lancones Trough (Morris and Aleman, 1975; Mourier, 1988).

Northern Peru and Eastern Ecuador

Along the coast, compressive deformation occurred during Late Albian and Cenomanian times, alternating with tensional periods (Bussel, 1983; Bussel and Pitcher, 1985). Intrusions of syntectonic gabbros and diorites continued (Patap unit, Cobbing et al., 1981; Pitcher et al., 1985), and they precede the emplacement of the coastal Batholith itself (Pitcher et al., 1985; Soler and Bonhomme, 1990). The subaerial nature of the overlying volcanic Pararin Formation (Myers, 1974) indicates that the Casma volcanic complex emerged. In the Lima area, marine conditions continued until Early Cenomanian times, as witnessed by the presence of *Mortoniceras* sp. and *Mantelliceras* sp. in the upper part of the Casma Group (Guevara, 1980). This compressional deformation was probably associated with important dextral wrenching motions (Myers, 1974; Bussel and Pitcher, 1985, Figure 10). The deformations observed in the western part of the Casma Group consist of mainly NW-trending and subordinate NE-trending open folds, commonly associated with a steep dipping axial plane cleavage (Myers, 1974; Mégard, 1987). In the eastern part, stronger deformations and cleavages are observed along the contact with the Western Trough.

In the northern part of the Western Trough, some slumpings, synsedimentary normal faults, large clastic dykes, and breccias in Late Albian and Early Cenomanian limestones are interpreted as the response to the compressive phase in this area (Jaillard, 1987, Figure 10). These features, as well as the pronounced differential subsidence in some

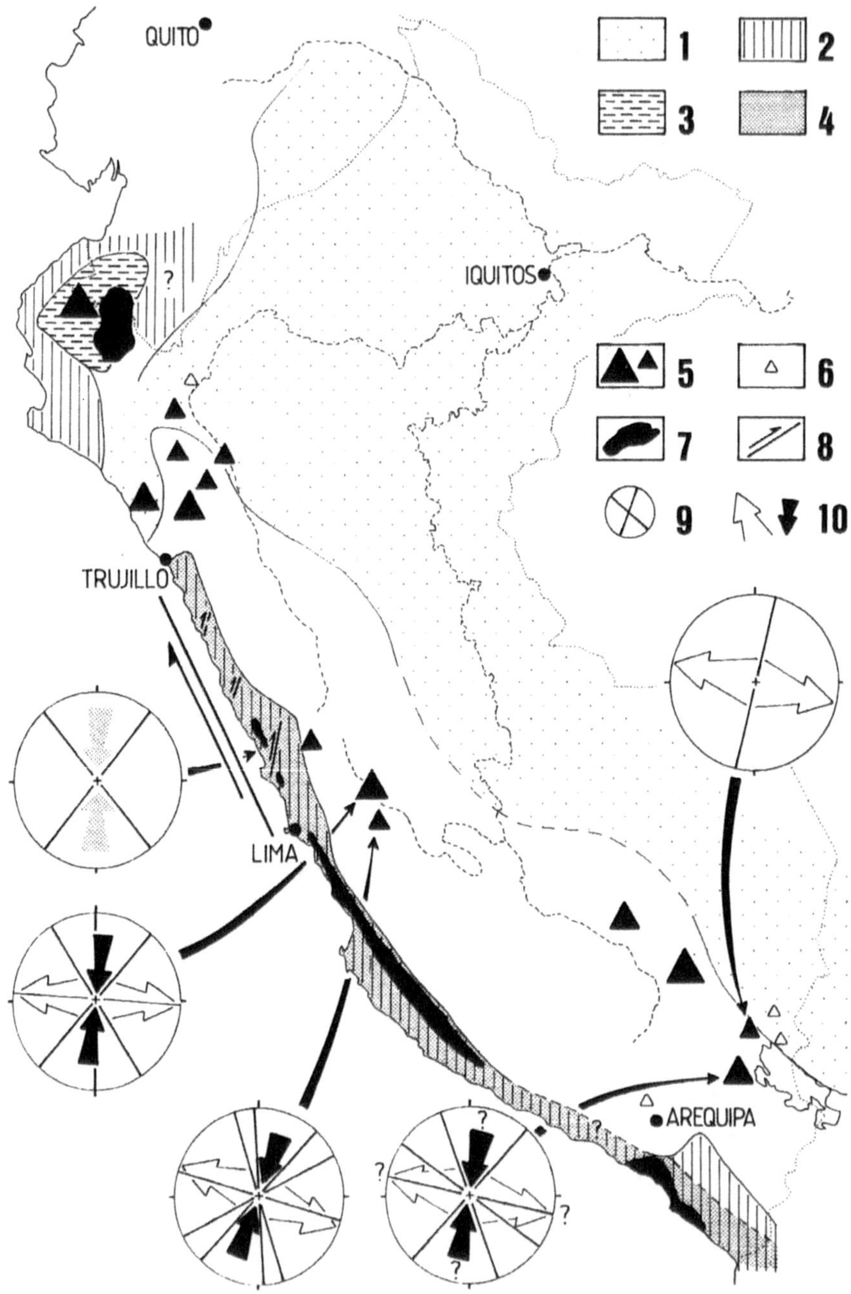
QUITO
IQUITOS
TRUJILLO
LIMA
AREQUIPA
1
2
3
4
5
6
7
8
9
10

areas (Jaillard, 1987), suggest a rather extensional paleo-stress. Northeast of Lima, beds of breccias (locally tens of metres thick) and synsedimentary normal faults and slumpings in the dolomitic lower part of the Jumasha Formation indicate a notable tectonic activity during the Early Cenomanian regression, and are regarded as a consequence of the Late Albian-Early Middle Cenomanian phase (Jaillard, 1987, Figure 10). The geometric analysis of these synsedimentary structures reveals the existence of two fracture systems, which trend NE-SW to ENE-WSW and NNW-SSE to N-S, respectively (Figure 11). A minor, NNE-SSW trending fracture group is associated with scarce reverse synsedimentary faults. Slump fold axes are mainly E-W trending and indicate a southward-dipping paleoslope (Figure 11). These features express a WNW-ESE tensional stress associated with a minor NNE-SSW compression (Figure 10).

In the Eastern Basin of Peru and Ecuador, the effects of the Late Albian-Early Middle Cenomanian phase are poorly known, because of unsatisfactory outcrop conditions. However, the westward progradation of the Late Albian-Cenomanian delta (Agua Caliente Formation of Peru, Rodriguez and Chalco, 1975; Soto, 1979; Middle sandstones of the lower Napo Formation of Ecuador, Tschopp, 1953), which seems mainly due to the Early Cenomanian eustatic regression, may have been accentuated by a mild uplift correlative with the tectonic event (Jaillard, 1987).

Southern Peru

Along the coast, no structural data are available on the Albian-Cenomanian volcanic series (Copara and Quilmana formations) which are directly capped by Tertiary deposits (Caldas, 1978; Anónimo, 1980). The plutonic activity seems to have been very low (Soler and Bonhomme, 1990), except in the Nazca area, where well-dated plutons were emplaced during Middle Albian-Middle Cenomanian times (101 to 94 Ma, Beckinsale et al., 1985; Mukasa, 1986).

In the western part of the Western Trough, the volcanic activity probably ceased, since the Senonian (?) Omoye Formation (García, 1978) disconformably overlies the already weathered Matalaque Formation, thus suggesting that a large interval of subaerial expo-

Fig. 10

◄ Sketch map of the late Albian-early middle Cenomanian deformations in Peru. 1: Maximum extent of the deltaic influence. 2: Emergent areas. 3: Turbiditic troughs. 4: Folded areas. 5: Synsedimentary tectonic features (size of triangles is roughly proportional to the importance of the deformation). 6: No observed deformation. 7: Late Albian-Cenomanian intrusions of the Coastal Batholith. 8: Dextral wrench faults (after Bussel, 1983; Bussel and Pitcher, 1985). 9: Late Albian-early middle Cenomanian joint fractures (data north of Lima from Bussel and Pitcher, 1985). 10: Late Albian- early middle Cenomanian interpreted stress (north of Lima: personal interpretation).

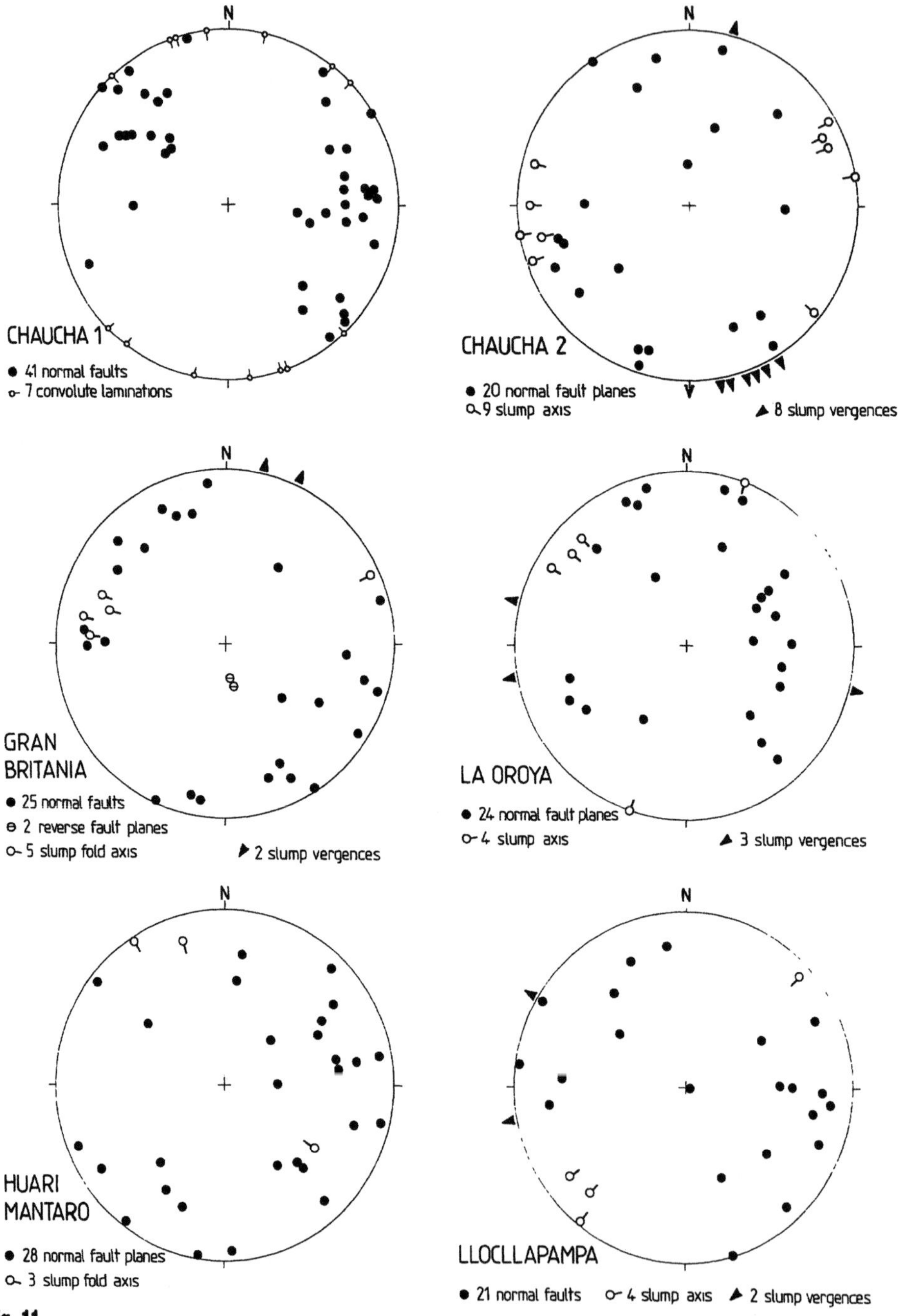

Fig. 11

Geometry of late Albian- early middle Cenomanian synsedimentary tectonic structures in the central Peruvian margin (Jumasha Formation).

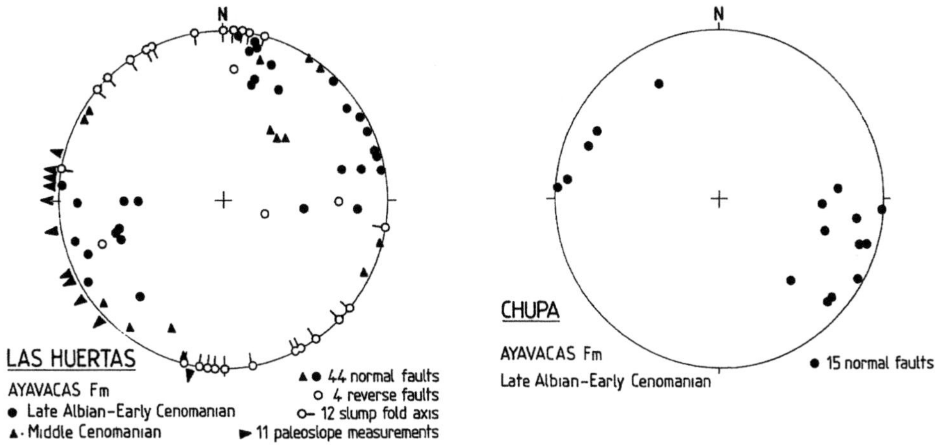

Fig. 12

Geometry of late Albian- early middle Cenomanian synsedimentary tectonic structures in the south Peruvian margin (Ayavacas Formation).

sure and alteration occurred before the Senonian (Figure 10). Moreover, the intercalations of volcanogenic sandstones, within the probably Early to Middle Cenomanian regressive horizons of the Arcurquina Formation, suggest that the Matalaque Formation was at least partly eroded at this time. However, no detailed studies have been made so far of the pre-Senonian deformation of the Matalaque Formation. In the eastern part of the Western Trough, no important synsedimentary deformations have been reported in the Arcurquina Formation, but neither are there any detailed studies available.

Near the Axial Swell and in the Eastern Basin, some noteworthy synsedimentary deformations are known in the Ayavacas limestones (Figure 10). In the Sicuani area, Audebaud (1971, 1973) describes large-scale mass slumping and collapses, angular unconformities and karstifications. In the Puno and Cuzco areas, synsedimentary breccias, slumpings, faults and gravity slides are also present (Portugal, 1974). In many cases the Mid-Cretaceous age of the deformation is attested to by the discordancy of overlying strata (upper Ayavacas, Hanchipacha and Vilquechico formations, Audebaud, 1971; Jaillard, unpublished). Geometrical analysis of the deformational structures evidences a dominant NNW-SSE direction of the normal faults and a WSW-ward associated paleoslope (Figure 12), which is supported by the westward slides observed by Audebaud (1971) and by the eastward thinning and increasing erosions of the coeval Miraflores limestone of Bolivia (Sempere, this volume). The presence of subordinate WNW-

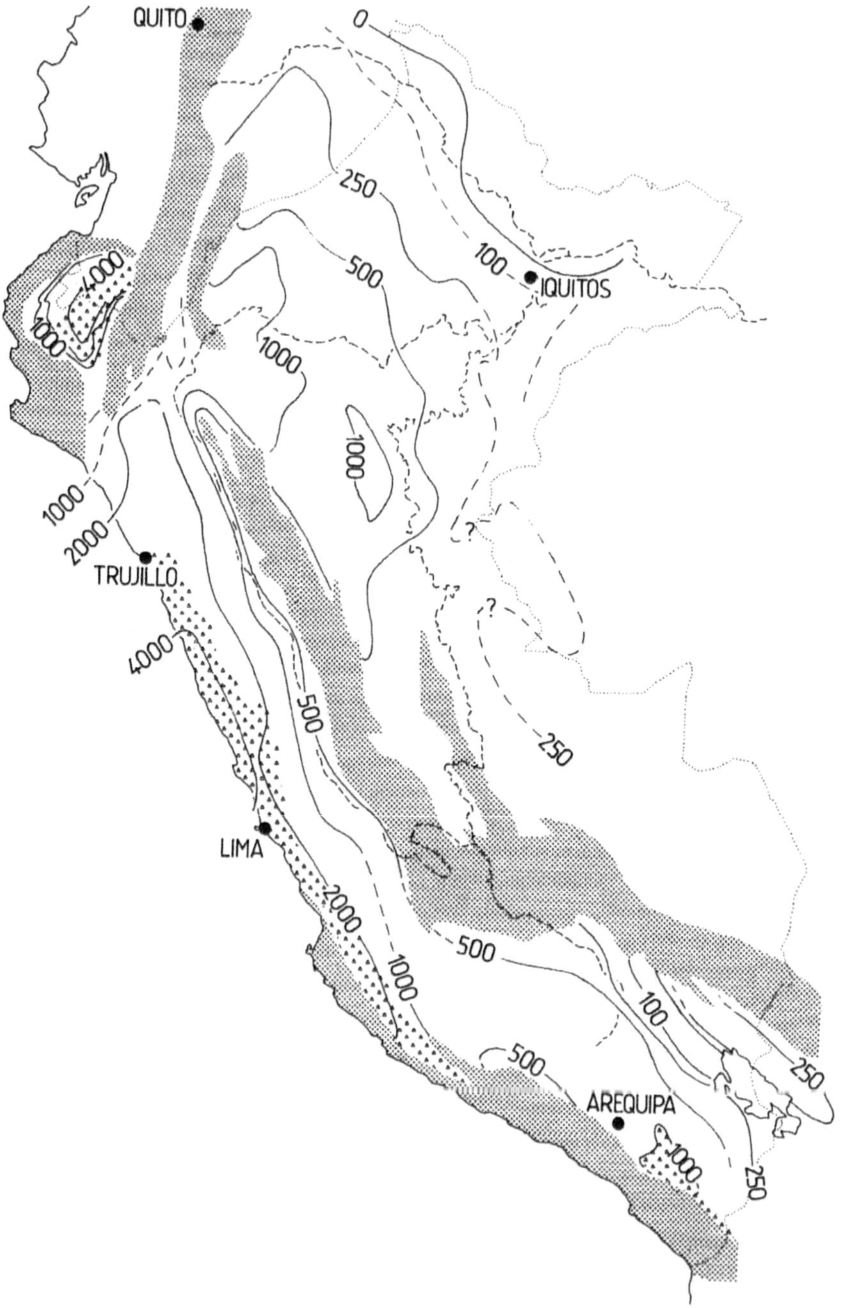

Fig. 13

Isopach map of late Aptian-late Turonian deposits. Shaded areas: no late Aptian-late Turonian deposits. Triangles: Albian-Cenomanian volcanic sequences.

trending normal faults might indicate a NE-SW tensional stress. However, scarce reverse faults (Figure 11) rather suggest a NE-SW compression, and more data would be necessary before reaching a final conclusion (Figure 12).

(Note that where the Ayavacas limestones are not capped by Cretaceous beds their chaotic aspect must be ascribed to superimposed Cretaceous and Tertiary deformations, rather than to Tertiary deformations only, as sometimes suggested (De Jong, 1974; Ellison et al., 1989).

Farther northeast, no data are available.

The Middle Cenomanian-Turonian Period

During this period a widespread marine transgression permitted the deposition of thick even carbonate shelves on most of the margin. Meanwhile, in the probably emergent coastal area, the coastal Batholith was emplaced.

Northwestern Peru, Northern Peru and Ecuador

In the Peruvian-Ecuadorian Lancones-Celica Trough, turbidite sedimentation continued to take place (Morris and Aleman, 1975, Figure 13).

In the coastal area, few radiometric data are available. However, a noticeable acidic to intermediate plutonic pulse is recorded during the Middle to Late Cenomanian (94-90 Ma, Mukasa, 1986; Soler and Bonhomme, 1990). The early intrusions of the coastal Batholith were emplaced along the western border of the Western Trough, whereas the Albian plutons were emplaced within the volcanic rocks (Mégard, 1984; Soler and Bonhomme, 1990). This suggests that a weak but real crustal shortening occurred during the Late Albian-Early Middle Cenomanian phase. The study of the joints and veins associated with these Early intrusions reveals NE-SW and NW-SE orientations, and wrenching motions seem to have continued to be important (Bussel, 1983; Bussel and Pitcher, 1985).

In the Western Trough, no synsedimentary tectonic manifestations have been observed. However, detailed studies evidence a change in the orientation of the isopach curves from NNW-SSE to WNW-ESE, and a very weak SW-proceeding detrital supply, thus suggesting minor paleogeographic changes (Jaillard, 1987). Moreover, the subsidence rate notably increased during this time-span. For instance, near Oyón, the Late Middle to Late Cenomanian limestones are 900 metres thick, thus expressing a very high sedimentation rate for this period (ca 300 m/Ma, without decompaction, Jaillard, 1987; Figure 13).

During the Early Turonian, the sea reached its maximum extent in the Eastern Basin of Peru and Ecuador and deposited fossiliferous marls (lower Chonta Formation, top of lower Napo Formation), but no tectonic features have been mentioned so far.

Southern Peru

In the coastal area, few radiometric data are available, and no plutonic event has been recorded so far.

In the Western Trough, very local NNW and ENE trending minor synsedimentary normal faults have been observed in the Turonian upper part of the Arcurquina Formation. They express a weak and probably local distensional instability.

Near the Axial Swell, and in the western part of the Eastern Basin, Turonian deposits are locally lacking, because of the paleotopography inherited from slidings of the Ayavacas limestones, or from deformation of the substratum (Figure 13). In this case, Senonian deposits disconformably overlie the often eroded or karstified Ayavacas limestones (Audebaud, 1971; Sempere, this volume). These features, however, do not indicate Middle Cenomanian or Turonian tectonic events.

Tectonic Interpretations

The Albian folding of the Casma Group was named the "Mochica" phase by Mégard (1984). We propose to extend this name to the various events recorded during the Late Aptian-Early Middle Cenomanian period.

The Late Aptian-earliest Albian Mochica 1 event (ca 110-107 Ma): Precursor tensional instability.

Evidence of a mild extensional instability, together with the scattered but frequent acidic or bimodal volcanic flows (Figure 9), indicate a tensional strain in the western part of the Peruvian continental margin. The direction of the strain is unknown. It is interpreted as a precursor event of the subsequent Albian "Mochica" deformations. The resumption of plutonic activity 113 to 111 Ma ago (Lancones-Celica and Ilo plutons) suggests a change in the subduction pattern.

The Early and Middle Albian Mochica 2 events (ca 107-100 Ma): Extension-compression in the arc or back-arc system.

In northern Peru, although most of the thick subduction-related coastal volcanic sequences yielded Mid-Albian fauna, it is possible that locally they began to overflow earlier (pre-113 Ma in the Lancones Trough). In southern Peru, their frequently disconformable basal contact suggests that the early volcanic effusions were associated with tectonic activity.

North of Lima, and as far as southwestern Ecuador (Figures 9 and 13), the thick volcanic marine flows exhibit a tholeiitic trend, and are regarded either as derived from volcanic arc activity, or as related to the opening of a marginal basin, depending on the author. In any case, the extensional setting of these very thick deposits is widely accepted, though strike-slip motions are probable. The dyke swarms recorded in the Western Trough indicate that an extensional phase occurred during Mid-Albian times. However, Early Mid-Albian (102 Ma) intrusions which cross-cut already folded rocks demonstrate that compressive pulses also occurred before this period (105 Ma?, original "Mochica" phase of Mégard, 1984). This indicates an alternation of compressive-tensional events, which can be due to the presence of strike-slip movements inducing local and/or sporadic compressions.

In contrast, south of Lima the effusive sequences are partly terrestrial and generally thinner (Figures 9 and 13). Moreover, they are generally interpreted as volcanic arcs, and have not yielded evidence so far of back-arc extension. These differences, as well as the lack of dyke swarms in the Western Trough of southern Peru, suggest first, that the Mid-Albian extensional stress was weaker than in northern Peru, and second, that the subduction regime was different.

The Late Albian-Early to Middle Cenomanian Mochica 3 phase (ca 100-94 Ma): Accretion of subduction-related volcanic centres.

In northwestern Peru, further studies are necessary, in order to specify the nature, importance and extent of the major regional tectonic event, embodied by the Late Albian-Cenomanian unconformity. However, it probably correlates with the compressional folding event responsible for the emergence of the coastal volcanic units of northern Peru. We interpret this event as the collision and accretion of deformed volcanic centres along the western edge of the continental margin.

Subsequent deformations probably modified the original geometry of the tectonic structures, and Tertiary volcanism obscured the structural relations with the present-day Western Cordillera. As a result of these poor outcropping conditions, the Mochica 3 phase has been interpreted either as a mild and relatively minor event (Mégard, 1984, 1987), or as an important tectonic phase responsible for a significant crustal shortening (Vicente, 1990). Whichever the case, it represents the earliest major compressional Cretaceous event in the Peruvian margin. Along the south Peruvian coast, no specific studies have been carried out on Mid-Cretaceous tectonic events. However, neither have any Mid-Cretaceous folds been observed so far in this area.

The N-S to NNE-SSW compression and the related E-W to WNW-ESE extension recorded in the Western Trough of central Peru apparently conflict with the expected ENE-WSW compression provoked by the closure of the NNW trending coastal basin (Figure 10). However, strong NNW-SSE to N-S dextral wrenching movements (Bussel and Pitcher, 1985) may have induced a NNE-SSW compression and a WNW-ESE extension, which could account for the observed features. These strike-slip movements together with the observed compression trend (Figure 10) suggest an oblique accretion, and a north to northeastward-trending convergence.

Whereas the coastal area of northern Peru underwent an intense continental stretching during Late Jurassic-Early Cretaceous times (Figure 13), the Albian-Cenomanian synsedimentary deformations recorded in the neighbouring Western Trough are weak (Figure 10). In contrast, these latter are important on the Axial Swell of southern Peru, whereas the Late Jurassic-Early Cretaceous crustal extension was less intense on the coast. The important slides in the Mid-Cretaceous limestones of southern Peru can be partly explained by the abundance of interbedded plastic red shales, or by their location on the possibly mobile Axial Swell. However, it is also probable that, in northern Peru, most of the Mochica shortening has been accommodated by tectonic inversion of the Albian extensional crustal structures of the coastal basin, whereas in southern Peru the paucity of such inherited extensional structures permitted transmission of the tectonic strain to the Eastern regions.

The Late Middle Cenomanian-Turonian period (ca 94-88 Ma): Tectonic remission and flexural subsidence?.

In most of the Peruvian margin this time-span is a quiescent magmatic and tectonic period. The Western Trough of northern Peru registered a noticeable increase in subsidence and sedimentation rates by Cenomanian times (Figure 13). This may be explained, if one accepts the eastward thrusting of the coastal units during the Mochica phase (Vicente, 1990), by crustal bending resulting from the thrust load. Nevertheless, the lack of structural data makes such an interpretation hypothetical. In southern Peru, the scarce and minor Late Turonian tectonic features probably represent precursors of the Senonian tectonic events.

Coniacian-Latest Campanian: The "Peruvian Phase"

Senonian times are still rather poorly understood, because they mostly gave place to unfossiliferous continental beds, and subsequent erosions have often removed part of the deposits. The progressive emergence of the Peruvian margin during the Senonian has been denominated the "Peruvian phase" (Steinmann, 1929).

The Turonian-Coniacian Boundary Event

The Turonian-Coniacian boundary is characterized by the irruption of red to brownish marine or continental shaly deposits.

Northwestern Peru, Northern Peru and Ecuador

In the Lancones-Celica Trough of northwestern Peru and southwestern Ecuador, turbiditic sedimentation continued (Morris and Aleman, 1975).
In the coastal area, Turonian and Early Santonian times (91 to 85 Ma) are marked by a striking plutonic quiescence (Beckinsale et al., 1985; Soler and Bonhomme, 1990). The tectonic regime seems to have been dominated by a high fault-slip rate, and a variable regional compression (Bussel and Pitcher, 1985).
In the Western Trough, marine Coniacian deposits conformably overlie the Turonian limestones (Celendín Formation, Benavides, 1956; Wilson, 1963; Jaillard, 1987). They are characterized by an abundant shaly and subordinate sandy detrital supply, and by a restricted depositional environment (Jaillard, 1987; Mourier et al., 1988b). In central Peru, the Celendín Formation is less calcareous and shows shallower, even tidal, depositional environments. Moreover, the 200 to 300 m-thick series deposited during the 4 to 6 Ma-long Early Senonian (Figure 14) contrast with the 1000 to 2000 m-thick limestones deposited during the 7 to 8 Ma-long Cenomanian-Turonian interval (Figure 13).
In the Eastern Basin of Peru and as far as eastern Ecuador, the Turonian limestones are capped by confined marine shales (upper Napo Formation, Tschopp, 1953; Bristow and Hoffstetter, 1977; upper Chonta Formation, Kummel, 1948; Ducloz and Rivera, 1956; Rodriguez and Chalco, 1975; Jaillard, 1987; Figure 15).

Southern Peru

Along the coast, very little is known about magmatic activity in the Batholith area.
In the Western Trough, in the Arequipa area, red shales and evaporites (Chilcane Formation) of probable Coniacian age conformably overlie the Turonian shelf limestones. South of Abancay, Coniacian deposits seem to be absent, since the Senonian continental Anta-Anta Formation disconformably overlies the Turonian limestones (Pecho, 1981).
On the Axial Swell, and in the eastern part of the Western Trough, sparse outcrops of red shales, overlying the pre-Senonian limestones, are disconformably overlain by Oligocene conglomerates, and may be tentatively ascribed to the Coniacian.
In the Eastern Basin, near the Lake Titicaca and in the Cuzco area, evaporite-bearing red shales conformably overlie the Mid-Cretaceous limestones or the undated Cotacucho sandstones (upper Yuncaypata Formation, Kalafatovitch, 1957, lower Vilquechico Formation, Jaillard et al., in press; Figure 15). In the Sicuani zone, red sands and shales (Hanchipacha Formation) unconformably overlie the deformed and partly eroded Ayavacas limestones (Audebaud, 1971, 1973). In most of the eastern area of southern Peru, the 300 to 600 m thick Coniacian-Santonian deposits (Figure 14) contrast with the much thinner (20 to 200 m) Late Aptian-Turonian series (Figure 13).

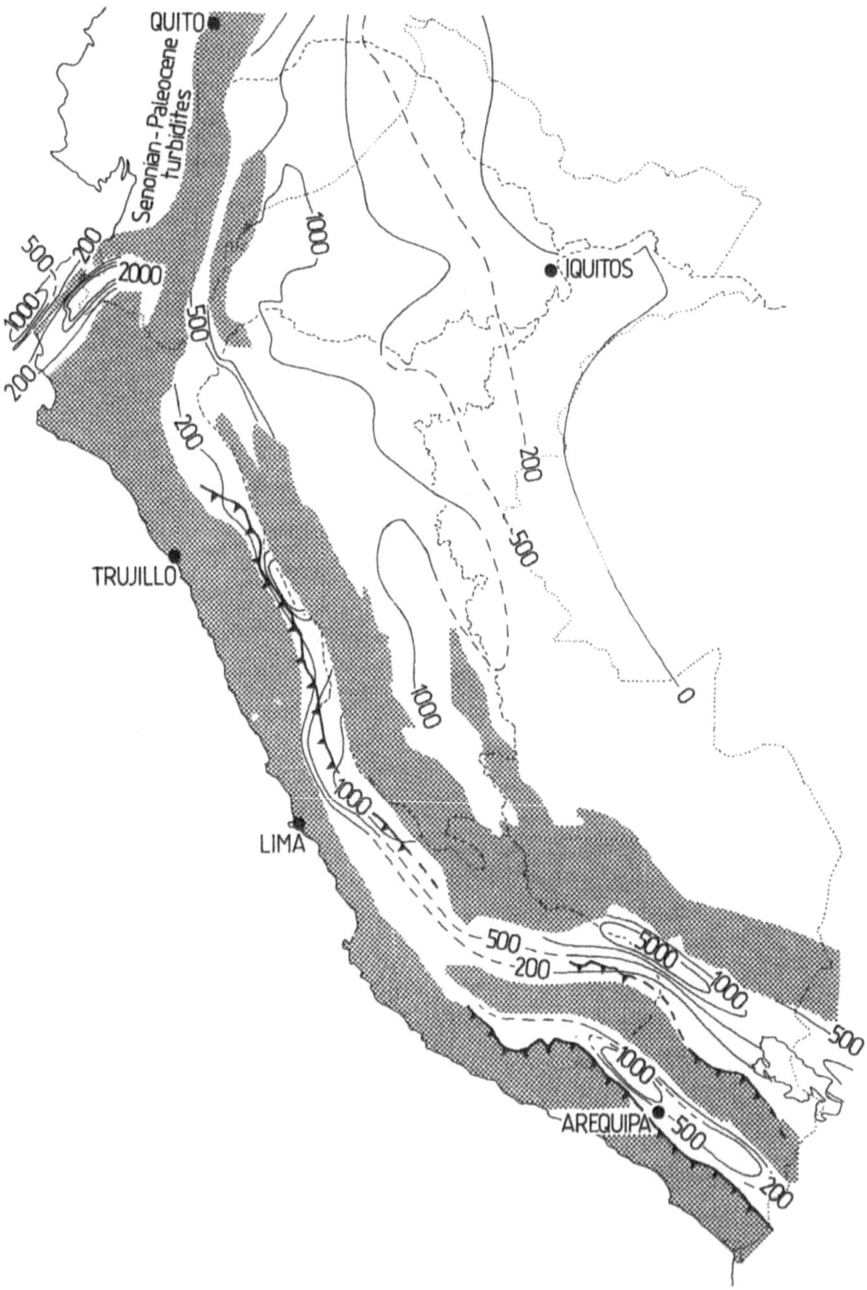

Fig. 14

Isopach map of Coniacian-Paleocene deposits, and probably active thrust faults. Shadowed areas: no Senonian-Paleocene deposits.

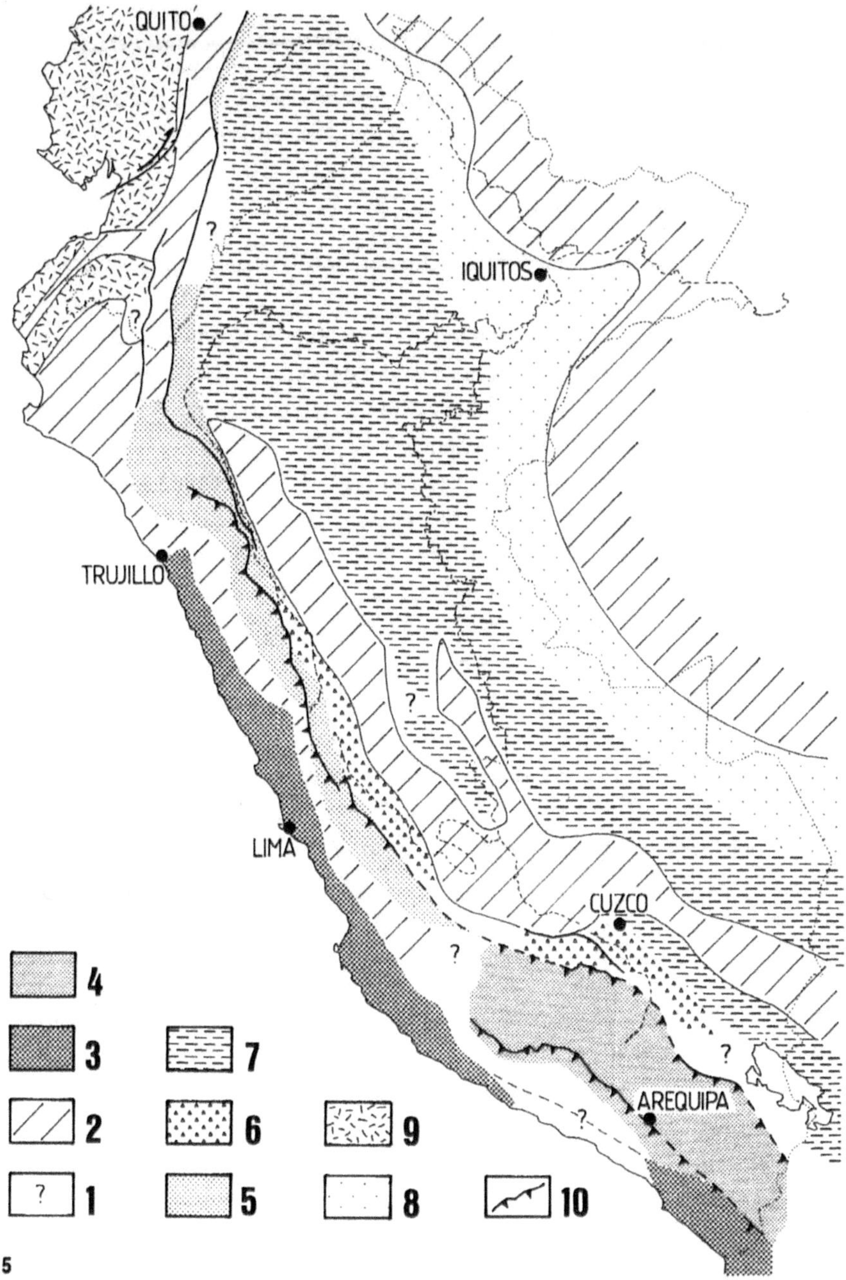

Fig. 15

Sketch map of the Senonian sedimentary facies in Peru and Eastern Ecuador. 1: No information. 2: No outcrops. 3: Emergent since Cenomanian times. 4: Emergent since the Coniacian, with a short-lived Santonian marine transgression, and with Campanian (?) foreland deposits. 5: Emergent since latest Santonian times. 6: Subsident red bed troughs of probable late Campanian-Maastrichtian age. 7: Late Campanian and Maastrichtian partly marine deposits. 8: Continental deposits throughout Senonian times. 10: Thrust faults probably active during the Senonian.

The Earliest Santonian (?) Event, and the Late Santonian-Campanian Period

This time-span is marked by the diachronous emergence of the margin, and by the beginning of compressional deformation within the western Trough.

Northwestern Peru, Northern Peru and Ecuador

In the Lancones-Celica Basin, turbiditic sedimentation continued to take place (Morris and Aleman, 1975).

Along the coast, plutonic activity resumed, with mainly granodioritic 85 to 76 Ma-old plutonic rocks (Beckinsale et al., 1985; Soler and Bonhomme, 1990). Near Chiclayo, synmetamorphic deformation is indicated by the K-Ar 82 Ma age (Early Campanian) yielded by a foliated gabbro (Mourier, 1988).

In the Western Trough, except in its easternmost part, the top of the Celendín Formation is of Early Santonian age (Benavides, 1956; Wilson, 1963; Mégard, 1978). By Late Santonian times most of the region was emergent (Figure 15). The overlying red beds have long been considered Santonian in age (Mégard, 1978; Romani, 1982). However, their charophyte assemblages most probably indicate a Middle to Late Campanian age (Mourier et al., 1988b; Jaillard et al., 1993).

In the Eastern Basin of Peru and Ecuador, marine sedimentation prevailed until Santonian times (upper Chonta and upper Napo formations, Kummel, 1948; Tschopp, 1953; Pardo and Zuñiga, 1976). Restricted conditions are expressed by the local occurrence of black shales. The lack of any Late Santonian to Early Campanian fauna suggests the existence of a sedimentary hiatus, probably due to of the progressive uplift of the whole area.

Southern Peru.

Along the coast, after the emplacement of the Tiabaya pluton, an important plutonic gap occurred between 84 and 70 Ma (Soler et al., 1989), though a 77-80 Ar-Ar age has recently been reported in southwesternmost Peru (Clark et al., 1990).

In the Western Trough near Arequipa, the Coniacian evaporites and shales grade upward into coarsening-upward, southwestward-flooding fluvial shales and sands (lower Querque Formation, Vicente et al., 1979; Figure 16). Farther southeast, the Mid-Cretaceous volcanic Matalaque Formation is unconformably capped by a coarsening-upward sequence of SW-flooding fluvial sandstones and conglomerates which exhibits internal disconformities (lower Omoye Formation, Figure 16). These latter indicate a progressive northeasterly bending of the substratum, thus suggesting a NE-SW compression (Figure 18). In both areas, these are capped by marine limestones locally dated as Santonian (Middle Querque Formation, Hosttas, 1967; Vicente, 1981; Middle Omoye

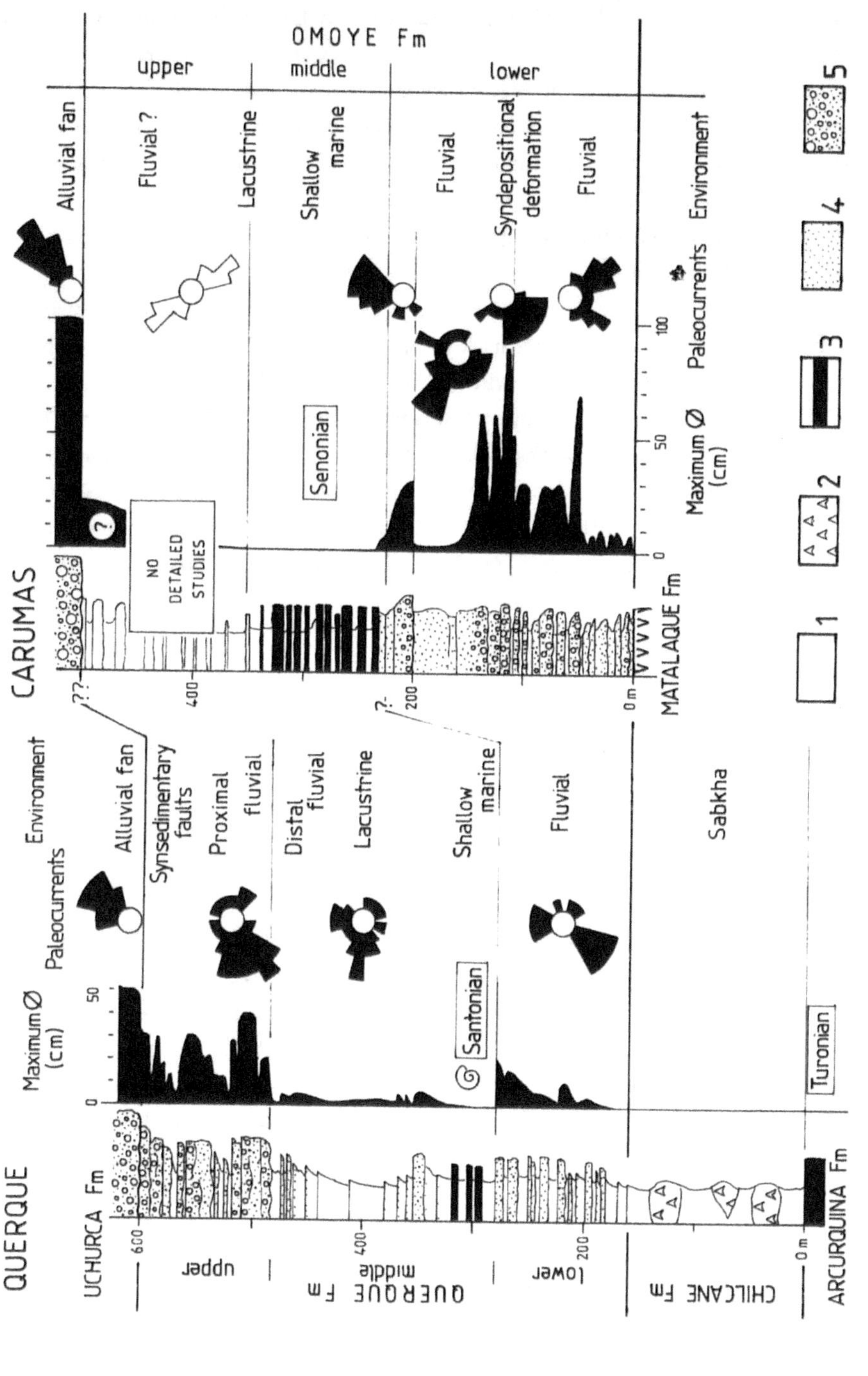

Fig. 16

Sedimentary-tectonic evolution of the Senonian deposits of the Arequipa area. 1 : Shale. 2: Gypsum. 3: Limestone. 4: Sandstone. 5: Conglomerate.

Formation, García, 1978), which probably correspond to the Early Santonian eustatic sea-level rise. They are overlain by lacustrine to fluvial coarsening-upward sequences of probably Late Santonian to Campanian age (upper Querque and upper Omoye Formation, Figure 16). Tectonic activity thus seems to have prevailed during the latest Coniacian-earliest Santonian (?), whereas the Late Santonian-Campanian time-span is a relatively quiet period, though a mild uplift suggests that compression continued.

In the eastern part of the Western Trough and on the Axial Swell, there are no known Santonian-Campanian deposits.

In the Eastern Basin, from Lake Titicaca up to the Cuzco area, shales and subordinate sands and limestones contain a well-defined marine horizon, which correlates with the Santonian transgression (lower and Middle Vilquechico Formation, Jaillard and Sempere, 1989; Jaillard et al., 1993, upper Yuncaypata Formation, Kalafatovitch, 1957; Carlotto et al., 1990; Hanchipacha Formation, Audebaud, 1973). Near Abancay, red shales and subordinate sands are also known (Anta-Anta Formation, Pecho, 1981), but their age is unsure. Farther east, in the Oriente, marine sedimentation went on till Senonian times (Dávila and Ponce de León, 1971), but marine Campanian deposits are unknown.

The Late Campanian Phase

The Late Campanian (and earliest Maastrichtian?) phase follows a quiet period, locally marked by a short-lived marine transgression of Mid-Campanian age. It probably represents the tectonic climax of the "Peruvian phase".

Northwestern Peru

In northwestern Peru, the Campanian Tablones conglomerates unconformably overlie Paleozoic to Senonian rocks (Olsson, 1944; Séranne, 1987; Morris and Aleman, 1975; Figures 14 and 15) and mark the end of turbiditic sedimentation in the western part of the Lancones Basin (Morris and Aleman, 1975). They express both the beginning of a marine transgression (Zuñiga and Cruzado, 1979; Séranne, 1987), and a noticeable uplift of the pre-Mesozoic crystalline massifs (Olsson, 1944; Morris and Aleman, 1975).

Northern Peru and Eastern Ecuador

In the coastal Batholith, a significant plutonic gap is recorded between 77 and 73 Ma (Soler and Bonhomme, 1990), and is followed by a significant magmatic pulse between 73 and 70 Ma (Beckinsale et al., 1985).

In the coastal zone, Mourier (1988) supposed that uplift continued, and that thrustings and ductile deformations occurred. However, in spite of one 82 Ma date (Mourier, 1988), the age of this deformation is actually poorly constrained (post-Turonian and pre-latest Paleocene), and at least part of it can result from the latest Paleocene Inca 1 tectonic phase.

Near the boundary between the Western and the Eastern Troughs, a short-lived Mid-Campanian marine transgression is locally recorded below the Late Campanian to Maastrichtian red beds (Mourier et al., 1988b). We may suppose that most of the continental red beds of the Western Trough are of post-Middle Campanian age. This is probably also the case with the thick red bed sequence of the Sihuas area (improperly named Chota Formation by Benavides, 1956 and Wilson et al., 1967), and of the La Oroya area, since the base of the latter contains charophytes (Mégard, 1978) similar to those yielded by the Mid-Campanian beds of northern Peru (Mourier et al., 1988b; Jaillard et al., 1993). Cretaceous foraminifera-bearing marine layers interbedded in the red beds (Mabire, 1961) could represent one of the Maastrichtian marine transgressions. They indicate that deformation had already begun at this time, since they are associated with conglomeratic beds (Mabire, 1961). In some areas, the thickness of the red bed sequence can reach as much as 3000 m, but the presence of Early Tertiary deposits is probable (Benavides, 1956; Jenks, 1961; Wilson, 1963; Mégard, 1978, Figure 14). On the Axial Swell, some plutons previously considered as Late Cretaceous are now known to be of Permian-Triassic age (Soler and Bonhomme, 1987).

In the Eastern Basin of Peru and Ecuador, the resumption of detrital sedimentation is expressed by conspicuous Late Campanian to Early Maastrichtian sandstones, which conformably overlie the Santonian marine deposits (Areniscas de Azúcar 1, Vivian and Tena formations, Tschopp, 1953; Koch and Blissenbach, 1962; Fyfe, 1962; Seminario and Guizado, 1976; Petroperú, 1989; Figure 15).

Southern Peru

No intrusions are recorded in the coastal Batholith between 78 and 70 Ma (Beckinsale et al., 1985; Soler et al., 1989).

In the Western Trough, undated coarsening-upward, southwestward-flooding fluvial sequences were deposited, probably during Late Campanian times (upper Querque Formation and upper Omoye Formation, Figure 16). A few synsedimentary minor faults indicate a NE-SW compression and a NW-SE extension. The upper Querque and Omoye formations are unconformably overlain by undated coarse-grained northeastward-flooding fanglomerates (García, 1978; Uchurca Formation, Vicente et al., 1979; Figure 16). They have been tentatively ascribed to the Early Tertiary on the basis of their abundant volcanic clasts, which would have derived from the Paleocene Toquepala Formation (Vicente et al., 1979). However, they may be alternatively interpreted as Late Cretaceous foreland deposits, related to the NE progression of the Lluta overthrust, since numerous

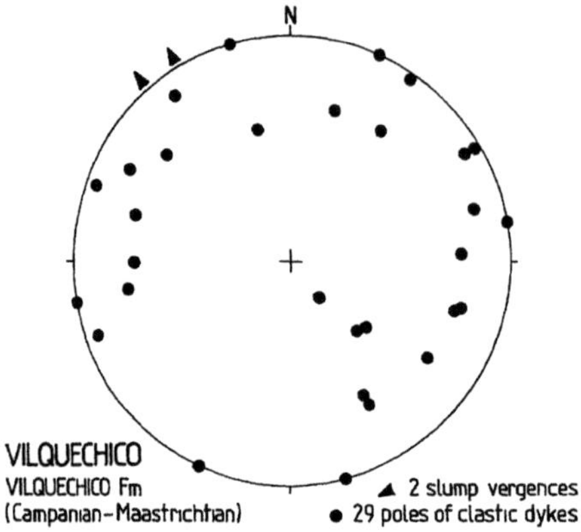

Fig. 17
Poles of clastic dykes in the middle and upper Vilquechico Formation (Middle Campanian-Late Maastrichtian).

volcanic clasts are also reported from the Late Cretaceous troughs of the Cuzco region (Noblet et al., 1987; Marocco and Noblet, 1990, see below). If this is the case, the onset of the Lluta thrust would not be of Cenomanian age (Vicente, 1990) but rather of Santonian-Campanian age (Figures 15 and 16). Nevertheless, the problem remains open, since the upper conglomerates are unfossiliferous. (Note that the contact of the Lluta thrust is intruded near Arequipa by plutons, part of which yielded Late Liassic U-Pb ages ! (188-184 Ma, Mukasa, 1986)). In the Abancay area, Late Senonian fluvial and lacustrine partly evaporitic red beds (Capas Rojas, Marocco, 1975; Anta-Anta Formation, Pecho, 1981) probably correlate with the red beds of central Peru and of the Cuzco area. On the Axial Swell, no Campanian-Maastrichtian deposits are known.

In the Eastern Basin, near Lake Titicaca, a tectonic quiescent period of probable Middle to Late Campanian age is expressed by a short-lived marine transgression (upper Middle Vilquechico Formation), and is followed by the resumption of detrital supply by latest Campanian and Early Maastrichtian times (base of upper Vilquechico Formation, Jaillard and Sempere, 1989; Jaillard et al., 1993). The orientation of open clastic dykes (Figure 17) indicates a NE to NNE-trending compressional strain, which agrees closely with the results obtained from the Cuzco and Sicuani areas. Farther northwest, the Hanchipacha Formation yielded Cretaceous marine fauna (Aptychus), and correlates with the Vilquechico Formation (Audebaud, 1973). In the Oriente region of southern Peru, the sandstones of the Vivian Formation are overlain by badly known monotonous red beds.

In the Cuzco and Sicuani areas, 4500 m-thick red bed sequences contain dinosaur tracks (San Jerónimo Group, Córdova, 1986; Noblet et al., 1987; López and Córdova, 1988). The creation of such very subsident troughs constitutes a major tectonic event of probably Late Campanian to earliest Maastrichtian age. The northward paleocurrents and the NE-SW synsedimentary compression (Figure 18) suggest that the creation and infilling of

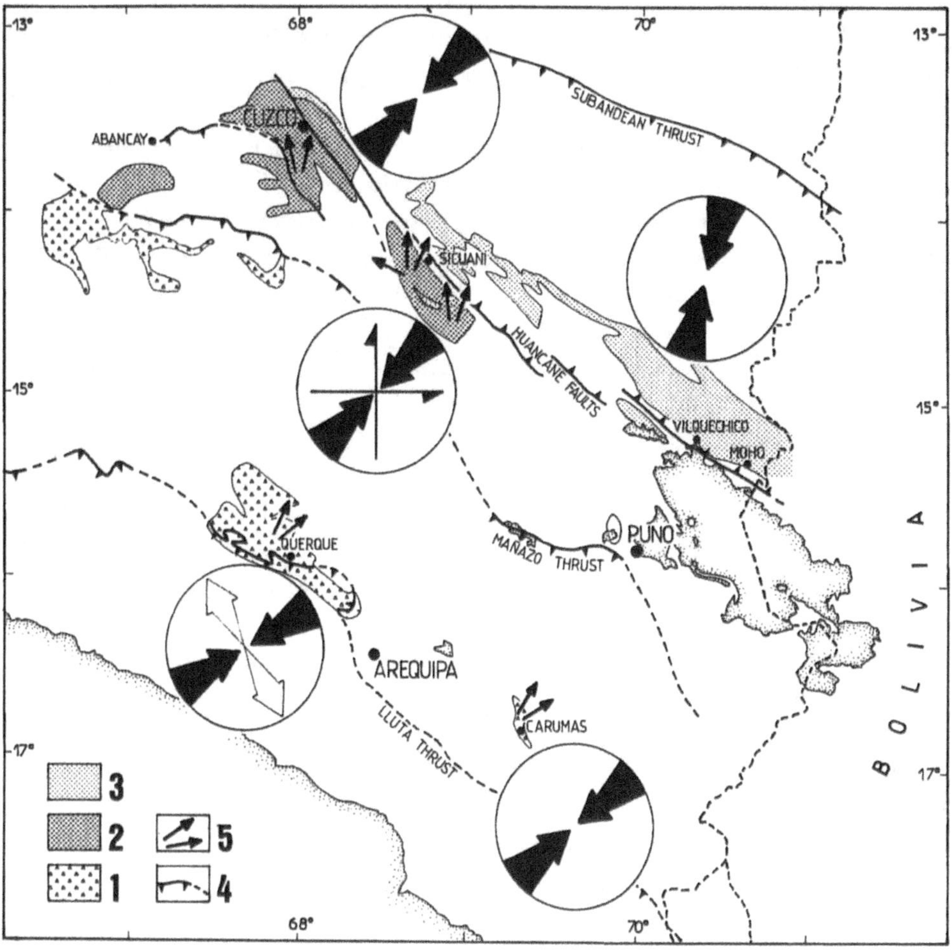

Fig. 18

Paleogeography and tectonic strain in southern Peru during Late Senonian times. 1: Western Facies: fanglomerate (of questionnable age) near Arequipa, shales and sands south of Abancay. 2: Cuzco Facies: thick fluvial red beds. 3: Eastern Facies: fine-grained, partly marine deposits. 4: Present-day major thrusts. 5: Paleocurrents.

these troughs are related to the onset of NE-trending overthrusts and to wrenching motions (Noblet, 1985; Córdova, 1986; Noblet et al., 1987; López and Córdova, 1988).

Tectonic Interpretations

During the Peruvian phase, compressional events (Turonian-Coniacian and Coniacian-Santonian boundaries, latest Campanian) alternate with relaxation episodes, which permitted marine transgressions to reach parts of the margin (Early Coniacian, Early Santonian, Middle Campanian).

The Turonian-Coniacian boundary Peruvian 1 phase (ca 89-88 Ma): Argillaceous detrital deposits, incipient coastal uplift and inversion of subsidence.

At the Turonian-Coniacian boundary, a major paleogeographic change caused the irruption of a fine-grained detrital supply over the entire margin (Sempere, this volume). The restricted environment of the Early Senonian deposits (evaporites, confined shales) suggests that the coastal area was uplifted, thus isolating the western areas from the open sea. This is supported by the emergence of southwestern Peru, by the weak subsidence of the Western Trough, and by the increasing subsidence of the Eastern Trough, giving an inverted subsidence pattern with respect to the Mid-Cretaceous (Sempere et al., 1988; Figures 13 and 14). No evidence of tectonic activity has been reported so far in the Coniacian deposits. Late Turonian and Coniacian times coincide with a striking plutonic gap in the coastal Batholith area.

The Late Coniacian-earliest Santonian Peruvian 2 phase (ca 87- 86 Ma): Incipient thrusting and progressive emergence.

In the Western Trough of southern Peru, emergence and tectonic activity are related to a NE-SW compressional strain (Figure 18). The Late Coniacian-earliest Santonian clastic sequences of the Arequipa area are interpreted as incipient foreland deposits resulting from the beginning of the Lluta overthrust, since the paleocurrents progressively rotate from SW to NW and then to the NE (Figure 16). In the Eastern Basin, the sedimentary response to the western deformations is possibly represented by the detrital deposits below the Santonian transgression.

More generally, the increasing compression of the coastal area, associated with lateral movements, caused the uplift of the margin and the progressive retreat of the sea during Late Santonian and Campanian times. The lack of Late Santonian-Campanian deposits in the eastern part of the Western Trough and near the Axial Swell can be due to incipient uplift or to subsequent erosion. This occurred earlier in southern Peru than in northern Peru, and in the western areas before the eastern ones (Figure 15). Thus, as for the Mochica phase, the response of southern Peru to compressional stress is clearly stronger and more rapid than that of the north Peruvian margin. However, the Early Campanian ductile deformation of gabbros indicate that deformation also occurred in the coastal area of Northern Peru (Mourier, 1988).

In the Arequipa area, after the (Early?) Santonian transgression, lacustrine deposits seem to correspond to a relaxation period. This is supported, in the Eastern Basin, by the deposition of fine-grained red shales and by a Middle Campanian marine transgression.
The Late Campanian Peruvian 3 phase (ca 76-73 Ma): Thrusting and foreland deposits.
In northwestern Peru, the deposition of the Tablones conglomerates indicates a marked paleogeographic change, characterized by the uplift and erosion of the Coastal Cordillera and the sinking of the coastal area, thus expressing the creation of the first Late Cretaceous forearc basin. In the coastal zone of northern Peru, uplift probably continued and thrusting and ductile deformation possibly occurred.
In southern Peru, NE-SW compressional stress clearly dominated (Figure 18). According to our interpretation, the coarse-grained deposits of the Arequipa area might represent the infilling of the foreland basins of the Lluta major thrust (Vicente et al., 1979; Vicente, 1990). Farther northeast, erosion probably occurred.
In the Western Trough (incipient Western Cordillera), the red bed basins are located just east of the western faults of the present-day Marañón and Mañazo thrust belts (Figure 14 and 15), thus strongly suggesting that they were related to the early interplay of these faults. As for the Cuzco red bed troughs (Noblet et al., 1987; López and Córdova, 1988), they are thus interpreted as the foreland basins of these incipient overthrusts (Figures 14 and 18). Moreover, their discontinuous shape (Figure 14), their very high subsidence rate, and the northward transport of the clastic supply suggest that they are pull-apart basins, and that (dextral?) wrenching movements occurred (Noblet, 1985). However, detailed studies of these basins are necessary in order to verify such a hypothesis.
Farther east, the Late Campanian (and earliest Maastrichtian?) phase is only recorded by the detrital supply, and by clastic dykes which trend normally to the NE vergence of the Lluta and Mañazo thrusts.

Maastrichtian-Paleocene

The Maastrichtian

Northwestern Peru and Southwestern Ecuador

In the Lancones-Celica Trough, the turbiditic sedimentation may have continued (Baldock, 1982). In the coastal Talara Basin, the Tablones conglomerates are overlain by marine fine-grained pro-delta deposits of Late Campanian and Maastrichtian age (Redondo Formation, Zuñiga and Cruzado, 1979; Séranne, 1987; Figure 15).

Northern Peru and Eastern Ecuador

Along the coast, the Batholith recorded a very important magmatic pulse associated with incipient strike-slip movements (72-64 Ma, Beckinsale et al., 1985; Bussel and Pitcher, 1985). With respect to the Campanian intrusions, the Maastrichtian plutons were intruded a few tens of kilometers eastward (Soler and Bonhomme, 1990). This could suggest that some shortening occurred during Late Campanian times, or that there was a change in subduction geometry.

In the Western Trough, Maastrichtian deposits might be represented by mainly fine-grained red beds (Jenks, 1961; Mabire, 1961; Mégard, 1978), since some of them yielded marine to continental Maastrichtian fauna near the transition to the eastern Basin (Bagua area, Fundo el Triunfo Formation, Mourier et al., 1988b).

In the Eastern Basin of Ecuador and Peru, Maastrichtian deposits consist of mainly continental fine-grained deposits, with thin intercalations of marine black shales (Areniscas de Azúcar 2 and 3, Cachiyacu, Huchpayacu and Tena formations, Tschopp, 1953; Koch and Blissenbach, 1962; Bristow and Hoffstetter, 1977; Petroperú, 1989; Figure 15). Ammonites have been mentioned, but their location or determination are uncertain (Rodríguez and Chalco, 1975; Vargas, 1988).

Southern Peru

In the Western Trough near Arequipa, intrusions and volcanic flows (Toquepala Formation) began to appear by Late Maastrichtian times (70-66 Ma, James et al., 1975; Beckinsale et al., 1985; Mukasa, 1986). Maastrichtian conglomerates could be represented by the Uchurca (and Jahuay) Formation (see discussion above). South of Abancay, Pecho (1981) describes fine-grained red beds which grade northward into thicker and coarser-grained red beds (Marocco, 1978).

In the Eastern Basin, fine-grained partly marine Maastrichtian shales are known in southern Peru (upper Vilquechico Formation, Dávila and Ponce de León, 1971; Jaillard et al., in press), in Bolivia, and as far south as Northern Argentina (El Molino and Yacoraite formations, Sempere et al., 1988; Marquillas and Salfity, 1988).

On the Axial Swell, mammal-bearing sandy red shales have been ascribed to the Maastrichtian (Grambast et al., 1967; Sigé, 1972). Comparable deposits have been observed in the Cuzco area, and might represent an eastern, coarser-grained equivalent of the Maastrichtian upper Vilquechico Formation (Jaillard et al., in press).

Near Cuzco and Sicuani, subsidence took place, and as much as 4500 m of coarse-grained fluvial red beds were deposited between latest Campanian and Late Maastrichtian times, as indicated by dinosaur tracks (Noblet et al., 1987; San Jerónimo Group, López and Córdova, 1989; Figure 14 and 15). The increasing abundance of volcanic clasts is regarded as indicating the presence and increasing erosion of a southern coeval volcanic arc (Marocco and Noblet, 1990), which could be the Toquepala system. Farther south, the emplacement of the Andahuaylas-Yauri pluton may have begun (Soler et al., 1989).

The Paleocene

Northwestern Peru, Northern Peru and Eastern Ecuador

The Talara Basin received a marine fine-grained (black shales) sedimentation (Séranne, 1987).

Along the coast, a noticeable plutonic event occurred during Early Paleocene times, but it rapidly vanished, and no intrusions are known between 60 and 54 Ma (Beckinsale et al., 1985; Soler and Bonhomme, 1990). High fault slip-rate and regional compression are recorded between 68 and 64 Ma (Bussel and Pitcher, 1985).

In the Western Cordillera (formerly Western Trough), the presence of Paleocene deposits is assumed in medium-grained fluvial red beds (Jenks, 1961; Wilson, 1963; Mégard, 1978), but no reliable stratigraphic data is available at present.

In the Eastern Basin of northern Peru, at the boundary with the Western Trough, Paleocene deposits are lacking (Mourier et al., 1988b; Naeser et al., 1991). In the Oriente of Ecuador, Paleocene deposits seem to be reduced or even absent beneath the unconformable Late Paleocene-Eocene sandstones and conglomerates (Tschopp, 1953; Bristow and Hoffstetter, 1977; Baldock, 1982). Farther southeast, the Paleocene is poorly characterized. Fine-grained red shales and silts with brackish intercalations (evaporites and marls) are ascribed to this stage (Sol and Yahuarango formations, Koch and Blissenbach, 1962; Fyfe, 1962; Feist et al., 1989; Petroperú, 1989).

Southern Peru

Along the coast and in the western part of the Western Cordillera (ex-Western Trough), numerous magmatic effusions and intrusions took place (Toquepala system, Beckinsale et al., 1985), recording a climax near the Danian-Thanetian boundary (57-62 Ma; Laughlin et al., 1968; Stewart et al., 1974; Bellon and Lefèvre, 1976; Vatin-Pérignon et al., 1982; Beckinsale et al., 1985; Clark et al., 1990). According to Bussel and Pitcher (1985), the compressive event recorded farther north is poorly expressed in southern Peru. In the rest of the western domain, no Paleocene deposits are recorded.

In the Eastern Basin, the Vilquechico Formation is conformably overlain by a few tens of meters of shales. This continental series of purple shales and thin-bedded sandstones becomes much thicker in Bolivia (Santa Lucía and Impora formations, Sempere, this volume), and farther northwest (Chilca Formation of the Sicuani area, Audebaud, 1973), where it yielded Paleocene fossils (Gayet et al., in press; Mourier et al., 1988). In the

Cuzco area, some hundreds of meters of fine-grained red shales overlie the Cretaceous deposits (base of Punacancha Formation, López and Córdova, 1988) and are followed by 2000 m of undated fluvial conglomerates (Punacancha Formation, Córdova, 1986; López and Córdova, 1988; Figure 14). In the Oriente of southern Peru, undated red beds were deposited. According to Soler et al. (1989), the Andahuaylas-Yauri batholith must have been formed at that time.

The Paleocene-Eocene Boundary Phase

This event closes the marine evolution of the Peruvian margin. It is marked throughout the margin by disconformable coarse-grained deposits, which often post-date tectonic deformations.

Northwestern Peru, Northern Peru and Eastern Ecuador

In the Talara basin, the Basal Salinas sands and conglomerates disconformably overlie the marine Paleocene shales (Marsaglia and Carozzi, 1990), and grade upward into Eocene fluviodeltaic sandstones (Séranne, 1987).

On the coast of central Peru, a conspicuous magmatic gap occurred between 54 and 50 Ma (Beckinsale et al., 1985; Soler and Bonhomme, 1990).

In the Western Cordillera of northern Peru, volcanic strata unconformably overlie folded sediments (Wilson, 1975), and fanglomerates interbedded with 49-50 Ma volcanic layers post-date a folding phase (Noble et al., 1990). Farther east, Maastrichtian red beds are disconformably capped by fluvial conglomerates associated with a 54 Ma tuff (Rentema Formation, Naeser et al., 1991).

In the Eastern Basin of Peru and Ecuador, the Paleocene or Cretaceous deposits are often disconformably overlain by Late Paleocene-Early Eocene sandstones and conglomerates (Tiyuyacu and Cuzutca formations, Tschopp, 1953; Koch and Blissenbach, 1962; Bristow and Hoffstetter, 1977; Basal Pozo Formation, Petroperú, 1989).

Southern Peru

Near the coast, the Toquepala magmatism ceased by 60 Ma, but hydrothermal phenomenae are then recorded during the Early Eocene (52 Ma K-Ar and Ar-Ar ages, Clark et al., 1990), and are possibly later than the tectonic phase. In the Western Cordillera, no information is available so far.

In the Eastern Basin, the Eocene fluvial conglomerates of the Muñani Formation (Audebaud et al., 1976; Feist et al., 1989) disconformably overlie the Maastrichtian upper Vilquechico Formation. Near Cuzco, the basal contact of the mainly volcaniclastic coarse-grained conglomerates of the upper Punacancha Formation (Córdova, 1986) could represent the discontinuity of the Paleocene-Eocene boundary.

Tectonic Interpretations

The Maastrichtian (ca 73-66 Ma): Tectonic remission, marine transgression and magmatic pulse.

The red bed sedimentation of the Western areas proceeds from erosion of the formerly created reliefs, but the lack of important sedimentary discontinuities indicates a rather stable tectonic regime. However, the very subsident basins of the Cuzco area suggest that continuous tectonic processes went on locally (Figure 14). This relative tectonic relaxation permitted a widespread marine transgression to invade the eastern parts of the entire central Andean margin and to deposit very homogeneous sediments (Figure 15). This non-compressive period favoured the occurrence of a significant magmatic pulse along the coast.

The Paleocene (ca 66-56 Ma): Reduced sedimentation and volcanic pulse.

During Paleocene times, the western areas and the western parts of the Eastern Basin recorded a reduced sedimentation or even a stratigraphic gap (northeastern Peru, Ecuador, Titicaca area), whereas the more easterly zones (eastern Peru, Bolivia) received rather thick red bed deposits. The latter exhibit a roughly coarsening-upward evolution (Sempere, this volume). Though no clear tectonic event can be identified, these features can be regarded as resulting from the resumption of a continuous mild compressional uplift of the Western Areas, responsible for the erosion or reduced sedimentation recorded in the western parts of the Eastern Area. After 60 Ma (Danian-Thanetian boundary), the disappearance of the intrusions in northern Peru and of the Toquepala volcanism of southern Peru could be related to the increasing compressional strain.

The Paleocene-Eocene boundary Inca 1 phase (ca 55-53 Ma): The first Tertiary Andean phase. The study of this phase is beyond the scope of this paper. It has only recently been identified in Peru, and its tectonic manifestations are still poorly documented. However, the irruption of unconformable widespread coarse-grained deposits, the uplift indicated by the change from marine shales to deltaic sands in the Talara basin, and local folding events express the importance of this tectonic phase, which is interpreted as the earliest well-defined Tertiary compressional phase of the central Andean margin.

Relationships with Geodynamic and Magmatic Events

Relationships between Tectonic Phases and Geodynamic Events

Three main tectonic periods can be recognized during the Kimmeridgian-Paleocene interval (Figure 19). We propose the name "Virú period" for the Kimmeridgian-Berriasian events. They are followed by the Early Albian-Early Middle Cenomanian "Mochica period", and the Early Coniacian-earliest Maastrichtian "Peruvian period". They are separated by periods of tectonic quiescence or relaxation. As no pre-Late Cretaceous plate reconstructions are available, the Kimmeridgian-Berriasian period will be analyzed separately.

The Kimmeridgian-Berriasian "Virú Period" (145-130 Ma)

It can be divided into three discrete phases. The Kimmeridgian event mostly affects southern Peru, and seems to be related to the Araucan phase of Argentina and Chile. The Tithonian phase is best regarded as a consequence of the collision of allochthonous terranes and the subsequent folding and thrusting of the western border of the Ecuadorian margin. The Berriasian event could be due to a general uplift originated in the east, and related to South Atlantic rifting. Since the location and origin of these events are quite separate, they can be ascribed either to a unique cause differently expressed because of local parameters (inherited tectonic features, heterogeneous structure of the crust, etc.), or to separate and independent geodynamic factors. The varied nature of these tectonic events supports the latter hypothesis, though interactions with local parameters are possible.

In most of the oceanic spreading centres of the world, Kimmeridgian-Tithonian times (145-140 Ma) are marked by a sharp decrease in the accretion rates (Olivet et al., 1984; Klitgord and Schouten, 1986; Savostin et al., 1986). These ridges still belong to the roughly E-W-trending Tethyan system, and it can be supposed that the NE-trending hypothetical Colombian-Tethyan ridge (Mooney, 1980) recorded the same sharp decrease in the accretion rate (Jaillard et al., 1990). Combining models proposed by Duncan and Hargraves (1984) and Aspden et al. (1987); Jaillard et al. (1990) proposed that, along the Peruvian margin, the roughly southeastward convergence induced by spreading activity of the Tethyan oceanic ridges would have been replaced by a roughly northeastward convergence induced by Pacific spreading centres after Kimmeridgian times. This model would explain the appearance of a Tithonian subduction-related volcanic system along the Peruvian margin, and the coeval dextral accretion of south-proceeding terranes along the Ecuadorian-Colombian margin (Mourier et al., 1988; Aspden et al., 1988).

Meanwhile, the rifting of the roughly N-S trending southern South Atlantic ocean began (147-136 Ma, Ojeda, 1982; Sibuet et al., 1985). Although it is still difficult to separate the role of both geodynamic events in the tectonic phases recorded in the Peruvian margin, they indicate that Kimmeridgian-Tithonian times mark the end of the Tethyan-dominant regime, and their replacement by an Atlantic-PaleoPacific (Farallón)-dominant system.

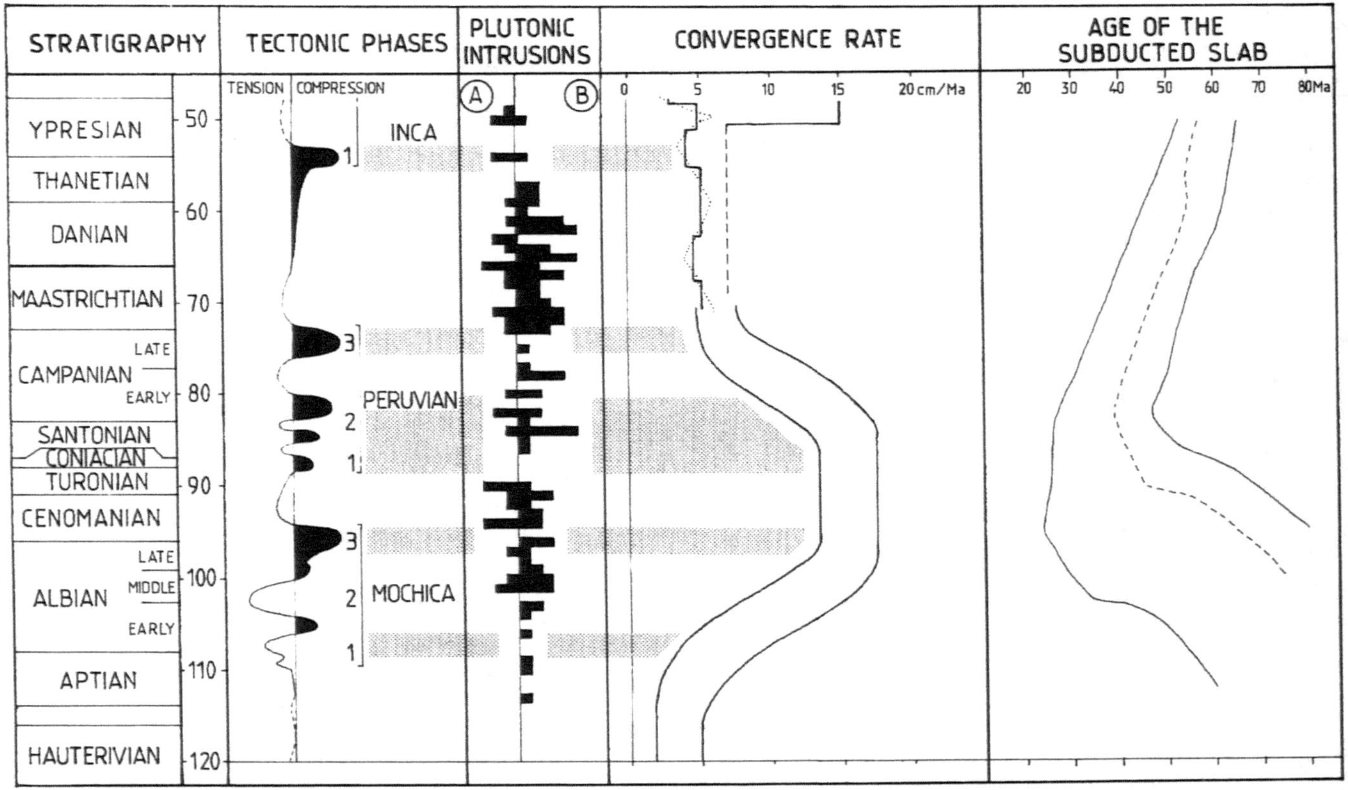

Fig. 19

Relations between tectonic events, magmatic pulses and Plate geodynamics, in the Peruvian margin, between Hauterivian and Paleocene times. Plutonic intrusions: A: On the coast of central Peru (after Soler and Bonhomme, 1990). B: Throughout the whole of coastal Peru (after Beckinsale et al., 1985; Mukasa, 1986). Convergence rate between the Farallón and the South American plates and relative age of the subducted slab (after Soler and Bonhomme, 1990).

By Berriasian times, the southern South Atlantic ocean began to open (135-130 Ma, Rabinowitz and La Brecque, 1979; Goodlad et al., 1982; Scotese et al., 1988). The coeval events (mainly uplift) of the Peruvian margin and the subsequent arrival of east-deriving mature sandstones (Valanginian) can therefore be ascribed to the rifting processes (doming?) affecting the northern South Atlantic ocean. The stable tectonic regime observed during most of Early Cretaceous times in Peru coincides with a spreading quiescence observed in the South Atlantic, before the opening of its northern segment by Late Aptian-Albian times (Lehner and De Ruiter, 1977; Rabinowitz and La Brecque, 1979).

The Middle and Late Cretaceous "Mochica" and "Peruvian" Periods (108-95 and 88-73 Ma)

Both phases will be analyzed together in this section, and compared with geodynamic events.

In a subduction setting, compressive stress in the upper plate is classically attributed to various independant factors which are (1) the young age of the subducted slab, which is then buoyant and induces a low-dipping subduction angle; (2) the subduction of oceanic obstacles; (3) the high convergence velocity of the subducted oceanic slab; and (4) the high absolute displacement rate of the overriding plate toward the trench (see Uyeda and Kanamori, 1979; Uyeda, 1982; Cross and Pilger, 1982; Mitrovica et al., 1989; Soler and Bonhomme, 1990).

Relationship with the age of the subducted plate

In spite of broad uncertainties in the Plate motion reconstructions for this period, the age of the subducted oceanic slab is known to have rapidly decreased between 115 and 100 Ma, to have been young (30 to 50 Ma) between Albian and Santonian times (100-80 Ma), and then to have progressively increased (40 to 65 Ma) during the Late Senonian and Paleocene (Soler et al., 1989; Soler and Bonhomme, 1990; Figure 19). Thus, the age of the oceanic lithosphere apparently cannot be directly brought in to explain the progressive increase of compressive deformation of the Peruvian continental margin during Late Cretaceous times (Soler and Bonhomme, 1990). Still less can it be used to explain the individual short-lived tectonic phases recorded by the Peruvian margin. However, the Mochica and Peruvian periods, as a whole, are coeval with the subduction of a rather young oceanic lithosphere, and these tectonic phases could be related to this long-term situation.

Relationship with convergence rates

During Middle Cretaceous to Paleocene times, the convergence rate along the Peruvian trench was in turn slow (Early Cretaceous), fast (Middle Cretaceous and Early Senonian), and slow again (Late Senonian and Paleocene)(Soler and Bonhomme, 1990). The period of fast convergence broadly correlates with two main periods of tectonic instability along the Peruvian margin, and the slow convergence periods coincide roughly with quieter intervals. Nevertheless, these correlations are unsatisfactory, since the high convergence period also includes extension or relaxation phases, and low convergence intervals include major compressive events (Peruvian 2 and 3, Incaic 1). In contrast, most of the tectonic events recorded in the Peruvian margin correlate closely with changes in the convergence rates (Mochica 1 and 3, Peruvian 2 and 3, Inca 1, Figure 19). So although high convergence rate periods broadly correlate with long-term compressive periods, the acceleration or deceleration of the convergent motion appear to be the greater determining factors in the generation of tectonic phases within the overriding continental plate. However, this observation is weakened by uncertainty with regard to stratigraphic data and plate motion reconstruction, especially during the Mid-Cretaceous quiet magnetic period.

Relationship with obliquity of convergence

Bussel and Pitcher (1985) emphasized the importance of strike-slip motions along the Peruvian coast during the Late Albian-Santonian (98-82 Ma) and the Early Paleocene (Danian, 68-64 Ma) periods. More recently (Soler, 1991a and b) proposed that strike-slip motions played a major part in the generation of the Mid-Albian to Early Cenomanian volcanic centers (see below). For the latest Cretaceous-Paleocene period, a strong dextral component along the Peruvian coast can be explained by the very oblique convergence of the Farallón oceanic plate, since the northward displacement of the latter was nearly parallel to the Peruvian continental margin (Pilger, 1984; Gordon and Jurdy, 1986; Pardo-Casas and Molnar, 1987; Mayes et al., 1990). This situation could also account for the pull-apart nature suggested for the Late Senonian red bed troughs (Noblet, 1985). If so, oblique oceanic subduction may largely have influenced the Middle and Late Cretaceous evolution of the Peruvian margin. Moreover, the change of rotation poles in plate motion can explain some of the short-lived tectonic phases, but no precise geometric reconstructions of the Pacific Plate are available for Middle and Early Late Cretaceous times.

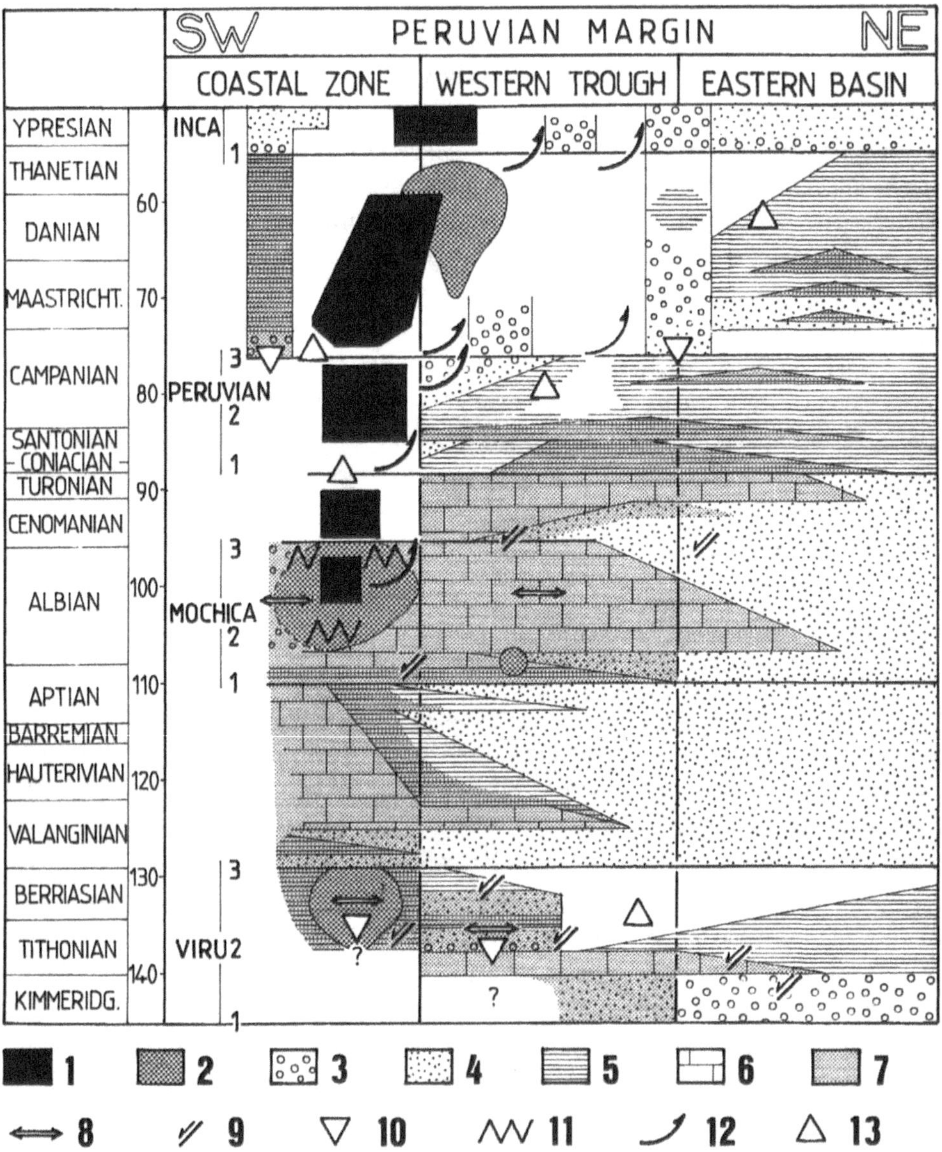

Fig. 20

Synopsis of the main tectonic-sedimentary and volcanic events of the Peruvian margin between Kimmeridgian and Paleocene times. 1: Intrusion. 2: Volcanism. 3: Conglomerate. 4: Sandstone. 5: Shale. 6: Limestone. 7: Marine deposits. 8: Extension. 9: Normal fault. 10: Tectonic subsidence. 11: Folding. 12: Thrusting. 13: Tectonic uplift.

Relationship between Tectonic Phases and Magmatism

In central Peru, Soler and Bonhomme (1990) identified four plutonic gaps in the Batholith emplacement (Figures 19 and 20). Such plutonic gaps are classically related either to compressive tectonic stress within the upper plate, which closes the crustal fractures channelling the magma (e.g. terrane accretion, Raymond and Swanson, 1980), or to the disappearance of the upper plate lithospheric wedge, which prevents magma generation. The latter situation occurs when oceanic obstacles are subducted (Nur and Ben Avraham, 1982), and/or when the oceanic slab possesses a low subduction angle.

Middle Albian times are characterized by both an extensional regime in the margin, and an intense volcanic activity in the arc-marginal basin system.

The first plutonic gap (97-94 Ma, latest Albian-Middle Cenomanian; Figure 19) coincides exactly with the Mochica 3 major phase, as recorded by the synsedimentary features in the Western Trough. It is followed by a plutonic pulse contemporaneous with the quiet Middle to Late Cenomanian period.

The second plutonic gap (90-84 Ma), more surprisingly, correlates both with a tectonic quiescent period (latest Cenomanian-Turonian) and with the beginning of the compressive Peruvian phase (Coniacian-Santonian). It is followed by a minor magmatic pulse of Early Campanian age (84-80 Ma), which correlates with the relatively quiescent interval between the Peruvian 2 and 3 events.

The third magmatic gap (80-73 Ma, Campanian, Figure 19) coincides with the Peruvian 3 major phase. A last plutonic and volcanic pulse (73-60 Ma) occurred during the Maastrichtian-Danian tectonic remission.

Finally, a plutonic gap, occurring between 60 and 54 Ma (Late Paleocene), corresponds to the incipient compression which culminated during the Incaic 1 phase of the Paleocene-Eocene boundary.

Such periods are also more or less apparent when compared with the radiometric data of the whole of coastal Peru (Figure 19). Thus, except for the latest Cenomanian-Coniacian period, the plutonic gaps broadly correlate with main compressive tectonic events.

No accretion of allochthonous terrane occurred along the Peruvian margin during Cretaceous times (Mégard, 1987; Beck, 1988). Moreover, Soler and Bonhomme (1990) assumed a rather steep-dipping angle of the subducted oceanic slab during Cretaceous times. Thus, the plutonic gaps must be due either to the subduction of oceanic obstacles, as assumed by Soler et al. (1989) for the Late Santonian-Early Maastrichtian interval in southern Peru, or, more generally, to compressive stress within the South American continental plate, the cause of which remains to be discussed.

A Tectonic-Geodynamic Model for the Cretaceous Andean Margin

In order to simplify comprehension, the Kimmeridgian-Paleocene evolution of the Peruvian margin can be divided into two major periods, separated by a period in which the general tectonic regime was inverted.

During the first period (Late Jurassic-Early Cretaceous), the Peruvian margin was dominated by an extensional regime, and from trench to continent reveals: (1) a possible forearc system, now disappeared; (2) a mainly effusive magmatic arc, the existence of which is now mostly attested to by its reworked products; (3) discontinuous marine or continental extensional basins filled both by autochthonous volcanic flows and by reworked volcanic products, which are interpreted either as ensialic marginal basins or as pull-apart basins located within the volcanic arc; (4) a mobile external margin, dominated by an extensional tectonic regime and by marine shelf deposits, and (5) a stable pericratonic internal margin, which underwent little subsidence and received mainly terrigenous sedimentation.

Such a margin pattern is classically related to an extensional margin submitted to the slow steep-dipping subduction of an old oceanic crust ("Mariana-type" of Uyeda and Kanamori, 1979; Uyeda, 1982).

During the second period (Late Cretaceous-Paleocene), the Peruvian margin underwent a northeastward propagating compressional strain, and is characterized by the following structures from the trench to the craton: (1) incipient marine to continental forearc basins; (2) a mainly intrusive emergent magmatic arc, incorporated into the deforming active margin; (3) an active thrust and fold belt, with sparse intermountain basins; (4) a belt of foreland basins; (5) a mobile zone, submitted to partly marine fine-grained detrital sedimentation and incipient compressional strain.

This pattern characterizes compressional margins submitted to the fast, low-dipping subduction of a young oceanic crust ("Chilean-type" of Uyeda and Kanamori, 1979; Uyeda, 1982).

Age of the Extensional and Compressional Periods

Extensional Period

The nature of the Kimmeridgian event is not well understood in the Peruvian margin. However, it is clearly associated with the play of normal faults in southern Peru. Extensional stress is then shown during the Tithonian phase by the opening of turbiditic and "within-arc" or "back-arc" basins and by structural analysis. Though the Berriasian event seems also to be related to Atlantic rifting, the existence of an extensional regime is suggested by the SW paleoslope incline and by the high subsidence rate of the Western Trough during Valanginian times. This situation then prevailed during Early Cretaceous

times, as shown by the noticeably differential subsidence registered on the north Peruvian margin.
The mostly extensional Virú tectonic period can therefore be regarded as a prelude to the tensional regime which was dominant during the Early Cretaceous.

Compressional Period

The beginning of compression is classically thought to be of Senonian age (ca 80 Ma, Peruvian phase) in the central Andean margin (Steinmann, 1929; Benavides, 1956; Mégard, 1978). However, the part played by the Mochica phase has recently been emphasized (Cobbing et al., 1981; Mégard, 1984; Jaillard and Sempere, 1989).
The Eastern Basin had undergone subsidence by Early Senonian times, and began to be uplifted in the Late Campanian, and more generally in Paleocene times (Figure 20). The beginning of compression there is of Late Senonian age, and roughly coincides with the Peruvian 3 phase.
In the Western Trough, the transition from subsidence to uplift took place between Turonian and Coniacian times, and the onset of syntectonic deposition is of earliest Santonian to Late Campanian age from the Southwest to the Northeast (Figure 20), expressing a clearly compressive regime (Figure 18). The beginning of compression in this region is therefore of Early Senonian age, and roughly coincides with the Peruvian 1 and 2 phases.
In the Coastal Zone, extension is thought to have been dominant during the Early Cretaceous; whereas compression began during the Early Albian, and clearly prevailed in the Late Albian and Early Middle Cenomanian (Figure 20).
Initial compression was therefore diachronous throughout the margin, and clearly began earlier toward the trench. Extrapolating these observations we may suppose that, at the trench itself, the transition from extensional to compressional stress must have occurred in the Late Aptian (ca 110 Ma), since it coincides with: (1) the arrival of a younger oceanic lithosphere in the trench; (2) the acceleration of the convergence rate (Figure 19); (3) the beginning of the oceanic accretion in the Northern South Atlantic, which led to the westward shift of the South American plate. If so, the actual beginning of compression in the Peruvian margin cannot be considered as of Senonian age. Interpretations of the consequences of this supposed Aptian-Albian compressional regime depend on the nature of the volcanic centres.

The Mochica Period

Opening and Closing of a Middle Cretaceous Marginal Basin? ...

Atherton et al. (1983, 1985) and Aguirre and Offler (1985) have suggested that the Tithonian-Berriasian and Albian volcanic centres must have belonged to the same back-arc extensional system. According to these authors, they probably represent aborted "ensialic" marginal basins, floored by continental crust. More recently, Atherton (1990) suggested that a true ocean-floored marginal basin, somewhat comparable to that of Patagonia (Dalziel et al., 1974; Aberg et al., 1984), developed in Northern Peru during Albian times. In these hypotheses, an extensional regime was dominant during Late Aptian-Middle Albian times. The marginal basin was created by crustal extensional stretching during Late Aptian or Early Albian times (Mochica 1 phase), and functioned and widened during the Middle Albian, thus explaining the high observed geothermal gradient and metamorphic pattern (Aguirre et al., 1989). Its closure occurred during Late Albian to Early Cenomanian times (Mochica 3 phase), and was associated with noticeable dextral motions (Bussel, 1983). If so, the early Mochica tectonic period would represent the time-span necessary for the trench to shift northeastward and reach the border of the continental margin proper.

This hypothesis is in keeping with the geological facts observed in the Western Trough, and has been adopted by most authors during the past few years (e.g. Mégard, 1987; Jaillard, 1987).

It remains difficult, however, to explain the compressive deformations observed during Early Albian times (105 Ma folding phase, Wilson, 1975, Mochica 2 phase), since an extensional tectonic regime is presumed to have existed at this time, and this would appear to contradict the geodynamic data, which seem to indicate a prevailing compressive tectonic regime after Late Aptian times. It would have been more likely for such a marginal basin to have opened during Early Cretaceous times, as a result of the oceanward shift of the trench-arc system, triggered by the continuous extensional regime. Moreover, in this hypothesis, it is difficult to explain why only the marginal basin is presently preserved, whereas no remnants of the volcanic arc are observed.

... Or Wrenching of a Pulled-Apart Volcanic Arc?

In contrast to the preceding hypothesis, Soler (1991a, 1991b), using the same geochemical data, in addition to some new findings, proposed that the whole range of volcanic and volcaniclastic formations could represent the remains and/or the products of the volcanic arc itself. The latter would have been active during Tithonian-Berriasian times, and then quiescent during most of the Early Cretaceous. After Late Aptian-Early Albian times, volcanic activity would have resumed, due to the new geodynamic pattern, and the volcanic arc would have been submitted to intense dextral strike-slip motions. This would explain the sporadic compressional deformation and the creation of the northern troughs filled by thick volcanic-derived products (Casma Group and Lancones Formation), these being interpreted as extensional pull-apart basins. This interpretation implies that oceanic convergence occurred toward the North, obliquely to the margin, during Albian times.

Such an interpretation of a pulled-apart volcanic arc is in agreement with the calc-alkaline geochemistry of the south Peruvian volcanic centres (Copara and Matalaque formations), as well as with the discontinuous shape of the volcanic belt (Figures 9 and 13). Moreover, the supposed northward convergence closely agrees with the compression direction registered during the Mochica phase on the Peruvian margin (Figure 10). In this hypothesis, the incipient compression would have been accommodated during the early Mochica period, by lateral displacements and wrenching deformations along the very edge of the continental margin.

Whichever the Case

The shift of the trench was achieved by the Late Albian, and culminated in the accretion of the volcanic arc or the remnants of the marginal basin with the Andean continental margin (latest Albian-Early Middle Cenomanian, Mochica 3 phase). It is worth noting that foldings and thrustings only involved the coastal zone during the Mochica period.

Though detailed geochemical and mineralogical studies have been carried out on these formations, their interpretation seems to remain debatable. Moreover, except in some areas (Myers, 1980; Webb, 1976), their overall lithologic and geometric features are so far poorly known, and further geologic surveys, as well as stratigraphic and structural field work are necessary before any conclusion can be reached.

The Peruvian Period

The Western Trough had undergone a strong crustal stretching by latest Jurassic-earliest Cretaceous times, and this controlled the Early Cretaceous subsidence (Jaillard, 1990). After the Cenomanian-Turonian quiescent period, the continental margin itself began to accommodate the ongoing compression, and the Mochica suture played again, while the extensional paleogeographic structures of the Western Trough began to invert.

The Turonian-Coniacian Peruvian 1 event seems to correspond to the re-utilization of the Cenomanian Mochica suture (uplift of the coastal zone). The Late Coniacian-earliest Santonian Peruvian 2 phase, materialized by the onset of the Lluta thrust of the Arequipa area, corresponds to the structural inversion of the southwestern border of the Western Trough. Finally, the Late Campanian Peruvian 3 phase, expressed by the creation of the red bed foreland basins, can be interpreted as the structural inversion of the northeastern border of the Western Trough (Figures 15 and 18). Therefore, the "Peruvian period" can be considered a period of tectonic inversion of the Western Trough extensional structures.

Because the westernmost areas were probably already partly inverted by the Mochica phase, and since southern Peru underwent a lesser amount of Mesozoic extension than northern Peru, the western and southern areas of the margin reacted and emerged earlier.

Conclusions

Tectonic Evolution of the Peruvian Margin

Detailed studies of the tectonic-sedimentary evolution of the Peruvian margin between Kimmeridgian and Paleocene times reveal the existence of three main tectonic periods, separated from each other by intervals of extensional regime or of tectonic relaxation. Each period can be subdivided in turn into various tectonic events (Figure 19).

The "Virú" period (Kimmeridgian-Berriasian) begins with the Kimmeridgian (?) Virú 1 event, marked by the resumption of clastic deposits in southern Peru. Although it is still poorly understood, it can be interpreted as a distal response of the Argentina-Chilean "Araucan" phase. The Tithonian Virú 2 phase is expressed by the creation of a deep turbiditic trough in northern Peru, by the opening of a back-arc basin in the coast of central Peru, and by the emergence of most of southern (and eastern?) Peru. It is interpreted as the consequence of both the accretion of allochthonous terranes in northwestern Peru and Ecuador, and an important extensional phase in Central and Northern Peru. The Berriasian Virú 3 tectonic event is a major paleogeographic change, mostly expressed by a regional unconformity beneath the Valanginian sandstones. It is related to the rifting of the northern South Atlantic ocean, and to ongoing extension at the margin edge.

The "Mochica" period (Late Aptian-Early Middle Cenomanian) starts with scattered precursory volcanic and tectonic extensional manifestations of Late Aptian-earliest Albian age (Mochica 1 phase). It continues during Early and Middle Albian times, with alternating compressional and extensional events within the arc or back-arc volcanic rocks (Mochica 2 phase). These are thought to express either the opening of a marginal basin, or the wrenching and pulling apart of a volcanic arc. The Mochica period ends with the accretion, folding and emergence of these volcanic centres by Late Albian to Early Middle Cenomanian times (Mochica 3 phase).

The "Peruvian" period (Coniacian-Campanian) begins at the Turonian-Coniacian boundary with a major paleogeographic change, probably due to the uplift of the coastal area, causing both an invasion of fine-grained detrital material over the entire margin, and a sharp decrease of subsidence in the Western Trough (Peruvian 1 phase). The Late Coniacian to Middle Campanian times are marked by the start of northeastward overthrusts in southern Peru (Peruvian 2 phase) and by the progressive emergence of the Peruvian margin. They are interpreted as the structural inversion of the southwestern edge of the Western Trough. The Peruvian period culminates with the Late Campanian Peruvian 3 phase, expressed by the development of a foreland basin belt located on the eastern side of the incipient Marañón and Mañazo thrusts. The latter are regarded as the expression of a structural inversion of the northeast border of the Western Trough.

The "Incaic" 1 phase (Paleocene-Eocene boundary) then marks the beginning of the Andean orogeny s.s.

Geodynamic Evolution of the Peruvian Margin

An extensional regime is thought to have been dominant at the trench during latest Jurassic and Early Cretaceous times, and to have induced the magmatic quiescence of the volcanic arc and an oceanward retreat of the oceanic slab.

A compressional regime is then believed to have begun at the trench in the Late Aptian, as the South American plate began to shift westward, and the subducted oceanic lithosphere became younger and arrived faster at the trench. The "Mochica period" is interpreted as the time-span necessary for the trench to migrate up to the border of the continental margin, leading to the consumption of the supposedly aborted marginal basin or to the deformation of the volcanic arc, up to their accretion on the continental margin.

The subsequent "Peruvian period" resulted from the northeastward propagation of the compressive strain within the South American Plate, which caused the structural inversion of the major pre-existing inherited structures, that is: the Cenomanian suture (Peruvian 1) and the western and eastern boundaries of the Western Trough (Peruvian 2 and 3, respectively).

Most of the tectonic events correlate with plutonic gaps, caused by compressive closure of the crustal features channelling the magma. The main tectonic, extensional or compressive phases coincide with quickening or slackening periods in the convergence between the oceanic plate and South America. Though little is yet known about them, dextral wrenching movements probably played an important part in the tectonic evolution of the Peruvian margin.

In more general terms, the determining factors in the tectonic evolution of a continental active margin seem to be: (1) the age of the subducted slab and the convergence rate at the trench, which determine the general strain regime of the trench-margin system; (2) pre-existent extensional or even compressional structures of the continental margin, which determine the tectonic expression and location of the propagating strain; (3) variations (quickening or slowing down) of the convergence rate, which bring about the major tectonic phases.

According to this model, the answers to the problems outlined at the beginning of this paper are:

During the Early Cretaceous, no chain formed along the active Peruvian continental margin, because subduction was low and involved an old oceanic slab. This induced a steep-dipping, "Mariana-type" subduction regime (Uyeda and Kanamori, 1979), characterized either by a magmatic quiescence or by the development of extensional back-arc basins.

After Late Aptian times, and in spite of a compressional tectonic regime due to an acceleration of the convergence rate and rejuvenation of the oceanic lithosphere at the trench -which brought about a possibly low-dipping, "Chilean-type" subduction (Uyeda and Kanamori, 1979)- neither was any chain formed during Middle Cretaceous times, because the shortening was first accommodated by deformation of the marginal basin or the volcanic arc, before the continental margin itself became involved in compressive deformation processes.

Acknowledgements

I am grateful to the Institut Français d'Études Andines for having financially supported most of the field work in Peru. Thanks are due to M. Batty, V. Carlotto, A. Córdova, J. Jacay and G. Santander for their enthusiastic participation in field work. Subsequent analyses and studies have been carried out at the Institut Français de Recherche Scientifique pour le Développement en Coopération (ORSTOM, UR 1H) and at the Institut Dolomieu of Grenoble. G. Mascle, T. Sempere and P. Soler greatly contributed to the improvement of the interpretations and conclusions. I thank the Argentine translator Mr Peter Edmunds, who helped me with my English.

REFERENCES

Åberg, G., Aguirre, L., Levi, B., and Nyström, J.O. (1984): Spreading-subsidence and generation of ensialic marginal basins: An example from the Early Cretaceous of Central Chile, *in* Marginal Basin Geology, (Eds.: Kokelaar, B.P., and Howells, M.F.), Geological Society Special Publication, v. 16, 185-193.

Aguirre, L. (1989): Metamorfismo pre-orogénico cretácico y marco geotectónico, Cordillera occidental de Colombia (perfil Buga-Buenaventura). Revista Geológica de Chile, v. 16, 123-144, Santiago.

Aguirre, L. (1990): Metamorphism of the Celica Formation, SW Ecuador: geotectonic implications. Symposium International Géodynamique Andine, Grenoble, Ed. ORSTOM, Colloques et Séminaires, 207-210.

Aguirre, L., and Offler, R. (1985): Burial metamorphism in the Western Peruvian Trough: Its relation to Andean magmatism and tectonics, *in* Magmatism at a Plate Edge: The Peruvian Andes, (Eds.: Pitcher, M.P., Cobbing, E.J., and Beckinsale, R.D.), London, Blackie Halsted Press, 59-71.

Aguirre, L., Levi, B., and Nyström, J.O. (1989): The link between metamorphism, volcanism and geotectonic setting during the evolution of the Andes, *in* Evolution of Metamorphic belts, (Eds.: J.S. Daly, J.S., Aliff, R., and Yardley, B.W.D.), Geological Society Special Publication, v. 43, 223-232.

Anderson, E.M. (1951): The Dynamics of Faulting. Edinburgh and London, Oliver, Boyd Ed.

Anonymous (1980): Sinopsis explicativo del mapa geológico del departamento de Ica. Boletín del Instituto Geológico Minero y Metalúrgico, (F), 2, 27 p., Lima.

Arthaud, F., Mégard, F., and Séguret, M. (1977): Cadre tectonique de quelques bassins sédimentaires. Bulletin des Centres de Recherche et d'Exploration-Production, Elf-Aquitaine, v. 1, 147-188.

Aspden, J.A., McCourt, W.J., and Brook, M. (1987): Geometrical control of subduction-related magmatism: the Mesozoic and Cenozoic plutonic history of Western Colombia. Journal of the Geological Society, v. 144, 893-905.

Aspden, J.A., Litherland, M., and Salazar, E. (1988): Una interpretación preliminar de la historia colisional del Centro y Sur del Ecuador y posibles controles para la geología cenozoica y de mineralización polimetálica. Politécnica, Monografía de Geología, 6, v. 13, 49-75, Quito.

Aspden, J.A., Rundle, C.C., Viteri, F., Bermudez, R., and Harrison, S.S. (1990): Edades radiométricas del Batolito de Zamora-rio Mayo. Boletín Geológico Ecuatoriano, v. 1, 85-88, Quito.

Atherton, M.P. (1990): The coastal Batholith of Peru: The product of rapid recycling of "new" crust formed within rifted continental margin. Geological Journal, v. 25, 337-349.

Atherton, M.P., and Webb, S. (1989): Volcanic facies, structure and geochemistry of the marginal basin rocks of Central Peru. Journal of South American Earth Sciences, v. 2, 241-261.

Atherton, M.P., Pitcher, W.S., and Warden, V. (1983): The Mesozoic marginal basin of Central Peru. Nature, v. 305, 303-305.

Atherton, M.P., Warden, V., and Sanderson, L.M. (1985): The Mesozoic marginal basin of Central Peru: A geochemichal study of within-plate-edge volcanism, *in* Magmatism at a Plate Edge: The Peruvian Andes, (Eds.: Pitcher, M.P., Cobbing, E.J., and Beckinsale, R.D.), London, Blackie Halsted Press, 47-58.

Audebaud, E. (1971): Mise au point sur la stratigraphie et la tectonique des calcaires cénomaniens du Sud-Est péruvien (formation Ayavacas). Comptes Rendus à l'Académie des Sciences, (D), 272, 1059-1062.

Audebaud, E. (1973): Geología de los cuadrángulos de Ocongate y Sicuani. Boletín del Instituto Geológico Minero y Metalúrgico, 25, 72 p., Lima.

Audebaud, E., Laubacher, G., and Marocco, R. (1976): Coupe géologique des Andes du Sud du Pérou de l'Océan Pacifique au Bouclier Brésilien. Geologische Rundschau, v. 65, 223-264.

Baldock, J.W. (1982) Geología del Ecuador. Quito, Dirección General de Geología y Minas, 66 pp.

Batty, M., and Jaillard, E. (1989): La sedimentación neocomiana (Jurásico terminal-Aptiano) en el Sur del Perú, *in* Contribuciones de los Simposios sobre el Cretácico de América Latina, A: Eventos y Registro Sedimentario, (Ed.: Spalletti, L.A.), GSGP-IGCP 242, 75-88, Buenos Aires.

Batty, M., Carlotto, V., Jacay, J., and Jaillard, E. (1990): The Kimmeridgian(?)-Early Valanginian tectonic events on the Peruvian margin. Symposium International Géodynamique Andine, Grenoble, Ed. ORSTOM, Colloques et Séminaires, 265-268.

Beck, M.E. Jr (1988): Analysis of Late Jurassic-Recent paleomagnetic data from active plate margins of South America. Journal of South American Earth Sciences, v. 1, 39-52.

Beckinsale, R.D., Sánchez-Fernandez, A.W., Brook, M., Cobbing, E.J., Taylor, W.P., and Moore, N.D. (1985): Rb-Sr whole-rock isochron and K-Ar age determinations for the Coastal Batholith of Peru, *in* Magmatism at a Plate Edge: The Peruvian Andes, (Eds.: Pitcher, M.P., Cobbing, E.J., and Beckinsale, R.D.), London, Blackie Halsted Press, 177-202.

Bellon, H., and Lefèvre, R. (1976): Données géochronométriques sur le volcanisme andin dans le Sud du Pérou. Implications volcano-tectoniques. Comptes Rendus à l'Académie des Sciences, (D), 283, 1-4.

Bellido, E. (1956): Geología del curso medio del río Huaytara, Huancavelica. Boletín de la Sociedad Geológica del Perú, v. 30, 33-47, Lima.

Benavides, V. (1956): Cretaceous System in Northern Peru. American Museum of Natural History Bulletin, v. 108, 352-494.

Benavides, V. (1962): Estratigrafía pre-Terciaria de la región de Arequipa. Boletín de la Sociedad Geológica del Perú, v. 38, 5-63, Lima.

Bristow, C.R., and Hoffstetter, R. (1977): Ecuador, (Second Edition), *in* Lexique Stratigraphique International, v. V: Amérique Latine (Dir.: Hoffstetter, R.), Fasc. 5 a 2, Paris, Centre National de la Recherche Scientifique, 412 pp.

Bussel, M.A. (1983): Timing of tectonic and magmatic events in the Central Andes of Peru. Journal of the Geological Society, v. 140, 279-286.

Bussel, M.A., and Pitcher, W.S. (1985): The structural control of batholith emplacement, *in* Magmatism at a Plate Edge: The Peruvian Andes, (Eds.: Pitcher, M.P., Cobbing, E.J., and Beckinsale, R.D.), London, Blackie Halsted Press, 167-176.

Caldas, J. (1978): Geología de los cuadrángulos de San Juan, Acari y Yauca. Boletín del Instituto Geológico Minero y Metalúrgico, (A) 30, 78 p., Lima.

Carlotto, V. (1989): Formación Huambutío: Nueva unidad estratigráfica, marcador del evento tectónico-sedimentario infraneocomiano. Workshop sobre el Cretáceo en el Perú, 1st, Lima, 5 p.

Carlotto, V., Jurado, J., Miranda, C., Carlier, G., and Jaillard, E. (1990): Evolución tectónica sedimentaria de la formación Yuncaypata (Aptiano superior?-Campaniano). Workshop sobre el Cretáceo en el Perú, 2nd, Lima, 4 p.

Chávez, A. (1982): Estratigrafía y tectónica del sector de Chapi. Thesis, Universidad Nacional de San Agustín, Arequipa, 101 pp.

Clark, A.H., Farrar, E., Kontak, D.J., Langridge, R.J., Arenas, M.J., France, L.J., McBride, S.L., Woodman, P.L., Wasteneys, H.A., Sandeman, H.A., and Douglas, D.A. (1990): Geologic and geochronologic constraints on the metallogenic evolution of the Andes of Southeastern Peru. Economic Geology, v. 85, 1520-1583.

Cobbing, E.J., Pitcher, W.S., Wilson, J.J., Baldock, J.W., Taylor, W.P., McCourt, W., and Snelling, N.J. (1981): The geology of the Western Cordillera of Northern Peru. London, Institute of Geological Sciences, Overseas Memoir, 5, 143 pp.

Córdova, A. (1986): Estudio estratigráfico y sedimentológico de las rocas del Cretáceo medio y superior en el Pongo de Rentema, Rio Marañón, departamento de Amazonas. Thesis, Universidad Nacional de San Marcos, Lima, 105 pp.

Córdova, E. (1986): Un bassin intramontagneux andin péruvien. Les Couches Rouges du bassin de Cuzco (Maastrichtien-Paléocène). Thesis Third Cycle, University of Pau, Pau, 272 pp.

Cross, T.A., and Pilger, R.H. (1982): Controls of subduction, location of magmatic arcs, and tectonics of arc and back-arc regions. Geological Society of America Bulletin, v. 93, 545-562.

Dalmayrac, B. (1978): Géologie des Andes Péruviennes. Paris, Travaux et Documents de l'ORSTOM, 93, 161 pp.

Dalziel, I.W.D., De Witt, M.J., and Palmer, K.F. (1974): A fossil marginal basin in the Southern Andes. Nature, v. 250, 291-294.

Dávila, J.J., and Ponce de León, V. (1971): La sección del río Inambari en la faja subandina del Perú, y la presencia de sedimentitas de la Formación Cancañiri (Zapla) del Silúrico. Revista Técnica de Yacimientos Petrolíferos Fiscales Bolivianos, v. 1, 67-85, La Paz.

De Jong, K.A. (1974): Mélange (Olistostrome) near the Titicaca lake, Peru. American Association of Petroleum Geologists Bulletin, v. 54, 729-741.

Ducloz, C., and Rivera, R. (1956): La Formación Chonta en la región del río Cahuapanas, Loreto. Boletín de la Sociedad Geológica del Perú, v. 30, 131-140, Lima.

Duncan, R.A., and Hargraves, R.B. (1984): Plate tectonic evolution of the Caribbean region in the mantle reference frame, in The Caribbean-South America Plate Boundary and Regional Tectonics, (Eds.: Bonini, W.E., Hargraves, R.B., and Shagam, R.), Geological Society of America Memoir, 162, 81-93.

Ellison, R.A., Klinck, B.A., and Hawkins, M.P. (1989): Deformation events in the Andean orogenic cycle in the Altiplano and Western Cordillera, southern Peru. Journal of South American Earth Sciences, v. 2, 263-276.

Feininger, T., and Bristow, C.R. (1980): Cretaceous and Paleogene history of coastal Ecuador. Geologische Rundschau, v. 69, 849-874.

Feist, M., Laubacher, G., and Sigé, B. (1989): Tentative correlations of Late Cretaceous and Lower Tertiary Andean Red Beds. European Union of Geosciences Congress, 5th, Strasbourg.

Fyfe, D. (1963): Estudio estratigráfico del Pozo Yurimaguas 2-1, Loreto, Perú. Boletín de la Sociedad Geológica del Perú, v. 37, 27-36, Lima.

Garcia, W. (1978): Geología de los cuadrángulos de Puquina, Omate, Huaitire, Mazo Cruz y Pizacoma. Boletín del Instituto Geológico Minero y Metalúrgico, (A), 29, 33 p., Lima.

Gayet, M., Marshall, L.G., and Sempere, T. (1991): The Mesozoic and Paleocene Vertebrates of Bolivia and their stratigraphic context: a review. *in* Fósiles y facies de Bolivia, (Ed., Suárez, R.), Revista técnica de Yacimientos Petrolíferos Bolivianos, v.12, 393-433, La Paz.

Geyer, O.F. (1983): Obertithonische Ammoniten-Faunen von Peru. Zentralblatt für Geologie und Paläontologie, v. 1 (3/4), 335-350.

Goodlad, S.W., Martin, A.K., and Hartnady, C.J.H. (1982): Mesozoic magmatic anomalies in the southern Natal Valley. Nature, v. 295, 686-688.

Gordon, R.G., and Jurdy, D.M. (1986): Cenozoic global plate motions. Journal of Geophysical Research, v. 91, 12389-12406.

Grambast, L., Martinez, M., Mattauer, M., and Thaler, L. (1967): *Perutherium altiplanense* nov. gen., nov. sp., premier mammifère d'Amérique du Sud. Comptes Rendus à l'Académie des Sciences, (D), 264, 707-710.

Guevara, C. (1980): El Grupo Casma del Perú central entre Trujillo y Mala. Boletín de la Sociedad Geológica del Perú, v. 49, 25-52, Lima.

Hall, M.L, and Calle, J. (1982): Geochronological control for the main tectonic-magmatic events of Ecuador, *in* Magmatic Evolution of the Andes, (Eds.: Linares, E., Cordani, U.G., and Munizaga, F.), Earth-Science Reviews, v. 18, 215-239.

Haq, B.U., Hardenbol, J., and Vail, P.R. (1987): Chronology of fluctuating sea levels since the Triassic. Science, v. 235, 1156-1167.

Hosttas, E. (1967): Estudio geológico del túnel terminal entre Huambo y Querque. Thesis, Universidad Nacional de San Agustín, Arequipa, 107 pp.

Injoque, J. (1985): Geochemistry of the Cu-Fe amphibole skarn deposits of the Peruvian central coast. Ph.D. Thesis, University of Nottingham, 392 pp.

Jaillard, E. (1987): Sedimentary evolution of an active margin during Middle and upper Cretaceous times: the North Peruvian margin from Late Aptian up to Senonian. Geologische Rundschau, v. 76, 677-697.

Jaillard, E., and Jacay, J. (1989): Les "couches Chicama" du Nord du Pérou: Colmatage d'un bassin né d'une collision oblique au Tithonique. Comptes Rendus à l'Académie des Sciences, (II), 308, 1459-1465.

Jaillard, E., and Sempéré, T. (1989): Cretaceous sequence stratigraphy of Peru and Bolivia, *in* Contribuciones de los Simposios sobre el Cretácico de América Latina, A: Eventos y Registro Sedimentario, (Ed.: Spalletti, L.A.), GSGP-IGCP 242, 1-27, Buenos Aires.

Jaillard, E., Soler, P., Carlier, G., and Mourier, T. (1990): Geodynamic evolution of the northern and central Andes during Early to Middle Mesozoic times: A Tethyan model. Journal of the Geological Society, v. 147, 1009-1022.

Jaillard, E., Cappetta, H., Ellenberger, P., Feist, M., Grambast-Fessard, N., Lefranc, J.-P., and Sigé, B. (1993): The Late Cretaceous Vilquechico Formation, Southern Peru. Sedimentology, Paleontology, Biostratigraphy, Correlations. Cretaceous Research, in press.

James, D.E., Brooks, C., and Cuyubamba, A. (1975): Early evolution of the Central Andean tectonic arc. Carnegie Institute Yearbook, v. 74, 247-250.

Jenks, W.F. (1961): Triassic to Tertiary stratigraphy near Cerro de Pasco, Peru. Geological Society of America Bulletin, v. 62, 203-219.

Kalafatovitch, C. (1957): Edad de las calizas de la Formación Yuncaypata, Cuzco. Boletín de la Sociedad Geológica del Perú, v. 32, 125-139, Lima.

Klinck, B.A., Ellison, R.A., and Hawkins, M.P. (1986): The geology of the Cordillera Occidental and Altiplano, West of the Lake Titicaca, Southern Peru. Lima, Instituto Geológico Minero y Metalúrgico, Preliminary report, 353 pp.

Klitgord, K.D., and Schouten, H. (1986): Plate kinematics of the Central Atlantic, *in* The Western North Atlantic Regions, (Eds.: Vogt, P.R., and Tucholke, B.E.), Geological Society of America Memoir, 351-378.

Koch, E., and Blissenbach, E. (1962): Las Capas Rojas del Cretáceo superior-Terciario en la región del curso medio del río Ucayali, Oriente del Perú. Boletín de la Sociedad Geológica del Perú, v. 39, 7-141, Lima.

Kummel, B. (1948): Geological reconnaissance of the Contamana region, Peru. Geological Society of America Bulletin, v. 59, 1217-1266.

Laubacher, G. (1978): Géologie de la Cordillère orientale et de l'Altiplano au Nord et Nord-Ouest du Lac Titicaca (Pérou). Paris, Travaux et Documents de l'ORSTOM, 95, 217 pp.

Laughlin, A.W., Damon, P.E., and Watson, B.N. (1968): Potassium-Argon dates from Toquepala and Michiquillay, Peru. Economic Geology, v. 63, 166-168.

Laurent, H. (1985): El Pre-Cretáceo en el Oriente peruano: su distribución y sus rasgos estructurales. Boletín de la Sociedad Geológica del Perú, v. 74, 33-59, Lima.

Lebrat, M., Mégard, F., Dupuy, C., and Dostal, J. (1987): Geochemistry and tectonic setting of pre-collision Cretaceous and Paleogene volcanic rocks of Ecuador. Geological Society of America Bulletin, v. 99, 569-578.

Lehner, P., and De Ruiter, P.A.C. (1977): Structural history of the Atlantic margin of Africa. American Association of Petroleum Geologists Bulletin, v. 61, 961-981.

Litherland, M., and Aspden, J. (1990): Evidence for pre-Cretaceous collision in the Ecuadorian Andes. Symposium International Géodynamique Andine, Grenoble, Ed. ORSTOM, Colloques et Séminaires, 199-202.

López, R., and Córdova E. (1988): Estratigrafía y sedimentación de la serie continental "Capas Rojas" (Maastrichtiano-Paleoceno) entre Cuzco y Ccorao. Boletín de la Sociedad Geológica del Perú, v. 78, 149-164, Lima.

Loughmann, D.L., and Hallam, A. (1982): A facies analysis of the Pucara group (Norian to Toarcian carbonates, organic-rich shales and phosphates) of Central and Northern Peru. Sedimentary Geology, v. 32, 161-194.

Loza, W. (1987): Aspectos litoestratigráficos mesozoicos del sector Sumbay-Arequipa. Thesis, Universidad Nacional de San Agustín, Arequipa, 138 pp.

Mabire, B. (1961): La "Série de Capas Rojas" Cretáceo-Terciarias en los Andes centrales del Perú. Boletín de la Sociedad Geológica del Perú, v. 36, 151-185, Lima.

Marocco, R. (1975): Geología de los cuadrángulos de Andahuaylas, Abancay y Cotabambas. Boletín del Instituto Geológico Minero y Metalúrgico, 27, 51 pp., Lima.

Marocco, R., and Del Pino, M. (1966): Geología del cuadrángulo de Ichuna. Boletín del Instituto Geológico Minero y Metalúrgico, 14, 57 pp., Lima.

Marocco, R., and Noblet, C. (1990): Sedimentation, tectonism and volcanism relationships in two Andean basins of southern Peru. Geologische Rundschau, v. 79, 111-120.

Marquillas, R.A., and Salfity, J.A. (1988): Tectonic framework and correlations of the Cretaceous-Eocene Salta Group, Argentina, *in* The Southern Central Andes, (Eds.: Bahlburg, H., Breitkreutz, Ch., and Giese, P.), Berlin-Heidelberg, Springer-Verlag, Lecture Notes in Earth Sciences, v. 17, 119-136.

Marsaglia, K.M., and Carozzi, A.V. (1990): Depositional environment, sand provenance, and diagenesis or the Basal Salina Formation (lower Eocene), northwestern Peru. Journal of South American Earth Sciences, v. 3, 253-267.

Mayes, C.L., Lawyer, L.A., and Sandwell, D.T. (1990): Tectonic history and New Isochron Chart of the South Pacific. Journal of Geophysical Research, v. 95, 8543-8567.

Mégard, F. (1978): Etude géologique des Andes du Pérou central. Paris, Mémoire de l'ORSTOM, 86, 310 pp.

Mégard, F. (1984): The Andean orogenic period and its major structures in Central and Northern Peru. Journal of the Geological Society, v. 141, 893-900.

Mégard, F. (1987): Cordilleran and marginal Andes: a review of Andean geology North of the Arica elbow (18° S), *in* Circum-Pacific belts and evolution of the Pacific ocean basin, (Eds.: Monger, J.W.H., and Francheteau, J.), American Geophysical Union, Geodynamic Series, v. 18, 71-95.

Miall, A.D. (1984): Principles of Sedimentary Basin Analysis. Berlin, Springer-Verlag, 490 pp.

Mooney, W.D. (1980): An East-Caribbean ridge during the Jurassic and the Cretaceous, and the Western Colombia, *in* The origin of the Gulf of Mexico and the opening of the Central North Atlantic Ocean, (Ed.: Pilger, R.J.), Louisiana State University, 55-73, Baton Rouge.

Morris, R.C., and Alemán, A.R. (1975): Sedimentation and tectonics of Middle Cretaceous Copa Sombrero formation in Northwest Peru. Boletín de la Sociedad Geológica del Perú, v. 48, 49-64, Lima.

Moulin, N. (1989): Facies et séquences de dépôt de la plate-forme du Jurassique moyen à l'Albien, et une coupe structurale des du Pérou central. Thesis, University of Montpellier, Montpellier, 287 pp.

Moulin, N., and Séguret, M. (1990): Rôle des variations du niveau marin dans la sédimentation fluvio-deltaïque du Néocomien des Andes du Pérou central (Formation Goyllarisquizga). Bulletin de la Société géologique de France, in press.

Mourier, T. (1988): La transition entre Andes marginales et Andes cordillèraines à ophiolites. Evolution sédimentaire, magmatique et structurale du relai de Huancabamba (3°-8°S, Nord Pérou-Sud Equateur). Dr Thesis, University of Paris XI, Paris, 275 pp.

Mourier, T., Mégard, F., Pardo, A., and Reyes, L. (1988a): L'évolution mésozoique des Andes de Huancabamba (3°-8°S) et l'hypothèse de l'accrétion du microcontinent Amotape-Tahuin. Bulletin de la Société géologique de France, (8), 4, 69-79.

Mourier, T., Bengtson, P., Bonhomme, M., Buge, E., Cappetta, H., Crochet, J.-Y., Feist, M., Hirsch, K.F., Jaillard, E., Laubacher, G., Lefranc, J.-P., Moullade, M., Noblet, C., Pons, D., Rey, J., Sigé, B., Tambareau, Y., and Taquet, P. (1988b): The Upper Cretaceous-Lower Tertiary marine to continental transition in the Bagua basin, Northern Peru. Paleontology, biostratigraphy, radiometry, correlations. Newsletters on Stratigraphy, v. 19, 143-177.

Mukasa, S.B. (1986): Zircon U-Pb ages of superunits in the Coastal Boatholith of Peru: Implications for magmatic and tectonic processes. Geological Society of America Bulletin, v. 97, 241-254.

Myers, J.S. (1974): Cretaceous stratigraphy and structure, Western Andes of Peru between latitudes 10°-10° 30'. American Association of Petroleum Geologists Bulletin, v. 58, 474-487.

Myers, J.S. (1975): Vertical crustal movements of the Andes in Peru. Nature, v. 254, 672-674.

Myers, J.S. (1980): Geología de los cuadrángulos de Huarmey y Huayllapampa. Boletín del Instituto Geológico Minero y Metalúrgico, (A), 33, 145 pp., Lima.

Naeser, C.W., Crochet, J.-Y., Jaillard, E., Laubacher, G., Mourier, T., and Sigé, B. (1991): Tertiary Fission-Track ages from the Bagua syncline (Northern Peru). Stratigraphic and tectonic implications. Journal of South American Earth Sciences, v. 4, 61-72.

Newell, N.D. (1949): Geology of the Titicaca region, Peru and Bolivia. Geological Society of America Memoir, 36, 111 pp.

Noble, D.C., McKee, E.H., Mourier, T., and Mégard, F. (1990): Cenozoic stratigraphy, magmatic activity, compressive deformation, and uplift in Northern Peru. Geological Society of America Bulletin, v. 102, 1105-1113.

Noblet, C. (1985): Analyse sédimentologique des Couches Rouges sud-péruviennes. ORSTOM-University of Pau, 50 pp., unpublished report.

Noblet, C., Marocco, R., and Delfaud, J. (1987): Analyse sédimentologique des "Couches Rouges" du bassin intramontagneux de Sicuani (Sud du Pérou). Bulletin de l'Institut Français d'Études Andines, v. 16, 55-78, Lima.

Nur, A., and Ben-Avraham, Z. (1982): Displaced Terranes and Mountain Building, *in* Mountain Building Processes, (Ed.: Hsü, K.J.), London, Academic Press, 73-84.

Odin, G.S., and Odin, C. (1990): Echelle numérique des temps géologiques. Mise à jour 1990. Géochronique, v. 35, 12-21.

Ojeda, H.A.O. (1982): Structural Framework, Stratigraphy, and Evolution of the Brazilian marginal Basins. American Association of Petroleum Geologists Bulletin, v. 66, 732-749.

Olchauski, L.E. (1980): Geología de los cuadrángulos de Jaqui, Coracora, Chala y Chaparra. Boletín del Instituto Geológico Minero y Metalúrgico, (A), 34, 71 p., Lima.

Olivet, J.-L., Bonnin, J., Beuzart, P., and Auzende, J.-M. (1984): Cinématique de l'Atlantique Nord et Central. Brest, Publication du Centre National pour l'Exploitation des Océans, Scientific and Technical Report, 54, 108 pp.

Olsson, A.A. (1944): Contributions to the Paleontology of Northern Peru. Part VII: The Cretaceous of the Paita region. Bulletins of American Paleontology, v. 28, 113 pp.

Pardo, A., and Zúñiga, F. (1976): Estratigrafía y evolución tectónica de la región de la Selva del Perú. Congreso Latinoamericano de Geología, 2nd, Caracas (1973), Memoria, II, Caracas, Dirección de Geología, Boletín de Geología, Publicación Especial, 7, 569-608.

Pardo-Casas, F., and Molnar, P. (1987): Relative motion of the Nazca (Farallón) and South America plates since Late Cretaceous times. Tectonics, v. 6, 233-248.

Paredes, J. (1982): Los yacimientos tipo "Hualgayoc" del sector norandino. BRGM-CEDIMIN, 15 pp., unpublished report, Lima.

Parize, O., Beaudouin, B., Fries, G., Pinault, M., and Pinoteau, B. (1989): Les filons gréseux sédimentaires: Des témoins privilégiés de la dynamique de remplissage des bassins sédimentaires. Notes et Mémoires, TOTAL CFP, 21, 221-233, Paris.

Pecho, V. (1981): Geología de los cuadrángulos de Chalhuanca, Antabamba y Santo Tomas. Boletín del Instituto Geológico Minero y Metalúrgico, (A), 35, 67 pp., Lima.

Pecho, V. (1983): Geología de los cuadrángulos de Pausa y Caravelli. Boletín del Instituto Geológico Minero y Metalúrgico, (A) 37, 125 pp., Lima.

Perez, V. (1981): Geología del sector de Pampacolca-Tipán-Unon. Prov. Castilla, Dpto. Arequipa. Thesis, Universidad Nacional de San Agustín, Arequipa, 129 pp.

Petroperú, (1989): Peruvian Petroleum, a renewed exploration opportunity. Lima, Petroperú S.A. Ed., 101 pp.

Pilger, R.H.Jr. (1984): Cenozoic plate kinematics, subduction and magmatism. Journal of the Geological Society, v. 141, 793-802.

Pitcher, M.P., Cobbing, E.J., and Beckinsale, R.D., Eds. (1985): Magmatism at a Plate Edge: The Peruvian Andes. London, Blackie Halsted Press, 323 pp.

Portugal, J. (1974): Mesozoic and Cenozoic Stratigraphy and Tectonic events of Puno-Santa Lucia Area, Department of Puno, Peru. American Association of Petroleum Geologists Bulletin, v. 58, 982-999.

Rabinowitz P.D., and La Brecque, J. (1979): The Mesozoic South Atlantic Ocean and evolution of its continental margins. Journal of Geophysical Research, v. 84, 5973-6002.

Raymond, L.A., and Swanson, S.E. (1980): Accretion and episodic plutonism. Nature, v. 285, 317-319.

Reyes, L., and Caldas, J. (1987): Geología de los cuadrángulos de Las Playas, La Tina, Las Lomas, Ayabaca, San Antonio, Chulucanas, Morropon, Huancabamba, Olmos y Pomahuaca. Boletín del Instituto Geológico Minero y Metalúrgico, (A), 39, 83 pp., Lima.

Riccardi, A.C. (1988): The Cretaceous System of Southern South America. Geological Society of America Memoir, 168, 161 pp.

Rivera, R. (1979): Zonas faunísticas del Cretáceo de Lima. Boletín de la Sociedad Geológica del Perú, v. 62, 19-23, Lima.

Rivera, R., Petersen, G., and Rivera, M. (1975): Estratigrafía de la Costa de Lima. Boletín de la Sociedad Geológica del Perú, v. 45, 159-196, Lima.

Rodríguez, A., and Chalco, A. (1975): Cuenca Huallaga, Reseña geológica y posibilidades petrolíferas. Boletín de la Sociedad Geológica del Perú, v. 45, 187-212, Lima.

Romani, M. (1982): Géologie de la région minière Uchucchacua-Hacienda Otuto, Pérou. Third Cycle Thesis, University of Grenoble, Grenoble, 176 pp.

Romeuf, N. (1990): Magmatisme et métamorphisme des coulées et des intrusifs jurassiques de la région côtière du Sud du Pérou. Diplôme d'Etudes Approfondies, 63 pp., unpublished, Marseille.

Rüegg, W. (1961): Hallazgo y posición estratigráfico-tectónica del Titoniano en la costa sur del Perú. Boletín de la Sociedad Geológica del Perú, v. 36, 203-208, Lima.

Savostin, L.A., Sibuet, J.-C., Zonenshain, L.P., Le Pichon, X., and Roulet, M.-J. (1986): Kinematic evolution of the Tethys belt from the Atlantic ocean to the Pamir since the Triassic. Tectonophysics, v. 123, 1-35.

Scotese, C.R., Gahagan, L.M., and Larson, R.L. (1988): Plate tectonic reconstructions of the Cretaceous and Cenozoic ocean basins, *in* Mesozoic and Cenozoic Plate Reconstructions, (Eds.: Scotese, C.R., and Sager, W.W.), Tectonophysics, v. 155, 27-48.

Seminario, F., and Guizado, J. (1976): Síntesis bioestratigráfica de la región de la Selva del Perú. Congreso Latinoamericano de Geología, 2nd, Caracas (1973), Memoria, II, Caracas, Dirección de Geología, Boletín de Geología, Publicación Especial, 7, 881-898.

Sempéré, T., Oller, J., and Barrios, L. (1988): Evolución tectosedimentaria de Bolivia durante el Cretácico. Congreso Geológico Chileno, 5th, Santiago, Actas, 3 (H), 37-65.

Séranne, M. (1987): Evolution tectono-sédimentaire du bassin de Talara (nord-ouest du Pérou). Bulletin de l'Institut Français d'Études Andines, v. 16, 103-125, Lima.

Sibuet, J.C., Hay, W.V., Prunier, A., Montadert, L., Hinz, K., and Fritsch, J. (1985): Early evolution of the South Atlantic Ocean: role of the Rifting episode. Initial Report DSDP, U.S. Government Printed Office, 75, 469-481, Washington.

Sigé B. (1972): La faunule de mammifères du Crétacé supérieur de Laguna Umayo (Andes péruviennes). Bulletin du Museum National d'Histoire Naturelle, (3), 99, 375-405, Paris.

Soler, P. (1989): Petrography and geochemistry of lower Cretaceous alkali basalts from the High Plateaus of central Peru and their tectonic significance. Zentralblatt für Geologie und Paläontologie, 1989, 1053-1064.

Soler, P. (1991a): El volcanismo Casma: cuenca marginal abortada o arco en ambiente distensivo?. Congreso Peruano de Geología, 7th, Lima, Anales, 2, 659-663.

Soler, P. (1991b): Contribution à l'étude du magmatisme associé aux zones de subduction. Pétrographie, géochimie et géochimie isotopique des roches intrusives sur un transect des Andes du Pérou Central. Implications géodynamiques et métallogeniques. Dr Thesis, University of Paris VI, Paris, 950 pp.

Soler, P., and Bonhomme, M. (1987): Données radiochronologiques K-Ar sur les granitoïdes de la Cordillère Orientale des Andes du Pérou central. Implications tectoniques. Comptes Rendus à l'Académie des Sciences, (II), 304, 841-845.

Soler, P., and Bonhomme, M. (1990): Relations of magmatic activity to plate dynamics in Central Peru from Late Cretaceous to Present, *in* Plutonism from Antarctica to Alaska, (Eds.: Kay, S., and Rapela, C.), Geological Society of America Memoir, 241, 173-192.

Soler, P., Carlier, G., and Marocco, R. (1989): Evidence for the subduction and underplating of an oceanic plateau beneath the south Peruvian margin during the Late Cretaceous: structural implications. Tectonophysics, v. 163, 13-24.

Soto, V. (1979): Facies y ambientes deposicionales cretácicos, área Centro-Sur de la cuenca Marañon. Boletín de la Sociedad Geológica del Perú, v. 60, 233-250, Lima.

Steinmann, G. (1929): Geologie von Peru. Heidelberg, Karl Winter Ed., 448 pp.

Stewart, J.W., Evernden, J.F., and Snelling, N.J. (1974): Age determinations from Andean Peru: A reconnaissance survey. Geological Society of America Bulletin, v. 85, 1107-1116.

Stipanicic, P.N., and Rodrigo, F. (1969): El diastrofismo jurásico en Argentina y Chile. Jornadas Geológicas Argentinas, 4th, Mendoza, Actas, 2, 353-368.

Tschopp, H.J. (1953): Oil explorations in the Oriente of Ecuador. American Association of Petroleum Geologists Bulletin, v. 37, 2303-2347.

Uyeda, S. (1982): Subduction zones: an introduction to comparative subductology. Tectonophysics, v. 81, 133-159.

Uyeda, S., and Kanamori, H. (1979): Back-arc opening and the mode of subduction. Journal of Geophysical Research, v. 84, B3, 1049-1061.

Vargas, J.M. (1988): Potencial petrolífero de la cuenca Huallaga, Oriente peruano. Simposio Bolivariano Exploración Petrolera de las Cuencas Subandinas, 3rd, Caracas, I, 195-225.

Vatin-Pérignon, N., Vivier, G., Sébrier, M., and Fornari, M. (1982): Les derniers événements andins marqués par le volcanisme cénozoïque de la cordillère occidentale sud-péruvienne et de son piémont pacifique entre 15°45 et 18° S. Bulletin de la Société géologique de France, (7), 24, 649-650.

Vicente, J.-C. (1981): Elementos de la estratigrafía mesozoica sur-peruana, *in* Cuencas Sedimentarias del Jurásico y Cretácico de América del Sur, (Eds.: Volkheimer, W., and Musacchio, E.), 1, 319-351, Buenos Aires.

Vicente, J.-C. (1989): Early Late Cretaceous overthrusting in the Western Cordillera of Southern Peru, *in* Geology of the Andes and its relations to energy and mineral resources, Santiago (1985), (Eds.: Ericksen, G.E., Cañas Pinochet, M.T., and Reinemund, J.A.), Circum-Pacific Council for Energy and Mineral Resources, Earth Science Series, v. 11, 91-117.

Vicente, J.-C., Sequeiros, F., Valdivia, M.A., and Zavala, J. (1979): El sobre-escurrimiento de Cincha-Lluta, elemento del accidente mayor andino al Nor-Oeste de Arequipa. Boletín de la Sociedad Geológica del Perú, v. 61, 67-99. Lima.

Vicente, J.-C., Beaudouin , B., Chávez, A., and León, I. (1982): La cuenca de Arequipa (Sur Peru) durante el Jurásico-Cretácico inferior. Congreso Latinoamericano de Geología, 5th, Buenos Aires, I, 121-153.

Vidal, C.E., Injoque-Espinoza, J., Sidder, G.B., and Mukasa, S.B. (1990): Amphibolitic Cu-Fe skarn deposits in the Central Coast of Peru. Economic Geology, v. 85, 1447-1461.

Wallrabe-Adams, H.-J. (1990): Petrology and geotectonic development of the Western Ecuadorian Andes: the Basic Igneous Complex. Tectonophysics, v. 85, 163-182.

Webb, S. (1976): The volcanic envelope of the coastal Batholith in Lima and Ancash, Peru. Ph.D. Thesis, University of Liverpool, Liverpool.

Wiedmann, J. (1981): El límite Jurásico-Cretácico: Problemas y soluciones. Congreso Argentino de Paleontología y Bioestratigrafía, 2nd, and Congreso Latinoamericano de Paleontología, 1st, Buenos Aires (1978), Actas, 5, Simposio Límite Jurásico-Cretácico, 103-120.

Wilson, J.J. (1963): Cretaceous stratigraphy of Central Andes of Peru. American Association of Petroleum Geologists Bulletin, v. 47, 1-34.

Wilson, J.J., Reyes, L., and Garayar, J. (1967): Geología de los cuadrángulos de Mollebamba, Tayabamba, Huaylas, Pomabamba, Carhuaz y Huari. Boletín del Instituto Geológico Minero y Metalúrgico, 16, 95 pp., Lima.

Wilson, P.A. (1975): K-Ar age studies in Peru with special reference to the emplacement of the Coastal Batholith. Ph.D. Thesis, University of Liverpool, Liverpool.

Winslow, M.A. (1983): Clastic dikes swarms and the structural evolution of the foreland fold and thrust belt of the Southern Andes. Geological Society of America Bulletin, v. 94, 1073-1080.

Zúñiga, F., and Cruzado, C. (1979): Bioestratigrafía del Noroeste Peruano. Boletín de la Sociedad Geológica del Perú, v. 60, 219-232, Lima.

Kimmeridgian? to Paleocene Tectonic Evolution of Bolivia

T. Sempere

Orstom (UR 1H), Convenio YPFB-Orstom, Santa Cruz, Bolivia; and CGGM, Ecole des Mines, 77305 Fontainebleau Cedex, France.

Present address: Consulting geologist, 13 rue Geoffroy l'Angevin, 75004 Paris, France.

Abstract

The Puca Group (Kimmeridgian?-Paleocene) of Bolivia recorded the external (distal) tectonic evolution of the Andean back-arc basin in these latitudes. Bolivia had been part of stable cratonic South America until Late Jurassic time, when it was captured by the Andean system during a large-scale extensional "Araucan"-age tectonic event. This episode seems to be related to the onset of large-scale extensional and transtensional conditions in northern Chile and coastal Peru. In Bolivia, it led to initiation of the tectonically-controlled, highly fragmented Potosí basin filled with unfossiliferous continental siliciclastic deposits (mostly red beds), with relation to the reactivation of the major transversal Khenayani-Turuchipa paleostructural corridor. Extension was locally accompanied by basic volcanism, and created a tilted-block structure, with half-grabens showing topographic downwarps and uplifts upon which younger fine-grained strata onlapped. The oldest and most important extensional episode took place during deposition of the lowermost part of the Puca Group (Condo conglomerates). A younger minor extensional episode developed locally, possibly in Late Neocomian and/or Aptian time. Albian? time saw a large-scale onlap of brown to violet-red mudstones over the previous deposits and, locally, on the Paleozoic basement, indicating a relative change in the tectonic setting, marked by a slow and gentle widening of the sedimentation area without any small-scale extensional manifestations.

Shallow-marine carbonates were deposited during the Cenomanian-Turonian interval, in an area of lesser extension than the Kimmeridgian?-Albian? basin. In this unit, thickness variations are very gentle, sedimentation rates seem very low and no direct indication of synsedimentary tectonics is known. It is thus assumed that this transgression was mainly of global-eustatic origin.

The Senonian-Paleocene sequence consists of a thick pile of mainly mudrocks beds. It was deposited in the external part of the wide, underfilled foreland basin of the paleo-Andes, which at that time were of small size and had produced only a minor flexure of the South American lithosphere. The base of the sequence records a noteworthy exten-

sional episode, accompanied by relatively widespread basaltic volcanism, which indicates coeval thickening of the asthenospheric wedge and thus an increase of the subduction angle, at least between lat 17° and 27°S. This tectonic upheaval marked the real onset of the compressional building of the Andes from the west. In Bolivia, it reactivated Late Jurassic-Early Cretaceous tensional structures and produced new normal faults. Coniacian playa-lakes formed in the low areas of the fragmented topography induced by this episode. Santonian-Campanian deposits (red mudstones and minor sandstones and evaporites, with two thin restricted-marine carbonate intercalations) show a clear and general onlap toward all external areas, which is interpreted as the distal foreland effect of the "Peruvian"-age deformation developing in the coastal regions at that time. Decrease of sedimentation rate and rapid progradation of immature sand from the west in the Middle and Late Campanian suggest relative coeval tectonic quiescence on the Pacific margin. Increase of sedimentation rate, along with arrival of clean sand from the external areas, near the Campanian-Maastrichtian boundary is interpreted to mark a notable reactivation of thrusting in the coastal regions, accompanied by uplift of a foreland forebulge. This episode is coeval with an important transgression from the north or northwest, which led to deposition of restricted-marine marl and carbonate facies during Maastrichtian and earliest Paleocene time. Lacustrine and alluvial strata were deposited during most of Paleocene time. No major tectonic event is perceptible during the Maastrichtian-Paleocene interval, but fairly steady subsidence suggests that tectonic loading and hence thrusting was going on in the coastal regions.
A major modification of the regional tectonic system occurred near the Paleocene-Eocene boundary and generated a widespread unconformity in Andean Bolivia.

Introduction

The mainly siliciclastic strata that accumulated in Bolivia between Kimmeridgian? and Paleocene times provide a distal (eastern) record of the tectonic evolution of the central Andean Pacific margin for this time span. Because in the Bolivian orocline a Late Oligocene-Recent deformation propagated considerably toward the east and northeast, through what had been the easternmost part of the Cretaceous Andean depositional area ("eastern domain" of Jaillard and Sempere, 1989; Sempere et al., 1989), the Altiplano and Cordillera Oriental of Bolivia provide numerous and excellent exposures of strata of that age (Figure 1), of better quality and easier access than the outcrops of coeval strata also deposited in that part of the eastern domain now located in central and northern Subandean Peru. Andean Bolivia is thus ideal for studying the distal (eastern) effects of the tectonic events which occurred in the coastal areas of Peru and northern Chile from latest Jurassic to Paleocene times.
Classic works on the "Cretaceous" Puca Group of Bolivia are by Lohmann and Branisa (1962), Russo and Rodrigo (1965), Kriz and Cherroni (1966), Lohmann (1970), Reyes

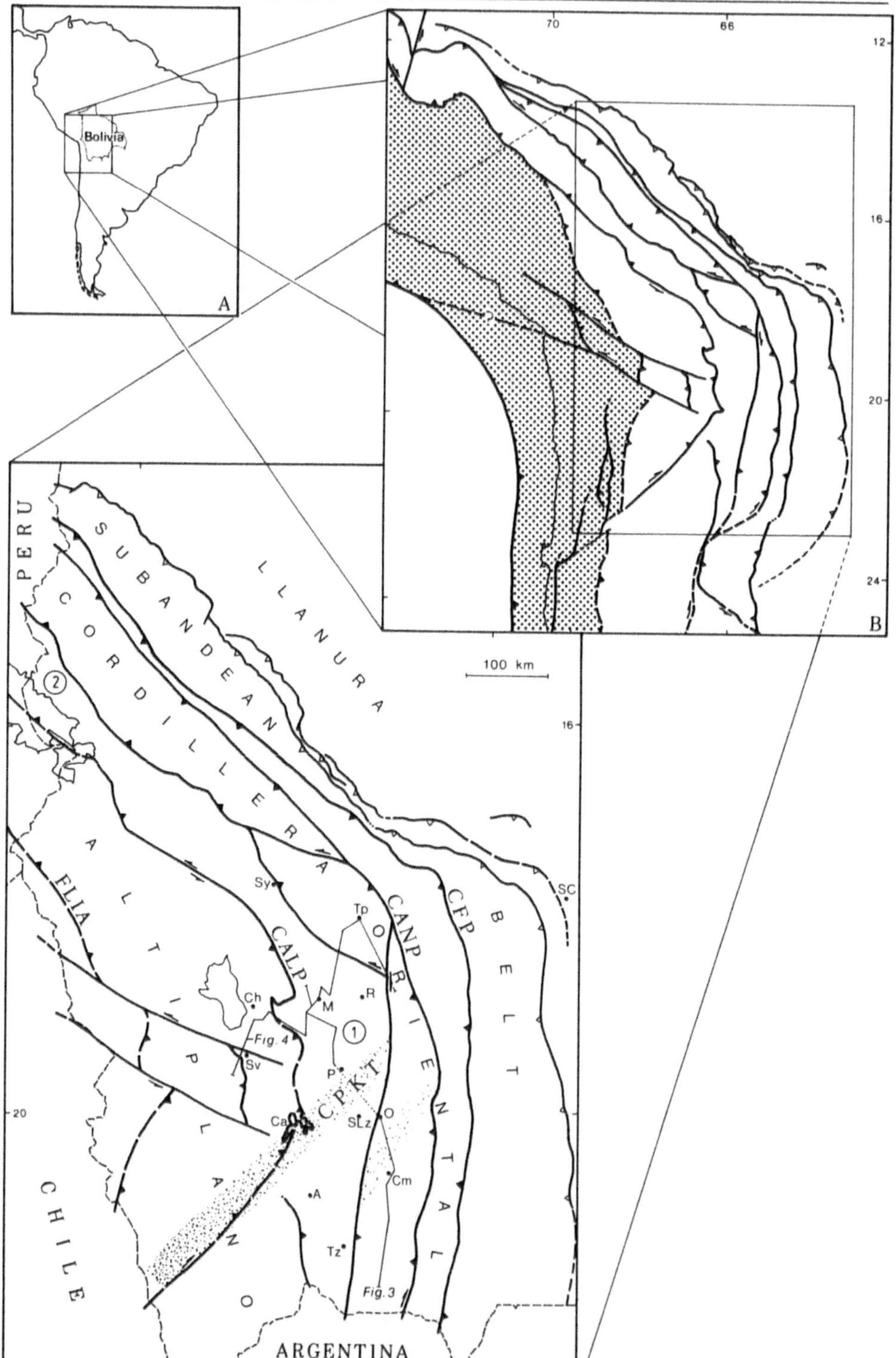
Bolivia
A
70
66
12
16
20
24
B
PERU
SUBANDEAN
CORDILLERA
LLANURA
ALTIPLANO
ORIENTAL
BELT
ELIA
CALP
CANP
CFP
CPKT
SC
Sy
Tp
Ch
M
R
Fig. 4
Sv
P
Ca
SLz
O
Cm
A
Tz
Fig. 3
100 km
16
20
CHILE
ARGENTINA
68
64
C

(1972) and Cherroni (1977). More recently, resumption of field studies, by Yacimientos Petrolíferos Fiscales Bolivianos and Orstom, and use of sequential methods and paleo-current analysis have led to studies of the sequence-stratigraphy and tectonic issues of the Puca Group (Sempere et al., 1988a; Jaillard and Sempere, 1989).

Paleostructural Setting

During most of the Silurian to Paleogene time lapse, Bolivia was located close to or at the southern tip of a longitudinal subsident domain which ran parallel to the Pacific margin of South America and generally deepened toward the north or northwest along its axis. This domain remained bounded southward by the Sierras Pampeanas area in northwestern Argentina during this time span. An important consequence of this geometry is that, at least from Silurian to Late Cretaceous times, the marine transgressions ingressed into Bolivia from the north or northwest. In the Late Cretaceous series of the eastern domain, more marine facies are generally to be found in a northerly direction.

The fact that the Devonian to Middle Jurassic stratigraphy of Bolivia closely resembles the coeval stratigraphy of the very wide Paraná basin of southern Brazil, eastern Paraguay, northwestern Uruguay and northeastern Argentina (Sempere, 1990) suggests that the Bolivian Cretaceous strata were deposited in a region which had previously been relatively stable and had behaved as a cratonic area (see Oller and Sempere, 1990).

Because thick siliciclastic deposits accumulated in only two regions of the Bolivian territory during the Kimmeridgian?-Early Cretaceous interval, two basins, -namely the Potosí and "Norte del Lago" basins- are usually distinguished (Sempere et al., 1988a). However, these areas were parts of the same depositional domain in subsequent times. The "Norte del Lago" basin is the southeastern tip of the Peruvian eastern Altiplano basin, to which it thus belongs tectonically (see Jaillard, this volume). The present study concentrates on the tectonic history of the Potosí basin, but it should be noted that both basins have had a parallel evolution in latest Jurassic-Early Cretaceous times, and were parts of the same basin throughout Late Cretaceous-Paleocene times.

Fig. 1

A -Location of Bolivia within South America. B - Structural sketch map of the Bolivian orocline area (after Sempere et al., 1990b, 1991); dotted region "western Andean belt" (Sempere et al., 1989), which includes the southern prolongations of Jaillard's (1991) Western Trough and Axial Swell. C - Traditional morphological provinces, main structural elements of the Bolivian Andes (after Sempere et al., 1990b, 1991) and localization of Figures 3 and 4. Main boundary faults: CFP- Main Frontal Thrust, CANP- Main Andean Thrust, CALP- Main Altiplanic Thrust, FLIA- Intra-Andean Boundary Fault. 1- Potosí basin, 2- "Norte del Lago" basin, CPKT- Khenayani-Turuchipa paleostructural corridor. Localities: A- Atocha, Ca- Calazaya, Ch- Challapata, Cm- Camargo, M- Macha, O- Otavi, P- Potosí, R- Ravelo, SC- Santa Cruz, SLz- San Lorenzo, Sv- Sevaruyo, Sy- Sayari, Tp- Tiupampa, Tz- Tupiza.

Only two regions of the present-day Subandean belt and Llanura received sedimentation during the interval studied; the sedimentation began in possibly Santonian time at the northwesternmost tip of the belt, and probably later (Campanian or Maastrichtian) in the Santa Cruz area. During Late Jurassic to Early Oligocene times, a large part of this wide domain was covered by silcrete-type paleosols affecting the Ichoa Formation of Jurassic age (Oller and Sempere, 1990; Sempere et al., 1990a). This rock unit consists of sandstones mostly of eolian origin. As will be documented below, the Ichoa and the silcretes which capped it were periodically gently uplifted when forebulges formed in response to shortening along the Pacific margin, and produced light-colored sands and clasts of silicified sandstone which were transported westward and northwestward into the basin.

The southern half of Andean Bolivia is crossed by a NE-striking major tectonic element, called the Khenayani-Turuchipa paleostructural corridor (CPKT; Figure 1), which appears to have largely controlled both Phanerozoic sedimentation and deformation in southwestern Bolivia (Sempere et al., 1991; see below). The Tupiza-San Lorenzo paleograben, the existence of which is inferred from spatial repartition of facies and thicknesses of some units (see below), appears to lie just west of the southern segment of the N-S-striking Aiquile-Tupiza fault (Sempere et al., 1987, 1988b), suggesting that both structural features are related and that the latter is an old tectonic element which was reactivated during some Cretaceous periods (see Figures 5 and 7 below).

The boundary faults of the tectonostratigraphic domains identified in Bolivia (Sempere et al., 1988b) have been included in the paleogeographic maps which illustrate this paper, because the domains they define were displaced and shortened, sometimes considerably, during the Andean orogeny (Sempere et al., 1990; Sempere et al., 1991). Thus two neighbouring points separated by one of these boundaries were generally farther apart at the time of Cretaceous sedimentation.

Stratigraphy

Figure 2 shows the stratigraphic units of the Puca Group presently used in Bolivia. Age attributions are from Jaillard and Sempere (1989, 1991) and Gayet et al. (1991), and are discussed in these papers. It must be mentioned that, unfortunately, due to some lack of experience, erroneous stratigraphic attributions sometimes occur in the literature (e.g., Martinez et al. (1990) for the Otavi syncline, and Okamoto et al. (1990) for the Camargo syncline).

Depending on locality, the Puca Group overlies sandy sedimentary rocks of fluvio-eolian origin and probable Middle Triassic-Jurassic age (Ravelo Formation; Oller and Sempere, 1990; Sempere, 1990), locally deformed Paleozoic strata (mostly Ordovician and Silurian, and locally Devonian to Permian), or the Precambrian basement (in the northwestern Altiplano; Lehmann, 1978). It must be noted that the thickness of the

Ravelo Formation may locally exceed 1000 m in the Andean domain, as at Sayari and Ravelo; conversely, this unit was been totally eroded before deposition of the Puca Group at many other locations.

When not in an erosion level, the Puca Group is normally overlain by the Early Eocene-age Cayara Formation, which marks the onset of the functioning of Andean Bolivia as a continental external "classic" foreland basin, probably resembling the present-day Chaco-Beni basin (Jaillard and Sempere, 1989; Sempere et al., 1989). The age of this unconformity is close to the Paleocene-Eocene boundary (Cirbián et al., 1986), and might be earliest Eocene (Sempere, 1990; Marshall and Sempere, 1991).

The "Cretaceous" stratigraphic units of Bolivia may be grouped, according to their characteristics, into three sets bounded by sedimentary discontinuities (Figures 2, 3 and 4):

-The "Puca A" set (Kimmeridgian?-Albian?) which is highly variable in thickness (0-1200 m), mostly consists of unfossiliferous red beds and local pink to yellowish sandstones, and displays many signs of synsedimentary tensional tectonics.

-The "Puca B" set (Cenomanian-Turonian) which shows a fairly regular thickness (20-30 m) inside the basin, and consists of marine fossiliferous limestones and subordinate marls.

-The "Puca C" set (Coniacian-Paleocene) which is formed by a thick sedimentary pile (200-1200 m) dominated by red mudstones, with some conspicuous units consisting of green marls and fossiliferous limestones, and evaporite bodies. The fauna listed for this set includes continental and marine taxa (Gayet et al., 1991, 1992; see below).

Strikingly enough, in map view, the Puca C basin contains the Puca A basin, which in turn contains the Puca B basin. The Puca C deposits show a clear onlap over areas previously devoid of Cretaceous sedimentation.

Kimmeridgian?-Albian?

The stratigraphy of the Puca A set (Kimmeridgian?-Albian?) is most complex, because of facies convergences and lateral changes, and because to date it has proved completely devoid of fossils. Bounded below by the unconformity at the base of the Puca Group, and above by the sharp lithologic discontinuity at the base of the Miraflores limestones, the Puca A set also shows two internal sedimentary discontinuities in some areas, such as northwest of Potosí. It is doubtful that the lower and upper boundaries of the Puca A are synchronous surfaces, because (1) it is likely that deposition extended diachronously on a substratum deformed by extensional tectonics of latest Jurassic-Early Cretaceous age, and (2) it is possible that the rapid transgression of the Miraflores Formation covered an area which was in part affected by a gentle erosional surface.

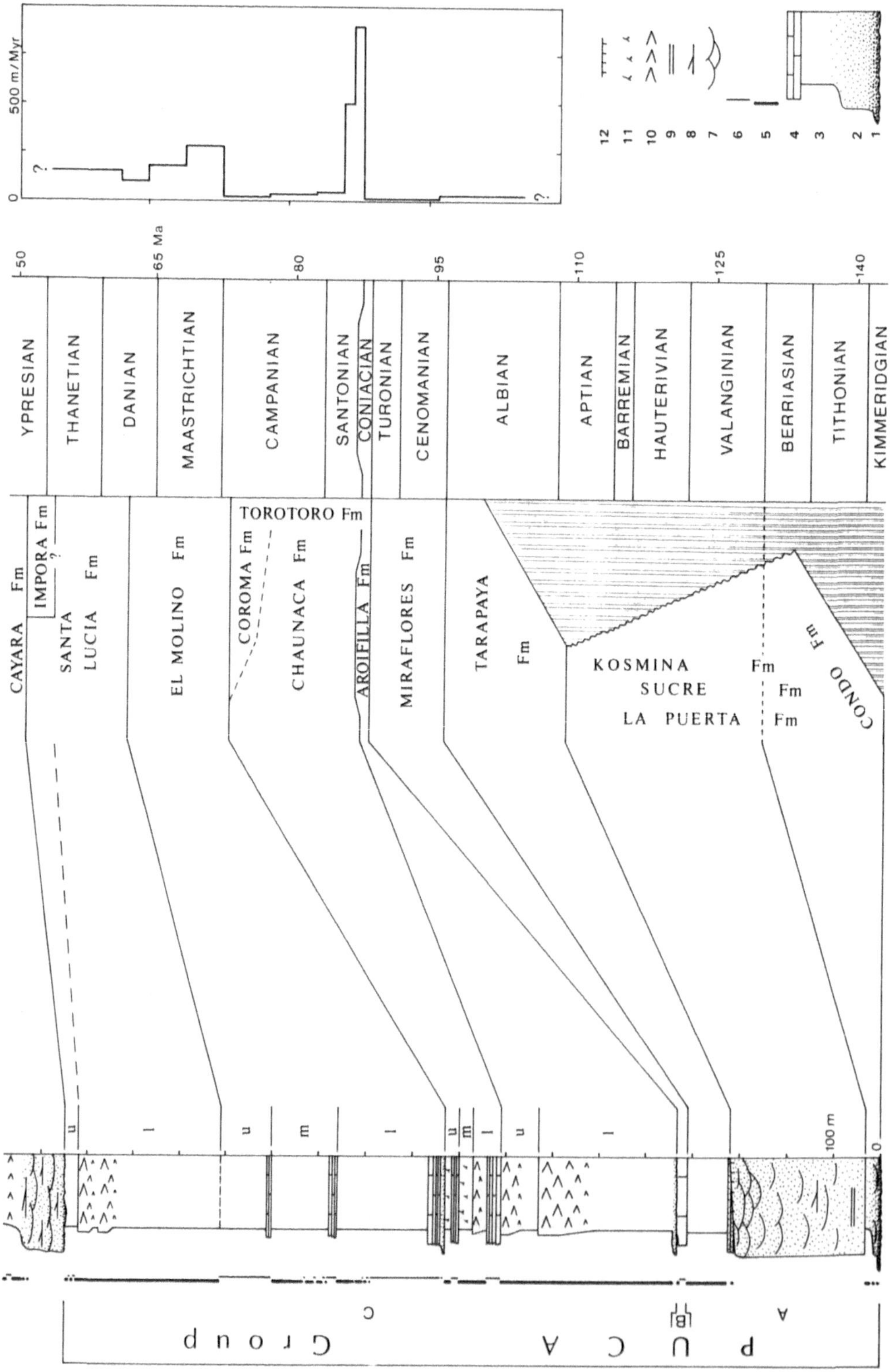
500 m / Myr
0
?
50
65 Ma
80
95
110
125
140
YPRESIAN
THANETIAN
DANIAN
MAASTRICHTIAN
CAMPANIAN
SANTONIAN
CONIACIAN
TURONIAN
CENOMANIAN
ALBIAN
APTIAN
BARREMIAN
HAUTERIVIAN
VALANGINIAN
BERRIASIAN
TITHONIAN
KIMMERIDGIAN
CAYARA Fm
IMPORA Fm
SANTA LUCIA Fm
EL MOLINO Fm
TOROTORO Fm
COROMA Fm
CHAUNACA Fm
AROIFILLA Fm
MIRAFLORES Fm
TARAPAYA Fm
KOSMINA Fm
SUCRE Fm
LA PUERTA Fm
CONDO Fm
100 m
P U C A Group
A
B
C
1
2
3
4
5
6
7
8
9
10
11
12

Fig. 2

◀ Type-section, stratigraphic chart and approximate sedimentation rates for the Potosí basin. Section is from the La Puerta locality near Potosí, where it overlies gently folded, thick, late Ordovician shales and subordinate quartzitic sandstones. Some lithologic types, particularly the limestone lithologies from the Chaunaca and El Molino Formations, are exaggerated for more clarity. Lithological symbols: 1- conglomerates (upward-fining), 2- sandstones (upward- fining), 3- predominantly mudrocks, 4- limestones, 5- red colour, 6- green (or grey) colour, 7- channels, 8- crossbedding, 9- sub- horizontal bedding in sandstones, 10- evaporites (mostly gypsum), 11- rooting, 12- oxidation surface. Stratigraphic abbreviations: Fm- Formation, l- lower, m- middle, u- upper. Chronological attributions modified from Jaillard and Sempere (1989) and Sempere (1990) by data published by Jaillard and Sempere (1991) and Gayet et al. (1991). See text for details about lateral changes of stratigraphic units and facies. Time scale from Odin and Odin (1990). Upper right diagram illustrates variations in time of approximate (maximum known) decompacted sedimentation rate.

Formation Review

The stratigraphic units presently identified within the Puca A set include the Condo, Kosmina, Sucre, La Puerta (s.s.) and Tarapaya formations. Definition of these units was mainly based on their facies. Stratigraphic relations between these units have been presented in detail elsewhere (Sempere et al., 1988a).

The Condo Formation consists of brown-red conglomerates and conglomeratic sandstones, which unconformably overlie -in locally angular fashion- Paleozoic or Mesozoic strata. The clasts are mostly subangular Paleozoic sandstone fragments; they average 1-5 cm in size, and some reach 20 cm. Facies indicate that the Condo Formation was deposited, depending on locality, in alluvial fan to proximal fluvial environments. It should be noted that the Condo conglomerates are not always present at the base of the Puca Group, which may begin either with sandstones or with fine-grained red beds.

The Kosmina Formation transitionally overlies the Condo Formation. It consists of intercalations of brownish to reddish mudstones, siltstones and generally subordinate sandstones, deposited in an alluvial (to lacustrine?) environment. The Condo and Kosmina Formations together form an upward fining and thinning succession.

The Sucre Formation consists of red, pink and whitish sandstones deposited in fluvial environments. The sandstones are commonly cross-bedded and frequently interbed with thin mudcracked red mudstone layers. The Sucre Formation appears to be a sandier lateral equivalent of the Kosmina Formation.

The La Puerta Formation (sensu stricto, i. e. as defined at the La Puerta locality close to Potosí and thus distinct from the Middle Triassic-Jurassic age Ravelo Formation, long mistaken for the La Puerta s.s.) mostly consists of yellowish cross-bedded sandstones, with no or very few mudstone intercalations. At La Puerta, thin conglomeratic sandstones, with subangular Paleozoic clasts, and red siltstones are present in the first 20 m of the formation and probably represent a lateral equivalent of the Condo and lower-

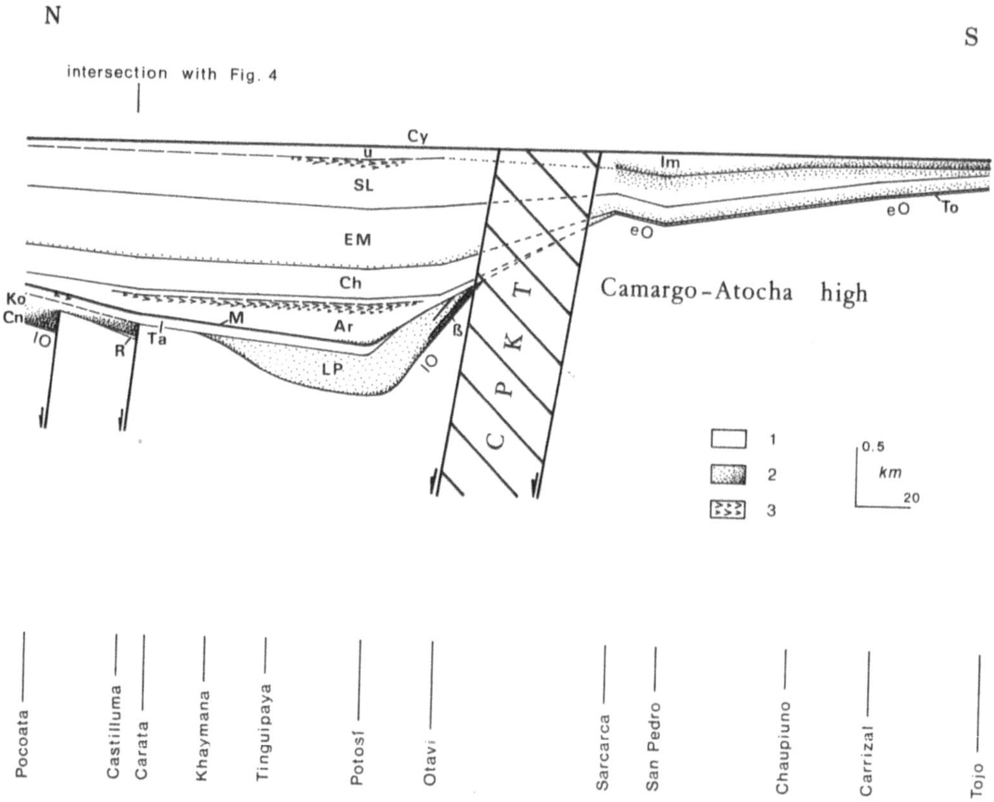

Fig. 3

Restored sedimentary cross- section from the subsident part of the Potosí basin to the Camargo-Atocha high (localization in Figure 1). Thicknesses from YPFB unpublished data and personal observations. Thicknesses from Sarcarca and San Pedro adapted from Okamoto et al. (1990). Thicknesses of non-outcropping rock units are hypothetical. Strata are decompacted below the early Eocene unconformity (base of the Cayara Formation) as reference level. Lithology: 1- mostly mudrocks (and limestones in the El Molino Formation), 2- mostly sandstones (and conglomerates), 3- evaporites. Abbreviations (also for Figure 4): eO- early Ordovician, lO- late Ordovician (Caradocian-Ashgillian), eS- early Silurian, lS- late Silurian, eD- early Devonian, lPz- late Paleozoic (late Triassic-Jurassic), R- Ravelo Formation (late Jurassic-early Cretaceous), Cn- Condo Formation, Ko- Kosmina Formation, LP- La Puerta Formation, Su- Sucre Formation, ß- basic lava flows, Ta- Tarapaya Formation (Cenomanian-Turonian), M- Miraflores Formation (Coniacian), Ar- Aroifilla Formation (Santonian to late Campanian), Ch- Chaunaca Formation, (Coniacian-late Campanian), To- Torotoro Formation (middle to late Campanian), Cr- Coroma Formation (latest Campanian-early Danian), EM- El Molino Formation (late Danian- ?earliest Eocene), SL- Santa Lucía Formation (u- upper) (?late Thanetian-?earliest Eocene), Im- Impora Formation (early Eocene), Cy- Cayara Formation, CPKT- Khenayani-Turuchipa paleostructural corridor.

most Kosmina. The La Puerta Formation appears as a sandier and thickly cross-bedded equivalent of the Sucre Formation, and hence of the Kosmina Formation. It was deposited in a relatively high-energy fluviatile environment.

The Tarapaya Formation consists of red to violet-red mudstones with minor sandstone intercalations. The upper boundary of the Tarapaya is defined as the base of the Miraflores limestones, which is always a discontinuity. At La Puerta, the Tarapaya overlies the La Puerta Formation with a sharp discontinuity, postdating normal faults that affect the latter (see below), but at some other localities the Tarapaya red mudstones transitionally grade downward into the Kosmina red beds (as in the Tawarreja-Thokori graben; Jaillard and Sempere, 1989). Because of its facies, the Tarapaya Formation was probably deposited in an alluvial (to lacustrine?) environment.

On the whole, three large-scale facies, vertically and laterally related, can be observed. In order of increasing frequency, they are: (1) a conglomeratic facies (the Condo conglomerate), generally with subangular, lithified sandstone clasts, and of very variable thickness (0-200 m), (2) a sandy facies (sand mud), interpreted as deposited in the main streams of an alluvial system, with channels, cross-bedding and some mudcracked thin overbank mudstones (Sucre and La Puerta Formations), and (3) an argillaceous facies (mud sand), interpreted as deposited in the flood plain (to lacustrine?) sub-environment of this alluvial system, or in a mud-dominated alluvial system, with thick, monotonous alternations of red mudstones, siltstones and sandstones (Kosmina and Tarapaya formations).

Sequence Stratigraphy

Sequential analysis has differentiated three sequences within the Puca A set (Sempere et al., 1988a). The second-degree sedimentary discontinuities that bound these sequences are best exposed at the Tres Cruces locality and can be followed between Potosí and Macha. However, they are absent at several other important localities of the Potosí basin.

The existence of these discontinuities, and hence of the sequences they define, have permitted correlations with the latest Jurassic-Early Cretaceous succession of Peru, and chronological assignments (Jaillard and Sempere, 1989). The ages of the three recognized sequences would thus be Kimmeridgian?-Berriasian, Valanginian-Late Aptian, and latest Aptian-Late? Albian, respectively.

The Kimmeridgian?-Berriasian sequence consists mainly of red beds of alluvial origin, with basal conglomerates and some sandstone bodies of more typical fluvial origin. Where distinguishable, the Valanginian-Late Aptian sequence typically consists of massive cross-bedded sandstones (as in most of the La Puerta Formation) of fluviatile origin.

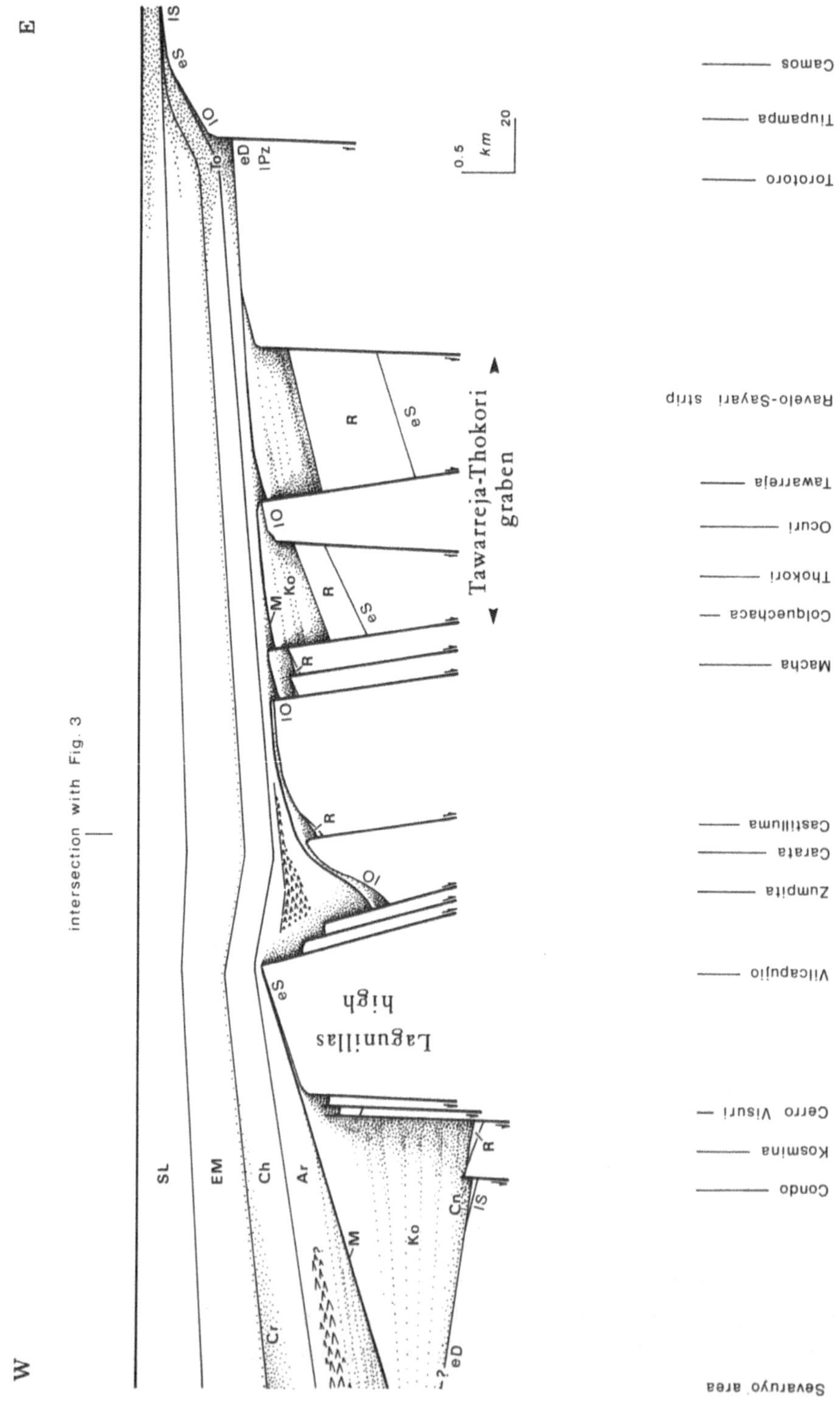
W
E
intersection with Fig. 3
SL
EM
Ch
Ar
Cr
M
Ko
Cn
lS
eD
R
eS
lO
eS
lS
lo
eD
IPz
Lagunillas high
Tawarreja-Thokori graben
0.5
km
20
Sevaruyo area
Condo
Kosmina
Cerro Visuri
Vilcapujio
Zumpita
Carata
Castilluma
Macha
Colquechaca
Thokori
Ocuri
Tawarreja
Ravelo-Sayari strip
Torotoro
Tiupampa
Camos

Fig. 4

◀ Restored sedimentary cross- section from Sevaruyo to Tiupampa and Camos (localization in Figure 1). Thicknesses from YPFB unpublished data and personal observations. Thicknesses from the Sevaruyo-Tambillo area adapted from Kriz and Cherroni (1966), but possibly exaggerated. Thicknesses are hypothetical in the zones where not all rock units crop out. Strata are decompacted below the early Eocene unconformity (base of the Cayara Formation) as reference level. Cretaceous distances are tentatively restored, assuming a gross 50% shortening and taking into account actual specific structures (but shortening may have been locally higher). Same abbreviations and lithological symbols as in Figure 3.

The latest Aptian-Late? Albian sequence equates with the Tarapaya Formation at its type locality, but this may well not be the case far from the Potosí syncline. It seems necessary to stress that, within the Puca A, there is no direct relationship between formations (which are large-scale facies) and sequences, and that the latter are not recognizable in all parts of the Potosí basin.

Synsedimentary Tectonic Manifestations

Unlike the overlying Puca B set, the Puca A shows strong and laterally rapid thickness variations, which point to important differential subsidence and/or subsequent differential erosion. Both phenomena suggest that tectonics was active during at least part of the Kimmeridgian-Albian interval.

Synsedimentary normal faults are known at two stratigraphic levels:

(1) Most of these faults affect the substratum and the lower part of the Condo Formation at many localities (especially in the Macha, Challapata and Calazaya areas), and are postdated by beds of the same Condo or Kosmina Formations (Chávez, 1987; Martinez and Vargas, 1988; Sempere et al., 1988a). They are commonly associated with per descensum clastic dykes in the Macha area, where both types of tensional structures testify to a local E-W extension (Sempere et al., 1988a). In the Challapata area, faults and clastic dykes suggest an NE-SW extension (Chávez, 1987). In the Calazaya area, the faults seem to strike roughly ENE and are located within the CPKT (Sempere et al., 1991). In all three areas, relicts of the Middle Triassic-Jurassic Ravelo Formation, some of them affected by normal faults postdated by Condo beds, are beveled to complete erosion and form asymmetric wedges which suggest that block-tilting accompanied normal faulting, forming half-grabens. At Macha, the Condo Formation overlies such a wedge of Early Mesozoic strata with a low-angle but clear angular unconformity. According to the correlations used (Sempere et al., 1988a; Jaillard and Sempere, 1989), this large-scale normal faulting would be of Kimmeridgian?-earliest Cretaceous age.

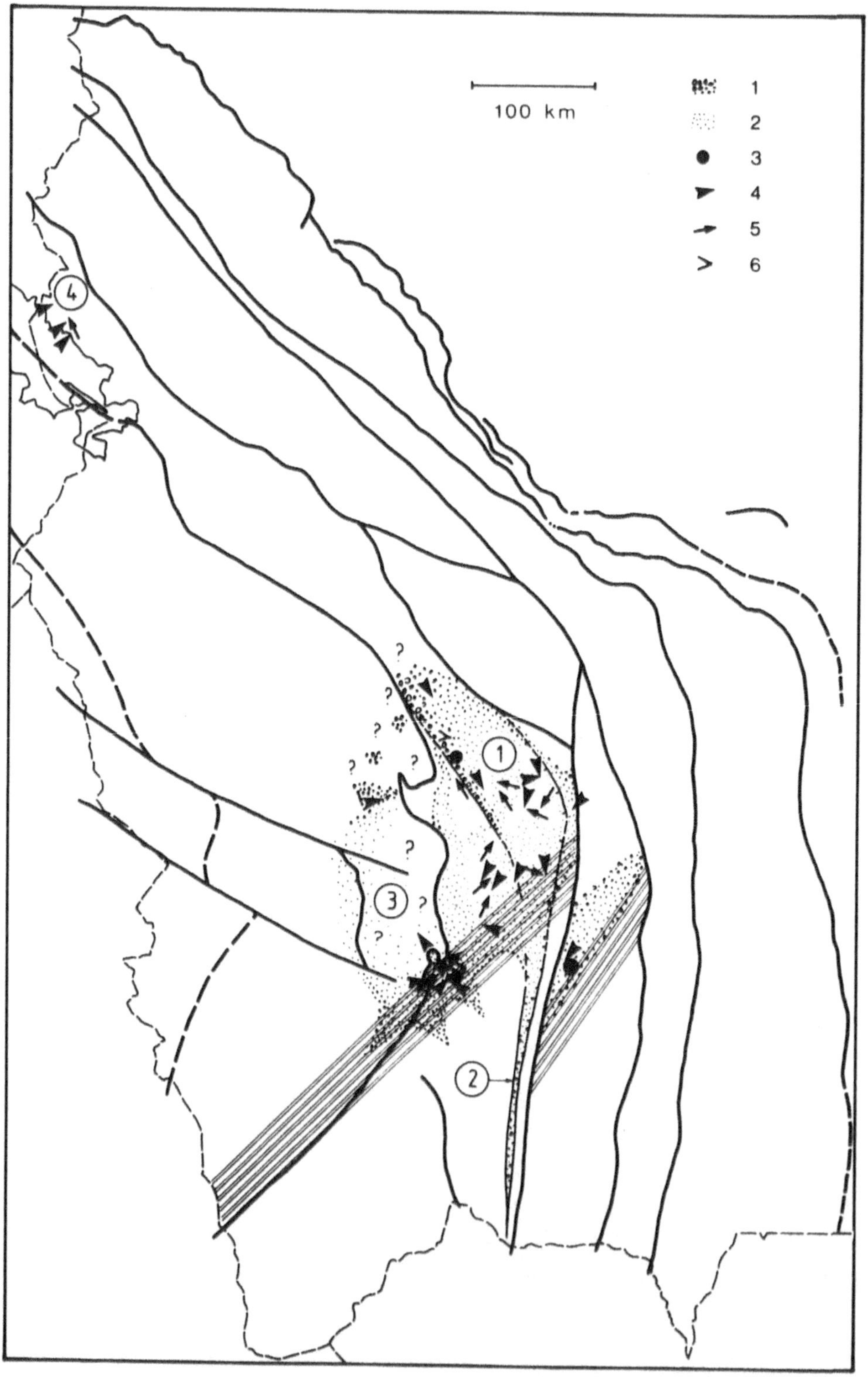
100 km
1
2
3
4
5
6
1
2
3
4

(2) At La Puerta and Tres Cruces, lying respectively 7 and 22 km northwest of Potosí, on the road to Oruro, two normal faults, striking N140°E and N60°E, affect the La Puerta Formation and are postdated by the Tarapaya Formation. The top of the La Puerta is irregularly overlain by very ferruginous silty deposits and, very locally, by what seems to be an altered basaltic arenite (YPFB laboratories, personal communication). The basal beds of the Tarapaya show a local onlap toward the southwest upon this discontinuity surface, which thus postdates minor normal faulting of possible Late Neocomian to Early Aptian age.

Both generations of faults are present in the northwestern part of the Otavi syncline, which is located on the northwestern edge of the CPKT, where normal faults indicating a NE-SW extension affect the Kimmeridgian?-Berriasian and/or Valanginian-Late Aptian sequences of the La Puerta Formation (see Martinez et al. (1990), although some of their stratigraphic attributions are wrong). Conglomerates and volcanic flows (see below) are present, mainly in the first sequence of the unit. The younger generation of faults is postdated by sandstone strata of the upper La Puerta, which in the area is capped by a calcareous crust (probably a lateral equivalent of the Cenomanian-Turonian-age Miraflores Formation), overlain in turn by the upper Aroifilla Formation. A Late Neocomian and/or Aptian age for this normal faulting is thus probable.

One case of synsedimentary reverse fault postdated by Condo beds was described 21 km north of Macha (Martinez and Vargas, 1988); but because of the very poor quality of the exposure this reference should be considered as quite dubious.

Fig. 5

◀ Schematic paleogeographic image for the Kimmeridgian?- early Cretaceous interval. Circled numbers: 1- Tawarreja-Thokori graben, 2- San Lorenzo-Tupiza graben, 3- Sevaruyo sub-basin (poorly known), 4- "Norte del Lago" basin. Ruled: CPKT. Symbols concerning the Kimmerigian?- Hauterivian? interval: 1 and 2 - (respectively) mainly conglomeratic and sandy facies, 3- basalt flows, 4- paleoccurrent data. Symbols concerning the Barremian?- Albian? interval: 5- paleocurrent data, 6- presence of gypsum in the Tarapaya Formation. Thicknesses are highly variable (see text and Figures 3 and 4). Local apparent inhomogeneities between the two sets of paleocurrent patterns as a result of integration of data collected at different stratigraphic levels, thus suggesting that paleotopography suffered local modifications from tectonic activity during the considered time interval; the data concerning the Barremian?-Albian? interval are thought to express reworking of previous deposits following modifications of the basin geometry and drainage system by extensional tectonics (see text). Note that the Tupiza-San Lorenzo inferred graben is parallel and close to the Andean-age Aiquile-Tupiza N-S-striking fault (Sempere et al., 1988b), suggesting that the latter was preexistent and influenced activity of the former (see also Figure 7). Most Valanginian-Barremian? paleocurrents from cross-bedded sandstones in the "Norte del Lago" basin are directed toward the Cordillera Real, where no Cretaceous outcrops exist; the usual interpretation favors these strata having been deposited in wave-dominated deltaic environments (see Sempere et al., 1988a) fed from the northeast. Another possibility would be that the deltas were indeed fed from the southwest and that more distal facies were deposited in the Cordillera Real area, but further studies are needed to clarify this issue.

Interpretation of the Condo Formation

It seems that there is a close relationship between the Condo conglomerates in their typical alluvial fan facies, and coeval synsedimentary normal faulting. The Condo Formation is coarser and thicker in the areas where synsedimentary extension is recorded; at the outcrop level it usually overlies preserved wedges of the Ravelo Formation, indicating relative local downwarping. Where the Condo is thin or absent, the Puca Group rests directly on the Paleozoic, indicating complete erosion of the Ravelo and thus relative local uplift. Exposures where the Condo is thick (and coarse) and where it is thin (or absent) commonly alternate over short distances in the same area.

The Condo conglomerates are interpreted as having been deposited in alluvial fans and proximal streams associated with normal faults. Normal faulting must have produced block-tilting: the thicker and coarser facies accumulated on the downwarped sides of the half-grabens, i.e. at the foot of fault escarpments, while the uplifted crests were submitted to erosion. The Condo Formation may therefore be considered a "trace" of nearby synsedimentary normal faulting (see Figure 4).

Synsedimentary Magmatic Manifestations

Some volcanic flows are interbedded locally within the Puca A. A basalt flow recorded in its base portion (Sucre Formation) 14 km north of Macha. In the Otavi syncline, located on the northwestern edge of the CPKT, several basalt flows associated with conglomerates and red beds occur in the Kimmeridgian?-Berriasian sequence, and an andesite flow is intercalated in the Valanginian-Late Aptian sequence. Basalts may be present in the Puca A strata of the Incapampa syncline, also located on the northwestern edge of the CPKT, but this needs confirmation.

These basalts are thought to be alkaline, and to indicate crustal extension (see Martinez, 1980).

Tectonic Interpretation of the Kimmerigian?-Albian? Interval

The base of the Puca Group records the large-scale onset of an extensional tectonic system in the Potosí basin, mostly northwest of the CPKT (Figure 5). Normal faulting developed dramatically at the beginning of the period (Condo Formation) and progressively waned, to resume on a much smaller scale in Late Neocomian and/or Aptian time, and subsequently die out. The extension had a NE-SW to E-W orientation, and was locally accompanied by volcanism.

Widespread erosion of the fluvial-eolian Ravelo Formation provided a large amount of relatively clean sand, which mainly accumulated in stream channels. Because sand accumulation is maximum along the northwestern edge of the CPKT, and because Cretaceous sedimentation on the Camargo-Atocha high southeast of the CPKT (except for the San Lorenzo-Tupiza graben; see Figure 5) only began in Senonian time (see Figure 3), it is suggested that the Ravelo was deposited on the Camargo-Atocha high but had been completely eroded by latest Jurassic?-Early Cretaceous times, and that the released sand was transported mostly toward the northwest and partly into the Tupiza-San Lorenzo graben. Paleozoic strata (shales and compact micaceous sandstones) were exposed and weathered in many areas due to block-tilting and provided sandstone clasts, an abundant clay fraction, and "dirty" sand.

The tectonic upheaval recorded at the base of the Puca Group ("Condo" event) has been related to the similarly outstanding Kimmeridgian-age "Araucan" event of central Chile and Argentina (Stipanicic and Rodrigo, 1969), which led to the age proposal (Sempere et al., 1988a) generally accepted. In central Chile (35°S), the "Araucan" event seems to correspond to thick-skinned compressional tectonics associated with important andesitic volcanism, as suggested by recent field work (R. Charrier and T. Sempere, 1990, unpublished). However, central Chile and Andean Bolivia are far from each other and do not belong to the same tectonostratigraphic domains. In northern Chile, transtensional conditions were established in Early Cretaceous (and latest Jurassic?) times (Naranjo et al., 1984; Flint et al., 1986; Muzzio et al., 1988). In coastal Peru, extensional and probably transtensional tectonics dominated latest Jurassic-earliest Cretaceous times (Jaillard, this volume).

As mentioned above, fine-grained red beds assigned to the Tarapaya Formation always disconformably underlie the Cenomanian-Turonian-age Miraflores Formation. The Tarapaya is of variable thickness: from over 50 m thick, where it overlies older Puca Group units, it may thin down to only a few meters where it directly overlies the Paleozoic basement. Although there is no absolute certainty as to whether these Tarapaya red beds are all synchronous and hence chronologically comparable, it seems that in the Potosí basin fine-grained sedimentation onlapped over most of the topographically high areas northwest of the CPKT in Late Early Cretaceous (Albian?) times. This general onlap of fine-grained continental strata indicates (1) that the extensional conditions which had led to the creation and tectonic fragmentation of the Potosí basin had ceased prior to sedimentation of the Tarapaya, and (2) that this unit was deposited in a new tectonic setting which created a more equably distributed subsidence and permitted a generalized deposition of mainly mudstone sequences. However, it seems difficult at present to propose a definite tectonic model for the Tarapaya time of deposition on the basis of the Bolivian data alone. Results obtained in Peru (Jaillard, this volume) suggest that the overall tectonic setting of the eastern domain was probably extensional and/or transtensional on a very large scale, but apparently without small-scale manifestations in Bolivia.

The synthetic paleocurrent data shown in Figure 5 testify that a notable paleogeographic modification occurred during the Kimmeridgian?-Albian? time span. It seems that drainage in the mostly fluviatile Potosí basin was initially controlled by normal faults and major fault systems (such as the CPKT, and the Tawarreja-Thokori and San Lorenzo-Tupiza grabens, which appear to be in some geometrical continuity; see Figure 5), and finally directed toward the Sevaruyo sub-basin, which might have been lacustrine in part. However, because Early Cretaceous outcrops are very scarce in the still poorly known Sevaruyo area, this remains hypothetical. Another possibility would be that the Potosí basin fluvial system extended to the west, and connected with the southern prolongation of the Arequipa basin where coeval deltaic deposits are known (Murco Formation; see Vicente, 1989, or Jaillard, this volume). Despite this, paleocurrents from the upper part of the sequence (Tarapaya Formation and upper La Puerta, Sucre and Kosmina Formations) in the central part of the Potosí basin, are consistent with another drainage pattern, directed toward the Tawarreja-Thokori graben and, within this structure, toward the northwest (Figure 5). This notable paleogeographic modification is probably due to "Late Neocomian-Aptian" tectonic movements (see above), the effects of which were to reactivate subsidence in the Tawarreja-Thokori graben and gently uplift the area that was located between it and the Sevaruyo sub-basin (the localities situated near the city of Potosí belong to this area). Consequently, the sediments deposited in this area were then partly reworked toward the Tawarreja-Thokori graben.

Cenomanian-Turonian

The Puca B set equates with the Miraflores Formation. This unit consists of four sequences which can be compared to and correlated with the more fossiliferous and better expressed Peruvian sequences (Jaillard and Sempere, 1991; assigned ages are approximate).

-The "black Miraflores" (Early Cenomanian) consists of very dark grey, strongly bioturbated limestones, and in the Macha area shows a conspicuous ostracod-bearing level at its top.

-The "grey Miraflores" (Middle Cenomanian) consists of three light grey, upward-thickening sub-sequences made of marly fossiliferous limestones, grading to micritic limestones at the top; this sequence yielded Neolobites cf. kummeli ammonites.

-The thin "yellow Miraflores" (Late Cenomanian), which consists of a yellow limestone bed, locally well-laminated, and some thin light-colored marls.

-The "red Miraflores" (Turonian), which consists of pink, laminated and mudcracked limestones, and red mudstones.

Thickness variations in the basin are very gradual, and the sequences can be easily followed (if not eroded below the Aroifilla). The sedimentation rate must have been very low, i.e. less than 12 m/m.y. (with decompaction), since less than 30 m now correspond to the Cenomanian-Turonian interval, which spans approximately 8 m.y. It may be noted that the coeval succession thickens toward the northwest and can reach 1500 m in central Peru (Jaillard, 1986, 1987; Jaillard and Sempere, 1991).

No direct evidence of synsedimentary tectonics has ever been observed. Moreover, the depositional domain of the Puca B lies within that of the Puca A in map view (Figure 6), which indicates that the marine deposits did not onlap onto the previous ones. These data strongly suggest that sedimentation of the Miraflores limestones suffered little tectonic control.

Hence, the main event recorded by the Puca B set seems to be of global-eustatic nature: the marine ingression extended over a very slightly subsident domain, bounded on the southeast by the CPKT, and on the west by the western Altiplano (?). The subsident area for this time span was probably located west of the Bolivian Altiplano (Arcurquina Formation of southern Peru; Jaillard, this volume).

Senonian-Paleocene

The Puca C set may rest on any of the Puca B sequences, and even on the Puca A deposits in some rare parts of the known depositional area of the Puca B, thus proving that the Miraflores limestones were locally submitted to erosion prior to deposition of the Puca C (Chávez, 1987; Jaillard and Sempere, 1991).

The Puca B / Puca C discontinuity is marked by the change from carbonates to pelitic mudrocks as predominant facies, which is accompanied by a sharp increase of the sedimentation rate (Sempere et al., 1988a). Because of the age assigned to the uppermost sequence of the Miraflores Formation, this change is likely to have taken place near the Turonian-Coniacian boundary.

The geographic distribution of the Puca C units indicates that sedimentation clearly onlapped toward external areas during this time span.

Coniacian-Late Campanian

This interval is represented in the Potosí basin by the Aroifilla and Chaunaca Formations and their time equivalents (Figure 2).

Formation Review

The Aroifilla Formation mainly consists of orange-red mudstones, subordinate fine-grained red to green laminated sandstones, often showing halite crystal casts, and evaporites. These evaporites are known to include gypsum beds and stratabound bodies,

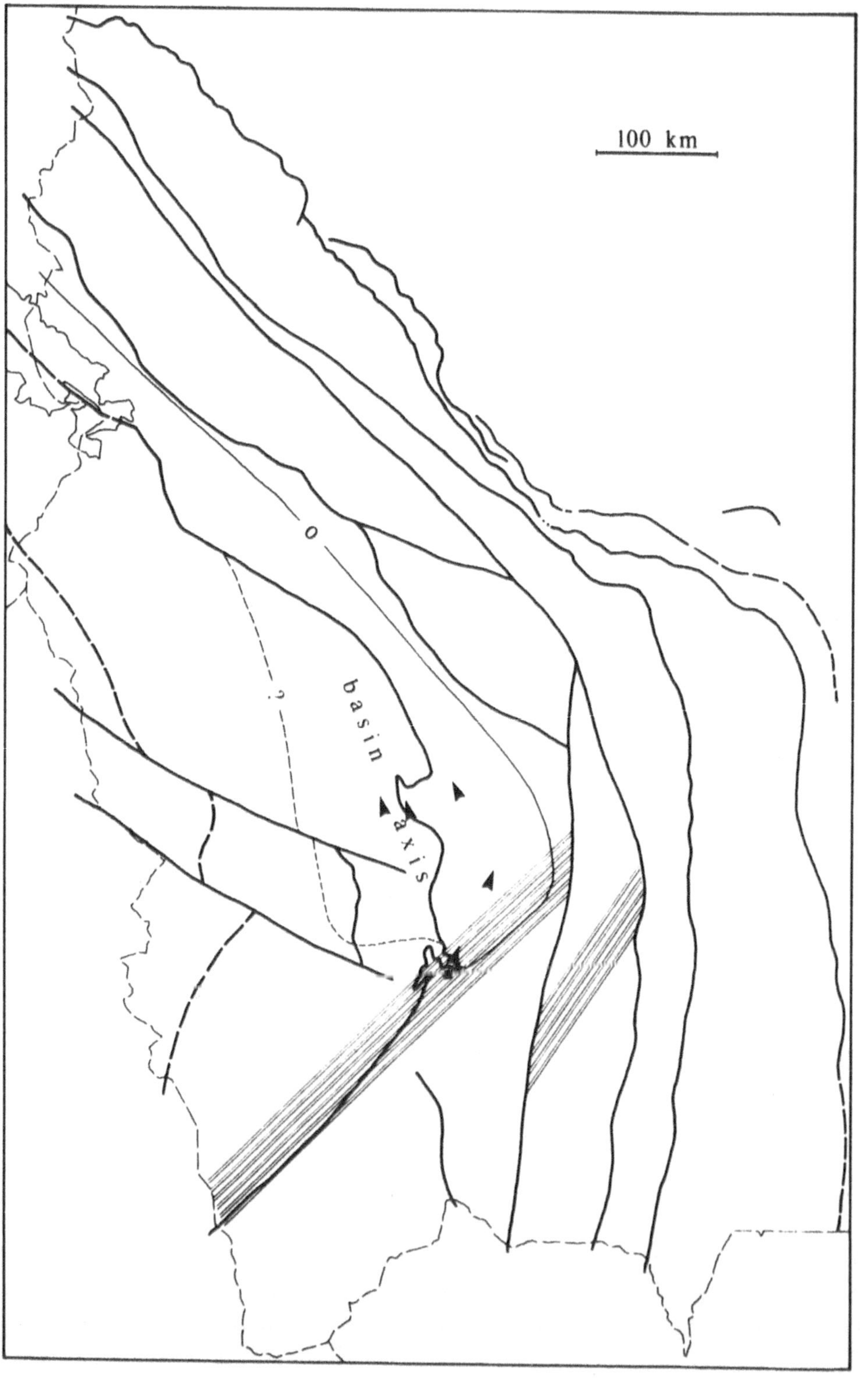
100 km
basin axis
?
0

pink-brown gypsiferous mudstones, anhydrite nodules altered to gypsum, and efflorescent halite (J.-M. Rouchy, personal communication) in outcrops; and a 2300 m thick halite mass (very probably a diapir) in an exploration well south of Lake Poopó (YPFB, unpublished). Halite-bearing gypsum diapirs intruding Cenozoic strata are known in several areas of the Bolivian Altiplano, and might have remobilized evaporites from the Aroifilla. Coarse sandstones or conglomerates (including mafic volcanic clasts) are locally present in the base of the Aroifilla. Basaltic flows are intercalated in the lower part of the Formation at several localities (Figure 7). Thin light-green to white tuffaceous beds are known in a few western sections. The Aroifilla was deposited in a distal alluvial to salt-lacustrine (playa-lake) environment. Its thickness is highly variable (Sempere et al., 1988a), unlike the overlying Chaunaca Formation.

The base of the Chaunaca Formation is conventionally the base of a 30 m thick guide-level, called the "basal limestone" of the Chaunaca, which consists of a fossiliferous layer of green marls, laminated grey limestones and black shales (locally bearing galena or sphalerite crystals, and/or malachite or azurite stains), and yellow dolomites. Palynologic samples from this level indicate a Santonian to earliest Campanian age (Pérez, 1987). It is overlain by red-brown mudstones which, in the basin axis, are interbedded in the lower half of the Formation with a level of pink-brown gypsiferous mudstones, and in its upper half with a 15 m thick level of green marls and thin limestones resembling the "basal limestone" (Figure 2). These intercalations are only present in a basin "axis" (Figure 8) since they thin out toward the west and east. Paleosols are quite frequent toward the top of the formation. As with the Aroifilla, the Chaunaca was mainly deposited in a distal alluvial to salt-lacustrine (playa-lake) environment. The two green levels, at the base and near the top of the unit, mark transgressions of water-bodies. Because of its bivalve (Brachidontes sp.; Branisa et al., 1966) and fish (Eotrigonodontidae; Gayet, Jaillard et Sempere, unpublished) fauna, the "basal limestone" could be of restricted-marine origin, and -in view of the similarity in facies- the upper green level probably as well. Black shales, such as those present in the lower portion of the "basal limestone", are commonly found at the base of marine transgressive deposits (Wignall, 1991). Although the question should be considered as still open, a restricted-marine en-

Fig. 6

◄ Paleogeographic extension of the marine Miraflores Formation (Cenomanian-Turonian; see Jaillard and Sempere, 1991). Ruled: CPKT. "Deepest" facies (which help to define the basin axis) are known in the Challapata area (see Figure 1), but because the Miraflores Formation was submitted to partial (and locally complete) erosion prior to deposition of the Aroifilla Formation in this area, the known maximum thicknesses are presently 25-30 m in the Macha area (see Figure 1). This includes the single locality where ammonites were found (Branisa et al., 1966). Arrows are paleocurrents obtained from orientation of gastropods and/or ripple-marks. Note that this map suggests that Cenozoic shortening has been higher in the NW-trending segment of the Bolivian orocline.

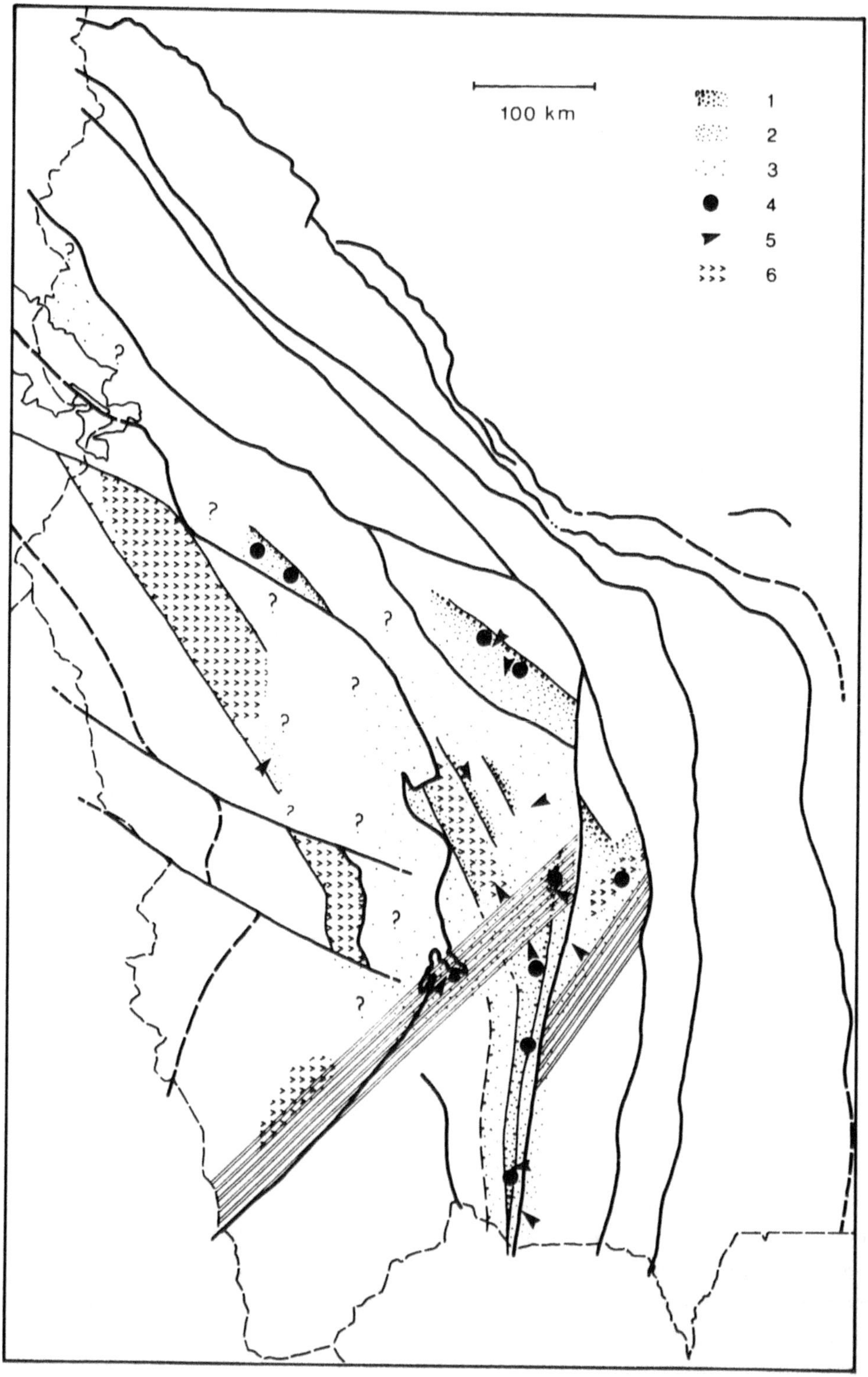
100 km
1
2
3
4
5
6
?

vironment is favored here on the basis of published data and of apparent correlation with similar marine guide-levels in southern Peru, one of which yielded ammonites (Vicente, 1981; Jaillard and Sempere, 1989; Jaillard et al., 1992; Jaillard, personal communication, and this volume). The thickness of the Chaunaca Formation is much more even than that of the underlying Aroifilla.

The Torotoro Formation is a sandy lateral equivalent -of alluvial origin- of the Aroifilla and Chaunaca in the northeastern and southeastern parts of the basin. It locally includes conglomerates (with mafic volcanic clasts) in its basal portion, and paleosols are frequent. This unit usually overlies Paleozoic strata, and marks the onlap of Cretaceous sedimentation toward the northeast and southeast.

The Coroma Formation transitionally overlies the Chaunaca in the central Altiplano. It is a lateral equivalent of the upper Chaunaca Formation of the classic Potosí region, since it underlies the El Molino Formation, the base of which is usually a discontinuity. The Coroma consists of an upward-thickening level of brownish, commonly rippled, fine to medium-grained sandstones and of red-brown mudstones identical to those of the underlying Chaunaca. It is coarser and thicker toward the west (Figure 4); preliminary paleocurent data also suggest a western source for these sands.

Sequence Stratigraphy

Sequential analysis applied to the Aroifilla and Chaunaca Formations (Figure 2) points to the following type-sequence: (0) sharp lithological contact overlain by (1) green marls, black shales and carbonates (more dolomitic toward the top) - (2) brown-red mudstones with rare paleosols and/or evaporites - (3') evaporite-bearing brown-red mudstones grading to evaporite beds or (3") brown-red mudstones with abundant rooted paleosols. The various occurrences of facies 3' or 3" in the interval studied are described below; they depended on paleoclimate. Facies 3' was probably deposited in playa-lakes, which would signify a dry climate, whereas facies 3" indicates vegetation, and thus a more humid climate.

Five sequences, some of them incomplete when compared to the type-sequence, are therefore distinguished in the Aroifilla-Chaunaca succession along the basin axis (thick-

Fig. 7

◄ Schematic paleogeographic image for the lower Aroifilla Formation depositional interval (early Coniacian). 1- presence of conglomeratic facies, 2- mainly red sandstones (i.e., Torotoro Formation), 3- mainly red mudstones, 4- basaltic flows, 5- paleocurrent data, 6- evaporite member at top of sequence. Ruled: CPKT. The gypsum and salt rock now forming large diapirs in the Altiplano are here considered to have been initially deposited during the Aroifilla (and Chaunaca?) depositional interval (yet another possibility is that some or all are of late Permian to Triassic or Jurassic age). The western Altiplano northwest of the CPKT is devoid of pre-Eocene outcrops in Bolivia and its Cretaceous history is thus very poorly known.

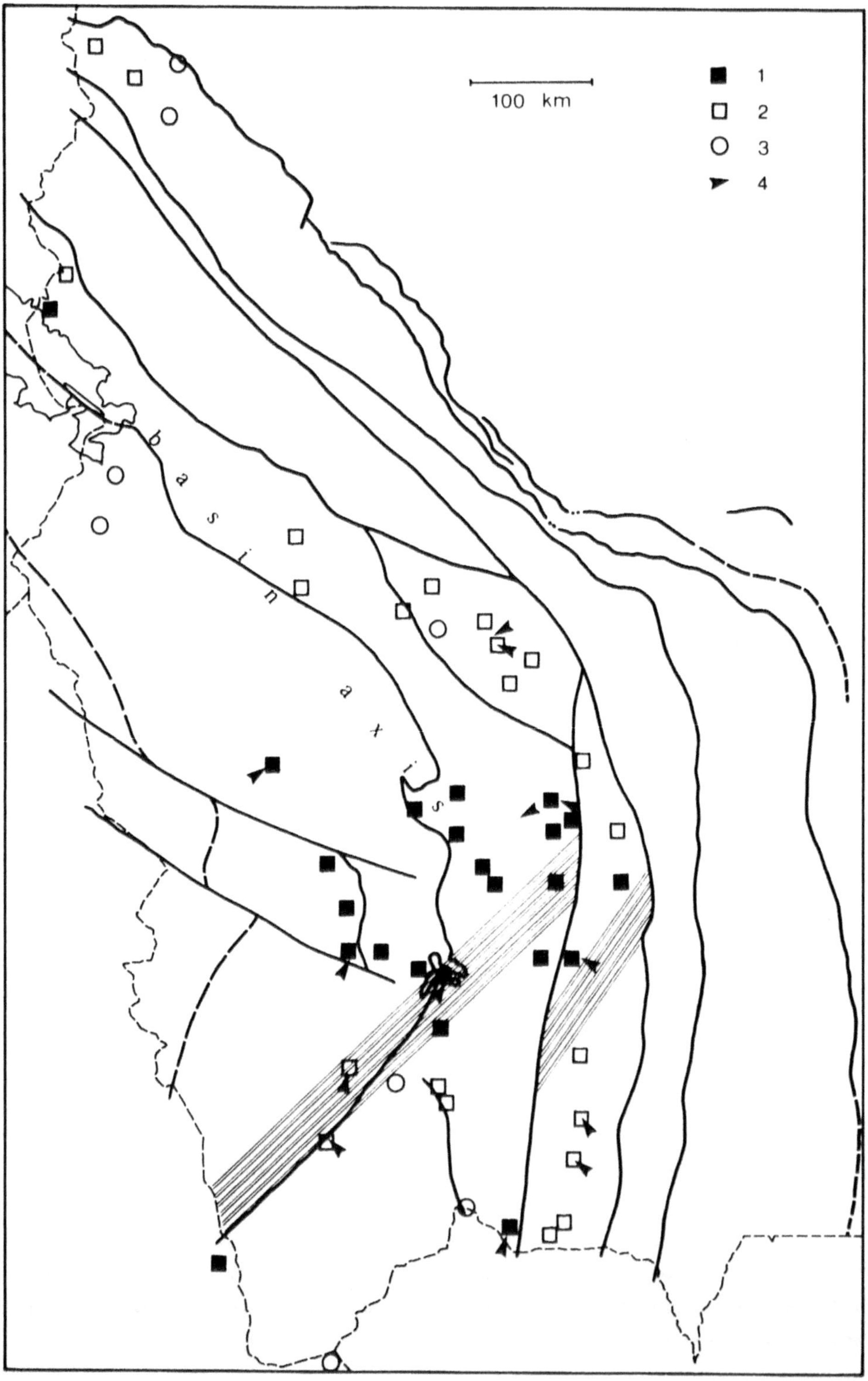
100 km
1
2
3
4
basin axis

nesses are from the La Palca section near Potosí; Figure 2). They are the lower Aroifilla (facies 2 - 3'), 300 m; upper Aroifilla (facies 2 - 3'), 80 m; lower Chaunaca (facies 1 - 2 - 3'), 60 m; middle Chaunaca (facies 2 - 3"), 30 m; and upper Chaunaca (facies 1 - 2 - 3"), 30 m. The lower Aroifilla sequence equates with sequence P31 of Sempere et al. (1988a) and with sequence K 30-32 of Jaillard and Sempere (1989). The upper Aroifilla to upper Chaunaca sequences constitute a 200 m thick grouping which equates with sequence P32 of Sempere et al. (1998a) and with sequence K 32-40 of Jaillard and Sempere (1989).

The type-sequence is interpreted as produced by (1) the transgression of a restricted-marine water body, and subsequent progressive regression, (2) establishment of very shallow lacustrine conditions, evolving toward (3') dessication of concentrated brines and hypersaline waters, or (3") emergence by progradation of distal alluvial facies. Each of the five recognized sequences is interpreted to have been deposited in all or some of these environments, according to the existing facies. The superposition of these five sequences gives some indications of the tectonic and/or eustatic history of the basin (see below).

Because long-distance correlations of the recognized sequences, and even of many 2 to 3 m thick sub-sequences within the "basal limestone", are easy to make in most parts of the basin, one has to keep in mind that during at least Santonian and Campanian times the paleolandscape was extremely flat.

It may be noted that the first three sequences end with the development of evaporites, whereas the last two end with the development of vegetated soils. It can therefore be proposed that the climate was rather dry during Coniacian and Santonian times, and became more humid in Campanian times.

Fig. 8

◀ Schematic paleogeographic image for the upper Aroifilla and Chaunaca Formation depositional interval (late Coniacian-late Campanian). 1- presence of the "basal limestone" of the Chaunaca Formation, within a sequence mostly composed of red mudstones and subordinate sandstones, 2- absence of the "basal limestone" of the Chaunaca from a sequence consisting of interbedded red sandstones and mudstones, 3- no deposits (because of non- deposition or pre-El Molino erosion), 4- paleocurrent data (from sandy facies, including in the west the Coroma Formation). Ruled: CPKT. The basin axis is defined by the occurrence of both green transgressive levels (see text) and deepest facies of the "basal limestone". Presence of the "basal limestone" of the Chaunaca north of Lake Titicaca is inferred from the presence of a similar limestone unit in nearby Peru (Jaillard et al., 1992). The sandy Coroma Formation only exists west of the basin axis and thickens toward the west. Presence of pre-Maastrichtian deposits in the northwesternmost Subandean belt is inferred from presence of red- colored sandstones at the base of published sections (Perry, 1963). Part or all of the dinosaur- bearing Cajones Formation of the central Subandean belt (see Figure 10) might have been deposited during this time interval. Data published by Marinovic and Lahsen (1984) demonstrate that Brachidontes sp. is present in a limestone unit, which is therefore an equivalent of the "basal limestone" of the Chaunaca, within the Lomas Negras Formation at its locality of definition in northern Chile. As in Figure 6, this paleogeographic image suggests that Cenozoic shortening was more important in the NW- trending segment of the Bolivian orocline.

The two green levels of the Chaunaca apparently correlate toward the northeast; with two guide-levels formed by coalescent rhizolith-rich paleosols, established on brown-red argillaceous sands deposited in an alluvial environment and belonging to the Torotoro Formation. Multiplication of rooting at these two particular levels suggests biostacy, and thus seems related to the establishment of highstand conditions in the basin axis.

Sequence Age and Sedimentation Rates

The Puca C set postdates the Miraflores Formation, the top of which is Turonian in age. The lower boundary of the Puca C is thus of latest Turonian to earliest Santonian age, and is considered to be approximately coincident with the Turonian-Coniacian boundary in this paper.

The "basal limestone" of the Chaunaca Formation correlates with the transgressive level which occurs at the base of the middle Vilquechico Formation of the southern Peruvian Altiplano, which in turn correlates toward northern Peru with an ammonite-bearing, Early Santonian-age, transgressive level (P. Bengtson in Mourier et al., 1988; Jaillard et al., 1992; E. Jaillard, personal communication). An Early Santonian age, which is in agreement with the biochronostratigraphic conclusions of Pérez (1987), is thus favored for this level. The age of the Aroifilla Formation, which is locally 800 m thick, would be mainly Coniacian (1 m.y. duration), and latest Turonian-earliest Santonian (about 3 m.y. duration) as a maximum.

The second transgressive green level of the upper Chaunaca correlates with a similar selachian-bearing level which occurs near the top of the middle Vilquechico Formation of southern Peru; this in turn most probably correlates with an ammonite-rich Middle to Late Campanian-age transgressive level in northern Peru (P. Bengtson in Mourier et al., 1988; Jaillard et al., 1992; E. Jaillard, personal communication). A Middle Campanian age is thus proposed here for the Bolivian level, which forms the base of the upper Chaunaca sequence.

The Chaunaca/El Molino discontinuity is Late Campanian to earliest Maastrichtian in age (see below), and is considered to be approximately of latest Campanian age in this paper.

Consequently, the five sequences identified in the Aroifilla-Chaunaca succession would correspond to the following time spans:

-lower Aroifilla: Early to Middle Coniacian;

-upper Aroifilla: Late Coniacian (and earliest Santonian?);

-lower Chaunaca: Santonian;

-middle Chaunaca: Early Campanian;

-upper Chaunaca: Middle to Late (but not latest) Campanian.

Average decompacted sedimentation rates for the Potosí area, which is located within the basin axis, would hence be very approximately:

-lower Aroifilla: about 900 m/m.y.
-upper Aroifilla: 350 to 700 m/m.y.
-lower Chaunaca: 45 m/m.y.
-middle Chaunaca: 30 m/m.y.
-upper Chaunaca: 20 m/m.y.

Although approximate, these figures show a clear decrease of the sedimentation rate with time. Because these units are similar in lithology (mostly mudstones), were deposited at paleodepths close to sea level, and were submitted to similar conditions of compaction, a noteworthy decrease of the sedimentation -and thus subsidence- rates can be deduced.

A question arises about the possible relationship of the two transgressive green levels of the Chaunaca with global-eustatic transgressions as listed by Haq et al. (1987). If the "basal limestone" is indeed of Early Santonian age, slight downwarping tectonic processes are necessary to explain the marine ingression, since only a minor transgression is known for that time. In contrast, a Middle Campanian age for the upper green level would suggest that its deposition was mainly brought about by the trascendent global coeval transgression.

Synsedimentary Magmatic Manifestations

In Bolivia, the Aroifilla Formation is the last Cretaceous unit to contain basalt flows, and the first to include thin felsic tuffaceous beds (Sempere et al., 1988a). The Chaunaca Formation seems devoid of volcanic intercalations.

The basalts are located in the lower portion of the Aroifilla, and the flows are usually several meters thick. One of these basalts was dated at 85.1 Ma (data by Evernden et al., 1977; age recalculated by McBride et al., 1983). This apparent age is Santonian by current standards, but the real age of the flow might be slightly older because of possible Ar loss. Because they bear olivine, the basalts were interpreted to be alkaline (see Martinez, 1980), but current studies by P. Soler will probably lead to greater precision with regard to their chemical affinity.

The tuffaceous beds are never thicker than 20 cm, and seem to occur in the middle and upper part of the Aroifilla.

Tectonic Interpretation of the Coniacian-Late Campanian Interval

Several phenomena have to be taken into account in order to give a coherent interpretation for this time span.

A first set of facts includes the often erosional unconformity at the base of the Aroifilla (Jaillard and Sempere, 1991), the abundant basic magmatism in the lower Aroifilla, the existence of narrow -apparently extensional- troughs in which the basalts are located

(Figure 7). The conjunction of these phenomena suggests that tectonic processes similar to rifting were active during the lower Aroifilla depositional time, i.e. Early Coniacian (and latest Turonian?; see above) times (Sempere et al., 1988a). The erosion which affects the Miraflores Formation in the Challapata area (Chávez, 1987) indicates that some parts of the basin were indeed gently uplifted during this event. Deposition of evaporites in at least two large playa-lakes (Sevaruyo-Chita area and Potosí-Agua Clara area; Figure 7) was approximately maximum during the Middle Coniacian. Both lakes are located in areas where the top of the Miraflores Formation shows no erosion below the Aroifilla, indicating that these NNW-elongated areas were relatively downwarped during the rifting event, and that the lakes were established in these topographic lows.

The highly variable thickness of the lower Aroifilla sequence suggests that it is due to differential extensional subsidence, and thus that the rifting period corresponded at least to the lower half of the sequence, and possibly to all of it, since the locally thick evaporites at its top accumulated in the center of some graben-like structures. In contrast, the first deposits to show onlap over the eastern external areas, indicating a relative change of tectonic setting, belong to the upper Aroifilla sequence. This change is also suggested by the marked contrast between the highly irregular thickness of the lower Aroifilla and the much more even thickness of the upper Aroifilla to upper Chaunaca set (Sempere et al., 1988a).

A second set of facts concerns the remainder of the succession (upper Aroifilla to upper Chaunaca sequences, Late Coniacian-Campanian) and includes progressive onlap over the external areas, a decrease in sedimentation and subsidence rates (see above), and the superposition of regressive sequences. The conjunction of these phenomena is reminiscent of aspects of some models of foreland basin evolution (see Tankard, 1986; Heller et al., 1988; Flemings and Jordan, 1989, 1990), suggesting that the Potosí basin was functioning as the distal part of the Andean foreland basin during that time, and thus that tectonic activity -including shortening and loading of the South American lithosphere- had started on the Pacific margin, and notably decreased with time, possibly reaching a state of relative tectonic quiescence in the Late Santonian-Campanian (see models by Flemings and Jordan, 1989, 1990). The superposition of the sequences described above was probably the result of minor subsidence reactivations in the basin axis, and hence of renewed shortening and tectonic loading along the margin. Detailed analysis of paleosol successions in the northeastern and southeastern highs might provide more data on this issue, since such reactivations would have produced slight uplifts of these areas.

Following the model by Flemings and Jordan (1990), a trend toward tectonic quiescence is suggested by the decrease in sedimentation and subsidence rates evidenced above, and is supported by the general onlap of fluviatile red beds in the eastern areas after Early Santonian times, and by the rapid invasion of the western part of the basin by eastward-prograding sandstones (Coroma Formation) in at least Middle to Late Campanian time.

To summarize, the tectonic history of the Coniacian-Late Campanian interval comprised two stages of very unequal duration:

-After a tectonically quiet period (Cenomanian-Turonian), widespread extensional conditions involving abundant basic magmatism were abruptly established near the Turonian-Coniacian boundary in the Potosí basin, generally reactivating Early Cretaceous structures. The duration of this extensional stage ("Vilcapujio" event; Chávez, 1987; Sempere et al., 1988a) was less than 3 m.y., and possibly about 1 m.y. Because basalts were erupted in the grabens and along reactivated faults, this extensional episode can be described as a regional "rifting" (Sempere et al., 1988a), as has been done in northwestern Argentina where a very similar evolution is recorded (Bianucci et al., 1981; Galliski and Viramonte, 1985; Salfity and Marquillas, 1986). This magmatic episode appears as volumetrically much more important than the Kimmeridgian?-Early Cretaceous one.

-Colmatation of tensional structures was accompanied or followed by an extension of sedimentation toward the external areas, some of which had remained totally devoid of Cretaceous deposits. Several series of data suggest that this Late Coniacian-Late Campanian stage, about 15 m.y. long, corresponds to the evolution of the Potosí basin as a relatively distal part of the Andean foreland basin. In this light, the "Vilcapujio" event probably represents the regional destabilization produced by this definite establishment of foreland conditions (see below). The sedimentary record suggests a progressive evolution toward relative tectonic quiescence during this time span.

Latest Campanian-Paleocene

This interval is represented by the El Molino, Santa Lucía and Impora Formations and their time-equivalents (Figure 2).

Formation Review

The fossiliferous El Molino Formation is commonly 500 m thick or more at the basin axis, where it mainly consists of green, violet, red, grey and black mudstones (marls and claystones), with frequently interbedded fine to coarse-grained limestones and very fine sandstones. White clean sandstones and carbonates, generally oolite-bearing, predominate in its lower part. The facies of this unit, or parts of it, or of its Argentine and Peruvian equivalents, have previously been described in detail (Castaños et al., 1975; Marquillas, 1984, 1985, 1986; Marquillas et al., 1984; Palma, 1986; Gómez Omil et al., 1989; Sempere et al., 1987; Sempere et al., 1988a; Jaillard and Sempere, 1989; Camoin et al., 1991; Jaillard et al., 1992). A recent summary of its rich fossil content (Gayet et al.,

1991, 1992) and other unpublished data suggest that the El Molino spans latest Campanian-earliest Paleocene time. The El Molino is equivalent to the Lecho, Yacoraite, Tunal and Olmedo Formations of northwest Argentina (Sempere et al., 1987).

The overlying Santa Lucía Formation consists of red-brown mudstones, with upward-thickening interbedded gypsum levels along the basin axis (Potosí), or of fine to coarse sandstones in its proximal portions. Northwest of the CPKT, the uppermost part of the unit consists of red-brown mudstones with some very thin green mudstone intercalations. The fauna from the lower Santa Lucía includes mammals and suggests that the unit is at least Early Paleocene in age (Gayet et al., 1991). The Santa Lucía is equivalent to the Mealla Formation of northwestern Argentina (Sempere et al., 1987). Charophytes ("Tectochara" ucayaliensis oblonga) indicating a Middle or Late Eocene age (Branisa et al., 1969) or a Paleocene-Eocene age (Feist, in Mourier et al., 1988) are reported from the Santa Lucía Formation near San Pedro in the Camargo syncline, but were in fact sampled from the lower Potoco Formation, which overlies the Cayara Formation and was long mistaken as a part of the Santa Lucía. This biochronological datum thus does not concern this unit.

The Impora Formation only exists southeast of the CPKT, in the Camargo syncline, where it overlies a major paleosol at the top of the Santa Lucía. The separate identity of the Impora is based on a significant difference of facies with the Santa Lucía, since the former consists of violet siltstones with carbonate concretions (grading to calcretes), green to violet mudstones, white to violet sandstones, and minor conglomerates (in the southern part of the syncline). The stratigraphic position of the Impora Formation relative to the Santa Lucía defined near Potosí is discussed below. The Impora is clearly equivalent to the Maíz Gordo Formation of northwestern Argentina, and is thus of Late Paleocene age (Sempere et al., 1988a).

The Campanian-Paleocene interval was recently studied in some detail by Okamoto et al. (1990) in the northern part of the Camargo syncline. However, these authors were strangely confused with regard to local stratigraphy, mistaking most of the El Molino Formation for their Torotoro Formation, and the Santa Lucía and Impora Formations for their El Molino. It seems necessary here to point out these discrepancies, in order to avoid uncritical use of misleading data.

Depositional Environments of the El Molino Formation

The question of the type of environment in which the El Molino and its equivalents were deposited is still being debated. A shallow marine origin was first suggested (Castaños et al., 1975) but more detailed sedimentological works insisted on the peculiar characteristics and restricted aspect of the paleoenvironment (Marquillas, 1985, 1986). Although the possibility of an oligohaline lacustrine environment has been proposed (Palma, 1986; Camoin et al., 1991), it seems that the El Molino basin was indeed frequently in connection with sea water because of the abundance and variety of marine fossils, which presently include foraminifers, dinoflagellates, marine thick-shelled bivalves and gastropods (all these groups remain virtually unstudied), an Aptychus, 11 species of selachians belonging to 3 families and 2 orders, and at least 6 species of marine and/or estuarine actinopterygians belonging to 5 families and 3 orders (Fritzsche, 1924; Benedetto and Sánchez, 1971; Méndez and Viviers, 1973; Cappetta, 1975; Audebaud et al., 1976; Gómez Omil et al., 1989; Cappetta, 1991; Gayet, 1991; Gayet et al., 1991, 1992; Jaillard et al., 1992). Many brackish-water ostracods from the El Molino are similar to taxa described from Early to Middle Maastrichtian rock units of the Neuquén basin in Argentina (Camoin et al., 1991), which were deposited in relation to an important marine transgression (Riccardi, 1988; Legarreta et al., 1989). However, the depositional environment of the El Molino Formation was never openly marine (Marquillas, 1985, 1986; Gómez Omil et al., 1989; Gayet et al., 1991, 1992), in part because ammonites and some other typically marine groups are apparently lacking (although this is probably due to shallow water depth) and because brackish to freshwater fauna and flora are also present (Fritzsche, 1924; Pilsbry, 1939; Gayet et al., 1991, 1992; Jaillard et al., 1992). In spite of its huge size (at least 500,000 km^2), the basin was markedly flat, and permanently very shallow to sub-emergent. This particular paleogeography, to which there is probably no present-day counterpart, would explain how a restricted-marine environment could have extended over such a large area. It may be necessary to remember that Andean Bolivia and northwestern Argentina were located in a "cul-de-sac" position at the extreme tip of the elongated Andean epicontinental basin, which probably only connected with the oceanic realm in present-day Venezuela, 3800 km more to the north (Figure 9): such a geometry is likely to have been responsible for the development of restricted conditions. Furthermore, abundance of sandy facies in the southernmost part of the basin in northwestern Argentina (see Gómez Omil et al., 1989) suggests that a major stream flowed flowed into the basin there from the south or south-east, and thus permanently provided a large amount of fresh water and mud to the flat and very shallow southern part of the basin (Figure 9). If confirmed, this is obviously another factor which would have helped to give the environment some peculiar characteristics. This stream probably occupied the axis of an underfilled, continental foreland basin at these latitudes. Other streams flowed from the east, and left sandy deposits in the Lomas de Olmedo area of northwestern Argentina and in the Santa Cruz to Cochabamba area of Bolivia (Figure 9).

Moreover, it may be noted that if a strictly lacustrine environment is favored for the El Molino, as do Camoin et al. (1991), it would be an amazing coincidence that a rapid "lacustrine transgression", bearing usually marine microfossils and a well-studied ichthyofauna of marine origin, took place in Bolivia in latest Campanian or earliest Maastrichtian time, i.e. at the time of a world-wide rise in sea level.

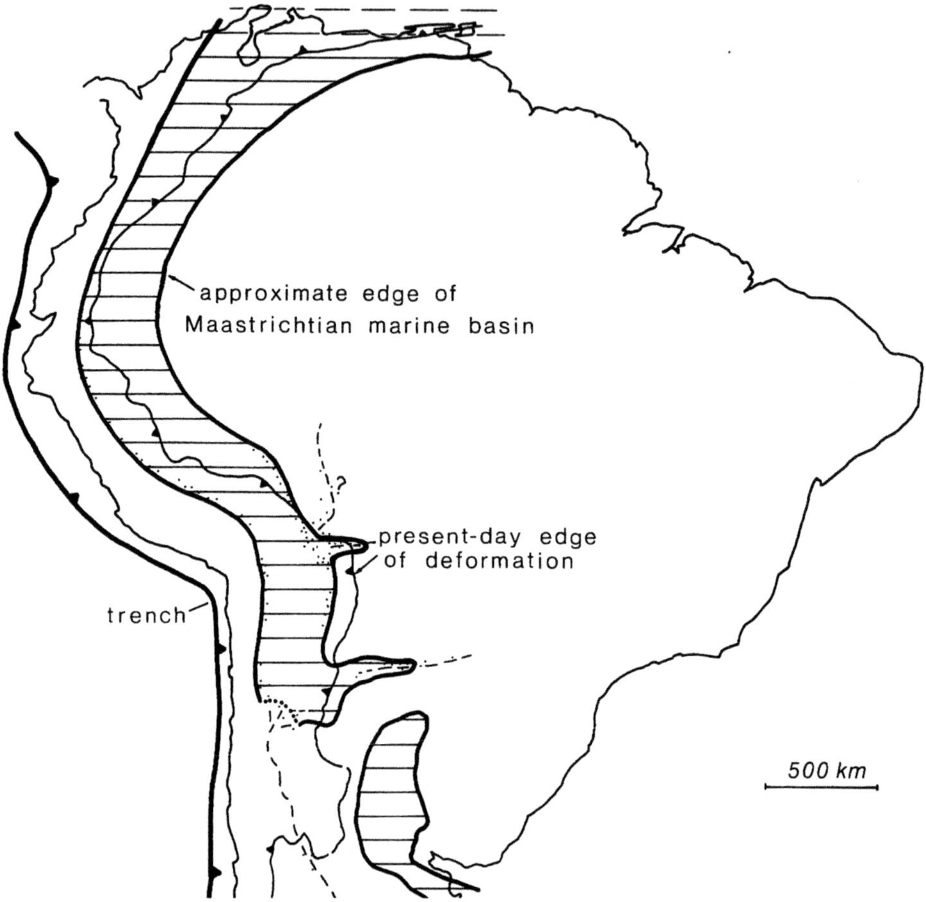

Fig. 9

Paleogeographic sketch-map of northern South America for Maastrichtian time showing the peculiar position of the Bolivian and northwestern Argentine regions at the southern tip of an elongated, partly marine, back- arc (foreland) basin connected to open sea in present-day Venezuela (see Rossi et al., 1987; Macellari, 1988; Chigne and Hernández, 1990).

In conclusion, the author here favors the idea that the El Molino Formation was mainly deposited in a wide, flat and very shallow area located at the southern tip of the elongate Andean back-arc basin, occupied by a restricted-marine environment, probably mesohaline and frequently in connection with northern sea water. However, more multidisciplinary studies are clearly needed in order to better define this interpretation.

Sequence Stratigraphy

On the basis of works by Sempere et al. (1987) and Jaillard et al. (1992), sequential analysis presently recognizes three major sequences in the El Molino Formation and two in the Santa Lucía Formation (Figure 2), as well as several sub-sequences within them. These five sequences are respectively termed the lower, middle and upper El Molino, and the lower and upper Santa Lucía. The lower El Molino practically equates to a grouping of the EM1 and EM2 sequences (Sempere et al., 1987), the P41 and P42 sequences (Sempere et al., 1988a), and the K 40-41 and K 41-42 sequences (Jaillard and Sempere, 1989).

Each of the lower, middle and upper El Molino sequences begins with a rapid marine transgression associated with black mudstones (see Gayet et al., 1991; Jaillard et al., 1992), reminiscent of the "basal limestone" of the Chaunaca and of current models of transgressive facies (Wignall, 1991). Although confirmation is needed, the available chronological data, which include unpublished Ar/Ar ages on single minerals obtained on a tuff from the lower El Molino, do suggest that the corresponding three transgressions correlate respectively with the Early Maastrichtian, Late Maastrichtian and Early Danian global-eustatic transgressions of Haq et al. (1987).

The lower and middle El Molino sequences end with red regressive facies, which may be of distal northeastern alluvial origin. Along the basin axis, the transgressive upper El Molino rapidly grades into the lacustrine lower Santa Lucía, and both sequences may be grouped in this area for interpretation. However, the lower Santa Lucía disconformably overlies the middle or upper El Molino in the eastern regions, where it shows onlap over a surface which cross-cuts these sequences.

The lower and upper Santa Lucía both begin with red-brown mudstones apparently deposited in a lacustrine environment. In the basin axis (Potosí), the top of the lower Santa Lucía consists of a progressively, upward-thickening interbedding of these mudstones with gypsum levels, which show chicken-wire or laminated structures (J.M. Rouchy, personal communication). In this area the upper El Molino-lower Santa Lucía set closely resembles some sequences and the type-sequence of the Aroifilla-Chaunaca succession (see above), and must be interpreted the same way -i.e., as an evolution from a transgressive restricted-marine environment toward playa-lacustrine conditions and dessiccation. Away from the basin axis, the lower Santa Lucía consists of upward-coarsening and thickening beds of alternating red mudstones and reddish sandstones, mainly deposited in a prograding fluviatile environment, which show onlap toward the external areas. The upper Santa Lucía, the top of which is truncated by the Early Eocene unconformity, was deposited in a lacustrine to alluvial environment.

The Impora Formation overlies a major paleosol at the top of the lower Santa Lucía in the Camargo syncline, southeast of the CPKT. It was deposited in a fluvial to lacustrine environment, which must have connected with the lacustrine depositional environment of the Late Paleocene-age Maíz Gordo Formation in northwestern Argentina because of strong facies similarities. The Impora is capped by the Early Eocene unconformity (Marocco et al., 1987). Two correlations are possible: (1) because of a similar stratigraphic position between the lower Santa Lucía and the Early Eocene unconformity, the upper Santa Lucía and the Impora are time-equivalents from respectively northwest and southeast of the CPKT; (2) the major paleosol at the top of the lower Santa Lucía southeast of the CPKT developed during the upper Santa Lucía deposition time, and the Impora constitutes a supplementary unit only present in the Camargo syncline. Because the Camargo syncline belonged to a permanently low-subsident domain (Camargo-Atocha high; Figure 3; see above), there is little chance that an additional unit would have been deposited and preserved there. For this reason, correlation (1) is preferred. The Impora-Maíz Gordo lacustrine basin thus evolved in apparently disconnected fashion from the remainder of the Potosí basin, where the upper Santa Lucía was being deposited at that same time. Separation of both basins in Late Paleocene time points to tectonic reactivation of the CPKT (see below). Marls and some carbonates were deposited in the first of these basins, while only siliciclastic sediments accumulated in the second, suggesting that some climatic gradient was also established in these latitudes during Late Paleocene time.

All five sequences show a regressive pattern organization, as do the Coniacian-Late Campanian ones. A type-sequence similar to the one evidenced for the Coniacian-Late Campanian interval can be deduced for the latest Campanian-Paleocene succession (Figure 2): three transgressions of restricted-marine water bodies occurred during the El Molino depositional time and were followed by progressive regressions, the last of which gave place to the development of an initial large lake system (lower Santa Lucía), which ended with almost total colmatation and dessiccation and was followed in turn by the contemporary development of two lake systems (Impora and upper Santa Lucía Formations).

These environmental interpretations are at odds with those reached near Camargo by Okamoto et al. (1990). As already noted above, these authors unfortunately mistook the Santa Lucía and Impora formations for the El Molino Formation, in spite of their characteristic red and purple colours. They moreover interpreted what happens to be the Santa Lucía as having been deposited in transitional-offshore, marginal bar and tidal channel (sand-wave) marine sub-environments, and what is the Impora Formation as having a transitional-marine to lagoonal and intertidal origin. Because paleosols and fluviatile channels are quite abundant in the corresponding outcrops of the Camargo area, particularly in the Santa Lucía Formation, Okamoto et al.'s interpretations would seem to be erroneous and misleading.

On the basis of chronological data and proposed correlations with global-eustatic transgressions, the five recognized sequences approximately correspond to the following time spans:

-lower El Molino: latest Campanian-Early Maastrichtian

-middle El Molino: Late Maastrichtian

-upper El Molino: Early Danian

-lower Santa Lucía: Late Danian-?Early Thanetian

-upper Santa Lucía (and Impora Formation): Late (and Early?) Thanetian-?earliest Eocene.

Decompacted average sedimentation rates may therefore be calculated for the basin axis as follows:

-lower El Molino: 180 to 380 m/m.y.

-middle El Molino: 110 to 250 m/m.y.

-upper El Molino: 60 to 130 m/m.y.

-lower Santa Lucía: 90 to 220 m/m.y.

These values are comparable because these units are mostly composed of mudrocks and were buried at a similar paleodepth. Thus, they suggest that average sedimentation and subsidence rates had a tendency to decrease during the El Molino depositional history, and to increase slightly near the El Molino-Santa Lucía boundary. However, there is no indication as to the evolution of the sedimentation and subsidence rates during each time interval, such as there is for the Coniacian-Late Campanian period.

Synsedimentary Magmatic Occurrences

White to green tuffaceous beds, partly reworked and altered and always less than 30 cm thick, occur at least in the lower part of the lower El Molino, in all the upper El Molino and in the base of the lower Santa Lucía at several localities throughout the basin. Their initial volcanic content is thought to have been emitted by the subduction-related magmatic arc of the Pacific margin.

Basalt flows intercalate with El Molino equivalents (Lecho, Yacoraite and Olmedo Formations) in northern Argentina (Bianucci et al., 1981; Salfity and Marquillas, 1986), mostly in the undeformed part of the basin; i.e., in an easternmost and quite external position.

Tectonic Interpretation for the Latest Campanian-Paleocene Interval

Figures 10 and 11 illustrate synthetic paleogeographies reconstructed for the lower El Molino and lower Santa Lucía Formations, respectively. Two tectonic events are particularly noteworthy within this time interval.

The first event, approximately of latest Campanian age, associates the abrupt increase in sedimentation rate at the Chaunaca/El Molino boundary (from about 20 m/m.y. to at least 180 m/m.y.) with the rapid translation of the basin axis toward the west (compare Figures 8 and 10) and the invasion of the lowest part of the El Molino by white sands transported from the east and southeast. These occurrences are interpreted as having been produced by a sharp increase in the rate of deformation and tectonic loading along the coastal areas. As described in current stratigraphic models of foreland basins (e.g., Tankard, 1986; Heller et al. 1988; Jordan et al., 1988; Flemings and Jordan, 1989, 1990), reactivation of tectonic loading increases subsidence, produces a forebulge uplift in the external areas (the source for the lower El Molino sands) and displaces the subsident axis toward the internal zones. Increased subsidence probably enhanced the effects of the global-eustatic earliest Maastrichtian marine transgression (see above).

The second event associates the Paleocene-age deformation described by Marocco et al. (1987) with the reactivation of the CPKT evidenced at the lower/upper Santa Lucía boundary (see above). According to the chronological data and hypotheses used in this paper, this event would have taken place in "Middle" Paleocene time. Because no increase in sedimentation rate, no migration of the basin axis, no grain size evolution and no onlap phenomena are perceptible, and because fault activations apparently occurred within the basin on a purely local scale (see Marocco et al., 1987), it is likely that this event is not directly related to a change in shortening velocity and/or intensity along the coastal areas. However, this event shortly predates the earliest Eocene (or latest Paleocene?) unconformity, which in Bolivia records a profound modification of the tectonic setting, including forebulge uplift (generating the light-colored Cayara Formation sands) and westward migration of the subsident axis (Sempere, 1990; Sempere, 1991, and unpublished; Marshall and Sempere, 1991). It might thus reflect a first stage in this major tectonic destabilization.

Between these events, the sedimentation rate remained relatively high and onlap developed toward the external areas (see above). Sedimentation and subsidence rates seem to have slightly increased in Early Paleocene time. However, the tectonic context remained basically the same from latest Campanian to "Middle" Paleocene times, and the basin functioned as the distal part of an underfilled foreland basin, continuously and somehow steadily. Hence it can be proposed that shortening along the Pacific margin of South America, at these latitudes, went on fairly regularly during this time interval. The latest Campanian event marked the definite re-initiation of relatively high sedimentation (and thus subsidence) rates in the external Bolivian part of the paleo-Andean foreland basin. In such a subsident context, it is not surprising that some of the contemporary global-eustatic variations were apparently well recorded.

Summary of the Senonian-Paleocene Evolution

The Senonian-Paleocene period appears to be the first Andean-age time interval during which foreland conditions, albeit distal, controlled sedimentation in Bolivia. Establishment of this foreland setting started approximately at the Turonian-Coniacian boundary and was recorded by the lower part of the Aroifilla Formation. A most prominent coeval phenomenon seems to be the abundant basic magmatism displayed by the lower Aroifilla sequence, in Andean Bolivia, and by sequences in northwestern Argentina. The geographic distribution of the lower Aroifilla basalts shows that old weakness zones such as the Tupiza-San Lorenzo graben and, to a lesser degree, the CPKT (Figure 7), were intensely reactivated at that time. Several rift geometries and features have been identified in the Argentine-Bolivian basin (Bianucci et al., 1981; Galliski and Viramonte, 1985; Salfity and Marquillas, 1986; Sempere et al., 1988a). These facts strongly suggest that the geometry of the asthenospheric wedge related to subduction was considerably modified in Coniacian time between latitudes 17°S and 27°S, generating a species of bulge which produced extensional phenomena and abundant basic magmatism at the surface. Such a modification of the asthenospheric wedge geometry is coeval with the onset of considerable shortening along the Pacific margin (Jaillard, this volume), which would indicate that this tectonic upheaval developed on a fairly major scale. Involvement of the asthenospheric wedge suggests that a change in subduction angle is likely to have occurred.

Because there are probable links between major global regressions and coeval major continental shortening in some parts of the world, the sharp increase in shortening marked by this central Andean event might possibly have been one of the factors causing a major global regression in latest Turonian time (Haq et al., 1987), in which case the exact age for the onset of this tectonic upheaval would be latest Turonian.

Once the foreland conditions were established, they were maintained by progressive -albeit unsteady- shortening along the Pacific margin, and the Bolivian basin evolved under this control (Figure 12). However, two main periods must be recognized in this evolution: the Coniacian (or latest Turonian?)-Late Campanian interval (~15 m.y.) and the latest Campanian-Paleocene interval (~18 to 22 m.y.). Both periods began with a tectonic event caused by (re)activation of tectonic loading along the margin, and marked in the Bolivian basin by a sedimentary discontinuity. Average apparent sedimentation and subsidence rates decrease after both discontinuities, but remain higher in the second period (Figure 2). The upper boundary of the Puca Group is another important event, of similar origin and of earliest Eocene or latest Paleocene age.

The Senonian-Paleocene interval is characterized by the occurrence of five unequal transgressions of very shallow restricted-marine environments, in Early Santonian, Middle Campanian, Early Maastrichtian, Late Maastrichtian and Early Danian times, respectively. The notable flatness of the basin and the varying subsidence conditions permitted and controlled these ingressions, along with the global eustatic level. The related deposits, which appear at the base of the respective sedimentary sequences, include green to black mudstones, and are interbedded with a sequence of predominantly red mudrocks.

Tectonic Interpretation of the "Cretaceous" Record, and Conclusions

The sedimentary record of Bolivia makes it possible to interpret the Cretaceous tectonic history of the Andean system between latitudes 17°S and 22°S. This history can be summarized as follows:

-Extensional and/or transtensional conditions were established in latest Jurassic and/or earliest Cretaceous time and initiated the functioning of the Potosí basin, which was highly fragmented during this structuration. Bolivia, which had until then been a part of stable South America, was captured by the Andean tectonic system at that time. Tensional expressions were evident until Late Neocomian or Early Aptian times.

-Because the extensional conditions decreased in intensity during the Neocomian, fragmentation of the Potosí basin had ceased by approximately Albian time (Tarapaya Formation depositional period), allowing sedimentation to onlap onto some high areas and subsidence to be more equally distributed. Although interpretation of the Tarapaya onlap remains unprecise, there is little reason to think that it was related to the establishment of foreland conditions according to what is known concerning the coeval evolution of Peru (Jaillard, this volume), and it was most likely due to a large-scale extensional and/or transtensional context.

Fig. 10

Schematic paleogeographic image for the lower sequence of the El Molino Formation (latest Campanian-early Maastrichtian). ➤ Arrows: paleocurrent data. Circle pattern: oolitic/oncolitic barrier areas. Black dots: main localities of the Cajones Formation (see below). Ruled: CPKT. Isopach curves are approximate, and thicknesses (undecompacted) are in meters; curve from Argentina is taken from Salfity and Marquillas (1986). The basin axis is defined by "deepest" facies, but its location is approximate, in part because of the notable flatness of the basin. The localities where the El Molino Formation directly overlies pre-Cretaceous rocks appear in Figure 8. The reddish, sandy and paleosol-rich Cajones Formation of the central Subandean belt was deposited in a fluviatile environment and is classically considered to be an equivalent of the El Molino Formation because of its scarce fish and dinosaur fauna (but it might also be of Campanian age in part).

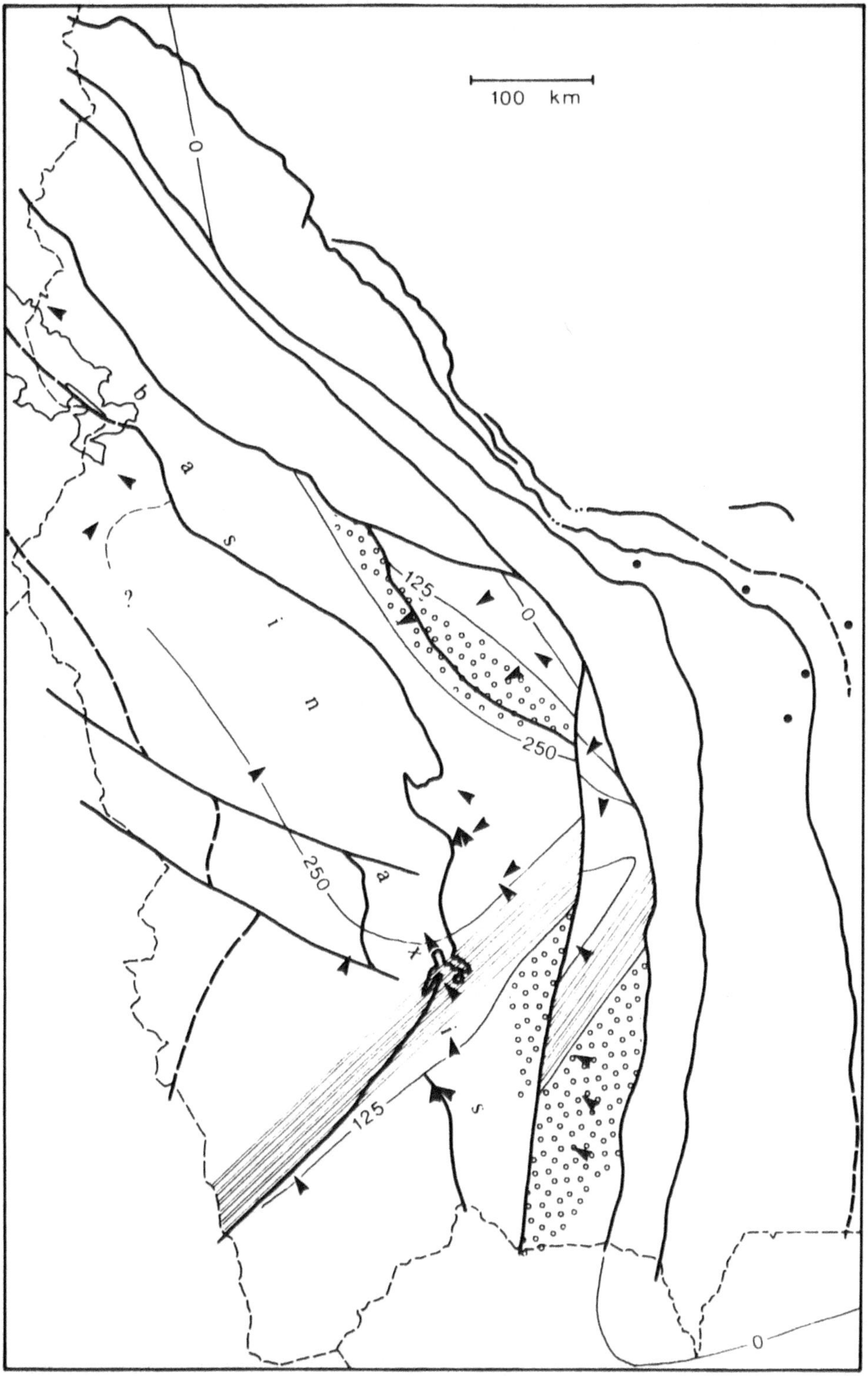
100 km
0
b
a
s
i
n
?
125
0
250
250
a
x
i
s
125
0

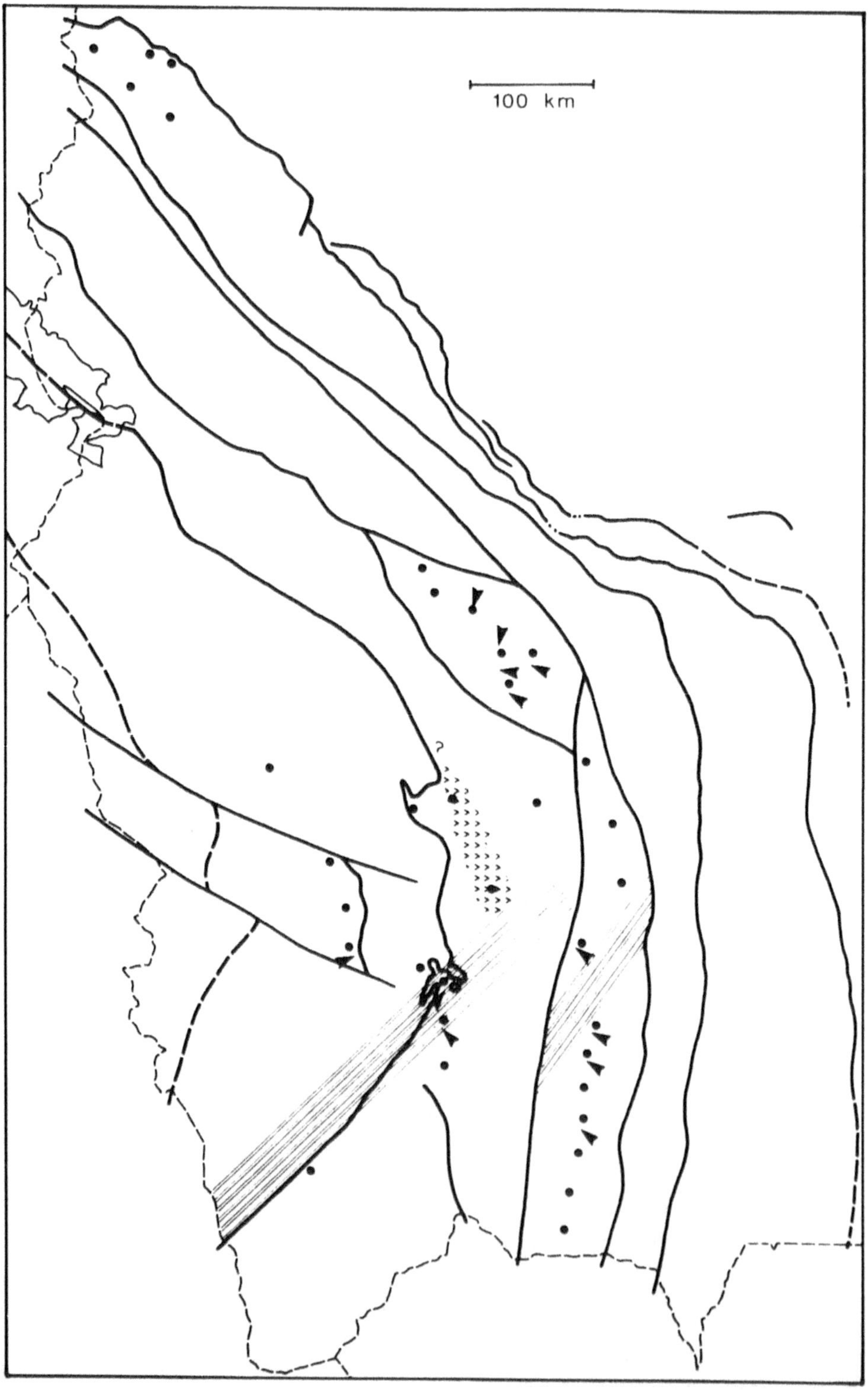
100 km

-Tectonic quiescence reigned during the Cenomanian-Turonian period. For the first time since the Permian, Bolivia was reached by a marine transgression, which in this case was due to the Cenomanian-Turonian global high sea level.
-A major tectonic upheaval started in latest Turonian or earliest Coniacian time, involving modification of the geometry of the asthenospheric wedge related to subduction and initiation of distal foreland conditions in Bolivia. This event points to a change in the subduction angle and to the onset of notable shortening along the Pacific margin. Another event of lesser magnitude, approximately in latest Campanian time, initiated a period of more rapid tectonic subsidence. The Puca Group depositional time ended with a third similar event, in earliest Eocene or latest Paleocene times. Ingression of very shallow restricted-marine waters occurred periodically in relation to global sea level highs, from the Santonian to the earliest Paleocene.
The Andean tectonic evolution began with the Kimmeridgian-age "Araucan" episode, which was coeval with, and therefore probably genetically related to, the initiation of rifting between South America and Africa. Until Late Turonian time, it was dominated by extensional and/or transtensional conditions interrupted along the Pacific coast by compressional and/or transpressional "phases" (Jaillard, this volume), and the corresponding tectonic regime was probably controlled by large-scale wrench-faulting that affected the internal margin and certainly built some "proto-Andean" reliefs along it. However, although latest Jurassic and/or Early Cretaceous thick red bed sequences are locally known in the costal areas (such as in northern Chile; Flint et al., 1986; Muzzio et al., 1988), no true foreland basins are known from this time span; i.e., before the Late Turonian. The Turonian-Coniacian boundary tectonic event thus marked a turning-point in the Andean evolution within the latitudes studied, since foreland conditions have prevailed in the eastern Andean domain from then on. Because the main Senonian to Recent sedimentary sequences of Bolivia were deposited as foreland basin fills, providing a record for substantial tectonic loading and hence shortening, it can be said that the compressional building of the central Andes really began at about the Turonian-Coniacian boundary.

Fig. 11

◄ Schematic paleogeographic image for the lower sequence of the Santa Lucìa Formation (late Danian-?early Thanetian). Dots: localities where at least part of the Santa Lucía Formation is present. Arrows: paleocurrent data. > pattern: evaporites (gypsum, anhydrite) in upper part of sequence. Ruled: CPKT. The basin axis would approximately coincide with the prolongation of the elongate evaporite playa- lake, the deposits of which crop out near Potosí. It is very likely that the Santa Lucía Formation was deposited in most of Andean Bolivia, but was partly or totally eroded before deposition of the Cayara-Potoco sequence (Eocene-early Oligocene).

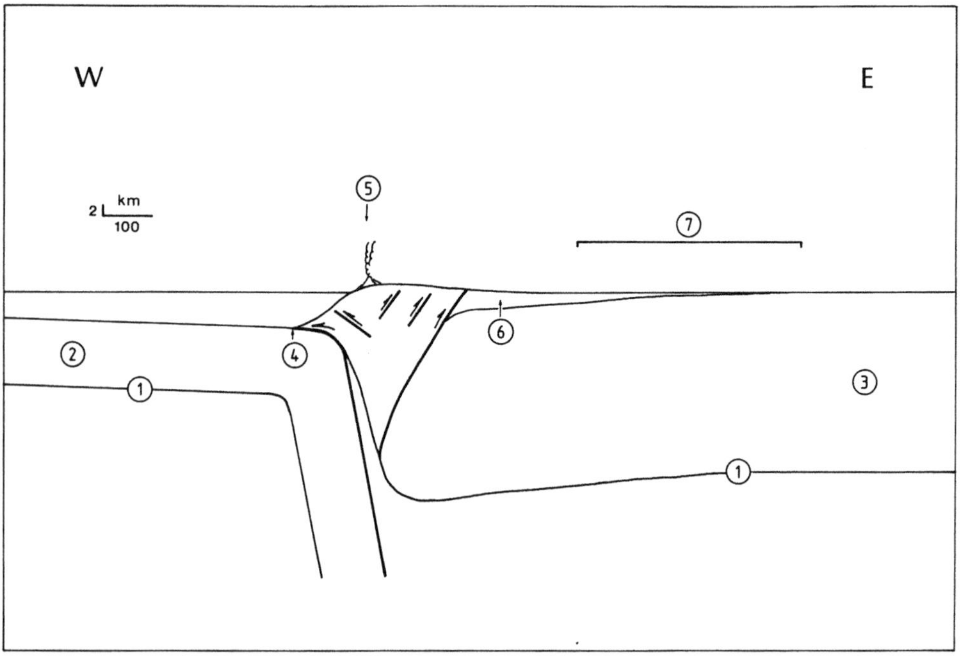

Fig. 12

Cartoon cross-section of the South American Pacific margin showing position of the Andean basin of Bolivia within the tectonic system induced by incipient compressional shortening of the margin in late Cretaceous time. Crustal heterogeneities, which are known to have existed, are not taken into account. Circled numbers: 1- Moho, 2- oceanic crust, 3- continental crust, 4- trench, 5- volcanic arc, 6- foreland basin, 7- approximate position of Bolivian part of foreland basin. In Bolivia, the Senonian-Paleocene foreland basin of the paleo-Andes was wide, gently subsident and filled with fine- grained sediments. According to the model by Flemings and Jordan (1989), such characteristics are favored by low flexural rigidities (e.g., about 10^{22} N.m) and/or high sediment transport coefficients (e.g., about 20000 km^2/m.y.). Low thrust velocities also tend to increase width of basin (albeit with a rather coarse-grained fill), whereas low erosional transport coefficients tend to increase proportion of fine-grained deposits in the basin (but to narrow its geometry). However, this model is based on a steady state hypothesis where the tectonic load applied to the passive (South American) lithosphere is held constant and relatively high. In Senonian-Paleocene times, thrusting had just begun along the Pacific margin of South America (see text), which had been weakened during Jurassic and early Cretaceous times: the tectonic load was therefore relatively low, as was the flexural rigidity of the margin. Flexure of the South American lithosphere is thus expected to have been localized at the tip of the plate; i.e., to have presented a low radius of curvature, which is schematically shown here.

Acknowledgments

Field work on the "Cretaceous" of Bolivia started in 1983 and was funded by Orstom (UR 1H) and Yacimientos Petrolíferos Fiscales Bolivianos (YPFB). L. Barrios, G. Beccar, J.

Blanco, G. Camoin, G. Carlier, H.C. Chávez, C. Cherroni, M. Cirbián, J. Dejax, M. Gayet, J. Jarandilla, M. López, R.A. Marquillas, R. Marocco, C. Martinez, G. Montemurro, M.A. Pérez, J.M. Rouchy, J.A. Salfity, P. Soler, H. Toledo, D. Zubieta and others participated in field trips and fruitful discussions. Special thanks go to J. Oller, head of the Division of Regional Studies at the Gerencia de Exploración of YPFB in Santa Cruz de la Sierra, for his incomparable knowledge of Bolivian stratigraphy and localities, to E. Jaillard, without whose aid reliable correlations with Peru and some conclusions of this work would have been impossible, to L.G. Marshall for fruitful discussions about the Maastrichtian-Paleocene period, to A. Iatzoura for valuable information concerning the Maastrichtian of Venezuela, and to B. Beaudoin for advice on appropriate decompaction techniques. The manuscript benefited from comments by E. Jaillard and L.G. Marshall.

References

Benedetto, J.L., and Sánchez, T. (1971): El hallazgo de peces Pycnodontiformes (Holostei) en la Formación Yacoraite (Cretácico superior) de la provincia de Salta, Argentina, y su importancia paleoecológica. Acta Geólogica Lilloana, v. 11, 151-175, Tucumán.

Bianucci, H., Acevedo, O., and Cerdán, J. (1981): Evolución tectosedimentaria del Grupo Salta en la subcuenca Lomas de Olmedo (provincias de Salta y Formosa). Congreso Geológico Argentino, 8th, San Luis, Actas, 3, 159-172.

Branisa, L., Hoffstetter, R., Freneix, S., Roman, J., and Sornay, J. (1966): Nouvelle contribution à l'étude de la paléontologie et de l'âge du Groupe Puca (Crétacé de Bolivie). Bulletin du Muséum d'Histoire Naturelle, 38/3, 301-310, Paris.

Branisa, L., Grambast, L., and Hoffstetter, R. (1969): Quelques précisions nouvelles, d'après les charophytes, sur l'âge du Groupe Puca (Crétacé-Paléocène, Bolivie). Comptes Rendus Sommaires des Séances de la Société Géologique de France, 8, 321-322.

Camoin, G., Rouchy, J.M., Babinot, J.F., Deconinck, J.F., and Tronchetti, G. (1991): Dynamique sédimentaire et évolution paléogéographique d'un bassin continental en position d'arrière-arc: exemple du Maastrichtien de la Cordillère Orientale (Bolivie). Comptes Rendus à l'Académie des Sciences de Paris, série II, 312, 1335-1341.

Cappetta, H. (1975): Sur quelques sélaciens nouveaux de Bolivie (Amérique du Sud). Geobios, v. 8, 5-24.

Cappetta, H. (1991): Late Cretaceous selachian faunas from Bolivia: new data and summary, *in* Fósiles y Facies de Bolivia, (Ed.: Suárez-Soruco, R.), Revista Técnica de Yacimientos Petrolíferos Fiscales Bolivianos, v. 12, 435-439, Santa Cruz.

Castaños, A., Pinedo, R., and Salfity, J.A. (1975): Nuevas consideraciones sobre la Formación Yacoraite del Cretácico superior del Norte argentino. Revista Técnica de Yacimientos Petrolíferos Fiscales Bolivianos, v. 4, 31-59, La Paz.

Chávez, H.C. (1987): Interpretación geodinámica de las formaciones sedimentarias y estructuras tectónicas de la zona de Challapata. Tesis de Grado, Universidad Mayor de San Andrés, La Paz, 163 pp.

Cherroni, C. (1977): El Sistema Ccretácico en la parte boliviana de la cuenca andina. Revista Técnica de Yacimientos Petrolíferos Fiscales Bolivianos, v. 6, 5-46, La Paz.

Chigne, N., and Hernández, L. (1990): Main aspects of petroleum exploration in the Apure area of southwestern Venezuela, 1985-1987, *in* Classic Petroleum Provinces, (Ed.: Brooks, J.), Geological Society Special Publication, 50, 55-75.

Cirbián, M., Marocco, R., Oller, J., and Sempere, T. (1986): Evidencia de una discontinuidad sedimentaria fundamental en el Eoterciario andino de Bolivia. Congreso Geológico Boliviano, 8th, La Paz, Resúmenes, p. 45.

Evernden, J.F., Kriz, S., and Cherroni, C. (1977): Potassium-argon ages of some Bolivian rocks. Economic Geology, v. 72, 1042-1061.

Flemings, P.B., and Jordan, T.E. (1989): A synthetic stratigraphic model of foreland basin development. Journal of Geophysical Research, v. 94, 3851-3866.

Flemings, P.B., and Jordan, T.E. (1990): Stratigraphic modeling of foreland basins: Interpreting thrust deformation and lithosphere rheology. Geology, v. 18, 430-434.

Flint, S., Clemmey, H., and Turner, P. (1986): The lower Cretaceous Way Group of northern Chile: an alluvial fan-fan delta complex. Sedimentary Geology, v. 46, 1-22.

Fritzsche, C.H. (1924): Neue Kreidefauna aus Südamerika (Chile, Bolivia, Peru, Columbia). Neues Jahrbuch für Mineralogie, Geologie und Paläontologie, v. 50, 1-56.

Galliski, M.A., and Viramonte, J.G. (1985): Un paleorift cretácico en el noroeste argentino: aproximación petrológica. Universidad de Chile, Departamento de Geología y Geofísica, Comunicaciones, 35, 89-91, Santiago.

Gayet, M. (1991): Holostean and teleostean fishes from Bolivia, *in* Fósiles y Facies de Bolivia, (Ed.: Suárez-Soruco, R.), Revista Técnica de Yacimientos Petrolíferos Fiscales Bolivianos, v. 12, 453-493, Santa Cruz.

Gayet, M., Marshall, L.G., and Sempere, T. (1991a): The Mesozoic and Paleocene vertebrates of Bolivia and their stratigraphic context: a review, *in* Fósiles y Facies de Bolivia, (Ed.: Suárez-Soruco, R.), Revista Técnica de Yacimientos Petrolíferos Fiscales Bolivianos, v. 12, 393-433, Santa Cruz.

Gayet, M., Sempere, T., and Cappetta, H. (1992): A propos de l'environnement marin restreint du bassin centro-andin au Maastrichtien. Comptes Rendus à l'Académie des Sciences de Paris, série II, 314, 223-228.

Gómez Omil, R.J., Boll, A., and Hernández, R.M. (1989): Cuenca cretácico-terciaria del Noroeste argentino (Grupo Salta) *in* Cuencas Sedimentarias Argentinas, (Eds.: Chebli, G.A., and Spalletti, L.A.), Universidad Nacional de Tucumán, Serie Correlación Geológica, 6, 43-64, Tucumán.

Haq, B.U., Hardenbol, J., and Vail, P.R. (1987): Chronology of fluctuating sea-levels since the Triassic. Science, v. 235, 1156-1167.

Heller, P.L., Angevine, C.L., Winslow, N.S., and Paola, C. (1988): Two-phase stratigraphic model of foreland-basin sequences. Geology, v. 16, 501-504.

Jaillard, E. (1986): La sédimentation crétacée dans les Andes du Pérou central: exemple de la Formation Jumasha (Albien moyen-supérieur à Turonien supérieur) dans la région d'Oyón (Département de Lima). Géodynamique, v. 1, 97-108.

Jaillard, E. (1987): Sedimentary evolution of an active margin during middle and upper Cretaceous times: the North Peruvian margin from Late Aptian up to Senonian. Geologische Rundschau, v. 76, 677-697.

Jaillard, E., and Sempere, T. (1989): Cretaceous sequence stratigraphy of Peru and Bolivia, *in* Contribuciones de los Simposios sobre el Cretácico de América Latina, A: Eventos y Registro Sedimentario, (Ed.: Spalletti, L.A.), GSGP-IGCP 242, 1-27, Buenos Aires.

Jaillard, E., and Sempere, T. (1991): Las secuencias sedimentarias de la Formación Miraflores y su significado cronoestratigráfico. Revista Técnica de Yacimientos Petrolíferos Fiscales Bolivianos, v. 12, 257-264, Santa Cruz.

Jaillard, E., Cappetta, H., Ellenberger, P., Feist, M., Grambast-Fessard, N., Lefranc, J.-P., and Sigé, B. (1992): The Late Cretaceous Vilquechico Formation, Southern Peru. Sedimentology, Paleontology, Biostratigraphy, Correlations. Newsletters on Stratigraphy, in press.

Jordan, T.E., Flemings, P.B., and Beer, J.A. (1988): Dating thrust-fault activity by use of foreland-basin strata, *in* New Perspectives in Basin Analysis, (Eds.: Kleinspehn, K.L., and Paola, C.), New York, Springer-Verlag, 307-330.

Kriz, S., and Cherroni, C. (1966): Diagramas correlativos de las formaciones cretácicas del Sudoeste de Bolivia. Servicio Geológico de Bolivia, Hoja Informativa, 2, La Paz.

Legarreta, L., Kokogián, D.A., and Boggetti, D.A. (1989): Depositional sequences of the Malargüe Group (upper Cretaceous-lower Tertiary), Neuquén Basin, Argentina. Cretaceous Research, v. 10, 337-356.

Lehmann, B. (1978): A Precambrian core sample from the Altiplano/ Bolivia. Geologische Rundschau, v. 67, 270-278.

Lohmann, H.H. (1970): Outline of tectonic history of Bolivian Andes. American Association of Petroleum Geologists Bulletin, v. 54, 735-757.

Lohmann, H.H., and Branisa, L. (1962): Estratigrafía y paleontología del Grupo Puca en el sinclinal de Miraflores, Potosí. Petróleo oliviano, v. 4, 9-16, La Paz.

Macellari, C. (1988): Cretaceous paleogeography and depositional cycles of western South America. Journal of South American Earth Sciences, v. 1, 373-418.

Marinovic, N., and Lahsen, A. (1984): Hoja Calama, Región de Antofagasta. Servicio Nacional de Geología y Minería, Carta Geológica de Chile, 58, 144 pp., Santiago.

Marocco, R., Sempere, T., Cirbián, M., and Oller, J. (1987): Mise en évidence d'une déformation paléocène en Bolivie du Sud. Sa place dans l'évolution géodynamique des Andes centrales. Comptes Rendus à l'Académie des Sciences de Paris, série II, 304, 1139-1143.

Marquillas, R.A. (1984): La Formación Yacoraite (Cretácico superior) en el Río Juramento, Salta: estratigrafía y ciclicidad. Congreso Geológico Argentino, 9th, Bariloche, Actas, 5, 186-196.

Marquillas, R.A. (1985): Estratigrafía, sedimentología y paleoambientes de la Formación Yacoraite (Cretácico superior) en el tramo austral de la cuenca, Norte argentino. Dr Thesis, Universidad Nacional de Salta, Salta, 139 pp.

Marquillas, R.A. (1986): Ambiente de depósito de la Formación Yacoraite (Grupo Salta, Cretácico-Eoceno), Norte argentino. Simposio PIGC-Proyecto 242 Cretácico de América Latina, 1st, La Paz, 153-169.

Marquillas, R.A., Boso, M.A., and Salfity, J.A. (1984): La Formación Yacoraite (Cretácico superior) en el Norte argentino, al sur del paralelo 24°. Congreso Geológico Argentino, 9th, Bariloche, Actas, 2, 300-310.

Marshall, L.G., and Sempere, T. (1991): The Eocene to Pleistocene vertebrates of Bolivia and their stratigraphic context: a review. *in* Fósiles y Facies de Bolivia, (Ed.: Suárez-Soruco, R.), Revista Técnica de Yacimientos Petrolíferos Fiscales Bolivianos, v. 12, 631-652, Santa Cruz.

Martinez, C. (1980): Structure et évolution de la chaîne Hercynienne et de la chaîne Andine dans le nord de la cordillère des Andes de Bolivie. Paris, Travaux et Documents de l'Orstom, 119, 352 pp.

Martinez, C., and Vargas, E. (1988): Sur les déformations synsédimentaires mésozoïques de la région de Macha-Pocoata-Colquechaca (nord de Potosí, Cordillère orientale de Bolivie). Géodynamique, v. 3, 107-115.

Martinez, C., Vargas, E., and Laubacher, G. (1990): Evolution tectono-sédimentaire dans le Crétacé du synclinal d'Otavi-San Lucas (Bolivie centre-sud). Symposium International Géodynamique Andine, Grenoble, Ed. ORSTOM, Colloques et Séminaires, 241-244.

McBride, S.L., Robertson, R.C.R., Clark, A.H., and Farrar, E. (1983): Magmatic and metallogenetic episodes in the northern tin belt, Cordillera Real, Bolivia. Geologische Rundschau, v. 72, 685-713.

Méndez, I., and Viviers, M.C. (1973): Estudio micropaleontológico de sedimentitas de la Formación Yacoraite (Provincias de Salta y Jujuy). Congreso Geológico Argentino, 5th, Carlos Paz (1972), Actas, 3, 467-470.

Mourier, T., Bengtson, P., Bonhomme, M., Buge, E., Cappetta, H., Crochet, J.-Y., Feist, M., Hirsch, K.F., Jaillard, E., Laubacher, G., Lefranc, J.-P., Moullade, M., Noblet, C., Pons, D., Rey, J., Sigé, B., Tambareau, Y., and Taquet, P. (1988b): The Upper Cretaceous-Lower Tertiary marine to continental transition in the Bagua basin, Northern Peru. Paleontology, biostratigraphy, radiometry, correlations. Newsletters on Stratigraphy, v. 19, 143-177.

Muzzio, G., Davidson, J., and Covacevich, V. (1988): Las formaciones Blanco y Atajaña del Cretácico inferior, Tarapacá, Chile: correlaciones regionales y paleogeografía. Universidad de Chile, Departamento de Geología y Geofísica, Comunicaciones, 39, p. 216, Santiago.

Naranjo, J.A., Hervé, F., Prieto, X., and Munizaga, F. (1984): Actividad cretácica de la falla Atacama al este de Chañaral: milonitización y plutonismo. Universidad de Chile, Departamento de Geología y Geofísica, Comunicaciones, 3, 57-66, Santiago.

Odin, G.S., and Odin, C. (1990): Echelle numérique des temps géologiques. Géochronique, v. 35, 12-21.

Okamoto, K., Oiso, Y., and Lema, J.C. (1990): Facies sedimentarias controladas por eventos transgresivos y regresivos en el Cretácico del sinclinal de Camargo (Cordillera Oriental de la zona andina, Departamento de Chuquisaca, Bolivia). Revista Técnica de Yacimientos Petrolíferos Fiscales Bolivianos, v. 11, 167-178, Cochabamba.

Oller, J., and Sempere, T. (1990): A fluvio-eolian sequence of probable Middle Triassic-Jurassic age in both Andean and Subandean Bolivia. Symposium International Géodynamique Andine, Grenoble, Ed. ORSTOM, Colloques et Séminaires, 237-240.

Palma, R.M. (1986): Interpretación ambiental de las formaciones del límite Cretácico-Terciario en el noroeste de Argentina. Zeitblatt für Geologie und Paläontologie, I, 1985, 1337-1350.

Pérez, M.A. (1987): Datos palinológicos del Cretácico superior de la sección estratigráfica de Carata (Potosí, Bolivia). Congreso Latinoamericano de Paleontología, 4th, Santa Cruz, Memorias, II, 739-756.

Perry, L.D. (1963): Flora Formation (Upper Cretaceous) of northern Bolivia. American Association of Petroleum Geologists Bulletin, v. 47, 1855-1861.

Pilsbry, H.A. (1939): Freshwater Mollusca and Crustacea from near El Molino, Bolivia. The Johns Hopkins University Studies in Geology, v. 13, 69-72.

Reyes, F.C. (1972): Correlaciones en el Cretácico de la cuenca andina de Bolivia, Perú y Chile. Revista Técnica de Yacimientos Petrolíferos Fiscales Bolivianos, v. 1, 101-144, La Paz.

Riccardi, A.C. (1988): The Cretaceous System of Southern South America. Geological Society of America Memoir, 168, 161 pp.

Rossi, T., Stephan, J.F., Blanchet, R., and Hernández, G. (1987): Etude géologique de la Serranía del Interior Oriental (Vénézuela) sur le transect Cariaco-Maturin. Revue de l'Institut Français du Pétrole, v. 42, 3-30.

Russo, A., and Rodrigo, L.A. (1965): Estratigrafía y paleogeografía del Grupo Puca en Bolivia. Boletín del Instituto Boliviano del Petróleo, v. 5, 5-51, La Paz.

Salfity, J.A., and Marquillas, R.A. (1986): Marco tectónico y correlaciones del Grupo Salta (Cretácico-Eoceno), República Argentina. Simposio PIGC-Proyecto 242 Cretácico de América Latina, 1st, La Paz, 170-184.

Sempere, T. (1990): Cuadros estratigráficos de Bolivia: propuestas nuevas. Revista Técnica de Yacimientos Petrolíferos Fiscales Bolivianos, v. 11, 215-227, Cochabamba.

Sempere, T. (1991): Cenozoic tectonic "phases" in Bolivia: Some needed clarifications. Congreso Geológico Chileno, 6th, Viña del Mar, Actas, 1, 877-881.

Sempere, T., Oller, J., Cherroni, C., Aranibar, O., Barrios, L., Branisa, L., Cirbián, M., and Pérez, M.A. (1987): Un ejemplo de cuenca carbonatada en un contexto distensivo de retroarco: paleogeografía del Cretácico terminal en la República de Bolivia (Formación El Molino y equivalentes). Congreso Geológico Argentino, 10th, Tucumán, Simposio PIGC-Proyecto 242 Cretácico de América Latina, 3rd, Resúmenes, 18-19, and unpublished manuscript.

Sempere, T., Oller, J., and Barrios, L. (1988a): Evolución tectosedimentaria de Bolivia durante el Cretácico. Congreso Geológico Chileno, 5th, Santiago, Actas, 3 (H), 37-65.

Sempere, T., Hérail, G., and Oller, J. (1988b): Los aspectos estructurales y sedimentarios del oroclino boliviano. Congreso Geológico Chileno, 5th, Santiago, Actas, 1 (A), 127-142.

Sempere, T., Hérail, G., Oller, J., and Baby, P. (1989): Geologic structure and tectonic history of the Bolivian orocline. International Geological Congress, 28th, Washington D.C., Abstract, 3, 72-73.

Sempere, T., Hérail, G., Oller, J., and Baby, P. (1990a): Estructura e historia tectónica del oroclino boliviano. Revista Técnica de Yacimientos Petrolíferos Fiscales Bolivianos, v. 11, 267-270, Cochabamba.

Sempere, T., Hérail, G., Oller, J., Baby, P., Barrios, L., and Marocco, R. (1990b): The Altiplano: a province of intermontane foreland basins related to crustal shortening in the Bolivian orocline area. Symposium International Géodynamique Andine, Grenoble, Ed. ORSTOM, Colloques et Séminaires, 167-170.

Sempere, T., Baby, P., Oller, J., and Hérail, G. (1991): La nappe de Calazaya: une preuve de raccourcissements majeurs gouvernés par des éléménts paléostructuraux dans les Andes boliviennes. Comptes Rendus à l'Académie des Sciences de Paris, série II, 312, 77-83.

Stipanicic, P.N., and Rodrigo, F. (1969): El distrofismo jurásico en Argentina y Chile. Jornadas Geológicas Argentinas, 4th, Mendoza, Actas, 2, 353-368.

Tankard, A.J. (1986): On the depositional response to thrusting and lithospheric flexure: examples from the Appalachian and Rocky Mountain basins, *in* Foreland Basins, (Eds.: Allen, P.A., and P. Horewood, P.), Special Publications of the International Association of Sedimentologists, Blackwell Publications, 8, 369-392.

Vicente, J.-C. (1981): Elementos de la estratigrafía mesozoica sur-peruana, *in* Cuencas Sedimentarias del Jurásico y Cretácico de América del Sur, (Eds.: Volkheimer, W., and Musacchio, E.), 1, 319-351, Buenos Aires.

Vicente, J.-C. (1989): Early Late Cretaceous overthrusting in the Western Cordillera of Southern Peru, *in* Geology of the Andes and its relations to energy and mineral resources, Santiago (1985), (Eds.: Ericksen, G.E., Cañas Pinochet, M.T., and Reinemund, J.A.), Circum-Pacific Council for Energy and Mineral Resources, Earth Science Series, v. 11, 91-117.

Wignall, P.B. (1991): Model for transgressive black shales? Geology, v. 19, 167-170.

Tectono-Sedimentary Evolution of the Cretaceous-Early Tertiary and Metallogenic Scheme of Northern Chile, Between 20° S and 26° S

T. Bogdanic C.
Cía Minera Carolina de Michilla S.A. Casilla 898. Antofagasta, Chile.

S. Espinoza R.
Departamento de Ciencias Geológicas, Universidad Católica del Norte, Casilla 1280, Antofagasta, Chile.

Abstract

A review of the Cretaceous geology and metallogenesis of the Chilean slope of the Southern Andes, between 20° and 26° south latitude (Antofagasta Region) is presented. An attempt is also made to interpret the geologic phenomena which occurred during the Late Jurassic-Early Tertiary time span.

The stratified rocks deposited during the period mentioned can be divided into three groups:

1) Sediments, mainly continental, resting on rocks from a fluvio- deltaic environment, which represent the final phase of infill of the Jurassic back-arc marine basin (Kimmeridgian-Barremian- Aptian?).

2) Andesitic-dacitic volcanic continental rocks, with sedimentary-volcanic intercalations (Albian-Cenomanian-Turonian?).

3) Sedimentary-volcanic continental rocks, with alluvial-fan development at the base, and volcanoclastic and volcanic rocks in the middle and upper part (latest Cretaceous-Late Eocene).

The intrusive rocks can also be divided into three groups, which are distributed in belts whose longitudinal axes migrate chronologically eastward. These groups are interpreted as being the result of the development of three magmatic arcs:

I) Late Jurassic-Early Cretaceous plutonic rocks; which correspond to the end of the Jurassic magmatic event.

II) Late Cretaceous intrusive rocks; end phase of the "Mid- Cretaceous" magmatic event.

III) Early Tertiary plutonic and sub-volcanic rocks; final phases of the latest Cretaceous-Eocene magmatism.

The end of the activity of each of these arcs is marked by the development of strike-slip fault systems (Atacama, San Cristóbal and West Fissure-Cordillera de Domeyko), and by folding, which can be considered an effect of the "Araucanian" (Kimmeridgian), "Peruvian" (Campanian) and "Incaic" (during the Eocene-Oligocene transition) tectonic phases.
In connection with these succesive periods of magmatic and tectonic activity a series of important metallogenic processes developed, both linked to volcanic and post-plutonic hydrothermal phases, which gave rise to the following metalliferous belts:
1) Coastal Cordillera Copper Belt (Late Jurassic-Early Cretaceous).
2) Central Depression Copper, Gold and Silver Belt (Late Cretaceous-Early Tertiary).
3) Pre-Andean Cordillera Porphyry-Copper Belt (Early Tertiary).

Introduction

Following the classic works of Brüggen (1950), Harrington (1961), Ruiz et al. (1965), García (1967) and Muñoz-Cristi (1973), and some others, which established the basic nomenclature for Chilean stratigraphy during the last few decades, there has been a great advance in knowledge of the evolution of the Andean chain. Numerous radiometric datings and a more detailed knowledge of the geologic units in the field from the wide range of studies available allow a far better understanding of the phenomena which occurred during Andean development.
The Cretaceous stratigraphic scheme for the area referred to differs considerably from the conventional model, especially with regard to its upper and lower limits, which in the field generally show an uninterrupted stratigraphic continuity.
The subdivisions of the Cretaceous System used here are those proposed by Haq and van Eysinga (1987) and those found in the Geologic Time Scale according to Harland et al. (1990).

Geomorphologic Framework

From a geomorphological point of view, this part of the Andean Cordillera has a very marked structural control, making it possible to identify various morphostructural blocks of individual shape and characteristics. In Northern Chile between 20°S and 26°S lat. (from west to east, see Figure 1), these entities are as follows:
a) *Coastal Cordillera*: This is a mountain chain running parallel to the coast, which stretches from Arica down to Central Chile, with a mean width of 50 km. On to the west, it rises sharply from the sea, with cliffs as high as 2,000 m; while to the east it slopes gently toward the Central Depression.

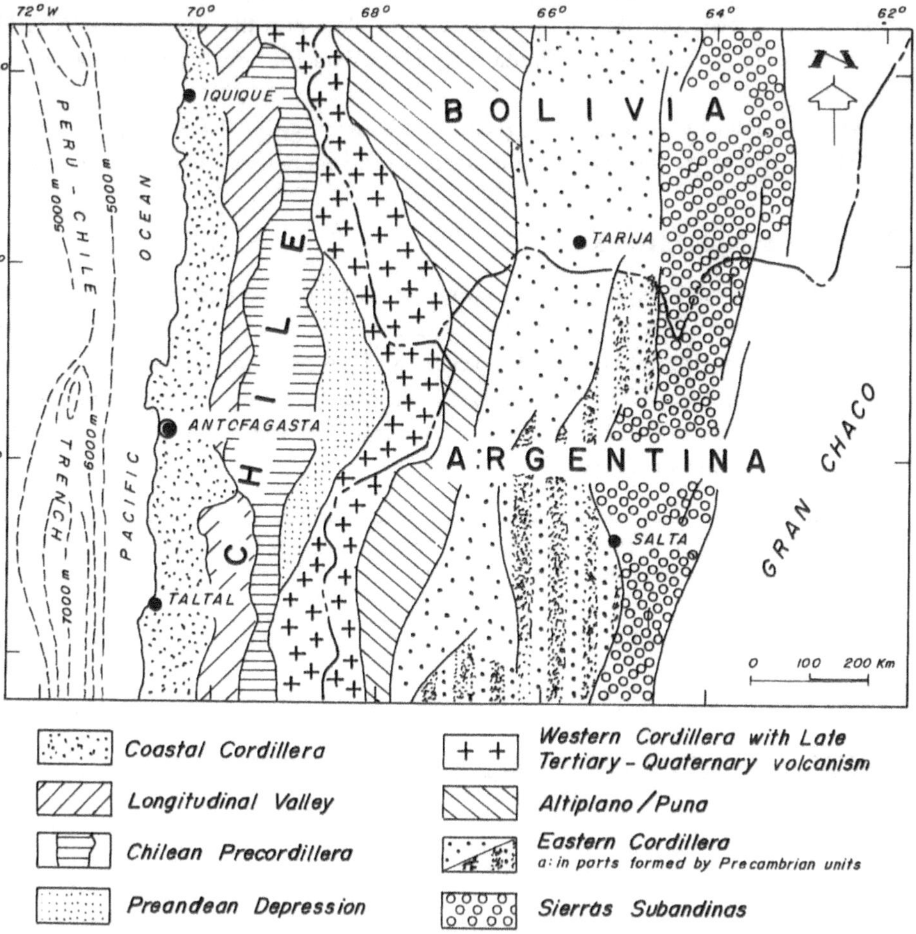

Fig. 1

Map of morphostructural units between 20°S and 26°S, after Reutter et al. (1986).

b) *Central Depression*: This is a geomorphologic feature that is well-defined between 18° and 23° S lat.; farther south, down to 27°S, it loses its physiographic identity and turns into a sequence of isolated hills and mountain ranges.

c) *Chilean Precordillera*: This is a mountain chain bearing NNE-SSW, with a mean altitude of 3,000 m asl which forms the eastern boundary of the Central Depression. The Precordillera is a highly complex structural massif which is cut by large E-W gullies ("quebradas"), and north of 21° S merges with the Western Cordillera.

d) *Preandean Depression*: This is a tectonic depression having a mean altitude of 2,000 masl and is located between the Chilean Precordillera and the High Andes (Cordillera Occidental). The Atacama and Punta Negra salt-flats ("salares") are emplaced in this depression.

e) *Western Cordillera*: This is the highest mountain range of the Andean chain (up to 6,500 masl) and forms the drainage divide; all the volcanoes now active are located in it.

Regional Geologic Framework

The area described contains a large variety of rocks whose ages range from Precambrian?-Paleozoic to Recent. The oldest rocks are a series of metamorphic rocks (migmatites, gneisses, amphibolites, mica-schists and greenschists), which outcrop as an elongate belt trending NNE-SSW at Sierra de Moreno, Cerros de Limón Verde and Sierra de Almeyda ranges in the Chilean Precordillera, and at the Península de Mejillones in the Coastal Cordillera

Location	Rock	Age (Ma)	Method	Interpretation
Sierra de Moreno *	migmatite gneiss	1.213 +28/-27	U/Pb in Zr	Maximum age
Limón Verde **	amphibolite	312-239	K/Ar in Hb	Metamorphic age
Mejillones ***	schist	530	Rb/Sr WR	Metamorphic age

* Damm et al. (1986); ** Herve et al. (1985); *** Díaz et al. (1985)

In these three localities the metamorphic rocks are intruded by granites and granodiorites whose ages correspond to the Ordovician-Silurian limit (U/Pb in Zr, between 502±7 and 450±11 Ma, Damm et al., 1986; K/Ar in muscovite, 431±10 Ma, Huete et al., 1977; Rb/Sr in rock, 452±4 and 441±8 Ma, Mpodozis et al., 1983).

Pelitic-psammitic sediments were locally deposited unconformably on these metamorphic sequences, and have been assigned to the Devonian-Carboniferous time span (Breitkreuz, 1986; Bahlburg et al., 1986; Niemeyer, 1989). In the Sierra de Moreno, Cerros Limón Verde and Sierra de Almeyda mountains these rocks are intruded by granodiorites and tonalites of ages ranging from Late Carboniferous to Early Triassic (322±5 to 225±7 Ma), according to dating by Huete et al. (1977), Rogers (1985), Herve et al. (1985), Damm et al. (1986), and Boric et al. (1990).

Later, from latest, Carboniferous up to Triassic times, a thick acid (dacite-rhyolite) volcanic sequence with clastic continental intercalations was laid down, having some marine and lacustrine levels in its upper part. Some datings from these volcanic rocks make it possible to asign them a 278±8 to 229±5 Ma age (Rb/Sr in rock, Rogers, 1985; K/Ar in biotite, Gardeweg, 1988; K/Ar in biotite, Breitkreuz, in press). These units apparently lie uncomformably on metamorphic and sedimentary rocks of the Early and Middle Paleozoic (Vergara and Thomas, 1984; Breitkreuz, 1986).

The basement formed by the rocks of the above-mentioned units lies in angular unconformity beneath the marine sediments of the latest Triassic-Jurassic marine basin. These marine sediments transgressed into the southern portion of the Antofagasta Region during Norian-Rhaetian times, spreading out widely during the Hettangian-Sinemurian and also locally during Bajocian time. These rocks are highly fossiliferous, with sporadic intercalations of volcanic rocks (mainly andesites) in the Bajocian and Callovian; whereas in the area of the present-day Coastal Cordillera these volcanic rocks are very abundant from Sinemurian time onwards (Gröschke et al., 1988). In Late Oxfordian time the marine basin shallowed out and evaporites (gypsum and anhydrite) were deposited therein. Finally, during Kimmeridgian time, the oceanic domain decreased and locally withdrew.

From Late Kimmeridgian-Tithonian? until Early Cretaceous (Barremian-Aptian?) times a thick fluvio-deltaic sequence, marine at the base, was laid down over the whole area. This sequence is in gradual and conformable contact with the underlying Jurassic marine deposits (Bogdanic, 1990).

In the Cretaceous, from Albian to Cenomanian-Turonian? times, volcanic rocks and andesitic lavas and breccias with sporadically interbedded sandstones and conglomerates, were deposited conformably, or with a slight erosional disconformity, on the fluvio-deltaic sequences. A Rb/Sr isochrone in an andesite from this sequence, at the Cerros de Montecristo area, gave a maximum age of 104.7±19 Ma (Rogers, 1985). Similarly, a K/Ar in biotite radiometric age for these rocks in the Taltal area gave 115±11, 111±35 and 106±3 Ma (Ulriksen, 1979). In the Sierra de Moreno, Cerritos Bayos and Sierra Gorda areas these fluvio- deltaic and volcanic rock sequences are intruded by granitic-granodioritic- monzonitic plutons, forming a continous outcrop belt. Dating of the plutonic rocks by various methods gave 110 to 80 Ma ages (Pichowiak and Hammerschmidt, 1989; Döbel, 1989; Boric et al., 1990). These results place the intrusive phenomena within the Albian-Santonian time span.

During the latest Cretaceous and Early Tertiary, thick coarse-grained clastic redbed sequences (mostly conglomerates) were laid down. These units are located unconformably over Jurassic marine formations (Late Jurassic), and are in turn conformably overlaid by Eocene volcanic rocks (Bogdanic, 1990). In the Quebrada Copaquiri and Quebrada Honda areas these sediments are intruded by subvolcanic bodies and cut by dikes showing ages of between 45.4±1 and 36.2±0.6 Ma (Ar/Ar in biotite and hornblende; Döbel, 1989), which makes it possible to place these rocks in pre-Eocene time.

Similarly, acid volcanic (rhyolite-dacites) and volcaniclastic (tuffs and agglomerates) rocks are present in the Early Tertiary, lying unconformably over Paleozoic, Jurassic and Cretaceous sequences. Radiometric ages of these volcanic and pyroclastic rocks place them in the Paleocene-Eocene (66 to 38 Ma) time span (Huete et al., 1977; Ramírez and Gardeweg, 1982; Naranjo and Puig, 1984; Döbel, 1989; Boric et al., 1990). Furthermore, there are intrusive bodies of an intermediate to acid composition associated with the above-mentioned volcanic rocks. Dating carried out on these rocks assigns them a 60 to 40 Ma age (Huete et al., 1977; Maksaev et al., 1988a, 1988b; Boric et al., 1990). Such intrusive bodies are responsible for the mineralization in the porphyry-copper type ore deposits such as Quebrada Blanca, El Abra, Chuquicamata and others. Radiometric dating of alteration and fault zone minerals has been carried out, thus giving some indications of mineralization and tectonic movement ages. These lie within the 39 to 30 Ma interval (Early Oligocene; Maksaev et al., 1988a, 1988b; Boric et al., 1990).

All the above units, including the volcanic and intrusive rocks of the Paleocene-Eocene interval, down to the basal Oligocene, are unconformably overlain by clastic continental sediments ("Atacama Gravels", Mortimer, 1973), with interbedded Oligocene- Early Miocene ashes and tuffs. These rocks fill the intermontane basins which are part of an ancient peneplain type relief.

Later, all these units were covered by Late Miocene ignimbrites, which in many cases cover the older landscape, including the Oligocene-Miocene gravels. K/Ar in biotite dating of these ignimbrites places them in the 9.5 to 8 Ma time span, that is to say in the Late Miocene (Huete et al., 1977; Baker, 1977). A typical feature of these rocks is their horizontal and subhorizontal attitude, covering the preexisting landscape, where even some minor quebradas and depressions were buried (Vergara, 1978).

During Pliocene and Quaternary times, large deposits of lacustrine sediments were formed (Loa Limestones, Naranjo and Paskoff, 1981); these are directly related to the tectonic activity which generated the present landscape. The basins of this type are of particular interest from an economic point of view, due to the non-metallic mineral deposits formed in the evaporitic or saline-lake? basins found in the Preandean Depression and the Western Cordillera (e.g., the Salar Atacama and Salar Punta Negra saline deposits).

The various morphologic-structural units present in the survey area are made up of and came about through structures formed by compressive tectonics. The Chilean Precordillera outcrops and is uplifted along the course of extensive overthrusts of the Precambrian-Paleozoic basement over highly folded Mid-Cenozoic rocks. These compressive structures are interpreted as "compressional pillars" by Godoy and Davidson (1976), or as large anticlines with overthrust in the flanks by Chong and Reutter (1985). In any event, these were the kinematic conditions of the tectonics under which the Chilean Precordillera was raised and formed during Late Eocene to Early Oligocene times.

Within this structural style, it has been recognized that N-S to NNE-SSW-striking overthrusts and strike-slip faults have been the main structures affecting all units up to Oligocene time. In this context, the most important regional tectonic features are the

Atacama and Western Fault Systems, in the Coastal Cordillera and Chilean Precordillera, respectively. The first of these systems corresponds to the uplift of crustal and intrusive blocks parallel to the orogen, along the trace of an older strike-slip fault, which had been sinistrally active during the Late Jurassic and up to Early Cretaceous time (Scheuber and Andriessen, 1990). Rb/Sr isochrones of the rocks associated with this structure (Scheuber et al., 1990) show that two periods of igneous activity (150 and 133-129 Ma), and two deformational periods (143 and 126 Ma) took place. The "Western Fault System", which limits the Chuquicamata ore body on the west, is a structure formed by an impressive vertical shear zone caused by a dextral type horizontal movement (Reutter et al., 1988; Reutter, 1990). Several important "porphyry-copper" type orebodies (the Quebrada Blanca, El Abra, Chuquicamata and La Escondida orebodies) are aligned along this fault zone, as well as large areas affected by hydrothermal alteration. Another significant structure is the large NNE-SSW-trending overthrust which cuts the Sierra de Moreno for about 120 kilometers, and places Precambrian-Paleozoic rock series over Mesozoic rock sequences. All the units which outcrop in the survey area, except for those of the Late Tertiary and Quaternary, are folded to a greater or lesser degree. In the areas proximal to large strike-slip or overthrust faults, the folding tends to be very tight and even overturned. The fold axes generally strike N-S, changing locally to the NE-SW.

Stratigraphy

The continental and volcanic sequences mentioned have been assigned different ages and given different formation names by various authors (Figure 2). In the present work, the sedimentary and volcanic rocks corresponding to this interval have been divided according to their lithologic characteristics, depositional environment and stratigraphic relationships into three periods, namely: (1) the Kimmeridgian-Barremian-Aptian?; (2) the Albian-Cenomanian-Turonian? and (3) the latest Cretaceous-Eocene.

Kimmeridgian-Barremian-Aptian? Period.

The description of the sedimentation during this period is based on observation of the Occidental Series (Bogdanic, 1990). This Series has a maximum thickness of 5,000 meters and outcrops in an almost continuous belt trending NNE-SSW, mainly along the western part of the Chilean Precordillera (Sierra de Moreno-Cordillera de Domeyko). There are also some isolated outcrops of this unit along the Coastal Cordillera, a brief description of which will also be given.

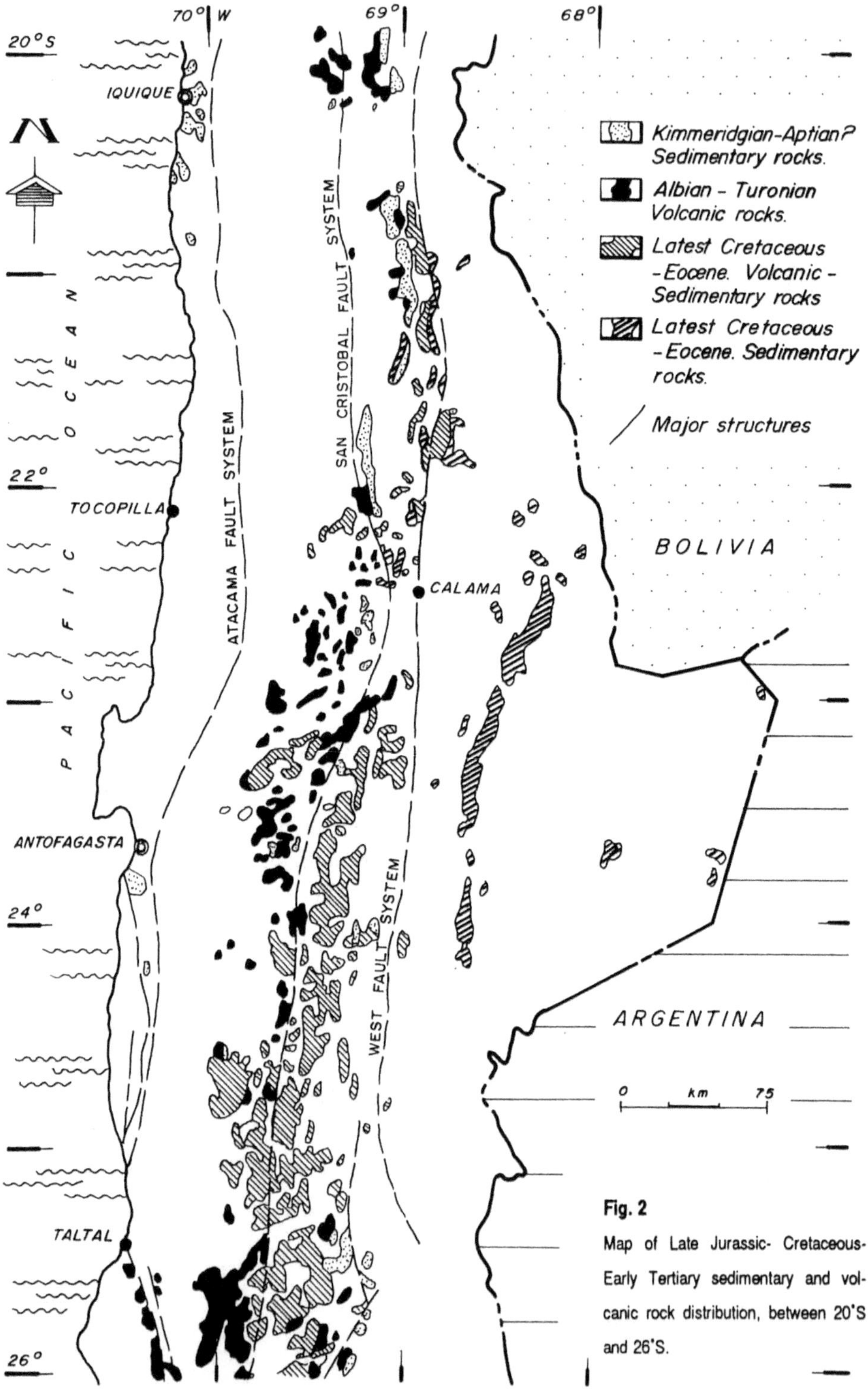

Fig. 2

Map of Late Jurassic- Cretaceous-Early Tertiary sedimentary and volcanic rock distribution, between 20°S and 26°S.

Outcrops Along the Western Part of the Chilean Precordillera.

Lithology.

The lithology for this period includes mainly fine to very fine- grained rocks, which range from sandstone through siltstone to claystone. The base of this formation includes some calcareous sandstones, and occasional intercalations of tuffs and andesitic lavas. Because of their lithology and sedimentary structures these rocks have been considered as having come from a fluviatile (braided and meandering river) and deltaic (fan-delta) type environment, according to Miall (1985) and Massari and Colella (1988), as seen in Figure 3. Paleocurrent measurements in tabular cross-bedding and asymmetric ripple-marks indicate an NE-SW flow direction and transport (Figure 4).

The fine-grained sandstones are generally present in decimetric beds, showing very even sorting and a wide lateral continuity. According to the classification scheme suggested by Pettijohn et al. (1987), these sandstones have been identified as quartz sandstones, subarkose, arkose, sublitharenite and lithic sandstones (Figure 5). The triangular plot shows an increase in lithic fragments in the areas located in the southern part of the study area.

Under the microscope, these sandstones show a clast-supported texture with a small amount of matrix (5%) or none at all, and are classified as mature to submature rocks. The sand grains are mainly of quartz (60-90%), of plutonic, metamorphic or -to a lesser extent -volcanic origin. Also present are minor quantities of plagioclase and sericitized alkaline feldspar grains (each 4- 6%), and subordinated muscovite crystals. Lithic clasts are very scarce (4%), and usually plutonic in origin. Some very scattered grains of claystone, siliceous aggregates, metamorphic rocks and heavy minerals (tourmaline, zircon and apatite) are also found.

The fine-grained rocks (siltstones and claystones) are generally dark gray and red-gray in color, and are very abundant in the lithology of this sequence, alternating with fine-grained sandstones. They are mainly highly fissile and friable. They are matrix-supported sediments with a very abundant matrix (60-70%) composed of argillaceous material (probably a mixture of kaolin, montmorrillonite and illite?). The argillaceous particles are sorted parallel to the stratification. Sand grains are scarce, and generally consist of quartz (of metamorphic origin) and plagioclase, with some secondary muscovite and heavy mineral fragments.

In the lower part of this unit, there are some very scarce decimetric beds of micritic and calcitic sandstones, and microsparite, according to Mount's (1985) classification. There are also sporadic and rare lenticular intercalations of tuffs and tuffaceous sandstones, bearing quartz, plagioclase and volcanic (andesitic-dacitic) rock fragments.

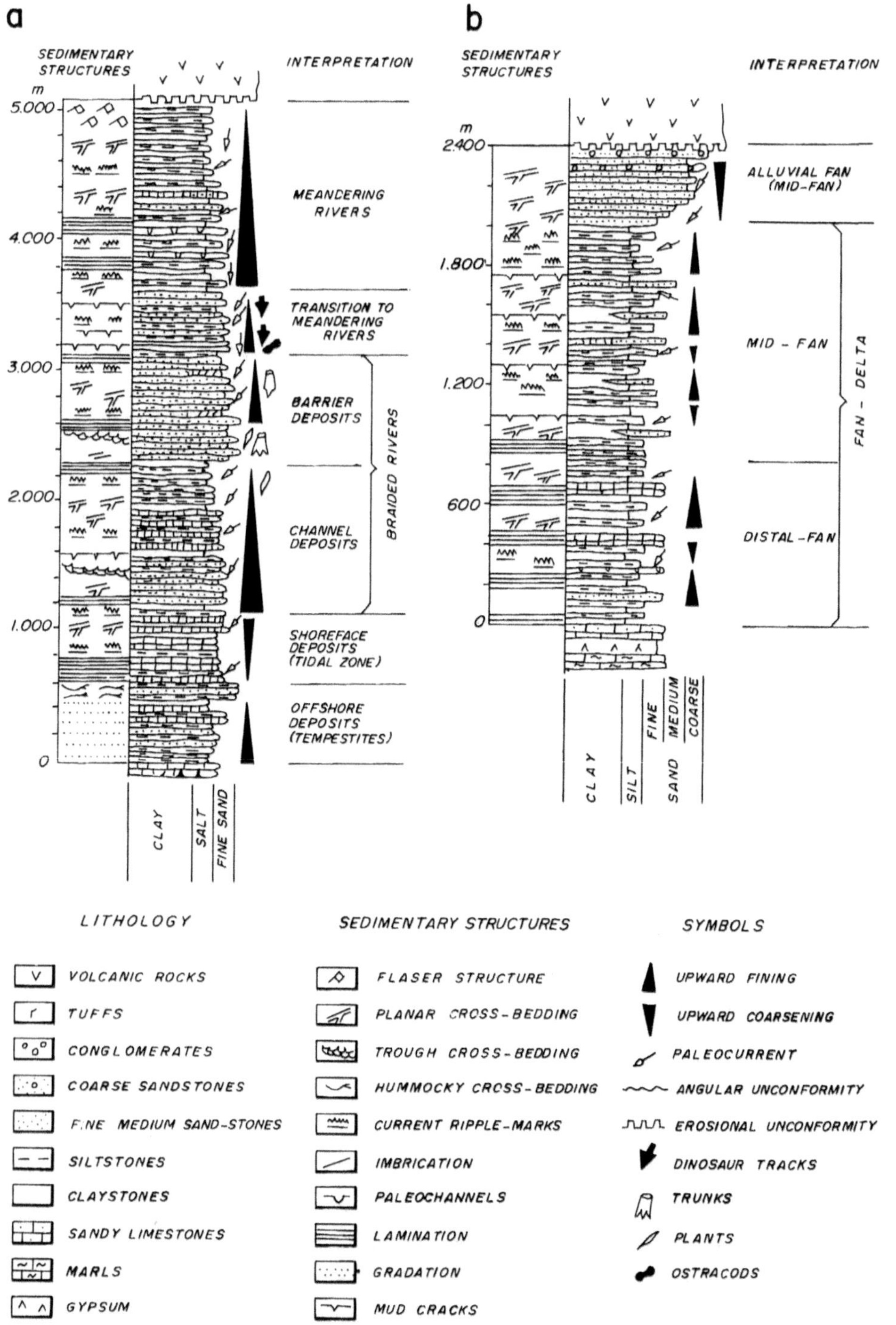

Fig. 3

Stratigraphic logs of the Kimmeridgian-Aptian ? sedimentary rocks. a: N sector (20°S-22°S), b: S sector (23°S-26°S).

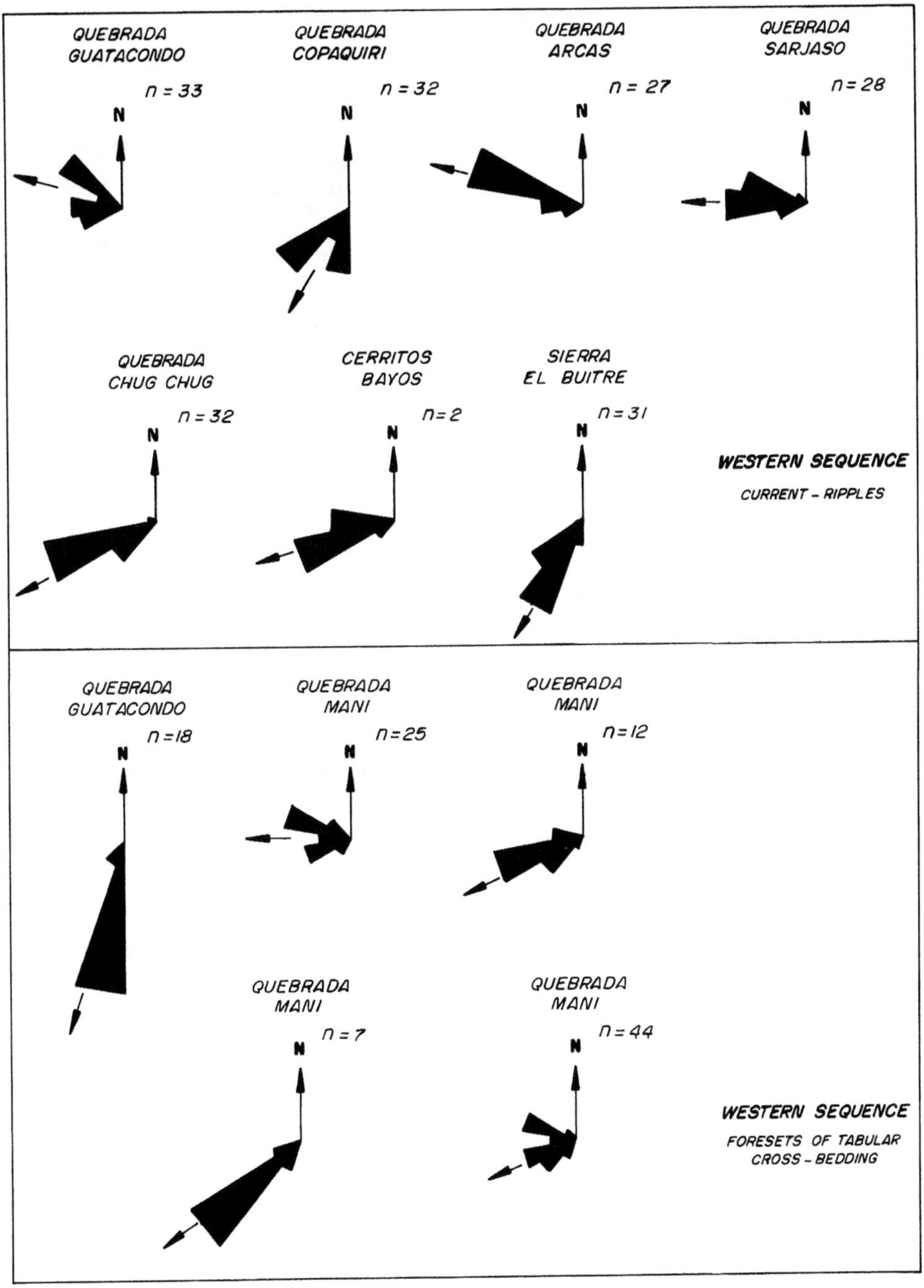

Fig. 4

Paleocurrent measurements of the Kimmeridgian-Aptian ? sedimentary rocks.

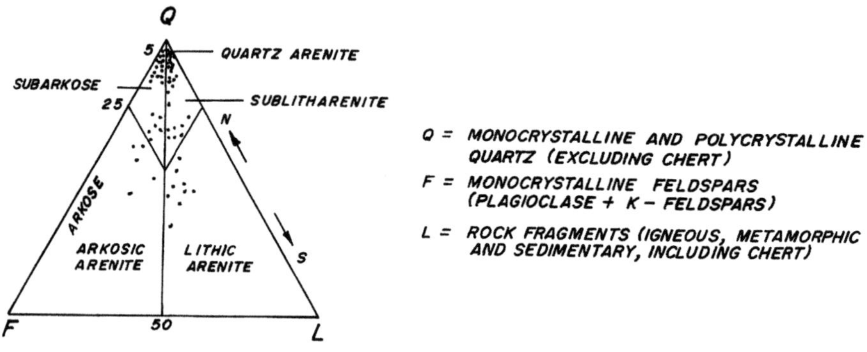

Fig. 5

QFL diagram after Pettijohn et al. (1987) for classification of the sandstones from Kimmeridgian-Aptian ? sedimentary rocks.

Provenance

Point-counter analysis in the sandstones from several sections (see Figure 6) were plotted in triangular diagrams, described by Dickinson (1985 and 1988), with the following results:

The Q_tFL_t plot shows that practically all the samples lie in the "Recycled Orogen" field, and the remainder in the "Continental Block Provenance" field. This means that the rocks from this unit clearly tend to proceed from a stable craton. Furthermore, with regard to the "Recycled Orogen" sector the greater part of the samples from the northern part of the study area (Sierra de Moreno) show an almost non-existent magmatic arc influence; whereas those from the south show the existence of a magmatic arc (La Negra Formation), which was undergoing erosion.

Two groups of samples can be observed in the Q_mFL_t triangle. The first of these, containing the majority of rocks, is located very near the Q_m-apex in the "Continental Block Provenance" field. As in the previous diagram, this shows there was a very strong stable craton influence. The second group shows a very reduced chert/Q_z ratio, due to the small amount of lithoclasts provided by the erosion of volcanic rocks from a nearby magmatic arc, thereby falling within the "Recycled Orogen" field.

The $Q_pL_vL_s$ plot shows that the Western Series sandstones contain a greater amount of sedimentary fragments than of volcanic fragments, the samples being plotted in the "Fold and Thrust-belt Sources" field and approaching the Q_p-apex, due to their polycrystalline quartz content. A large number of samples lie outside this field in the "Mixed Orogenic Sands" field, which might indicate that these sandstones were formed and deposited during the transitional period between two tectonic regimes, probably from an island arc to a continental margin tectonic environment, according to Mack (1984).

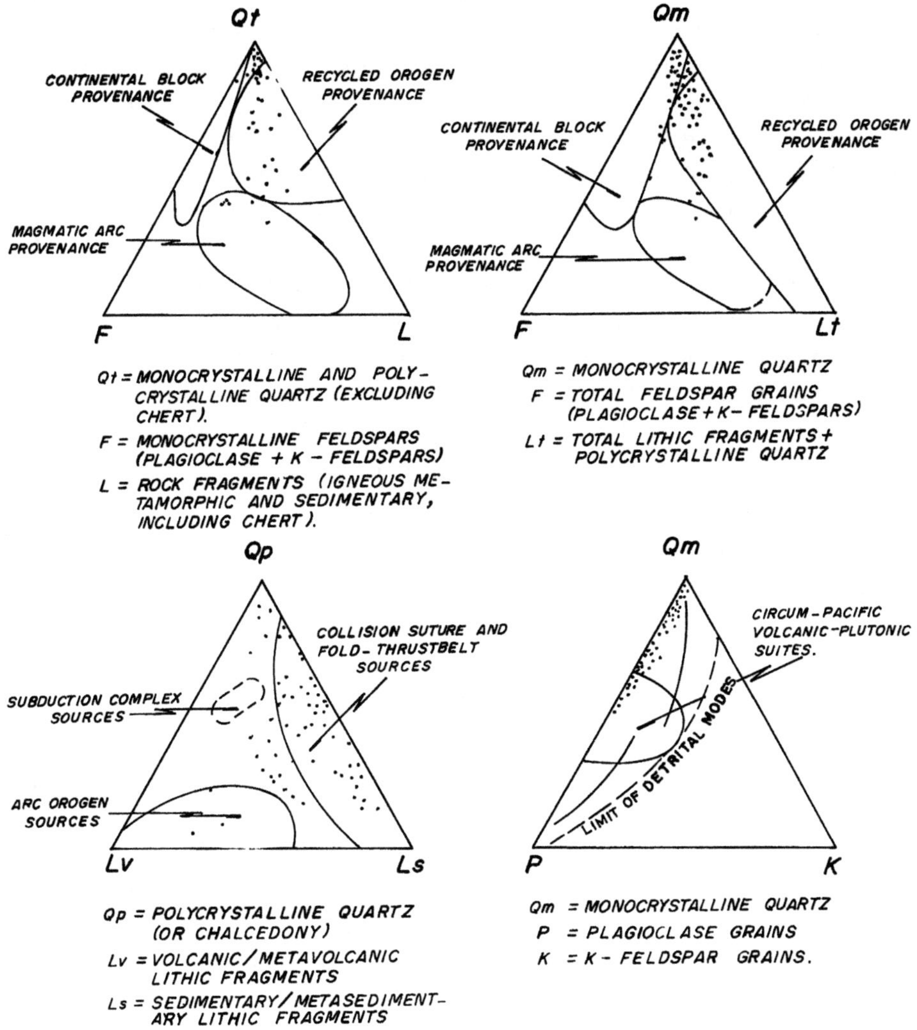

Fig. 6

Framework modes of sandstones pertaining to the Kimmeridgian-Aptian? sedimentary rocks, plotted after Dickinson (1985).

Finally, in the QmPK diagram it can be seen that samples from the northern part of the study area plot near the Qm-apex, which means that the minerals they contain are very stable. Conversely, samples from the southern part of the area are less mature and stable, as deduced from an increase in plagioclase fragments resulting from a greater volcanic influence.

A careful study of these triangular diagrams (Figure 6) indicates the Western Series sandstones as having a "recycled orogen" provenance. This signifies that the Western Series provenance is from folded metamorphic rocks. These units could certainly have been found on the eastern flank of the Jurassic- Early Cretaceous back- arc basin, according to the paleogeography outlined by Herve et al. (1987) and Mpodozis and Ramos (1989). These rocks show almost no trace of having been affected by the active magmatic arc located on the western side of this basin during Jurassic time.
Sedimentary structure and facies studies, as well as the trends of E-W-oriented paleocurrents, indicate that the Western Series originated in fluviatile systems, with the sediments being transported from the E over long distances for final deposition in this backarc basin. This topographic high or positive relief with orogen relicts was essentially made up of rocks of the Precambrian-Paleozoic basement, as well as by Late Paleozoic intrusive rocks and Permian-Triassic volcanic rocks. This can be seen in present-day outcrops in the Sierra de Moreno-Cordillera de Domeyko area.

Age, Stratigraphic Relationships and Correlations.

As concerns its stratigraphic relationships, the base of the Western Series prograded conformably over marine sediments and/or gypsum and anhydrite of Late Jurassic age, and contains Late Oxfordian and Early Kimmeridgian fossils (Guatacondo, Häberer and Reutter, 1986; Cerritos Bayos, Gröschke et al., 1988). Its top lies conformably and/or with a slight erosional unconformity beneath volcanic rocks and andesitic breccias of the Andesitic Volcanic Series (see below), assigned to the Albian-Cenomanian- Turonian? time span.
In the Quebrada Arcas, Cerritos Bayos, Sierra San Cristóbal and Sierra Candeleros areas these Western Series sediments are intruded by monzonites, granodiorites and granites, as well as by andesitic dikes, with ages assigned to the 110 to 80 Ma time span (Ulriksen, 1979; Rogers, 1985; Döbel, 1989; Pichowiak and Hammerschmidt, 1989). Stratigraphic relationships and the radiometric ages of associated magmatic rocks place this unit in the Late Kimmeridgian-Early Cretaceous (Aptian?). Furthermore, in the study area it is possible to correlate the following formations or parts of formations with this Series:

1) The middle and upper part of the Chacarilla Formation (Galli and Dingman, 1962).
2) The Copaquiri, Aquiuno, Majala and Guatacondo Formations (García, 1967).
3) The Punilla and Los Tambos Members of the Quehuita Formation (Vergara, 1978).
4) The lower portion of the Quehuita Formation Upper Member (after Vergara and Thomas, 1984).
5) The Upper Member of the Quinchamale Formation (after Skarmeta and Marinovic, 1981).
6) The Continental Sequence of the Cerritos Bayos (Ferraris, 1978).

7) The Cerritos Bayos Formation (García, 1967; Montaño, 1976).
8) The Early Cretaceous continental sediments (Baeza, 1976). 9) The San Manuel Strata and the Llanura Colorada Formation (Muñoz, 1989).
10) Member 1 of the Profeta Formation (Chong, 1973).
11) The Santa Ana Formation (Naranjo and Puig, 1984).

Cretaceous Outcrops in the Coastal Cordillera.

Some 15 km south of Antofagasta, at the Quebrada El Way, a series of limestones - known as the El Way Formation (Hoffstetter et al., 1957)- outcrops. These limestones rest conformably on a sequence of siltstones, red sandstones and alluvional conglomeratic breccias known as the Caleta Coloso Formation (Brüggen, 1950). These are the only known Cretaceous marine sediments in the Coastal Cordillera within the Antofagasta Region. Their possible relationships with similar rocks deposited south of Copiapó, in the Chañarcillo Basin (Segerstrom and Ruiz, 1962; Corvalán, 1974), through a connection between both basins is still open to discussion, since there are sedimentary rocks of Late Neocomian age (Candeleros, Chong, 1976, and Santa Ana Formations, Naranjo and Puig, 1984) at the latitude of Taltal (Sierra Candeleros). Studies of the fossil fauna (Alarcón and Vergara, 1964; Jurgan, 1974) place it in the Hauterivian-Barremian time span, possibly reaching the Aptian?. Later, studies were made by Flint et al. (1986) of the El Way Basin, especially of the subaerial alluvial facies and their transition to a fan-delta environment. Establishing a sedimentary supply from the NW for this unit, these authors considered the Upper Member of the Caleta Coloso Formation as another formation, and unjustifiably renamed it the El Way Formation.

Albian-Cenomanian-Turonian? sedimentation.

This unit crops out in a continous belt along the western flank of the Chilean Precordillera (Sierra de Moreno-Cordillera de Domeyko), as in the above case, and also along a part of the Central Depression.

Lithology.

The lithology of this unit consists mostly of lavas and andesitic brecciated agglomerates, with subordinate interbedded tuffs, sandstones and conglomerates. It has a minimum thickness of between 1,000 m and 1,500 m, with the volcanic rocks and breccias lying in beds 2 to 5 m thick. These strata have marked lateral continuity and are lithologically very homogeneous (Figure 7).
The andesitic lavas are usually gray-green to gray-red in color, massive in appearance and porphyritic in texture. Under the microscope they show an ophitic to subophitic

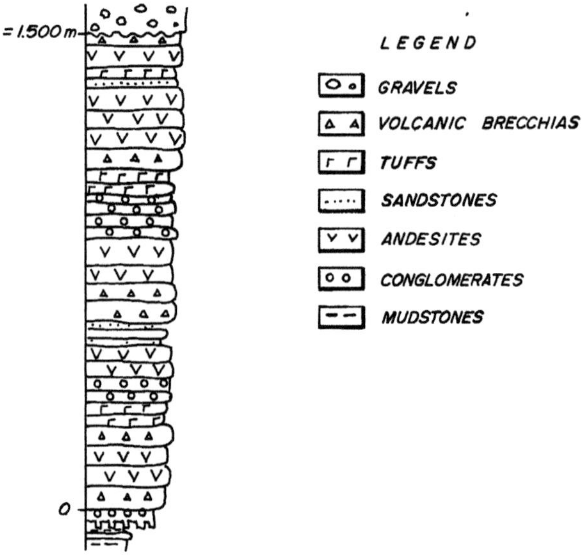

Fig. 7

Simplified stratigraphic log of Mid-Cretaceous volcanic rock unit.

and intergranular texture, with plagioclase phenocrysts (An_{30-50}) which are subhedral, twinned and zoned. Clinopyroxene (augite) and hornblende (usually surrounded by an opaque halo) phenocrysts are occasionally observed. The groundmass is formed by a plagioclase microlite aggregate, with lesser amounts of augite and hornblende. The groundmass is also found in cryptocrystalline or vitrophyric form surrounding the phenocrysts. Alteration minerals found in these rocks are calcite, sericite, chlorite, epidote and opaque minerals.

The andesitic brecciated agglomerates are usually green to grey- red in color, with angular fragments ranging from 10 to 50 cms in diameter. These are mainly composed of volcanic rock (dacite and rhyolite) clasts, with a lesser proportion of intrusive rock fragments (granite, granodiorite and rhyolitic porphyry), and ocasional metamorphic rock clasts (migmatite and schists). The matrix of these breccias is made up of gray-green andesite, having characteristics similar to those of the lavas described above. In the middle part of the unit there are occasional intercalations of lenticular 2 to 3 m thick beds of andesitic tuffs, conglomerates and red sandstones. The conglomerates are clast-supported, with mainly rounded to subrounded clasts, 2 to 40 cm in size, mostly volcanic in origin with lesser amounts of sedimentary and intrusive rock fragments. The matrix is of medium to coarse-grained red sandstone, with abundant quartz grains. This same sandstone appears in layers tens of centimeters thick, interbedded with the conglomerates. The tuffs are red to pink and green in color, lapilli or ash in clast size, with some quartz and plagioclase phenocrysts in a vitreous-pummicitic groundmass.

Age, Stratigraphic Relationships and Correlations.

This volcanic rock unit was laid down conformably or with a slight erosional disconformity over Kimmeridgian-Aptian? sediments (the Western Series). Unconformable gravel beds assigned an Oligocene-Miocene age overlie its top, which is also overlain by conglomerates assigned a latest Cretaceous-Paleocene age (Eastern Series), on the western flank of the Sierra de Moreno.

A Rb/Sr in rock isochrone in this unit at the Cerros de Montecristo area gave a 104±19 Ma age (Rogers, 1985). Other radiometric datings (K/Ar in rock) made in the volcanic rocks of this series in the Taltal area gave 115±11, 111±35 and 106±3 Ma (Ulriksen, 1979). In addition to this, these rocks have been intruded by diorite-monzonite-granodiorites in the Cerros de Montecristo, Cerritos Bayos and Sierra San Cristóbal mountains, and in the area E of Taltal. These plutonic rocks show ages in the 115 to 80 Ma range (Pichowiak and Hammerschmidt, 1989; Boric et al., 1990).

This unit's stratigraphic relationships, as well as the radiometric ages both in the volcanic rocks and in the plutons intruding them, make it possible to place this series in the Albian-Cenomanian-Turonian? time span.

With regard to correlations with similar series described for the area under study, the rocks in this interval can be correlated with all or part of the following formations:

1) Cerro Empexa Formation (Galli and Dingman, 1962).

2) Arca Formation and Upper part of La Negra Formation (after Skarmeta and Marinovic, 1981).

3) Upper part of La Negra Formation (after Ferraris, 1978).

4) Continental cycle (Baeza, 1976); upper part of volcanic rocks.

5) Río Seco Strata and Quebrada Mala Formation (Muñoz, 1989).

6) Upper part of Aeropuerto Formation (Ulriksen, 1979).

7) El estanque Strata (Naranjo and Puig, 1984).

Latest Cretaceous-Eocene Sedimentation.

As for the Kimmeridgian-Aptian? time span, the description and evolution of this unit is based on the Eastern Series as defined by Bogdanic (1990).

This Series has a maximum thickness of 1,500 m, although it varies greatly in thickness throughout the sections surveyed (see Figure 2). This unit is present mainly in the Sierra del Medio and to the E in the belt formed by the Kimmeridgian-Aptian? rocks, but also outcrops on the western flank of the Sierra de Moreno, in the Central Depression (Baquedano area), on the Cordillera de Domeyko (Cerros de Purilactis, Sierra de Varas and Sierra Candeleros areas), and on the Western Cordillera (Loma Negra and the Cerro Zapaleri area). These outcrops occur mainly in a NS-trending continuous belt and in many places are controlled by large NNE-SSW-trending faults.

Lithology

The lithology of this unit consists basically of coarse-grained red-colored clastic rocks (conglomerates and coarse sandstones). There are also some interbedded layers of sandstone at its base and of evaporites in its upper portion (Figure 8), and also scattered lenticular intercalations of tuffs and tuffaceous sandstones, especially in the base. The coarse-grained clastic rocks referred to pertain to proximal and intermediate alluvial-fan facies, according to Bull (1977) and Nielsen (1982). Paleocurrents have been measured in this unit on the basis of clast imbrication, cross-bedding and paleochannels that indicate a W-E-flowing sedimentary supply for the unit (Figure 9).

Conglomerates are the predominant type of sedimentary rock, and according to Folk et al. (1970) are polymictic igneorudites (Figure 10). They are generally present in beds from 3 to 8 m thick, with good lateral continuity, although occasionally lenticular in shape, well compacted, occasionally showing an imbricate grain pattern, and have a clast-supported texture. The constituent fragments are rounded to subrounded, 2 cm to

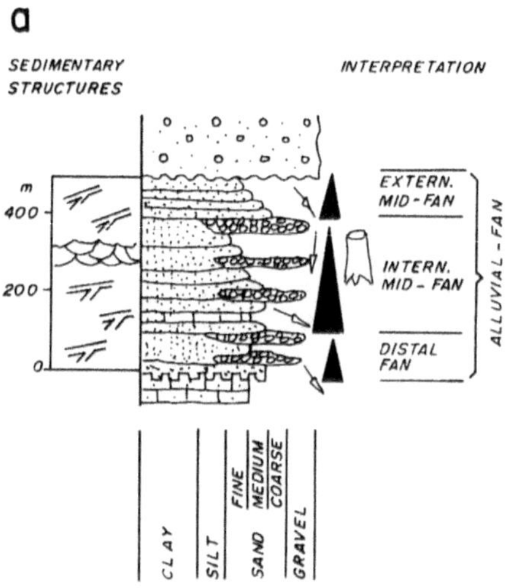

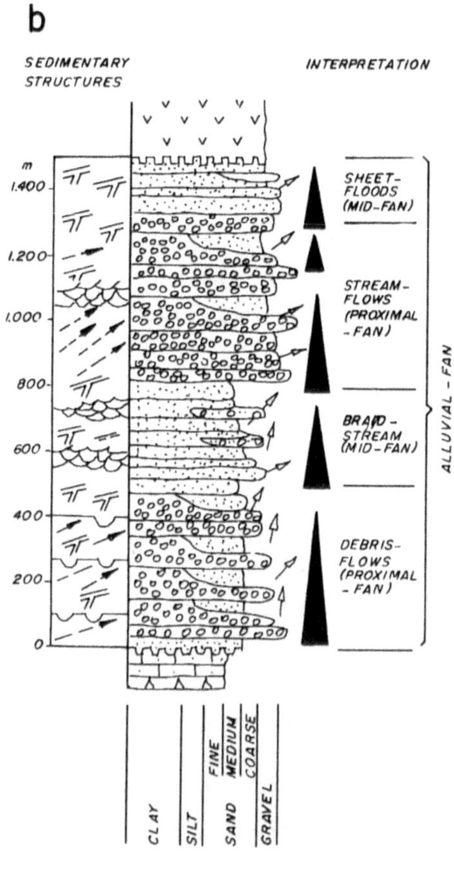

Fig. 8

Stratigraphic logs of latest Cretaceous-Eocene sedimentary rocks. a: 20°S-21°S, b: 22°S-23°S.

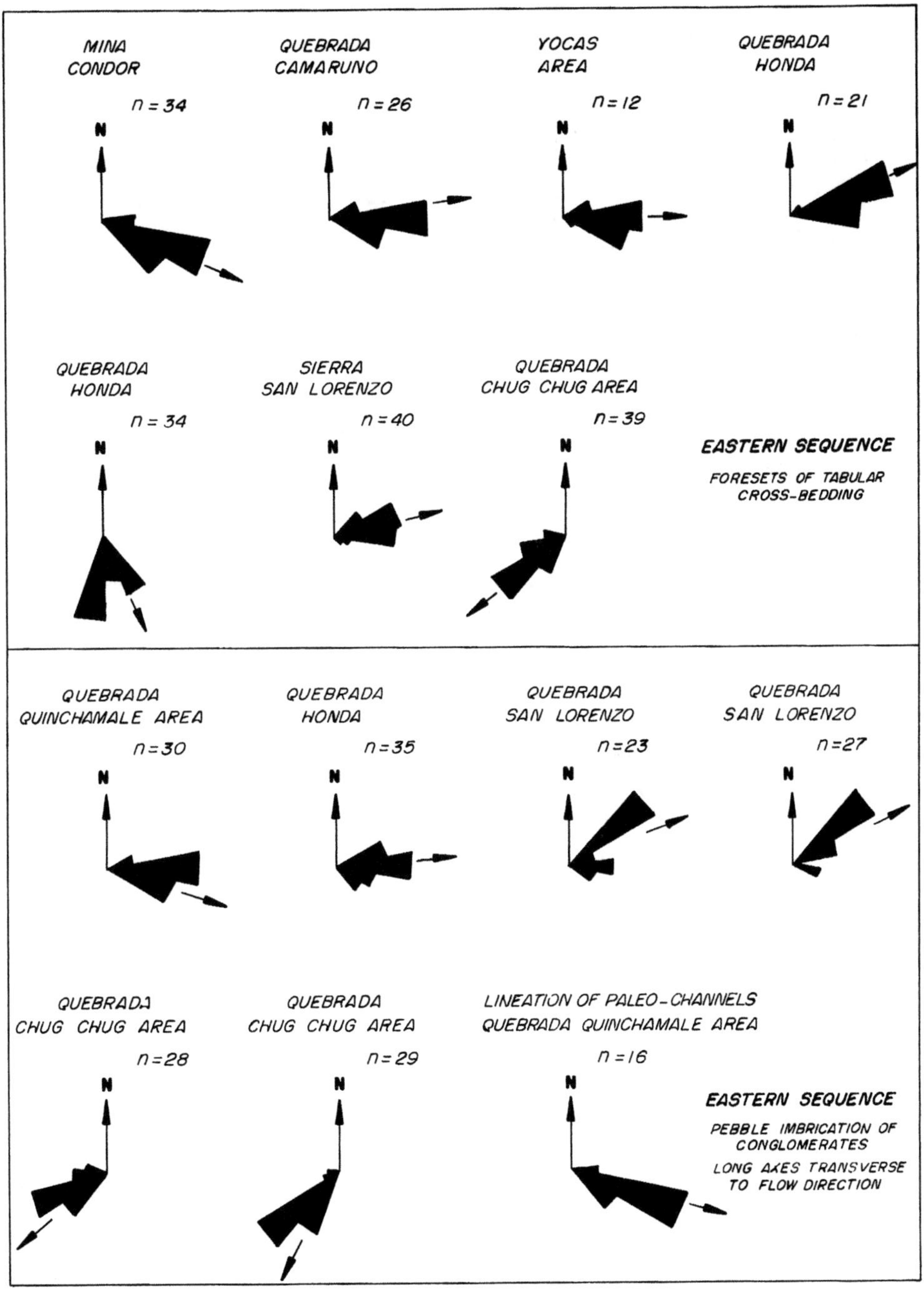

Fig. 9

Paleocurrent measurements of the latest Cretaceous-Eocene sedimentary rocks.

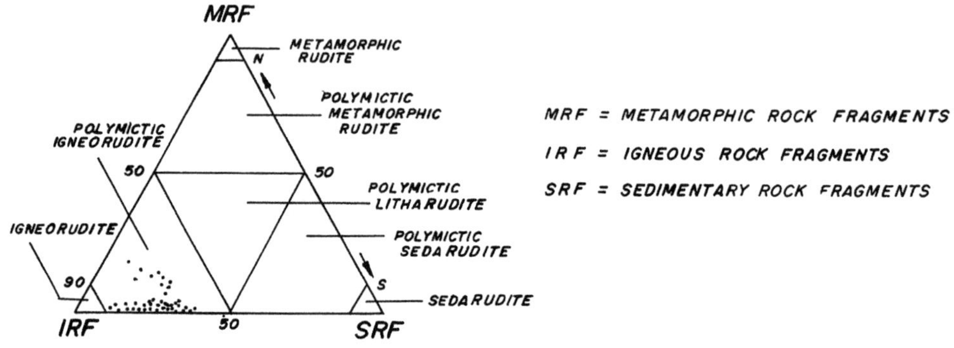

Fig. 10

Triangular diagram after Folk et al. (1970) for classification of conglomerates from latest Cretaceous-Eocene sedimentary rocks, based on clast lithology.

1 m in clast size, and show normal grading. There are also very scattered metamorphic (schist and gneiss) and sedimentary (sandstone, calcarenite, and Jurassic fosil-bearing calcareous nodule) rock fragments. The matrix of these conglomerates is formed of medium to coarse-grained sandstones, and according to Pettijohn et al. (1987) is an inmature lithic sandstone, with a large proportion of volcanic and intrusive porphyry grains; it has a clast-supported texture. The cement is usually coarse- grained calcite (sparry calcite), and also a hematite, limonite and calcite composite which generally surrounds the clasts.

The sandstones outcrop mainly in the northern part of the area (Copaquiri-Quebrada Choja), and are found in beds tens of centimeters thick with good lateral continuity. They can be classified as quartz sandstones, sublitharenite and lithic sandstones (Figure 11). Their main components show average contents of 67% (total quartz) Q, 8% (total feldspars) F and 25% (lithic grains) L.

Under the microscope, they show a clast-supported texture, with a 7% to 10% matrix formed by an aggregate of sericite, chlorite and clay minerals. The cement in these sandstones is hematite, calcite or a mixture of both. The grains are abundant and mostly of volcanic origin (andesite, dacite and rhyolite), with lesser amounts of sedimentary and metamorphic grains. Within the sand grains there is a large proportion of volcanic quartz, euhedral with dipyramid shapes and embayments. There are also lesser quantities of plagioclase (An_{10-30}) and alcaline feldspar grains present.

The tuff and tuffaceous sandstone levels are tens of centimeters thick and usually lenticular in shape. They are formed of crystalline tuffs with a porphyritic texture and a microcrystalline to microfelsic groundmass.

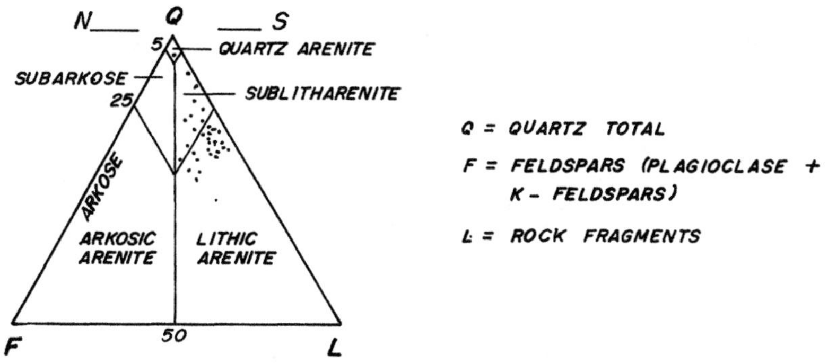

Fig. 11

Framework modes of sandstones pertaining to latest Cretaceous-Eocene sedimentary rocks, plotted after Pettijohn et al. (1987).

Provenance.

Point-counter petrographic analysis of the sandstones in this Series using Dickinson's (1985, 1988) triangular diagrams, reveals the following characteristics (Figure 12):

The Q_tFL_t diagram shows that the sandstones from this unit fall within the "Recycled Orogen" field, but have a larger lithic content than those of the Western Series (Kimmeridgian-Aptian?). This indicates that these rocks were more highly influenced by a volcanic-sedimentary supply, since the majority of the sand grains are volcanic.

The Q_mFL_t triangle also shows a "Recycled Orogen" type provenance, with a large proportion of chalcedony and quartz particles, which again points to an important volcanic rock supply (andesite, dacite, rhyolite), as can be seen from the composition of the lithic grains.

In the $Q_pL_vL_s$ plot nearly all the samples fall within the "Arc Orogen Sources" field, as a consequence of the high percentage of volcanic and pyroclastic lithic grains over as compared to those of polycrystalline quartz. There are three samples which plot outside this field but near the L_s apex, which means they have a high percentage of sedimentary grains. However, under the microscope these samples appear to be of mixed origin, part sedimentary and part pyroclastic.

Finally, the Q_mPK plot shows that the sedimentary rocks of the Eastern Series have a similar feldspar content to those of the Western Series, although the sections in the southern part of the Sierra del Medio show a marked increase in plagioclase content, perhaps due to greater proximity to the magmatic arc.

From observation of these diagrams (Figure 12) it can be concluded that the clastic portion of the Eastern Series comes from a "Recycled Orogen", but having a strong volcanic influence from a magmatic arc. These rocks come from basement units (according to the

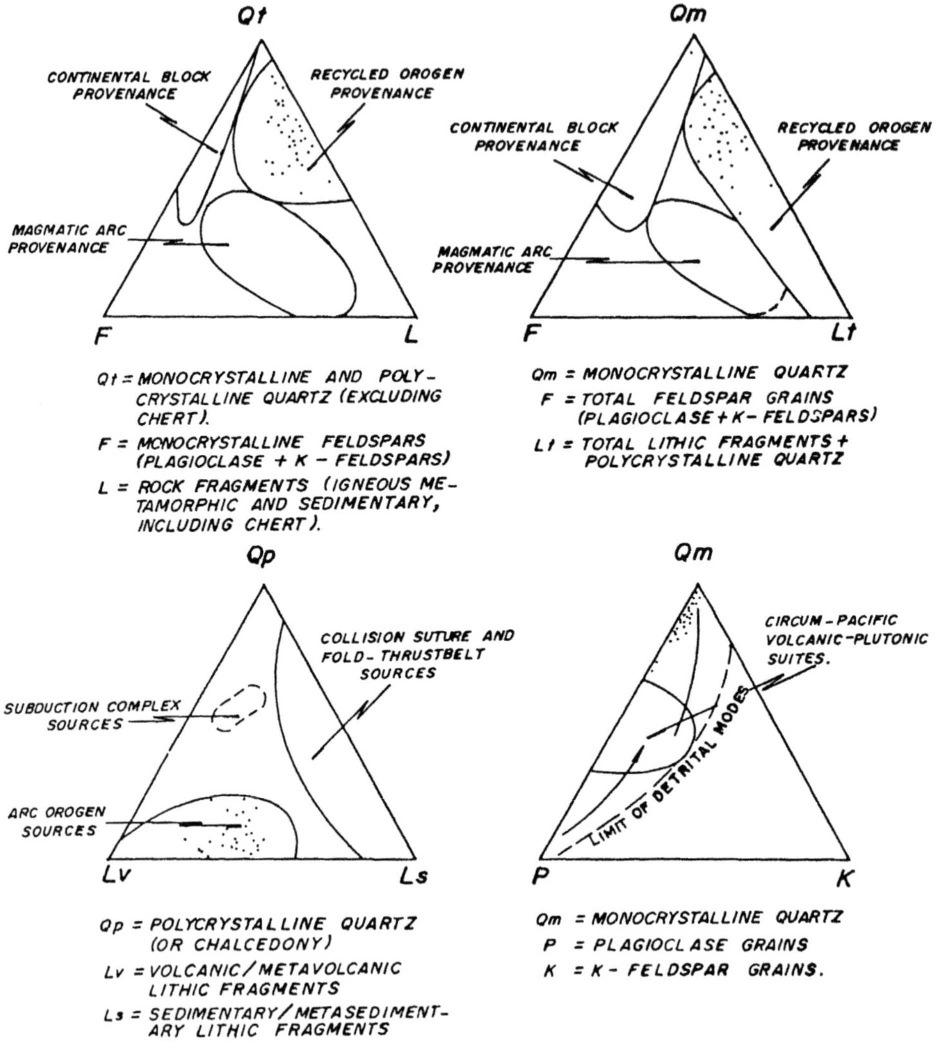

Fig. 12

Framework modes of sandstones pertaining to latest Cretaceous-Eocene sedimentary rocks, plotted after Dickinson (1985).

$Q_pL_vL_s$ plot) which have been intensely deformed and uplifted as a result of strong volcanic and magmatic activity (sensu Dickinson, 1988). However, this statement may be only relatively true, since the geography of the area does not rule out the possibility that part of the volcanic material may have come from the Permian-Triassic volcanic rocks lying under the Mesozoic overburden. It is also possible that these lithoclasts derive from volcanic rocks of the same age as or slightly older than the Eastern Series sediments.

Furthermore, the magmatic arc which was active during latest Cretaceous to Early Tertiary times was located in the present-day Chilean Precordillera (Cordillera de Domeyko), according to Reutter et al. (1988), but also extended to as far N as Calama, to the Sierra del Medio and to the western side of the Western Cordillera (Döbel, 1989). Therefore, the redbeds of the Eastern Series were deposited in intra-arc basins which were partly interconnected, giving place to large back-arc deposition areas in eastern Bolivia and northwestern Argentina, showing similar facies and depositional environments.
As already mentioned, south of Calama there are volcanic rocks from a synchronous volcanism with sedimentation from latest Cretaceous time onward. These volcanic rocks are calc- alkaline in nature, and are made up of lavas and pyroclastic deposits. In this type of sequence there is usually an increase in volcanic and pyroclastic interbedding from base to top, ranging from a clastic-volcanic unit in the base to a completely volcanic unit at the top.

Age, Stratigraphic Relationships and Correlations.

The age of the clastic part of this Eastern Series N of Calama is based mostly on its stratigraphic relationships and on some radiometric datings of the dikes and intrusives which cut it.
According to this, its minimum age is pre-Eocene, since it is conformably overlain by volcanic and pyroclastic rocks of the Icanche Formation (48-35 Ma; Huete et al., 1977; Döbel, 1989), and its maximum age would be post-Cenomanian-Turonian?, in accordance with its unconformable relationship over Mid- Cretaceous volcanic rocks (Bogdanic, 1990). Furthermore, the rocks of this Series are intruded by granitoid bodies aged 37-30 Ma (Maksaev et al., 1988a, 1988b), which gives this unit a maximum Late Cretaceous or Paleocene age.
The information used for assigning an age to the part of this Series lying S of Calama is based mainly on radiometric datings of volcanic and pyroclastic rocks, found both in its lower and its upper part. Its stratigraphic relationships are equally in agreement with the geochronological data thus obtained. Its maximum age would be post-Cenomanian-Turonian?, due to its unconformable relationship with underlying Mid-Cretaceous volcanic rocks. Its minimum age would be given by its unconformable relationship with the overlying Gravels, which have been assigned an Oligocene-Early Miocene age. K/Ar and Ar/Ar datings in the volcanic rocks -in biotite, in clinopyroxene and in rock- give ages between 64 and 40 Ma (Ramírez and Gardeweg, 1982; Döbel, 1989; Flint et al., 1989), which would place this unit between latest Cretaceous and Eocene times.
With regard to correlations, the clastic rocks of the Eastern Series, in the sector N of Calama, are lithologically similar to a series of units exposed along the Sierra de Moreno, Sierra del Medio and the Central Depression, as follows:
1) Quehuita Formation (Vergara, 1978); upper part of the Los Tambos Member.

2) Macata Formation (Vergara, 1978); Upper Member.
3) Quehuita Formation (after Vergara and Thomas, 1984); upper part of the Upper Member.
4) Quinchamale Formation (Maksaev, 1978); Upper Member.
5) Tolar Formation (Maksaev, 1978).
6) Sierra San Lorenzo Strata (after Marinovic and Lahsen, 1984).
7) Late Cretaceous red conglomerate deposits (Niemeyer et al., 1985).
8) Tambillo Formation (Skarmeta and Marinovic, 1981).
9) Cerritos Bayos Sequence (after Ferraris, 1978); part of the Continental Sediments.
10) Pajonales Formation (Naranjo and Puig, 1984).

The Eastern Series rocks which outcrop S of Calama (Cordillera de Domeyko and the Central Depression) can be correlated lithologically and stratigraphically with the following units:

1) Icanche Formation (Maksaev, 1978).
2) Upper part of the Purilactis Group (Purilactis Formation s.s.), sensu Charrier and Reutter (1990).
3) Quebrada Mala Formation (Muñoz, 1989).
4) Volcanic and volcanoclastic rocks (Baeza, 1976).
5) Quebrada Seca Formation (Montaño, 1976).
6) Augusta Victoria Formation (after Muñoz, 1989).
7) Cinchado Formation (Montaño, 1976).
8) Chile-Alemania Formation (after Naranjo and Puig, 1984).

A general correlation and redistribution of all the units described in the survey area (formerly defined by a number of authors), covering the Late Jurassic-Cretaceous-Early Tertiary time span, is shown in Figure 13.

Intrusive Rocks

The Cretaceous intrusive rocks, which cover the Late Jurassic- Early Tertiary time span, extend throughout the area in NNE-SSW- trending discontinuous elongate belts. They vary greatly in composition, from gabbros to granites, and their porphyritic varieties are those of hypabyssal bodies. The ocurrence of various of these plutons and hypabyssal bodies is controlled by major structures having an approximately N-S bearing (e.g., the Atacama Fault System), and they are generally linked to important mineral deposits and/or hydrothermal alteration zones.

Based mainly on the geographic position of the outcropping belts and on the geochronological ages established for them, the plutons of this interval can be divided into three large groups (see Figure 14):

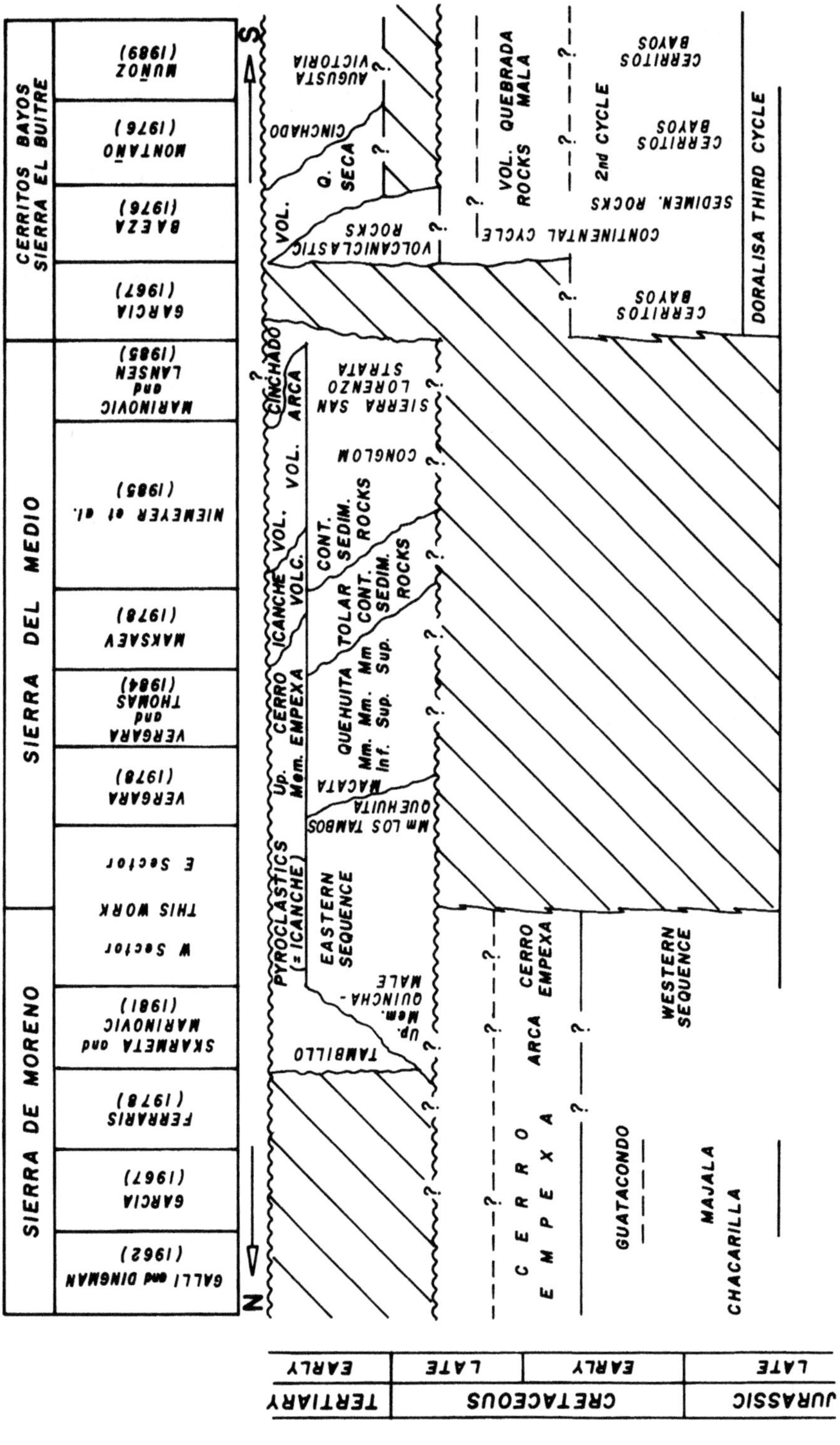

Fig. 13
Stratigraphic correlation chart of Late Jurassic-Cretaceous-Early Tertiary units from the survey area, modified from Bogdanic (1990).

Late Jurassic-Early Cretaceous Plutons.

These rocks form an elongate NS continuous belt, with batholithic characteristics, along the whole length of the Coastal Cordillera. Their contact relationships do not enable a relative chronology to be made, as they have intruded Paleozoic metamorphic rocks (Paposo Strata, Ferraris, 1978), Triassic volcaniclastic and sedimentary rocks (Permian-Triassic sediments, Scheuber, 1987), and Jurassic volcanic rocks (La Negra Formation, sensu Muñoz et al., 1988). The outcrops of these rocks are closely linked to and controlled by the Atacama Fault System, which runs parallel to the Coastal Cordillera. The age determinations made in these intrusive rocks by different methods (K/Ar, Ar/Ar, Rb/Sr and fission tracks; Herve et al., 1985; Herve and Marinovic, 1989; Pichowiak and Hammerschmidt, 1989; Scheuber and Andriessen, 1990), both in rock and in selected minerals (biotite, hornblende, apatite), place them within the 160 to 120 Ma time span.
From the lithologic point of view these plutons are made up of gabbros, monzonitic gabbros, diorites, monzonites, tonalites, granodiorites and granites, according to Herve and Marinovic (1989). From the foregoing characteristics these rocks can be divided into two groups or intrusive pulses, according to their intrusive relationships and radiometric ages:

Late Jurassic Plutons.

Located in a belt running parallel to the west side of the Coastal Cordillera. They are tonalites and granodiorites which are considered to be closely related to the Jurassic volcanism of the La Negra Formation. The ages of these rocks (K/Ar, Rb/Sr and U/Pb; Berg and Breitkreuz, 1983; Herve and Marinovic, 1989; Pichowiak and Hammerschmidt, 1989) fall within the 160 to 145 Ma time span; i.e., the Oxfordian-Kimmeridgian. Meanwhile a first deformational stage brought about by the Atacama Fault System is placed at 143 Ma by Scheuber et al. (1990).

Early Cretaceous Plutons.

These are located in a belt running parallel to the Atacama Fault System, resting directly on its eastern flank, and near this structure they show ductile deformation (Scheuber and Andriessen, 1990). Lithologically speaking they are mainly made up of granodiorites, monzonites and tonalites, which in this case are related to the final stages of Jurassic magmatic activity and to the sinistral movement of the Atacama Fault System (Scheuber and Reutter, 1988). Age determinations in these rocks by the K/Ar and Rb/Sr methods (Ulriksen, 1979; Naranjo and Puig, 1984; Herve et al., 1985; Herve and Marinovic, 1989) show ages between 133 and 126 Ma; i.e., in the Valanginian-Aptian time span. As in the former case, a second deformational period caused by activity of

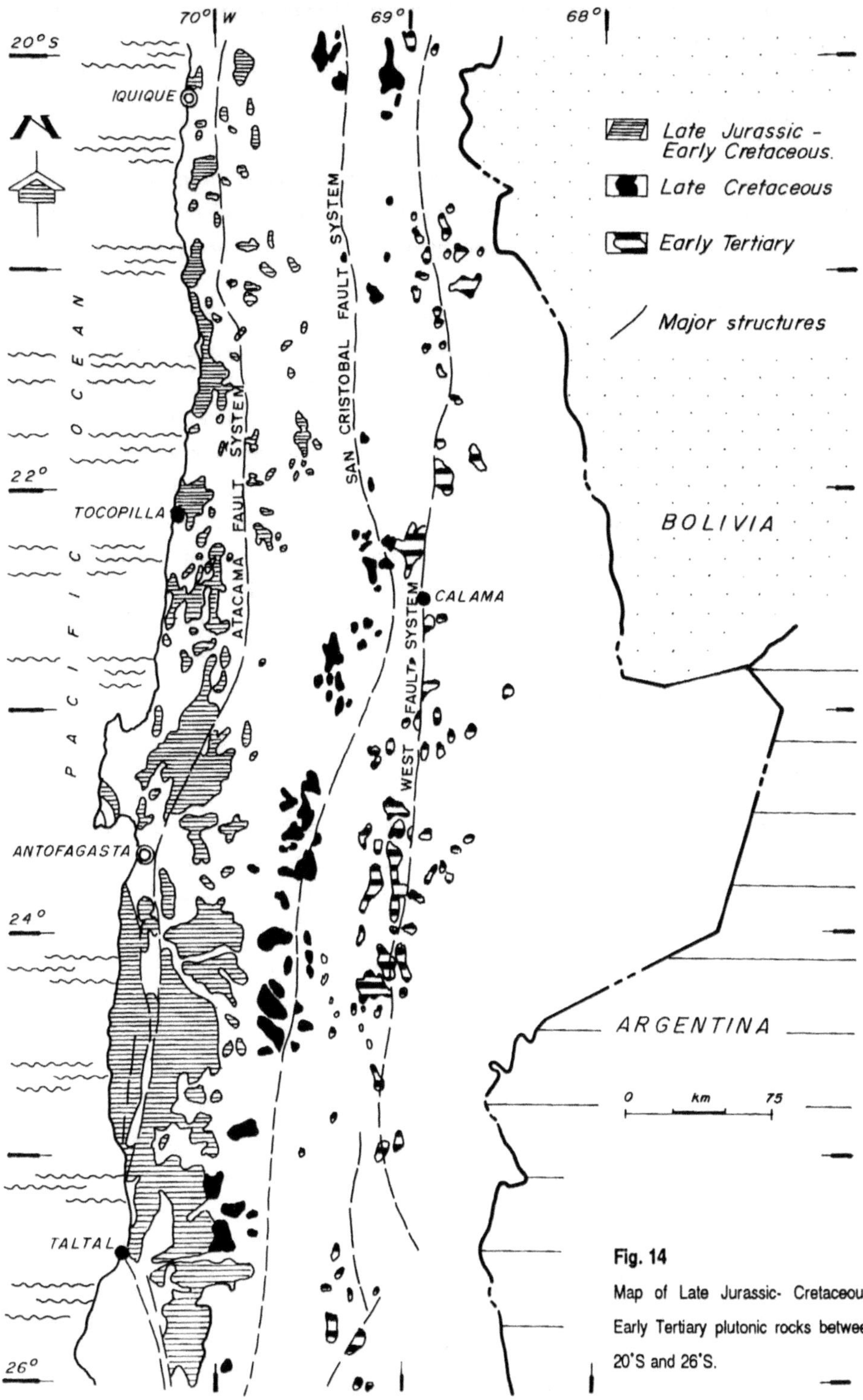

Fig. 14

Map of Late Jurassic- Cretaceous-Early Tertiary plutonic rocks between 20°S and 26°S.

the Atacama Fault System is also observed here; this has been given a 126 Ma dating by Scheuber et al. (1990).

Late Cretaceous Plutons.

Intrusive rocks from this period outcrop in a discontinuous belt on the west flank of the Sierra de Moreno, in the Central Depression (between Calama and Augusta Victoria), on the west flank of the Cordillera de Domeyko, and on the eastern side of the Coastal Cordillera.

Their contact relationships can be seen in the Coastal Cordillera, the Central Depression and the west flank of the Chilean Precordillera, where they intrude Jurassic and Early Cretaceous sedimentary rocks, Albian-Cenomanian volcanic rocks, and intrusive bodies of the Late Jurassic-Early Cretaceous period. Furthermore, these plutonic rocks are closely related to the tectonic activity of major faults; an example of which can seen on the eastern flank of the Sierra San Cristóbal (Quebrada Río Seco), where the activity of such a fault caused dynamic metamorphism in Mid-Cretaceous volcanic rocks.

Lithologically, these plutonic rocks consist of tonalites, monzonites, monzo-diorites and granodiorites. Age determinations carried out on these rocks by K/Ar, Ar/Ar and Rb/Sr methods (Rogers, 1985; Döbel, 1989; Herve and Marinovic, 1989; Pichowiak et al., 1990; Boric et al., 1990) gave results which lie in the 108 to 78 Ma time span, indicating a Cenomanian-Campanian age.

Early Tertiary Plutons.

The outcrops of these rocks occur in an almost continuous NNE-SSW-striking belt in the Central Depression and the Chilean Precordillera, although they also reach the east flank of the Coastal Cordillera in the Paposo-Taltal area. It is also quite possible that they may be present in parts of the Western Cordillera.

These rocks have intruded all the older rocks emplaced before them, belonging to the Precambrian-Paleozoic basement and the Mesozoic overburden. In turn, these intrusive rocks are overlaid by detrital Oligocene-Early Miocene rocks ("Atacama Gravels"). Field observations show that these rocks vary widely in composition, ranging from diorites through monzonites and granodiorites to granites. Porphyritic stocks of quartz-rhyolitic composition are also seen. Observation also showed contact and intrusive relationships between these plutonic rocks, indicating that they were emplaced in more than one intrusive event.

From their contact relationships and a large number of radiometric datings these intrusive bodies can be separated into two large groups:

Paleocene-Eocene Porphyries.

These are rocks ranging in composition from diorites to granites, and are generally found in association with large structures and paleo-calderas on the western edge of the Cordillera de Domeyko (Puig et al., 1988). Radiometric datings place the ages of these intrusive rocks between 66 and 41 Ma (Huete et al., 1977; Naranjo and Puig, 1984; Puig et al., 1988; Padilla, 1988; Maksaev et al., 1988a, 1988b).
The above-mentioned stocks are genetically related to epithermal precious metal ore bodies and/or associated hydrothermal alteration areas, such as the Sierra Gorda, Cerro Zanelli, El Soldado, Cachinal de la Sierra, El Guanaco and Sierra Exploradora Districts.

Early Oligocene Porphyries.

These intrusive rocks are quartz-rhyolitic in composition and are emplaced in a N-S elongate belt along the Cordillera de Domeyko-Sierra del Medio ranges. They are directly associated with the West Fissure Fault System. Radiometric ages place these bodies in the 38-32 Ma interval (Maksaev et al., 1988a, 1988b; Boric et al., 1990).
Intensive hydrothermal alteration and important cupriferous porphyry type mineralization are associated with these porphyries and with the West Fissure Fault System, as in the Quebrada Blanca, El Abra, Chuquicamata, La Escondida and El Salvador mineral deposits.

Paleogeography and Tectonics

The development of the Andean Tectonic Cycle (from Jurassic or perhaps Late Triassic times onward) in the study area took place in the framework of an active continental margin. This evolved within the convergent plate dynamic setting of an active subduction system. This pattern can be subdivided, after Dickinson and Seely (1979), into the following tectonic areas:
1) a fore-arc zone, 2) a magmatic arc, and 3)a back-arc zone.
This scheme can be applied from the beginning of the Andean Cycle. Various succesive parallel magmatic arcs developed from W to E during this cycle, are expressed as several parallel N-S belts of volcanic and intrusive rocks. These rocks, which are coeval, represent a subduction-erosion pattern in an active subduction zone setting (Scholl et al., 1980).
Since Jurassic time, four stages of development can be recognized within this setting: (a) the Coastal Cordillera stage in the Jurassic, (b) the Central Depression stage during the Cretaceous, (c) the Chilean Precordillera stage during the Early Tertiary, and (d) the Western Cordillera stage from the Miocene to the present (Reutter and Scheuber, 1988). During development, each of these magmatic arcs had a fore-arc zone to the west and a

back- arc zone to the east, but owing to the eventual eastward migration of the system, only the present-day magmatic arc remains active.

The back-arc regions of the three former stages have been preserved, or at least can be recognized and reconstructed from their sediments. The younger back-arc basins overlap the older ones. Due to the eastward migration of the younger arcs, these succesively covered the older back-arc basins with volcanic rocks and penetrated them with intrusive rocks, giving the whole complex a characteristic magmatic-arc tectonic setting.

In the study area, outcrops of Cretaceous and Early Tertiary sedimentary rocks with associated volcanic and intrusive rocks are frequent. These rock formations also occur beyond this particular area: in Central Chile (~ 34° S), southern Peru, western Bolivia and northwestern Argentina similar rocks series cover extensive areas. In northern Chile, correlations between the Cretaceous units which occur in the aforementioned areas are often uncertain, since their links have either been broken by Late Tertiary magmatism and by tectonic disruptions, or their outcrops have been covered by recent volcanic activity in the Western Cordillera. A correlation table for the units mentioned above is given in Figure 15.

A tentative description of the evolution of the geologic units studied is shown below:

Kimmeridgian-Barremian-(Aptian?) Period

Sedimentary deposition for this period started after the "Araucana" tectonic phase (after Riccardi, 1988), exemplified by an angular unconformity between the La Negra and Caleta Coloso Formations in the Coastal Cordillera. However, there is no unconformity in the Chilean Precordillera, and a marine regression and the start of the Western Sequence clastic continental sedimentation are observed instead.

The angular unconformity in the Coastal Cordillera was due to intense tectonic activity, which included important sinistral movements along the Atacama Fault System (Scheuber and Andriessen, 1990), and the uplifting of blocks of continental crust to the surface. This phenomenon provoked a rapid subsidence of the back-arc zone, and gave rise to a speedy filling of the Jurassic marine basin with clastic material. In turn, this activity caused a regression which is only evident in the Precordilleran area.

The paleogeographic map in Figure 16a clearly shows an elongate N-S basin, whose sediments represent the Western Sequence described in the section on Stratigraphy, and can be compared and correlated with the upper part of the Chacarilla Formation (Galli and Dingman, 1962). This unit can be found and correlated, as far as the south of Peru, with the upper part of the Labra and Murco Formations in the Arequipa Basin, and the Huancané Formation in the Altiplano Basin (after Batty and Jaillard, 1989). Similarly, these units are comparable to the Megasequence 1 described by Marocco (1988) (see Figure 15).

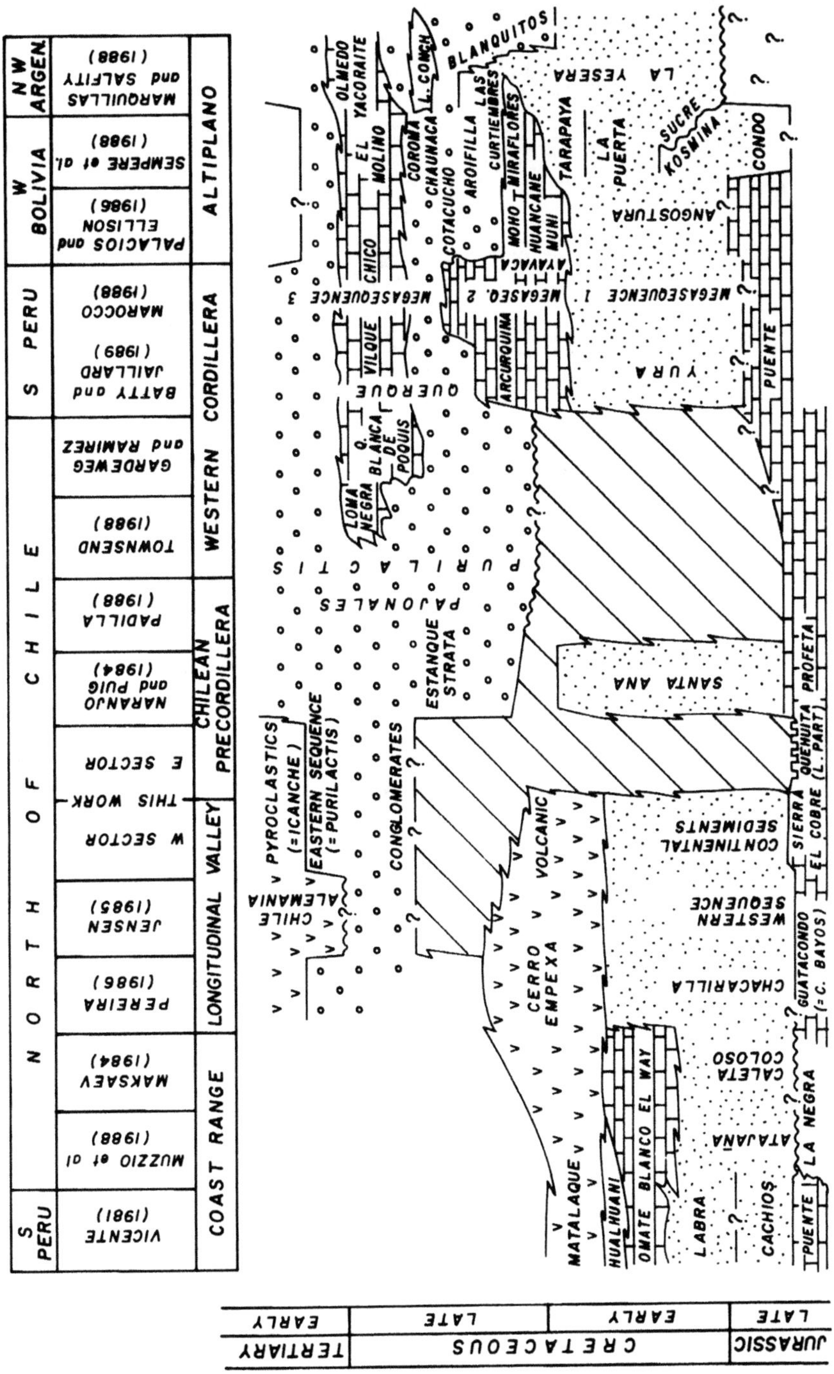

Fig. 15
Stratigraphic correlation chart of Late Jurassic-Cretaceous-Early Tertiary units from areas in N Chile, S Perú, W Bolivia and NW Argentina (16°S-26°S and 67°W-71°W), modified from Bogdanic (1990).

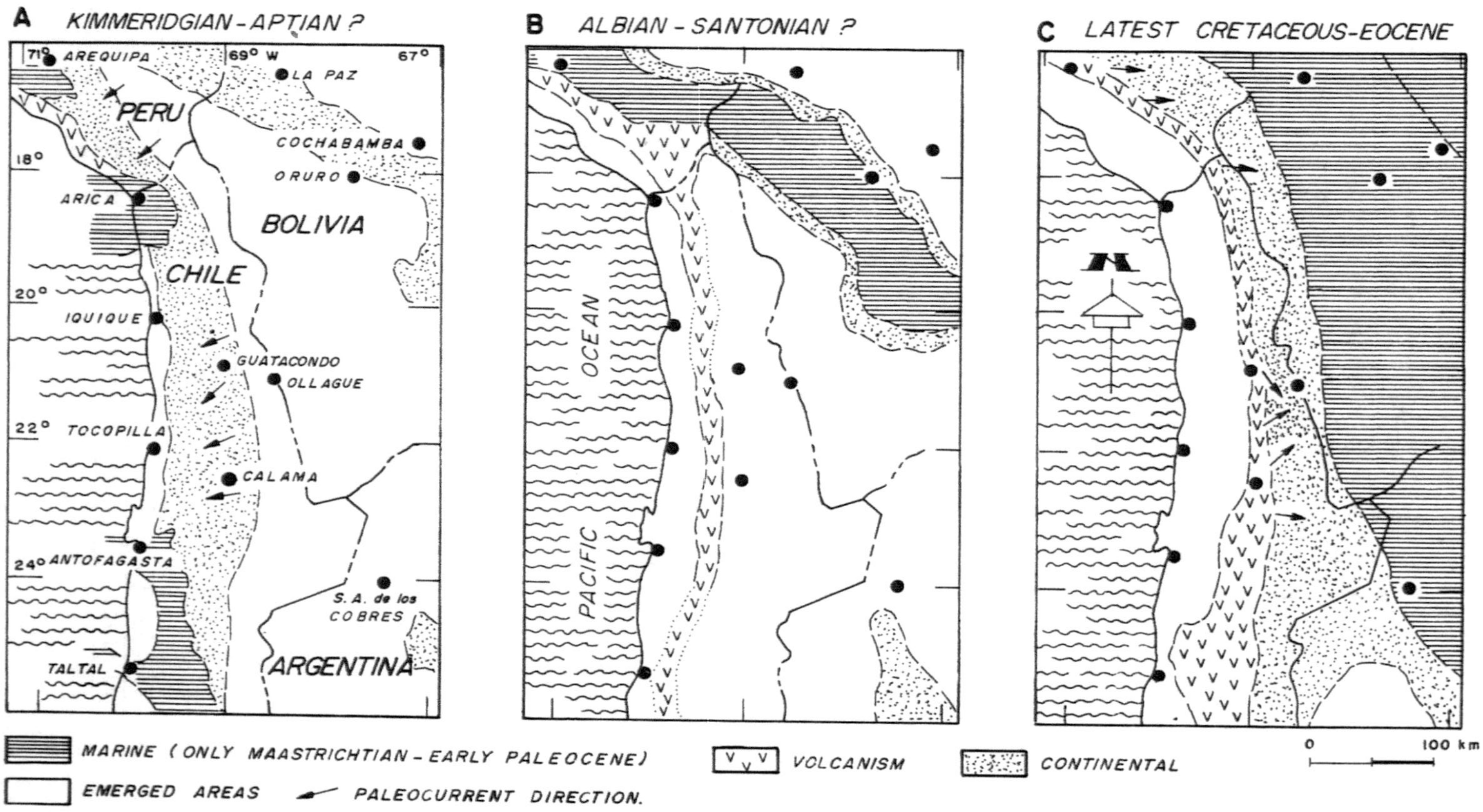

Fig. 16

Paleogeographic map of Andean segment between 16°S-26°S and 67°W-71 °W, modified from Bogdanic (1990).

The sedimentary rocks from this period show transport directions from the east and northeast, where there would have been a topographic high. It is not possible to recognize the eastern border of this basin; this may be due to Late Cretaceous erosion. However, the western limit of the basin was defined by its elevation by the magmatic arc, which was active during Jurassic times in the Coastal Cordillera. This sector of the basin received deposits of coarse-grained red clastic sediments, corresponding to the Atajaña (after Muzzio et al., 1988) and Caleta Coloso (after Maksaev, 1984) Formations. The latter showed transport directions from the west and northwest, caused by the extinct magmatic arc in the Coastal Cordillera. These continental-clastic sedimentary conditions on the western downslope of the basin continued until at least Valanginian times; whereas on the eastern slope, continental sedimentary deposition continued up to Barremian or perhaps Aptian times. Concurrently, there were local marine ingressions in the Coastal Cordillera from the Hauterivian until Barremian (Aptian?) times, with marine sediments being deposited conformably and transitionally over clastic redbeds. These marine sediments can be seen in the Blanco (after Muzzio et al., 1988) and El Way (after Maksaev, 1984) Formations, located respectively south of Arica and south of Antofagasta. These formations are made up of littoral and bioclastic limestones with abundant fauna, and are representative of a very shallow continental platform sea.

The true extent of this transgression is not known, but there are no marine sediments in the clastic rocks north of the area surveyed by Bogdanic (1990) (Guatacondo, 21° S), whereas in the Cerritos Bayos area (~22° S) and as far as the Sierra El Cobre area (~23° S) a transitional marine-continental sequence is observed (fan-delta, after Massari and Colella, 1988). Furthermore, to the south of the survey area, in the Central Depression (east of Paposo, ~24°30′ S), marine rocks with Early Cretaceous fauna outcrop and are similar to those in the El Way Formation (Marinovic, 1988, (personal communication). Farther south, in the Domeyko Cordillera (Sierra Candeleros, ~25°30′ S), the sea reached more easterly areas -as far as the west flank of the Precordillera- as is the case with the Santa Ana Formation (Naranjo and Puig, 1984).

The effects of this transgression in the south of Peru are evidenced by the presence of littoral limestones from a very shallow marine environment. However, in this sector the effects of transgression were manifest earlier, during the Tithonian and up to Berriasian times (Batty and Jaillard, 1989). According to these authors such transgressive units are found in the Gramadal Formation and in sedimentary rocks of the Tiabaya area, in the Arequipa Basin, and in the Sipin and Muni Formations of the Altiplano Basin.

Albian-Cenomanian-Turonian? Period

During this interval ("Mid" Cretaceous) a magmatic arc was emplaced in the back-arc basin formed by the Jurassic arc. This is the first easterly migration of the magmatic arc system (Figure 16b).

The rocks from this magmatic episode belong almost exclusively to an andesitic volcanic series, which lies with some slight erosional disconformity over the fluvio-deltaic sediments of the Kimmeridgian-Aptian? interval. These volcanic rocks (Cerro Empexa Formation, after Galli and Dingman, 1962; Aeropuerto Formation, after Naranjo and Puig, 1984) have been dated in the Cerros de Montecristo and Taltal areas. Such Rb/Sr and K/Ar datings place them in the 115 to 104 Ma interval (Ulriksen, 1979; Rogers, 1985). These rocks lie in a NNE-NNW to N-S discontinuous elongate belt along the west flank of the Sierra de Moreno, and extend southward across the Cerritos Bayos-Sierra San Cristóbal zones, the Augusta Victoria area and the vicinity of Taltal, and the Copiapó area (~27° S).

Because of their lithological characteristics, stratigraphic position and age, these volcanic rocks can be correlated and compared to similar units which outcrop in the coastal area of southern Peru, such as the Matalaque Formation (after Vicente, 1981; Batty and Jaillard, 1989). This last unit is made up of a thick series of andesite-dacite flows, volcanic agglomerates and associated acid tuffs, assigned to the Albian.

In addition, an important event in the south of Peru, the west of Bolivia and northwestern Argentina is a marine transgression which occurred during Albian time and lasted until at least the Early Santonian, but is not observed in the north of Chile. The marine transgression sediments point to the existence of a very shallow sea, and are locally interbedded with volcanosedimentary units, mainly in southwest Peru. These marine sediments are found in the Arcurquina Formation and the Moho Group, assigned to an Albian-Cenomanian age (Palacios and Ellison, 1986; Batty and Jaillard, 1989).

These volcanic and marine units formed a new magmatic arc with a back-arc basin to the east in the south of Peru, western Bolivia and northwestern Argentina. However, so far no sediments from the back-arc basin of this period have been discovered in Chile, between the Precordillera (on the west) and the Western Cordillera (on the east). These facts would indicate that this back-arc basin was only a locally developed episode, or that its sediments were eroded during latest Cretaceous-Eocene times.

The magmatic activity of this new arc is also related to tectonic activity along a series of large structures with sinistral movements located on the San Cristóbal Fault System (~69°30′ W). Indeed, the angular disconformity which separates these volcanic rocks from the Augusta Victoria-Chile Alemania Formation constitutes evidence of a deformation episode associated with such tectonic activity. This unconformity, judging by its stratigraphic relationships and radiometric datings, would indicate a tectonic phase which occurred in Santonian-Campanian times. This correlates well with an important deformation and orogeny which took place in Peru, known as the "Peruvian" or "Subhercinic" Phase (85-80 Ma; Megard, 1984, 1987; Ellison et al., 1989)

Latest Cretaceous-Eocene Period

During this lapse, red continental clastic sediments and the volcanic and pyroclastic rocks of the Eastern Sequence (this work) were deposited. The volcanic rocks are associated with hypabyssal intrusive rocks. The magmatism and sedimentation are due to the development of a magmatic arc during this period. These rocks are now exposed in the Chilean Precordillera and partially on the eastern flank of the Central Depression (Figure 16c).

This magmatic arc shows a different pattern of development south of Calama (22° S). From the latest Cretaceous to the Eocene these volcanic rocks outcropped as far as the western border of the Central Depression. In this region they are called the Augusta Victoria/Chile Alemania Formation (after Boric et al., 1990). A great number of radiometric datings on the volcanic rocks of this series place them in the 66-40 Ma interval (Maksaev, 1984; Naranjo and Puig, 1984; Puig et al., 1988). A large quantity of abyssal and hypabyssal granitic-granodioritic stocks also outcrop in this area and are associated with the above rocks; they show radiometric ages lying in the 66 to 41 Ma range (Huete et al., 1977; Ramírez and Gardeweg, 1982; Maksaev et al., 1988a, 1988b; Puig et al., 1988; Padilla, 1988). They are also related genetically to the mineralization and hydrothermal alteration of precious metal ore bodies; as for example, in the Cachinal de la Sierra district.

In contrast, in the Precordillera north of Calama, magmatic activity was restricted to the Eocene-Early Oligocene, exclusively. The ages of the volcanic rocks here (Arca and Icanche Formation, after Bogdanic, 1990) range from 50 to 38 Ma (Huete et al., 1977; Döbel, 1989). Intrusive abyssal and hypabyssal rocks of rhyolitic-quartzitic composition also outcrop here, whose ages lie in the 38 to 32 Ma time lapse (Maksaev et al., 1988a, 1988b). These have genetic relationships with mineralization and hydrothermal alteration processes of a porphyry-copper type. To the north of Calama the pre-Eocene interval is recorded only in red clastic sediments which do not crop out south of lat 22° S, except locally in the Domeyko Cordillera (e.g., Quebrada de Mulas and west flank of the southern part of the Salar Punta Negra). The age of these sediments falls within the development period of the magmatic arc; the sedimentary deposition, judging by its characteristics, took place in one or more small intermontane basins, which may have been partially interconnected. The outcrops of rocks pertaining to this interval (latest Cretaceous-Paleocene) extend north and eastward into southern Peru, western Bolivia and northwestern Argentina, and southward as far as Central Chile (~33° S).

The latest Cretaceous-Eocene magmatic arc must have had a fore- arc area to the west and a back-arc area to the east. Sediments from the fore-arc zone are not known at present, although the red conglomerates of the Quebrada Culebrón (E of Baquedano, ~23° S) could belong to it. The back-arc sediments are better known and documented, and they frequently outcrop in northwestern Argentina, western Bolivia and southern

Peru, were they can be correlated with the Salta Group sediments in northwestern Argentina and the Puca Group sediments in Bolivia and Peru. The latter have a Late Cretaceous-Eocene age, and include the Balbuena Subgroup (see Marquillas and Salfity, 1988) and the P_4 Sequence (see Sempere et al., 1988), respectively.

In the latest Cretaceous (Maastrichtian) a marine transgression occurred in the back-arc area, represented by the carbonaceous sediments of the Yacoraite Formation in northwestern Argentina (Marquillas and Salfity, 1988), and of the El Molino Formation in Bolivia and Peru (Palacios and Ellison, 1986). In Chile these units can be correlated lithologically with the Loma Negra (after Townsend, 1988) and the Quebrada Blanca de Poquis (Gardeweg and Ramirez, 1985) Formations, on the Chilean-Bolivian-Argentine border, respectively. During this period, extensional tectonic events and alkaline basalt volcanic events took place locally in the back-arc basin area of northwestern Argentina. These rocks define the back-arc setting of this depositional environment (paleorift; sensu Galliski and Viramonte, 1988). According to Marquillas and Salfity (1988) the marine transgression could have come from the east (Atlantic); or it is possible that this basin could have been conected to the west (Pacific) through the Central Peru area (Mourier et al., 1988). In any case, a connection of some kind between this back-arc basin and the west across the north of Chile can be discarded, because the transport direction of red clastic sediments (Eastern Sequence) indicates there was a topographic high to the west of the region during the latest Cretaceous.

On the western side of the Salar de Atacama basin the back-arc sediments are interbedded with erosional products from the magmatic arc outcropping in the Precordillera, represented by the Purilactis Group (sensu Charrier and Reutter, 1990). There are no outcrops of the Yacoraite Formation on the western flank of the Salar de Atacama area, although some sandstones with calcareous cement appear in the basal portion of the Purilactis Group (Reutter, 1990, personal communication) which could be correlated with the Yacoraite Formation. The start of deposition of the Purilactis Group can be dated as latest Cretaceous, due to the correlation of the base of the calcareous sandstone beds and the age of an interbedded andesite at the bottom of the unit (Ar/Ar, 60 Ma; Flint et al., 1989). Tuffs and lavas from the middle portion of the Purilactis Group have an age of 40 Ma (Ramirez and Gardeweg, 1982; Döbel, 1989). The sedimentary transport was from the west, where the magmatic arc must have been located. Interbedding between sediments and volcanic rocks occurs throughout the Purilactis Group sequence. This makes it possible to state that this unit was deposited in an intermediate area lying between the magmatic arc and the back-arc basin (Charrier and Reutter, 1990).

The Eastern Sequence sediments which outcrop north of Calama, can only be correlated with the Purilactis Formation sensu strictu (upper part of the Purilactis Group). For this reason, it is very probable that this red sediment has connections with the large sedimentary back-arc basin in Bolivia and northwestern Argentina.

As in the previous period studied, it can be seen that at the climax of magmatic activity here there was also intense tectonic activity, associated in this case with dextral movements along the Western Fault System (Reutter, 1990), as well as a folding phase which affected all the pre-Oligocene units. Radiometric datings and stratigraphic relationships in the area give evidence of a tectonic phase occurring at the Eocene-Oligocene boundary (34-32 Ma), which would correspond to the "Incaic" Phase known also in Peru, Bolivia and Argentina (Megard, 1984, 1987).

Metallogenesis

With regard to the formation of metal-bearing bodies, if a strictly chronological point of view is taken, the Cretaceous period appears to have been of minor importance. However, as indicated above, geologic boundary markers in the field are lacking due to the existence of gradual transitions between the Jurassic and the Cretaceous periods, as well as between the Cretaceous and the Tertiary. Also, there are ambiguities with regard to chronological time and duration of the mineralizing processes.

If the metallogenic characteristics along the Andes are analyzed, it is possible to identify well-defined segments containing ore deposits of clearly distinctive distribution, type, abundance, and metallic-ore characteristics. The area covered in this work roughly coincides with one such segment (21° to 26° S), the Antofagasta Region.

In contrast to the stretch of country south of lat 26° S, which is polymetallic in nature, the Antofagasta Region is notable for the development of ore belts having a markedly cupriferous nature and of almost entirely magmatic origin: the ore deposits whose mineralization processes can be ascribed to sedimentary, diagenetic or metamorphic phenomena are few and of little importance.

These metallogenic characteristics are the result of a subducting convergent plate tectonic setting. Changes in the rate and angle of convergence have controlled the development of magmatic cycles, which has caused the migration of magmatic activity progressively towards the east, in the same direction as the subducted oceanic plate. The magmatism, of a calc-alcaline type, falling within the magnetite series defined by Ishihara (1985), is presumed to have been generated mainly by partial melting of the oceanic crust. Within this setting certain depth and hydration conditions presumably favoured copper and sulphur segregation, thus generating the copper-bearing ore deposits. The possibility of metallic heritage from Paleozoic deposits is a complementary hypothesis to explain the origin of metallogenic belts. However, the scarcity of Paleozoic outcrops makes it difficult to study this theory.

Ore Belts

In the Antofagasta Region, several metalliferous ore belts can be distinguished (Figure 17); the most notable of which are, from west to east and, as explained above, in decreasing age order:

1) The Coastal Cordillera Copper Belt
2) The Central Depression Gold-Silver-Copper Belt
3) The Domeyko Cordillera Silver Belt
4) The Pre-Andean Cordillera Porphyry Copper-Molybdenum Belt
5) Andean Volcanic Belt.

Although these belts can be traced without the aid of any interpretative judgement, they coincide with morphostructural features that can be associated with the development of the Jurassic-Early Cretaceous magmatic arc, which brought about the formation of the Coastal Cordillera Copper Belt. Part of it, as we shall see, is emminently volcanic in origin, with sulphide deposits disseminated in lavas, breccias, and deep veins emplaced in plutonic rocks.

The Gold-Silver-Copper, Silver and Porphyry Copper-Molibdenum Belts, which comprise part of the Central Depression and of the western flank of the Domeyko Cordillera are related to the start of a new magmatic cycle, whose axis moved eastward. This metallogenic phase culminated in the Early Tertiary with the Porphyry Copper Belt, which developed under the structural control of major faults.

Mineralization related to the development of the La Negra Magmatic Arc (Jurassic-Early Cretaceous): Coastal Cordillera Copper Belt

The development of the La Negra magmatic arc originated in intense basaltic-andesitic volcanic activity during the Jurassic, and caused the deposition of a series of volcanic rocks having a measured total thickness of approximately 8,000 meters (Muñoz et al., 1988). These rocks, known in the Antofagasta Region as the La Negra Formation, occur in the Coastal Cordillera from Tocopilla to Taltal. They are intruded by various gabbro-dioritic bodies which developed during the Late Jurassic-Early Cretaceous interval (see chapter on Intrusive Rocks) in a magmatic arc setting. The Coastal Cordillera, which is the remnant of this arc, is a metalliferous ore province containing various types of ore bodies, but with a very marked copper specialization. Using an occurrence criterion, the following groups of ore deposits can be defined:

(a) volcanic rock ore deposits, (b) plutonic rock ore deposits, and (c) sedimentary rock ore deposits.

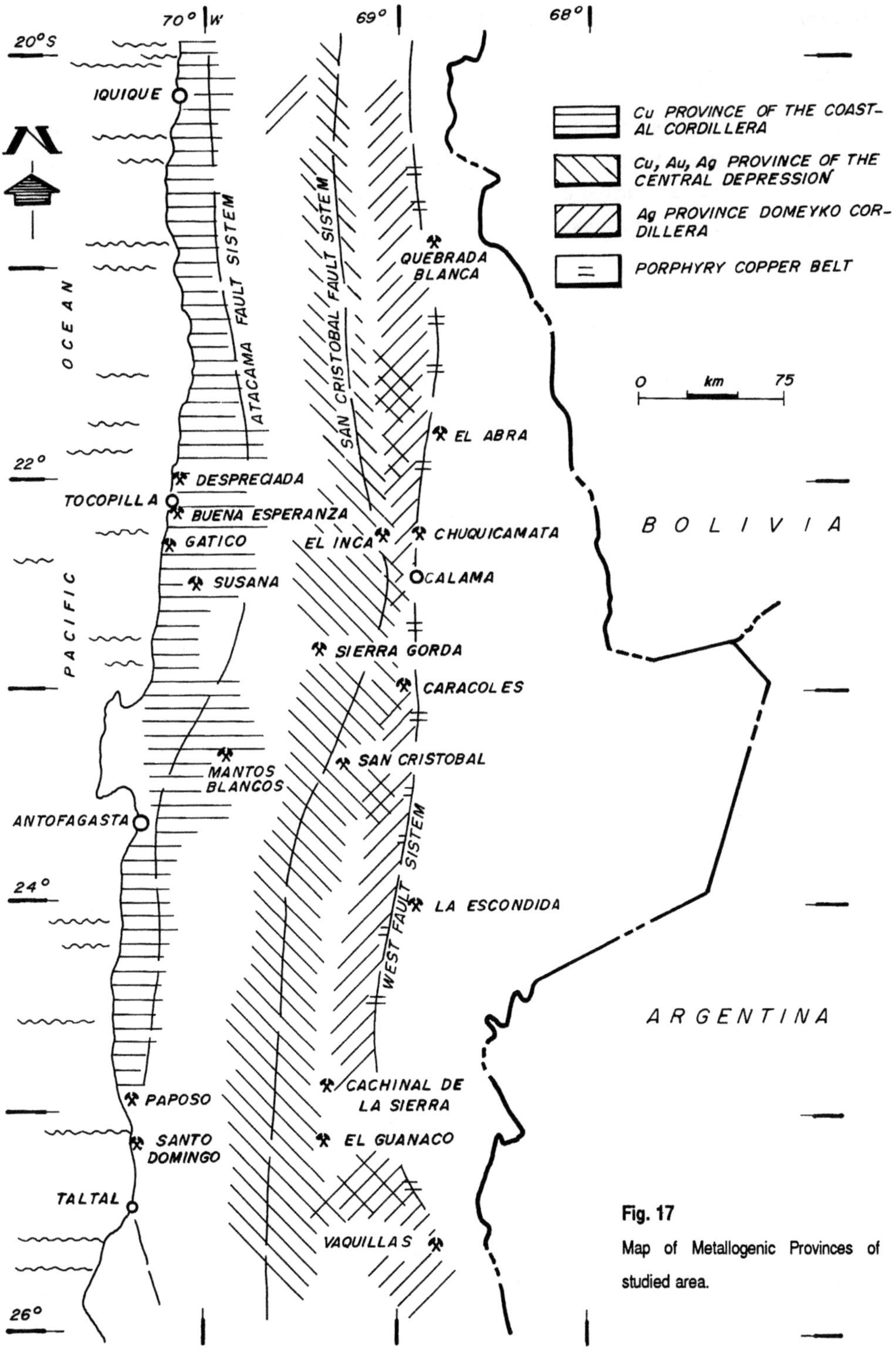

Fig. 17

Map of Metallogenic Provinces of studied area.

Volcanic Rock Ore Deposits (La Negra Formation)

Carolina de Michilla Vein Type

These are veins of copper oxidation minerals (chrysocholla, atacamite), which eventually present chalcocite, bornite, chalcopyrite and pyrite. These structures cut the volcanic strata of the La Negra Formation. Their origin seems to be related to events of a subvolcanic nature and could be genetically linked to the disseminated Buena Esperanza Type ore deposits.

Buena Esperanza Type Ore Bodies:

These are the most important ore deposits, both in terms of grade and tonnage, and the only ones of this type being actively mined at present. Various advantages, such as their proximity to highways and ports, the metallurgical quality of the ore, and the size of the ore bodies (tens of millions of tons with a better than 2% copper content), make these ore bodies the best copper- producing deposits in the region at this time, after the porphyry copper deposits.

The best examples of this type of ore body are, from south to north, Santo Domingo (Definis, 1985), Mantos Blancos (Chavez, 1983), Susana (Espinoza, 1981; Soto and Dreyer, 1985; Wolf et al., 1990), and Buena Esperanza (Losert, 1973; Palacios and Definis, 1981; Palacios et al., 1986; Espinoza and Orquera, 1988).

These ore deposits consist of intrusive-hydrothermal breccias and of chalcocite, bornite, chalcopyrite and pyrite disseminated mineralization in vesicular or brecciated volcanic lavas. In general terms, this type of orebody can be considered a form of stratabound or "manto" deposit, as historically denominated in Chile. However, some distinctive features which differentiate them from the massive sulphides, Kuroko or other deposits with which they can be compared, make it possible to define them under a different model, such as that proposed by Espinoza (1982) under the name "Disseminated Volcanic Copper Sulphides". This model is based on the idea of an early hydrothermal mineralization linked to the same volcanic phenomena which originated the wallrocks, and is compatible with the characteristics of the ore deposits and the data so far available. Thus, it provides an explanation, for instance, for the zoning of both ore and alteration minerals, the breccia bodies, and the constant presence of subvolcanic intrusives bodies. These subvolcanic intrusive bodies are mineralogically identical to the lavas they intrude; also, their textures are similar. Their attitudes show a marked vertical tendency with regard to the stratification, for which reason they have been interpreted as feeders, seams or cylindrical bodies related to the volcanic sequences (Espinoza and Palacios, 1982), Dating of these intrusive bodies has confirmed this supposition: Susana microgabbro, K/Ar (WR): 140±4 Ma (Astudillo, 1983) and 154-133 Ma (Boric et al., 1990); Buena Esperanza gabbro neck, K/Ar: 150±5 Ma (WR, Espinoza and Orquera, 1988) and 168±5

Ma (Plag., Boric et al., 1990); Mantos Blancos andesitic dike, K/Ar: 149-147 Ma (WR, Chavez, 1983).

Regional alteration of the host rocks, gave rise to some interesting speculation on the source of the copper mineralization. Losert (1973) in a classic study, defined this alteration as consisting of abundant epidotization, variable silicification (chalcedony), and discrete development of albite, zeolites, phrenite, chlorite, pumpellyte and sericite, considered to be the products of load metamorphism. In this regard, copper remobilization from host volcanic rocks by means of load metamorphism has been proposed by Sato (1984). However, the presence of hydrothermal breccias which are contemporaneous with intrusive subvolcanic bodies, such as in the "Susana Breccia" (Soto and Dreyer, 1985), leaves no doubt as to the origin of the mineralization being volcanic (Jurassic) activity. There is also further evidence, such as sulphur isotope ratios (Sasaki et al., 1984), and paleotemperature measurements in fluid inclusions of quartz and calcite at Buena Esperanza, which yielded values of 65° to 195°C, and a calculated depth of 1,000 to 1,500 meters (Nisterenko et al., 1974); all of which points to hydrothermal processes in a subvolcanic environment.

The foregoing does not, of course, exclude the possibility of a hydrothermal regional alteration, caused by reheated ground water, which would have brought about the removal of copper from the volcanic rocks and the transportation of it towards more favourable sites.

Ore Deposits Emplaced in Plutonic Rocks:

Despreciada Copper Vein Type:

These are deep veins of chalcocite associated with magnetite and actinolite, and eventually having a small content of uraninite, cobaltite, danaite and molybdenite, which are emplaced in coarse- grained diorites of the Coastal Batholith. Examples of these deposits are the Minita-Portezuelo-Despreciada and Tres Puntas Group in Tocopilla (Boric et al., 1990), the Gatico Mine District (Boric et al., 1990), the Naguayán Mine District (Kuntz, 1928), and the Julia-Montecristo Mine District and Paposo Mines (Ruiz et al., 1965). All of these were worked at the turn of the century; in some cases the workings reached depths greater than 500 meters.

Copper Stocks and Chimneys:

In some of the intrusive rocks of the Coastal Cordillera there are vertical breccia bodies, sometimes tabular and sometimes irregular, with oxidized copper ore mineralization, which at depth changes to veinlet mineralization (closely packed ore veinlets).

Boric et al. (1990) described the Chuminga Deposit, emplaced in Mid-Jurassic gabbros with chalcopyrite, bornite and pyrite as primary minerals. Molybdenite, polybasite and gold were also found, together with quartz and tremolite-actinolite. Other examples of this type of ore deposit are the Sorpresa Mine ores at Caleta El Cobre, which are emplaced in a Late Jurassic diorite-granodiorite, and the Puntillas Deposit, sited north of the Oficina Prosperidad Nitrate Mine, which is emplaced in a Late Jurassic-Early Cretaceous dacite porphyry and which shows abundant hydrothermal alteration.

Non-Chalcophile Veins:

Minor veins occur in plutonic rocks, which include iron ore bodies in the Naguayán area and gold (copper) veins in the southern portion of the Sierra Esmeralda (south of Taltal).

Copper Ore Deposits Emplaced in Late Jurassic-Early Cretaceous Sedimentary Rocks (Quebrada El Way):

These ore deposits are embedded in red conglomerates which form the Lower Member of the Caleta Coloso Formation (Brüggen, 1950). The copper deposits are stratiform and occur also in minor veins or complex bodies, with minerals such as atacamite, gypsum and djurleite. Diagenetic atacamite is found in sandstone lenses and evaporite levels in this unit. It has been proposed (Espinoza, 1983) that these deposits were formed by ground water which contained copper sulphate and copper chloride in solution. The source of these solutions would have been ore bodies emplaced in lavas (La Negra Formation), which were being eroded nearby, upstream from these deposits. This transported (or exotic) mineralization could have been transformed to djurleite during diagenesis, in response to a more reducing environment. Flint and Turner (1990) have suggested that this mineralogic transformation was due to the action of an extra-basinal hydrothermal fluid.

Intermediate Depression and Domeyko Cordillera Copper, Gold and Silver belt:

This stretch has a great number of veins, breccia bodies and small porphyry deposits containing copper, copper with gold, gold, and silver. These ore bodies are emplaced in rocks whose ages fall within the late Jurassic-Cretaceous-Early Tertiary time span. Ore deposits of this zone could be classified in various ways using different criteria if more knowledge was available, but the information at hand is poor and insufficient to allow a coherent overall pattern to be established. We shall subdivide it in a somewhat speculative way into two belts: (1) a western copper-gold-silver belt, which from north to south would include the Sierra Gorda, Lomas Bayas, San Cristóbal, Pan de Azúcar and Cachinal de la Sierra districts, and (2) an eastern silver-gold belt, which from north to south would comprise the Caracoles, El Inca, Chimborazo, Argomedo and Vaquillas districts.

The ore deposits in the western belt are mainly epithermal and associated with latest Cretaceous-Eocene volcanism; while those in the eastern belt are associated with rhyolite porphyry type intrusives. Puig et al. (1988) found that the El Soldado (Ag), Cachinal de la Sierra (Ag) and El Guanaco (Au, Cu) deposits were controlled by the structures of a Paleocene-Early Eocene caldera system, associated with siliceous domes and fumarole activity. Rivera and Stephens (1988), while studying Early Tertiary paleogeothermal fields, found that the majority of these were located on the edges of the Intermediate Depression (San Cristóbal, Cerro Búfalo, Sierra Overa) or in raised blocks within it.

The economic importance of this belt is relatively small, as far as gold is concerned. In the Antofagasta Region, a large amount of gold is produced as a by-product of exploitation of the large porhyry copper-molybdenum deposits. All the gold mines, with the exception of the El Inca, San Cristóbal (Rivera, 1980) and Guanaco mines (Cuitiño et al., 1988; Puig et al., 1988), are relatively small, however, in comparison with those contained in the important gold belts, which are found south of lat 25° S (Camus, 1985, 1990).

Conclusions

The development of the Andean Tectonic Cycle (Frutos, 1972), included the development through time of four magmatic arc systems which migrated from west to east, from the Coastal Cordillera to the Western Cordillera (Figure 19). These arcs developed in the Coastal Cordillera during the Jurassic (La Negra Arc), in the Central Depression during the Mid-Cretaceous (San Cristóbal Arc), in the Chilean Precordillera during the latest Cretaceous-Eocene interval (Sierra del Medio-Cordillera de Domeyko Arc), and in the Western Cordillera (Puna Arc) from Miocene time until the present day. Within this context, it can be said that the Kimmeridgian-Aptian? fluvio-deltaic sedimentary sequence was formed at the end of the first and the start of the second of these magmatic arcs. Similarly, the latest Cretaceous- Eocene sedimentary-volcanic sequence was associated with the third magmatic arc.

In the study area (20°-26° S), the lithologic sequences of continental sandstones or of conglomerates and sandstones with interbedded lavas and pyroclastic rocks are widely distributed. Detailed study and observation of these units have shown that they consist of two Sequences, which are clearly different in age, facies and outcrop distribution. These two continental units have been named the Western and the Eastern Sequences (Bogdanic, 1990).

The Western Sequence, of Kimmeridgian-Aptian? age, developed continuosly and transitionally over Late Jurassic marine sediments. These sediments originated in a back-arc basin over a continental crust, which included a magmatic arc to the west, whose extrusive and intrusive materials are now found in the Coastal Cordillera (La Negra Formation).

In Kimmeridgian time a N-S marine regression occurred in this basin, as a result of clastic material fill. The deposition lasted throughout magmatic activity in the Coastal Cordillera region, which was further affected by strong tectonic movements at the Jurassic-Cretaceous boundary, and a sinistral strike-slip movement along the Atacama Fault

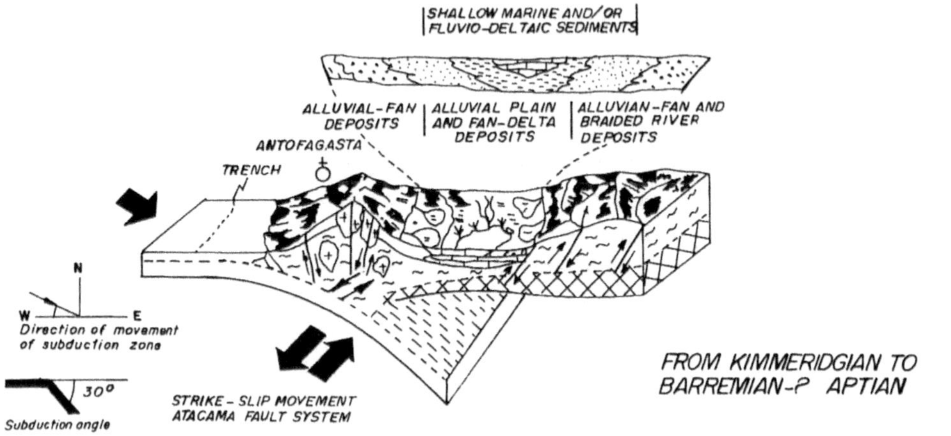
SHALLOW MARINE AND/OR FLUVIO-DELTAIC SEDIMENTS
ALLUVIAL-FAN DEPOSITS
ALLUVIAL PLAIN AND FAN-DELTA DEPOSITS
ALLUVIAN-FAN AND BRAIDED RIVER DEPOSITS
ANTOFAGASTA
TRENCH
N
W
E
Direction of movement of subduction zone
30°
Subduction angle
STRIKE-SLIP MOVEMENT ATACAMA FAULT SYSTEM
FROM KIMMERIDGIAN TO BARREMIAN-? APTIAN

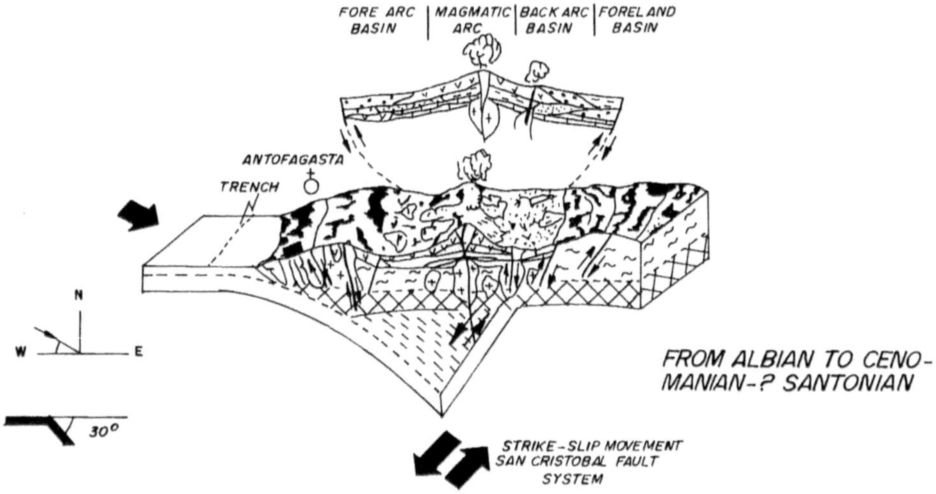
FORE ARC BASIN
MAGMATIC ARC
BACK ARC BASIN
FORELAND BASIN
ANTOFAGASTA
TRENCH
N
W
E
30°
FROM ALBIAN TO CENO-MANIAN-? SANTONIAN
STRIKE-SLIP MOVEMENT SAN CRISTOBAL FAULT SYSTEM

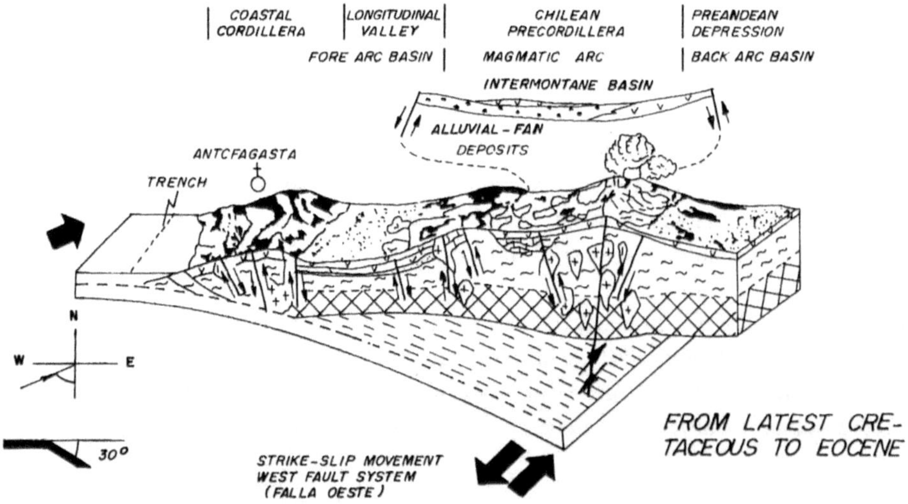
COASTAL CORDILLERA
LONGITUDINAL VALLEY
CHILEAN PRECORDILLERA
PREANDEAN DEPRESSION
FORE ARC BASIN
MAGMATIC ARC
BACK ARC BASIN
INTERMONTANE BASIN
ALLUVIAL-FAN DEPOSITS
ANTCFAGASTA
TRENCH
N
W
E
30°
STRIKE-SLIP MOVEMENT WEST FAULT SYSTEM (FALLA OESTE)
FROM LATEST CRE-TACEOUS TO EOCENE

System. This tectonic activity was also genetically linked to a phase of volcanogenic copper type mineralization.

The sandstone facies of the Western Sequence correspond to a "Recycled Orogen" (Dickinson, 1985). Furthermore, the sedimentary structures show that the constituents of the sandstones laid down in the back-arc basin, came from a magmatic arc to the west, and a continental block to the east, where the predominant facies in the basin were of fluvio-deltaic type (Figure 18a). Plotting these sandstones in a $Q_pL_vL_s$ triangular diagram shows that these sediments lie within the "Mixed Orogenic Sands" field. This indicates that the source material came from a transitional period between two tectonic regimes (Mack, 1984). This tectonic instability was marked by the Araucana Phase (after Riccardi, 1988), which ocurred at the start of the Kimmeridgian age in parts of southern Peru, central and southern Chile and western Argentina.

From Albian to Turonian? times a new volcanic chain was emplaced in the back-arc basin of the previous period, in the area in which the Central Depression now lies (see Figure 18b).

The products of this second magmatic arc were mainly andesitic in nature. They lie conformably, or with a slight erosional unconformity, on the Western Sequence. The volcanic rocks have ages between 115 and 104 Ma; whereas the genetically associated intrusive rocks which cut them have ages ranging from 100 to 80 Ma. The end of the active stage of this magmatic arc was marked by strong tectonic sinistral strike-slip movements along the San Cristóbal Fault System, and folding. These correspond to the Peruvian or Subherzinian Tectonic Phase (sensu Megard, 1984; Ellison et al., 1989).

However, the Western Sequence sediments are missing from the Sierra del Medio, as are the Mid-Cretaceous volcanic and intrusive rocks. The Jurassic marine deposits are covered pseudo-conformably by red clastic sediments of the Eastern Sequence, or else by Eocene volcanic rocks. These sediments and volcanic rocks were produced by a third magmatic arc, which was located in what is now the Chilean Precordillera and which formed very extensive intermontane basins north of lat 23° S (Figure 18c).

The tectonic setting again plays an important role in the formation of the sandstone facies -which in the classification proposed by Dickinson (1985) also came from a "Recycled Orogen"- but in this case with a direct and far more marked volcanic influence than the Western Sequence deposits. In the $Q_pL_vL_s$ diagram it can be seen that the sandstone source material came from an "Arc Orogen", which was very probably the magmatic arc active during this period. Another feature here is that the latest Cretaceous-Eocene deposits show paleocurrents from the west and southwest, which is in complete contrast to the paleocurrents of the Late Jurassic-Early Cretaceous sediments.

Fig. 18

◄ Geodynamic development of the active continental margin in the North Chilean Andes (between 20°S and 26°), modified from Bogdanic (1990).

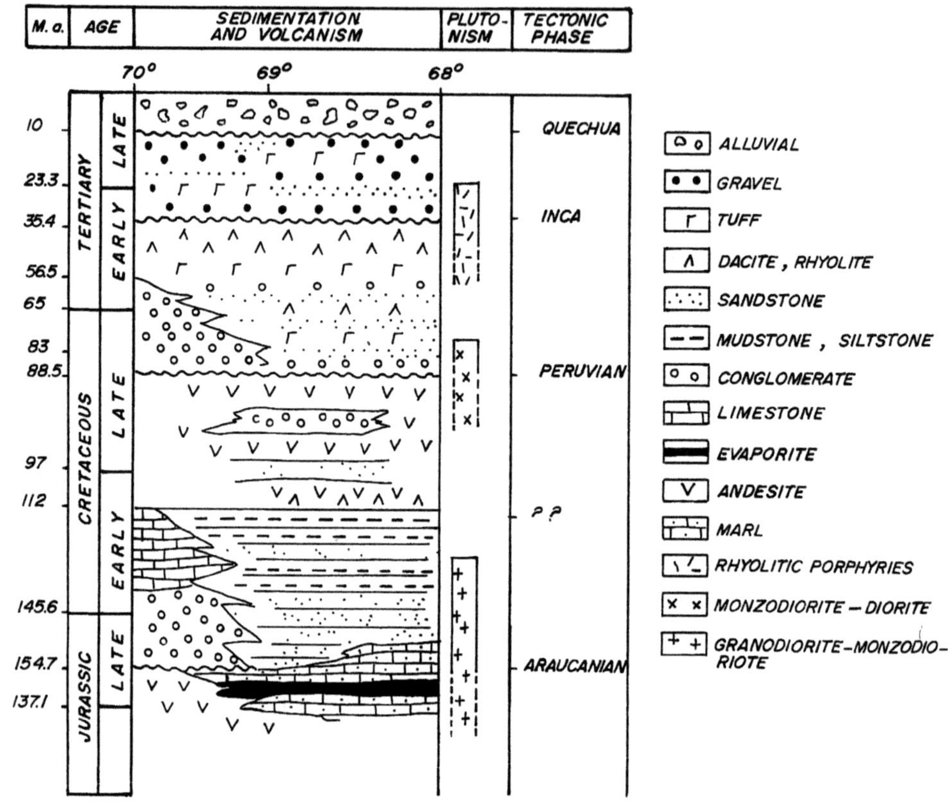

Fig. 19

Tectonostratigraphic chart of studied units between 20°S and 26°S.

In the southern part of the study area (south of 22° S), the volcanism and sedimentation developed from latest Cretaceous time onward, whereas north of 22° S, volcanism was confined exclusively to the Eocene.

The Eocene-Oligocene boundary marked the peak of the folding and dextral strike-slip tectonic movement (Western Fault System), which affected all pre-Oligocene units and finally brought about the uplifting of the Chilean Precordillera. This tectonic activity was also genetically linked to a phase of mineralization and hydrothermal alteration of a porphyry copper type. This phenomenon is known as the "Incaic" Tectonic Phase, and according to Megard (1987) was the most important tectonic event in the Andes, and was marked by a strong angular unconformity which is present in Peru, Bolivia (Ellison et al., 1989), Argentina (Salfity et al., 1984) and the rest of Chile (Hervé et al., 1987).

With regard to metallogenesis, it can be concluded that most of the metalliferous ore deposits of the 2nd Region are hydrothermal and magmatogenic, formed in mesothermal to epythermal environments.

Spatially, they are arranged in metallogenic provinces whose development would be related to the tectonic evolution of the magmatic arcs. Thus, from W to E we have 1) the Cupriferous Province of the Coastal Range (Jurassic-Early Cretaceous), originated during the development of the La Negra magmatic arc; 2) the Central Depression Polymetallic Province (Middle Cretaceous),formed during the development of the San Cristóbal arc, and 3) the Argentiferous province of the Domeyko Range and the Porphyry Copper-Molibdenum Belt (Cretaceous- Eocene),originated during the development of the Sierra del Medio-Cordillera de Domeyko arc.

The Antofagasta Region has a marked cupriferous specialization and hosts one of the largest known concentrations of copper in the world. The most prolific periods were the Jurassic and Tertiary. The Cretaceous period had in this region and -in strictly chronological terms- a somewhat polymetallic nature.

Acknowledgements

This paper is partly based on the first author's Ph.D. thesis, directed by Prof. Dr. K.-J. Reutter, and presented at the Free University of W-Berlin in 1990.

Our special thanks go to Dr. J.A. Salfity of the Universidad Nacional of Salta, Argentina, for his encouragement and assistance in carrying out this work.

Our special gratitude and regards go to Dr. J. Arias of the Universidad Católica del Norte, for his valuable suggestions and revision of the English manuscript.

Thanks are due to our colleague, G. González of the Universidad Católica del Norte, for his suggestions and critical revision of the manuscript. The drawings and tables presented in this paper were prepared by L. Jofré (Univ. Católica del Norte), to whom we are sincerely thankful.

References

Alarcón, B., and Vergara, M. (1964): Nuevos antecedentes sobre la geología de la Quebrada El Way. Universidad de Chile, Facultad de Ciencias Físicas y Matemáticas, Anales, v. 20-21, 97-130. Santiago.

Astudillo, O. (1983): Geología y metalogénesis del distrito minero Carolina de Michilla, Antofagasta, II Región, Chile. Memoria de Título, Universidad del Norte, Departamento de Geociencias, Antofagasta, 131 pp.

Baeza, L. (1976): Geología de Cerritos Bayos y áreas adyacentes entre los 22°30'-22°45' lat. S y los 68°55'-69°25' long. W; II Región-Antofagasta, Chile. Memoria de Título, Universidad del Norte, Departamento de Geociencias, Antofagasta, 155 pp.

Bahlburg, H., Breitkreuz, C., and Zeil, W. (1986): Paläozoische Sedimente Nordchiles. Berliner geowissenschaftliche Abhandlungen, A 66, I, 147-168.

Baker, M.C.W. (1977): Geochronology of Upper Tertiary volcanic activity in the Andes of North Chile. Geologische Rundschau, v. 66, 455-564.

Batty, M., and Jaillard, E. (1989): La sedimentación neocomiana (Jurásico terminal-Aptiano) en el Sur del Perú, *in* Contribuciones de los Simposios sobre el Cretácico de América Latina, A: Eventos y Registro Sedimentario, (Ed.: Spalletti, L.A.), GSGP-IGCP 242, 75-88, Buenos Aires.

Berg, K., and Breitkreuz, C. (1983): Mesozoische plutone in der Nordchilenischen Küstenkordillere: Petrogenese, Geochronologie, Geochemie und Geodynamic mantelbetonter Magmatite. Geotektonische Forschungen, 66, 107 pp.

Bogdanic, T. (1990): Kontinentale Sedimentation der Kreide und des Alttertiärs im Umfeld des subduktionsbedingten Magmatismus in der chilenischen Präkordillere (21°-23° S). Berliner geowissenschaftliche Abhandlungen, A, 123, 117 pp.

Boric, R., Díaz, F., and Maksaev, V. (1990): Geología y yacimientos Metalíferos de la Región de Antofagasta. Santiago, Servicio Nacional de Geología y Minería, Bol. 40, 246 pp.

Breitkreuz, C. (1986): Das Paläozoikum in der Kordilleren Nordchiles (21°-25°S). Geotektonische Forschungen, 70, 88 pp.

Brüggen, J. (1950): Fundamentos de la Geología de Chile. Santiago, Instituto Geográfico Militar Ed., 378 pp.

Bull, W.B. (1977): The alluvial-fan environment. Progress in Physical Geography, v. 1, 222-270.

Camus, F. (1985): Yacimientos epitermales de oro de Chile. y posible modelo idealizado, *in* Geología y Recursos Minerales de Chile, (Eds.: Frutos, J., Oyarzún, R., and Pincheira, M.), Concepción, Editorial de la Universidad de Concepción, II, 653-689.

Camus, F. (1990): The geology of hydrothermal gold deposits in Chile. Journal of Geochemical Exploration, v. 36, 197-232.

Charrier, R., and Reutter, K.-J. (1990): The Purilactis Group of Northern Chile: Link between arc and backarc during Late Cretaceous and Paleogene. Symposium International Géodynamique Andine, Grenoble, Ed. ORSTOM, Colloques et Séminaires, 249-252.

Chávez, W. (1983): The Geologic Setting of Disseminated Copper Sulfide Mineralization of the Mantos Blancos Copper-Silver District, Antofagasta Province, Chile. AIME Annual Meeting, Atlanta, Geology Section, Pre-print 193, 20 pp.

Chong, G. (1973): Reconocimiento geológico del área Catalina-Sierra de Varas y estratigrafía del Jurásico del Profeta. Memoria de Título, Universidad de Chile, Departamento de Geología, Santiago, 284 pp.

Chong, G. (1976): Las relaciones de los sistemas Jurásico y Cretácico en la zona preandina del norte de Chile. Congreso Geológico Chileno, 1st, Santiago, Actas, 1 (A), 21-42.

Chong, G., and Reutter, K.-J. (1985): Fenómenos de tectónica compresiva en las Sierras de Varas y Argomedo. Congreso Geológico Chileno, 4th, Antofagasta, Actas, 4 (2), 219-238.

Corvalán, J. (1974): Estratigrafía del Neocomiano marino de la región al sur de Copiapó, Provincia de Atacama. Revista Geológica de Chile, v. 1, 13-36, Santiago.

Cuitiño, L., Díaz, S., and Puig, A. (1988): Aspectos de la mineralogía, geoquímica y geotermometría de los yacimientos epitermales Guanaco y Cachinal de la Sierra, Antofagasta, Chile. Congreso Geológico Chileno, 5th, Santiago, Actas, 1 (B), 273-298.

Damm, K., Pichowiak, S., and Todt, W. (1986): Geochemie, Petrologie und Geochronologie der Plutonite und des metamorphen Grundgebirges in Nordchile. Berliner geowissenschaftliche Abhandlungen, (A) 66, I, 73-146.

Definis, A. (1985): Antecedentes geológicos del yacimiento de cobre Santo Domingo, Taltal, y discusión acerca de su relación con un sistema de filones gabrodioríticos. Congreso Geológico Chileno, 4th, Antofagasta, Actas, 2 (3), 204-215.

Díaz, M., Coedani, U., Kawashita,K., Baeza, L., Venegas, R., Hervé, F., and Munizaga, F. (1985): Edades radiométricas preliminares de la Península de Mejillones, Norte de Chile. Universidad de Chile, Departamento de Geología y Geofísica, Comunicaciones, 35, 59-67, Santiago.

Dickinson, W.R. (1985): Interpreting provenance relations from detrital modes of sandstones, *in* Provenance of Arenites, (Ed.: Zuffa, G.G.), Dordrecht, Reidel, 333-361.

Dickinson, W.R. (1988): Provenance and Sediment Dispersal in Relation to Paleotectonics and Paleogeography of Sedimentary Basins, *in* New Perspectives in Basin Analysis, (Eds.: Kleinspehn, K.L., and Paola, C.), New York, Springer-Verlag, 3- 25.

Dickinson, W.R., and Seely, D.R. (1979): Structure and Stratigraphy of Forearc Regions. American Association of Petroleum Geologists Bulletin, v. 63, 2-31.

Döbel, R. (1989): Geochemie und Geochronologie alttertiärer Vulkanite aus der Präkordillere Nordchiles zwischen 21°00′ und 23°30′ S. Inaug.-Diss., Fachbereich. Geowiss. Freie Universität Berlin, Berlin, 178 pp.

Ellison, R.A., Klinck, B.A., and Hawkins, M.P. (1989): Deformation events in the Andean orogenic cycle in the Altiplano and Western Cordillera, southern Peru. Journal of South American Earth Sciences, v. 2, 263-276.

Espinoza, S. (1981): Esbozo metalogénico del distrito de Michilla, II Región, Chile. Coloquio sobre Volcanismo y Metalogénesis, 1st, Antofagasta, Departamento de Geociencias, Universidad del Norte, 71-79.

Espinoza, S. (1982): Definición del tipo diseminado infravolcánico de sulfuros de cobre. Congreso Geológico Chileno, 3rd, Concepción, Actas, 2 (E), 201-219.

Espinoza, S. (1983): Geología y génesis de la mineralización cuprífera del sector Caleta Coloso, al sur de Antofagasta. Revista Geológica de Chile, v. 10, 81-91, Santiago.

Espinoza, S., and Orquera, W. (1988): El yacimiento de cobre Buena Esperanza, geología y alteración en zona de superficie. Nuevas evidencias sobre edad y génesis. Congreso Geológico Chileno, 5th, Santiago, Actas, 3 (G), 1-20.

Ferraris, F. (1978): Hoja Tocopilla, Región de Antofagasta. Instituto de Investigaciones Geológicas, Mapas Geológicos Preliminares de Chile, 3, 32 pp., Santiago.

Flint, S., and Turner, P. (1990): The Conglomerate- Hosted Copper Deposits at Caleta Coloso, Chile, *in* Stratabound Ore Deposits in the Andes, (Eds.: Fontboté, L., Amstutz, G.C., Cardozo, M., Cedillo, E., and Frutos, J.), Berlin Heidelberg, Springer-Verlag, 339-352.

Flint, S., Clemmey, H., and Turner, P. (1986): The Lower Cretaceous Way Group of Northern Chile: an Alluvial Fan-Fan Delta Complex. Sedimentary Geology, v. 46, 1-22.

Flint, S., Hartley, A.J., Rex, D.C, Guise, P., and Turner, P. (1989): Geochronology of the Purilactis Formation, Northern Chile: an insight into Late Cretaceous/Early Tertiary basin dynamics of the Central Andes. Revista Geológica de Chile, v. 16, 241-246, Santiago.

Folk, R., Andrews, P., and Lewis, D. (1979): Detrital sedimentary rock classification and nomenclature for use in New Zealand. New Zealand Journal of Geology and Geophysics, v. 13, 937-968.

Frutos, J. (1972): Ciclos tectónicos sucesivos y direcciones estructurales sobreimpuestas en los Andes del Norte Grande de Chile. Conferencia sobre Problemas de la Tierra Sólida, Buenos Aires (1970), v. II, Simposio Sobre los Resultados de Investigaciones del Manto Superior con Enfasis en América Latina, 473-483.

Galli, C., and Dingman, R.J. (1962): Cuadrángulos Pica, Alca, Matilla y Chacarilla, Provincia de Tarapacá. Instituto de Investigaciones Geológicas, Carta Geológica de Chile, v. III, 2-5, 125 pp., Santiago.

Galliski, M.A., and Viramonte, J. (1988): The Cretaceous paleorift in north-western Argentina: A petrologic approach. Journal of South American Earth Sciences, v. 1, 329- 342.

García, F. (1967): Geología del Norte Grande de Chile. Symposium sobre el Geosinclinal Andino, Santiago (1962), Sociedad Geológica de Chile, 3, 138 pp.

Gardeweg, M. (1988): Petrología y Geoquímica del Complejo Volcánico Tumisa, Altiplano de Antofagasta, Andes del Norte de Chile. Congreso Geológico Chileno, 5th, Santiago, Actas, 3 (I), 183-208.

Gardeweg, M., and Ramírez, C.F. (1985): Hoja Río Zapaleri, II Región de Antofagasta. Servicio Nacional de Geología y Minería, Carta Geológica de Chile, 66, 122 pp., Santiago.

Godoy, E., and Davidson, J. (1976): Pilares en compresión de edad Mioceno Superior en los Andes del Norte de Chile (22°30' lat.S). Congreso Geológico Chileno, 1st, Santiago, Actas, 1 (B), 87-103.

Gröschke, M., Hillebrandt, A.v., Prinz, P., Quinzio, L.A., and Wilke, H.-G. (1988): Marine Mesozoic Paleogeography in Northern Chile between 21°-26°S, *in* The Southern Central Andes, (Eds.: Bahlburg, H., Breitkreutz, Ch., and Giese, P.), Berlin-Heidelberg, Springer-Verlag, Lecture Notes in Earth Sciences, v. 17, 105-117.

Häberer, H., and Reutter, K.-J. (1986): Strukturgeologische und Stratigraphische Untersuchungen in Bereich der Quebrada Guatacondo (21°S), Nord-Chile. Berliner geowissenschaftliche Abhandlungen, (A) 66, I, 225-230.

Haq, B.U., and van Eysinga, W.B. (1987): Geological Time Table. Fourth revised enlarged and updated edition. Amsterdam, Elsevier Sci. Publishers.

Harland, W., Armstrong, R., Cox, A., Craig, L., Smith, A., and Smith, D. (1990): A Geological Time Scale. Cambridge, New York, Port Chester, Melbourne, Sydney, Cambridge University Press, 263 pp.

Harrington, H. (1961): Geology of parts of Antofagasta and Atacama provinces of Northern Chile. American Association of Petroleum Geologists Bulletin, v. 45, 169-197.

Hervé, F., Marinovic, N. (1989): Geocronología y evolución del Batolito Vicuña Mackenna, Cordillera de la Costa, sur de Antofagasta (24°-25° S). Revista Geológica de Chile, v. 16, 31-49, Santiago.

Hervé, F., Marinovic, N., Mpodozis, C., and Perez de Arce, C. (1985): Geocronología K/Ar de la Cordillera de la Costa entre los 24°y 25° latitud Sur. Antecedentes preliminares. Congreso Geológico Chileno, 4th, Antofagasta, Resúmenes, 4.19, p. 158.

Hervé, F., Munizaga, F., Marinovic, N., Hervé, M., Kawashita, K., Brook, M., and Snelling, N. (1985): Geocronología Rb/Sr y K/Ar del basamento cristalino de Sierra Limón Verde, Antofagasta, Chile. Congreso Geológico Chileno, 4th, Antofagasta, Actas, 3 (4), 235-253.

Hervé, F., Godoy, E., Parada, M.A., Ramos, V., Rapela, C., Mpodozis, C., and Davidson, J. (1987): A general view on the Chilean-Argentine Andes, with emphasis on their early history, *in* Circum-Pacific belts and evolution of the Pacific ocean basin, (Eds.: Monger, J.W.H., and Francheteau, J.), American Geophysical Union, Geodynamic Series, v. 18, 97-113.

Hoffstetter, B., Fuenzalida, H., and Cecioni, G. (1957): Chile, *in* Lexique Stratigraphique International, v. V: Amérique Latine (Dir.: Hoffstetter, R.), Fasc. 7, Paris, Centre National de la Recherche Scientifique, 444 pp.

Huete, C., Maksaev, V., Moscoso, R., Ulriksen, C., and Vergara, H. (1977): Antecedentes geocronológicos de rocas intrusivas de la Cordillera de los Andes comprendida entre la Sierra de Moreno y el Río Loa y los 21° y 22° lat. sur, II Región, Chile. Revista Geológica de Chile, v. 4, 35-41, Santiago.

Ishihara, S. (1985): Concentration of heavy metals during granitoid magmatic processes, *in* World Resources and the Development of the Earth's Surface, (Eds.: Sasaki, A., Ishihara, S., and Seki, Y.), London, John Wiley & Sons, 24-27.

Jurgan, V.H. (1974): Die marine Kalkfolge der Unterkreide in der Quebrada El Way-Antofagasta, Chile. Geologische Rundschau, v. 63, 490-516.

Kuntz, J. (1928): Monografía Minera de la Provincia de Antofagasta. Santiago, SONAMI, 77 pp.

Losert, J. (1973): Genesis of copper mineralization and associated alterations in the Jurassic volcanic rocks of Buena Esperanza mining area. Santiago, Universidad de Chile, Departamento de Geología, Publicación, 40, 104 pp.

Mack, G.H. (1984): Exceptions to the relationship between plate tectonics and sandstone composition. Journal of Sedimentary Petrology, v. 54, 212-220.

Maksaev, V. (1978): Cuadrángulo Chitigua y Sector Occidental del Cuadrángulo Cerro Palpana, Región de Antofagasta. Instituto de Investigaciones Geológicas, Carta Geológica de Chile, 31, 55 pp., Santiago.

Maksaev, V. (1984): Mesozoico a Paleogeno de la Region de Antofagasta. Seminario Actualización de la Geología de Chile, Apuntes, Santiago, Servicio Nacional de Geología y Minería, Miscelánea, 4, C1-C20.

Maksaev, V., Boric, R., Zentilli, M., and Reynolds, P. (1988a): Significado metalogénico de dataciones K/Ar, $^{40}Ar/^{39}Ar$ y trazas de fisión de zonas mineralizadas en el Norte Grande de Chile. Congreso Geológico Chileno, 5th, Santiago, Actas, 1 (B), 65-86.

Maksaev, V., Boric, R., Zentilli, M., and Reynolds, P. (1988b): Geocronología $^{40}Ar/^{39}Ar$ de Depósitos de tipo Pórfido Cuprífero del Norte Grande de Chile. Congreso Geológico Chileno, 5th, Santiago, Actas, 1 (B), 109-133.

Marinovic, N., and Lahsen, A. (1984): Hoja Calama, Región de Antofagasta. Servicio Nacional de Geología y Minería, Carta Geológica de Chile, 58, 144 pp., Santiago.

Marocco, R. (1988): Dinámica y Paleogeografía Cretácicas del Perú Central y Meridional. Litoestratigrafía Secuencial. Universidad de Chile, Departamento de Geología y Geofísica, Comunicaciones, 39, p. 214, Santiago.

Marquillas, R., and Salfity, J. (1988): Tectonic framework and correlations of the Cretaceous-Eocene Salta Group, Argentine, *in* The Southern Central Andes, (Eds.: Bahlburg, H., Breitkreutz, Ch., and Giese, P.), Berlin-Heidelberg, Springer- Verlag, Lecture Notes in Earth Sciences, v. 17, 119-136.

Massari, F., and Colella, A. (1988): Evolution and types of fan-delta systems in some major tectonic setting, *in* Fan Deltas: Sedimentology and Tectonic Settings, (Eds.: Nemec, W., and Steel, R.J.), Glasgow and London, Blackie & Sohn, 103- 122.

Mégard, F. (1984): The Andean Orogenic Period and its major structures in central and northern Peru. Journal of the Geological Society, v. 141, 893-900.

Mégard, F. (1987): Cordilleran and marginal Andes: a review of Andean geology North of the Arica elbow (18° S), *in* Circum-Pacific belts and evolution of the Pacific ocean basin, (Eds.: Monger, J.W.H., and Francheteau, J.), American Geophysical Union, Geodynamic Series, v. 18, 71-95.

Miall, A.D. (1985): Architectural-Element Analysis: A new Method of Facies Analysis Applied to Fluvial Deposits. Earth-Science Reviews, v. 22, 261-308.

Montaño, J.M. (1976): Estudio geológico de la zona de Caracoles y áreas vecinas, con énfasis en el Sistema Jurásico, Provincia de Antofagasta, II Región, Chile. Dr Thesis, Universidad de Chile, Departamento de Geología, Santiago, 168 pp.

Mount, J. (1985): Mixed siliciclastic and carbonate sediments: a proposed first-order textural and compositional classification. Sedimentology, v. 32, 435-442.

Mourier, T., Bengtson, P., Bonhomme, M., Buge, E., Cappetta, H., Crochet, J.-Y., Feist, M., Hirsch, K.F., Jaillard, E., Laubacher, G., Lefranc, J.-P., Moullade, M., Noblet, C., Pons, D., Rey, J., Sigé, B., Tambareau, Y., and Taquet, P. (1988b): The Upper Cretaceous-Lower Tertiary marine to continental transition in the Bagua basin, Northern Peru. Paleontology, biostratigraphy, radiometry, correlations. Newsletters on Stratigraphy, v. 19, 143-177.

Mpodozis, C., and Ramos, V. (1989): The Andes of Chile and Argentina, *in* Geology of the Andes and its relations to energy and mineral resources, Santiago (1985), (Eds.: Ericksen, G.E., Cañas Pinochet, M.T., and Reinemund, J.A.), Circum-Pacific Council for Energy and Mineral Resources, Earth Science Series, v. 11, 59-90.

Mpodozis, C. Hervé, F., Davidson, J., and Rivano, S. (1983): Los granitoides de Cerro de Lila, manifestaciones de un epidodio intrusivo y termal del Paleozoico Inferior en los Andes del norte de Chile. Revista Geológica de Chile, v. 10, 3-14, Santiago.

Muñoz, N. (1989): Estudio Geológico Estratigráfico de las Hojas Baquedano y Pampa Unión. II Región, Antofagasta- Chile. Memoria de Título, Universidad de Chile, Departamento de Geología y Geofísica, Santiago, 152 pp.

Muñoz, N., Venegas, R., and Tellez, C. (1988): Formación La Negra: nuevos antecedentes estratigráficos en la Cordillera de la Costa de Antofagasta. Congreso Geológico Chileno, 5th, Santiago, Actas, 1 (B), 283-311.

Muñoz-Cristi, J. (1973): Geología de Chile. Prepaleozoico-Paleozoico y Mesozoico. Santiago, Editorial Andrés Bello, 209 pp.

Muzzio, G., Davidson, J., and Covacevich, V. (1988): Las Formaciones Blanco y Atajaña del Cretácico Inferior, Tarapacá, Chile. Correlaciones Regionales y Paleogeografía. Universidad de Chile, Departamento de Geología y Geofísica, Comunicaciones, 39, p. 216, Santiago.

Naranjo, J., and Paskoff, R. (1981): Estratigrafía de los depósitos cenozoicos de la región de ChiuChiu-Calama, Desierto de Atacama. Revista Geológica de Chile, v. 8, 78-85, Santiago.

Naranjo, J., and Puig, A. (1984): Hojas Taltal y Chañaral, Regiones de Antofagasta y Atacama. Servicio Nacional de Geología y Minería, Carta Geológica de Chile, 62-63, 140 pp., Santiago.

Niemeyer, H. (1989): El Complejo Igneo-Sedimentario del Cordón de Lila, Región de Antofagasta: significado tectónico. Revista Geológica de Chile, v. 16, 163-181, Santiago.

Niemeyer, H., Venegas, R., Baeza, L., and Soto, H. (1985): Reconocimiento geológico del sector sur-occidental del Cuadrángulo Cerro Yocas, ubicado en la zona de falla Quebrada Blanca-Chuquicamata, Región de Antofagasta. Congreso Geológico Chileno, 4th, Antofagasta, Actas, 4 (1), 629-653.

Nilsen, T.H. (1982): Alluvial Fan Deposits, *in* Sandstone Depositional Environments, (Eds.: Scholle, P.A., and Spearing, D.R.), American Association of Petroleum Geologists Memoir, 31, 49-86.

Nisterenko, G., Losert, J., Chávez, L., and Naumov, V. (1974): Temperaturas y presiones de formación de algunos yacimientos cupríferos de Chile. Revista Geológica de Chile, v. 1, 74-80, Santiago.

Padilla, H. (1988): Eventos intrusivos y deformaciones en la Cordillera de Domeyko a la latitud del Salar de Punta Negra. Antecedentes Geocronológicos K/Ar. Congreso Geológico Chileno, 5th, Santiago, Actas, 3 (I), 229-243.

Palacios, C., and Definis, A. (1981): Petrología del yacimiento Buena Esperanza, II Región, Chile. Coloquio sobre Volcanismo y Metalogénesis, 1st, Antofagasta, Departamento de Geociencias, Universidad del Norte, 48-67.

Palacios, C., Hein, U., and Auski, P. (1986): Behaviours of rare earth elements during hydrothermal alteration of the Buena Esperanza copper-silver deposit, Northern Chile. Earth and Planetary Science Letters, v. 80, 208-216.

Palacios, O., and Ellison, R. (1986): El Sistema Cretácico en la Región del Lago Titicaca (Perú). Simposio PIGC-Proyecto 242 Cretácico de América Latina, 1st, La Paz, 32-51.

Pettijohn, F.J., Potter, P.E., and Siever, R. (1987): Sand and Sandstone (Second Edition). Berlin, Heidelberg, New York, Springer-Verlag, 553 pp.

Pichowiak, S., and Hammerschmidt, K. (1989): Age relations and magmagenesis for a suite of early Andean gabbroic to granitic rocks-evidences from trace-element and Sr isotope systematic. 67 Jahrestagung der Deutschen Mineralogischen gesellschaft, Kurzfassungen der Beiträge und Poster, p. 32, Berlin.

Pichowiak, S., Hammerschmidt, K., and Friedrichsen, H. (1990): Early Andean Magmatism in North Chile -Chemical and Isotopic Constrains. Final Workshop, Berlin, Structure and Evolution of the Central Andes in Northern Chile, Southern Bolivia and Northwestern Argentina, Abstract Volume, p.93.

Puig, A., Díaz, S., and Cuitiño, L. (1988): Sistemas hidrotermales asociados a calderas en el arco volcánico paleógeno de la Región de Antofagasta, Chile: distritos El Guanaco, Cachinal de la Sierra y El Soldado. Revista Geológica de Chile, v. 15, 57-81, Santiago.

Ramírez, C.F., and Gardeweg, M. (1982): Hoja Toconao, Región de Antofagasta. Servicio Nacional de Geología y Minería, Carta Geológica de Chile, 54, 122 pp., Santiago.

Reutter, K.-J. (1990): Implication of Magmatic Arcs in the Deformation History of the Central Andes. Final Workshop, Berlin, Structure and Evolution of the Central Andes in Northern Chile, Southern Bolivia and Northwestern Argentina, Abstract Volume, p.64.

Reutter, K.-J., and Scheuber, E. (1988): Relation between tectonics and magmatism in the Andes of Northern Chile and adyacent areas between 21° and 25° S. Congreso Geológico Chileno, 5th, Santiago, Actas, 1 (A), 345-363.

Reutter, K.-J., Schwab, K., and Giese, P. (1986): Oberflächen- und Tiefenstrukturen in den zentralen Anden. Berliner geowissenschaftliche Abhandlungen, (A) 66, I, 247-264.

Reutter, K.-J., Giese, P., Götze, H.-J., Scheuber, E., Schwab, K., Schwarz, G., and Wigger, P. (1988): Structures and crustal development of the central Andes between 21° and 25° S, *in* The Southern Central Andes, (Eds.: Bahlburg, H., Breitkreutz, Ch., and Giese, P.), Berlin-Heidelberg, Springer- Verlag, Lecture Notes in Earth Sciences, v. 17, 231-261.

Riccardi, A.C. (1988): The Cretaceous System of Southern South America. Geological Society of America Memoir, 168, 161 pp.

Rivera, S. (1980): Geología del Distrito Aurífero San Cristóbal, II Región. Memoria de Título, Universidad del Norte, Departamento de Geociencias, Antofagasta, 202 pp.

Rivera, S., and Stephens, A.J. (1988): Campos geotermales fósiles de edad terciaria inferior y mineralización asociada en la Región de Antofagasta. Congreso Geológico Chileno, 5th, Santiago, Actas, 1 (B), 39-64.

Rogers, G. (1985): A geochemical traverse across the North Chilean Andes. Ph.D. Thesis, Open University, Department of Earth Sciences, Milton-Keynes, 333 pp.

Ruiz, C., Aguirre, L., Corvalán, J., Klohn, C., Klohn, E., and Levi, B. (1965): Geología y yacimientos metalíferos de Chile. Santiago, Instituto de Investigaciones Geológicas, 305 pp.

Salfity, J., Gorustovich, S.A., and Moya, M.C. (1984): Las fases diastróficas en los Andes del norte argentino. International Symposium, Central-Andean Tectonics and Relations with Natural Resources, La Paz, Preprint, 23 pp.

Sasaki, A., Ulriksen, C., Sato, K., and Ishihara, S. (1984): Sulfur Isotope Reconnaissance of Porphyry Copper and Manto Type Deposits in Chile and the Phillippines. Bulletin of the Geological Survey of Japan, v. 35, 615-622.

Sato, K. (1984): Manto Type Copper Deposits in Chile. A review. Bulletin of the Geological Survey of Japan, v. 35, 565-582.

Scheuber, E. (1987): Geologie der nordchilenischen Küstenkordillere zwischen 24°30' und 25°S - unter besonderer Berücksichtigung duktiler Scherzonen im Bereich des Atacama-Störungssystems. Ph.D. Thesis, Freie Universität Berlin, Berlin, 180 pp.

Scheuber, E., and Andriessen, P. (1990): The kinematic and geodynamic significance of the Atacama fault zone, Northern Chile. Journal of Structural Geology, v. 12, 243-257.

Scheuber, E., Hammerschmidt, K., Teufel, S., and Friedrichsen, H. (1990): Rb/Sr Isotopic Data from Mylonites from the Atacama Fault Zone -Preliminary Results. Final Workshop, Berlin, Structure and Evolution of the Central Andes in Northern Chile, Southern Bolivia and Northwestern Argentina, Abstract Volume, p.97.

Scholl, D.W., von Huene, R., Vallier, T.L., and Howell, D.G. (1980): Sedimentary masses and concepts about tectonic processes at underthrust ocean margins. Geology, v. 8, 564-568.

Segerstrom, K., and Ruiz, C. (1962): Cuadrángulo Copiapó, Provincia de Atacama. Instituto de Investigaciones Geológicas, Carta Geológica de Chile, III, 1, 115 pp., Santiago.

Sempere, T., Oller, J., and Barrios, L. (1988): Evolución tectosedimentaria de Bolivia durante el Cretácico. Congreso Geológico Chileno, 5th, Santiago, Actas, 3 (H), 37-65.

Skarmeta, J., and Marinovic, N. (1981): Hoja Quillagua, Región de Antofagasta. Instituto de Investigaciones Geológicas, Carta Geológica de Chile, 51, 63 pp., Santiago.

Soto, H., and Dreyer, H. (1985): Geología de "Mina Susana". Un yacimiento novedoso en Carolina de Michilla. Congreso Geológico Chileno, 4th, Antofagasta, Actas, 2 (3), 354-382.

Townsend, F. (1988): Exploración petrolera en la Cuenca del Salar de Atacama, Región de Antofagasta, Chile. Universidad del Norte, Facultad de Ingeniería y Ciencias Geológicas, Vertiente, v. 4, 45-55, Antofagasta.

Ulriksen, C. (1979): Regional geology, geochronology and metallogeny of the Coastal Cordillera of Chile between 25°30' and 26°00' south. M.Sc. Thesis, Dalhousie University, Halifax, 221 pp.

Vergara, H. (1978): Cuadrángulo Quehuita y Sector Occidental del Cuadrángulo Volcán Miño, Región de Tarapacá. Instituto de Investigaciones Geológicas, Carta Geológica de Chile, 32, 44 pp., Santiago.

Vergara, H., and Thomas, A. (1984): Hoja Collacagua, Región de Tarapacá. Servicio Nacional de Geología y Minería, Carta Geológica de Chile, 59, 79 pp., Santiago.

Vicente, J.-C. (1981): Elementos de la estratigrafía mesozoica sur-peruana, *in* Cuencas Sedimentarias del Jurásico y Cretácico de América del Sur, (Eds.: Volkheimer, W., and Musacchio, E.), 1, 319-351, Buenos Aires.

Wolf, F., Fontboté, L., and Amstutz, G.C. (1990): The Susana Copper (-Silver) Deposit in Northern Chile. Hydrothermal Mineralization Associated with a Jurassic Volcanic Arc, *in* Stratabound Ore Deposits in the Andes, (Eds.: Fontboté, L., Amstutz, G.C., Cardozo, M., Cedillo, E., and Frutos, J.), Berlin Heidelberg, Springer-Verlag, 319-338.

Tectonic and Sedimentary Evolution of the Cretaceous-Eocene Salta Group Basin, Argentina

J.A. Salfity, R.A. Marquillas
Universidad Nacional de Salta-Conicet, Buenos Aires 177, 4400 Salta, Argentina.

Abstract

The present work summarizes the geologic history of the Salta Group basin. The information available for carrying out the work is based on personal information which, with the data provided by the bibliography, adds up to some three hundred surface sections of the outcropping of the basin, in addition to the subsurface data published.

The nature of the pre-Cretaceous basement is first analyzed, together with the structural frame in which sedimentation commenced.

The sedimentary history of the Salta Group, the eruptive episodes and the structural processes which took place during and after the Salta Group deposition are then summarized, in the following order:

a) The synrift accumulations of the Pirgua Subgroup redbeds during Kimmeridgian?-Early Cretaceous to Campanian times, which filled and levelled the rift troughs throughout the whole basin.

b) The postrift accumulations, in the shape of the Balbuena Subgroup ingression and the final fill provided by the Santa Bárbara Subgroup, which occurred between Maastrichtian and Mid- Eocene times.

c) Facies evolution and the flooding processes interbedded in the red bed sequences of the Salta Group.

d) The eruptive processes which occurred in the basin, expressed in three magmatic cycles.

e) The tectonic episodes which took place after the Salta Group deposition and the basin inversion during Late Tertiary time.

Introduction

The main morphostructural and geological characteristics of the Central Andes in northwest Argentina (Figure 1) are the following;

a) The southernmost ending of the South American Subandean chain outcrops in this region.
b) The Cambrian and Early Ordovician marine outcrops, typical of the Bolivian-Argentine Eastern Cordillera, are found only north of the El Toro lineament.
c) The Precambrian and Paleozoic metamorphic and igneous rocks of the northern stretch of the Argentine Pampean Ranges outcrop to the south of the Eastern Cordillera.
d) In the western part of the region lies the Puna plateau, whose west flank is formed by the Principal Cordillera or Cenozoic volcanic arc.
e) The eastern part of the region corresponds to the Chaco Plain.

The Salta Group is present in nearly all the geologic provinces of the region, north of the latitude of the city of Tucumán (Figure 2). The best exposed outcrops are those of the northern Pampean Ranges, the Subandean Ranges (east of the city of Jujuy) and the Eastern Cordillera (south of the Bolivia border). The Puna outcrops are scarcer and more isolated.

The geologic history of the Salta Group began in the Kimmeridgian?-Early Cretaceous, probably as a consequence of the Araucanian movements (Figure 3).

The typical marine Jurassic accumulations of the Andes in the north of Chile are not present in northern Argentina. It is probable that the presence of the San Pablo and Traspampean arches prevented their extension into Argentina (Figure 5).

The Salta Group basin therefore developed entirely over a Precambrian and Paleozoic basement (Figures 3 and 6).

At the start of the Cretaceous, a series of rift troughs opened in northwest Argentina, at first singly and then in interconnected fashion. The development of these troughs also affected the territories of Bolivia (Cherroni,1977; Sempere, this volume), Paraguay (Clebsch, 1991) and probably Chile (Bogdanic 1990) (Figure 5).

The conjoint set of these troughs or grabens in Argentine territory form a part of the Salta Group basin, also known as the Northwest Basin (Moreno, 1970) and the Andean Basin (Reyes, 1972).

The positive structural elements which governed the Salta Group sedimentation (Figure 5) were the Michicola arch (Vilela, 1965), the Quirquincho Arch (Salfity, 1980), the Pampean and Traspampean arches (Padula and Mingramm, 1968), and the San Pablo and Salta- Jujuy structural highs (Reyes, 1972).

The troughs which make up the Salta Group basin follow three preferential structural strikes: north-south, northeast-southwest and northwest-southeast. The most outstanding is the northeast- southwest strike, which was definitely preponderant in controlling the Late Paleozoic sedimentary basins (Bianucci et al., 1984; Salfity, 1985), whose structures were reactivated during the Cretaceous.

The grabens or depocenters identified have been named Tres Cruces, Lomas de Olmedo, Metán, Alemanía, Cerro Hermoso, El Rey and El Charco or Sey (Reyes, 1972; Salfity,

1980; Schwab, 1984, 1985) (Figures 7 and 8). The Huaitiquina high (Salfity et al., 1985) must have separated the Purilactis basin in the north of Chile from the Salta Group basin in pre-Maastrichtian time (Figure 5).

The filling of the basin was an essentially continuous episode (Groeber, 1953), only altered by regional or local discontinuities of sedimentary or volcanic origin. The Salta Group has no interior regional angular unconformities.

The Salta Group deposition was interrupted during the Eocene, due to the effects of the Incaic diastrophism. Later, probably from Oligocene time onward, new foreland type basins were formed in the Subandean chain, and new intramontane to intra-arc basins were formed in the Puna.

The Cretaceous deposits of northwest Argentina have been known since the end of the last century (Brackebusch, 1883, 1891). The nature of the sedimentary fill and environment, the contemporary volcanism, and the general tectonic framework of the basin were studied and interpreted by various researchers during the first half of the present century, such as Bonarelli (1913, 1921), Frenguelli (1930, 1936), Schlagintweit (1937) and Groeber (1953), among others.

The above-mentioned authors concluded that three successive types of terrane had accumulated in this basin. These were:

(a) Very thick conglomeratic, sandy and argillaceous redbeds, with interbedded magmatic rocks, lying on a Precambrian or Paleozoic basement, and which filled and levelled the depocenters.

(b) Above these redbeds lay -conformably- beds of white sandstones, limestones and black and green shales, whose environmental settings (lacustrine or marine) and ages were under discussion for many years.

(c) Finally, red and green shales and mudstones, with interbedded limestones and sandstones, were also laid down conformably, thus completing the basin fill.

These authors pointed out that the youngest sedimentary episodes successively overlapped the older episodes, and were the cause of basin expansion over the structural highs then in existence; this phenomenon was very well described with regard to the Salta-Jujuy high.

Later investigation made it possible to adjust and perfect the previously acquired knowledge of this basin. The Salta Group (Turner, 1959) has been formally divided into three sub-groups (Figures 3 and 4): the Pirgua Subgroup (Reyes and Salfity, 1973), which corresponds to the basal red beds; the Balbuena Subgroup (Moreno, 1970), made up of

Fig. 1

Morphostructural regions of Northwest Argentina and neighbouring areas. 1 Pampean Plain, 2 Subandean Ranges, 3 Eastern Cordillera, 4 Puna, 5 Principal Cordillera, 6 Northern Pampean Ranges, 7 Traspampean Ranges, 8 Domeyko Cordillera, 9 Coastal Cordillera, 10 Quaternary valley. ➤

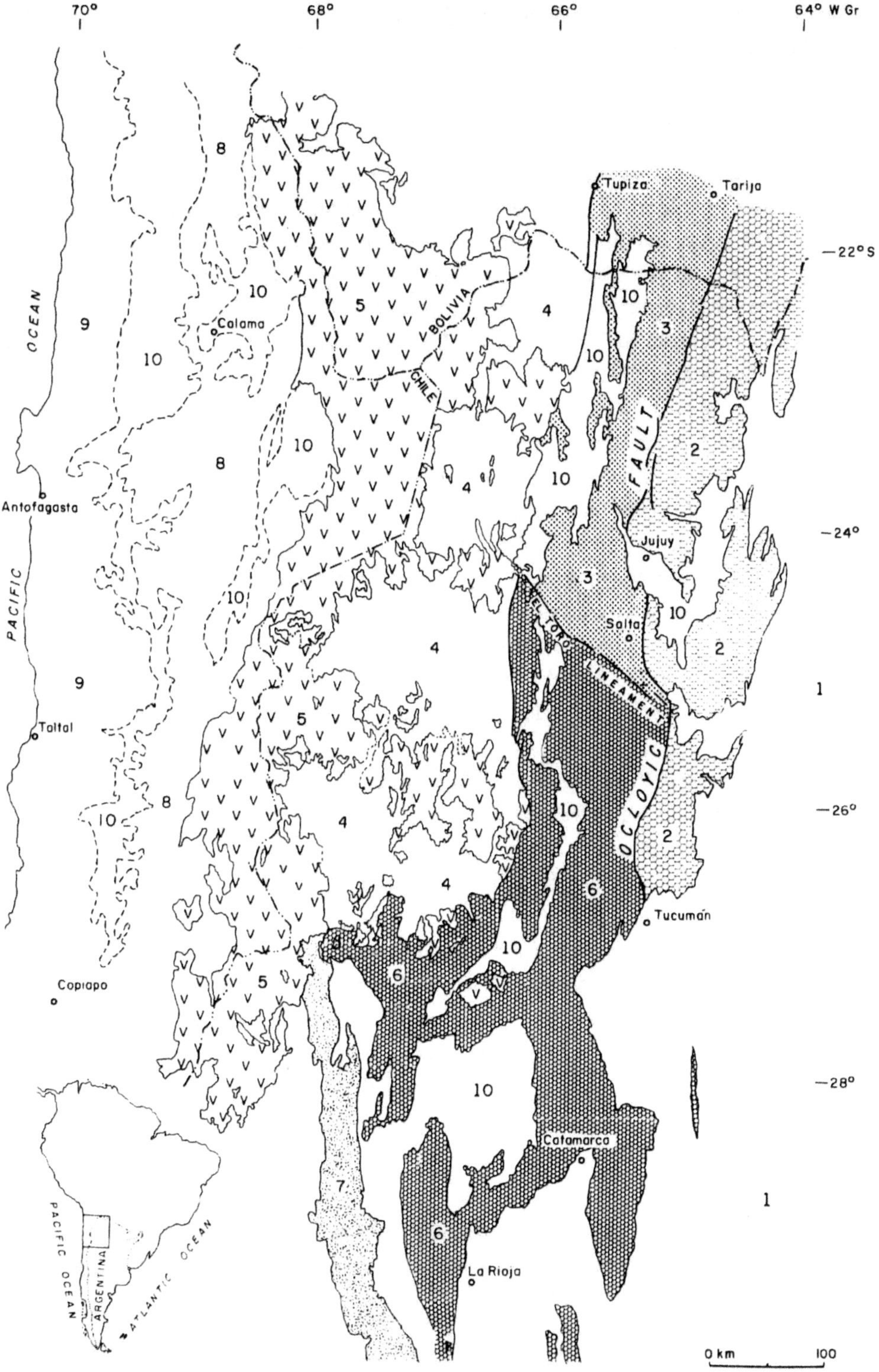
70°
68°
66°
64° W Gr
22° S
24°
26°
28°
OCEAN
PACIFIC
Antofagasta
Taltal
Copiapo
Calama
Tupiza
Tarija
Jujuy
Salta
Tucumán
Catamarca
La Rioja
BOLIVIA
CHILE
FAULT
OCLOYIC
EL TORO
LINEAMENT
PACIFIC OCEAN
ATLANTIC OCEAN
ARGENTINA
0 km
100
1
2
3
4
5
6
7
8
9
10

sandstone, limestone and shale; finally, the Santa Bárbara Subgroup (Moreno, 1970), formed of varicolored pelitic rocks, with limestone and sandstone.

The Pre-Cretaceous Basement

The base of the Salta Group is seen in a large number of surface sections and was logged in some exploration wells, both within the synrift troughs (Figures 7 and 8) as well as on the structural highs overlapped by postrift sequences (Figures 9 and 10).

Figure 6 shows the pre-Cretaceous geology, based on available field and subsurface data, and on overlays of paleogeographic reconstructions of the pre-Cretaceous basins (Salfity, 1979; Bianucci and Homovc, 1982; Carlé et al., 1989; Salfity and Marquillas, 1989; Mon and Hongn, 1991).

The age of the basin basement, as well as that of the structural highs, ranges between the Precambrian and the Late Paleozoic.

Precambrian - The southern portion of the Salta Group basin (Figure 6) had Precambrian formations of flysch, limestone and granitoids, which are part of the northern end of the Pampean Ranges, as a basement. During the Cretaceous, the El Toro lineament bounded these Precambrian terranes, separating them from the Cambrian and Ordovician terranes of the Eastern Cordillera, a situation which is still in evidence (Figures 1, 15 and 17F). Moreover, the postrift deposits of the Yacoraite Formation along the El Toro lineament rest on Precambrian granites (the La Quesera Granite), which shows that the plutonic rocks lodged in the Precambrian basement were exposed during the Late Cretaceous.

Cambrian - The postrift deposits of the Yacoraite Formation lie on Precambrian and Cambrian sediments north of the city of Jujuy (Figure 6) (Lencinas and Salfity, 1973). This is illustrative of the erosion undergone during the Cretaceous by the thick Ordovician deposits on the northern stretch of the Salta-Jujuy ridge (Moya and Salfity, in press).

Ordovician - During the Cretaceous, the Ordovician basement cropped out on the Eastern Cordillera and on the northern Puna; that is, to the west of the Ocloyic fracture front, and to the north of the El Toro lineament (compare Figures 1 and 6). The Ordovician sediments of the Puna form the host rock for the intrusives of the Eastern Puna eruptive belt (Méndez et al. 1973), which also underlie the Salta Group.

Fig. 2

General geologic map of Northwest Argentina. 1 Pre-Cretaceous basement (Precambrian and Paleozoic), 2 Salta Group (Cretaceous-Eocene), 3 Cretaceous granitoids, 4 Late Cenozoic sediments, 5 Late Cenozoic volcanics, 6 Quaternary, 7 Salar. ELT: El Toro Lineament, EOF: Eastern Ocloyic Fault. ➤

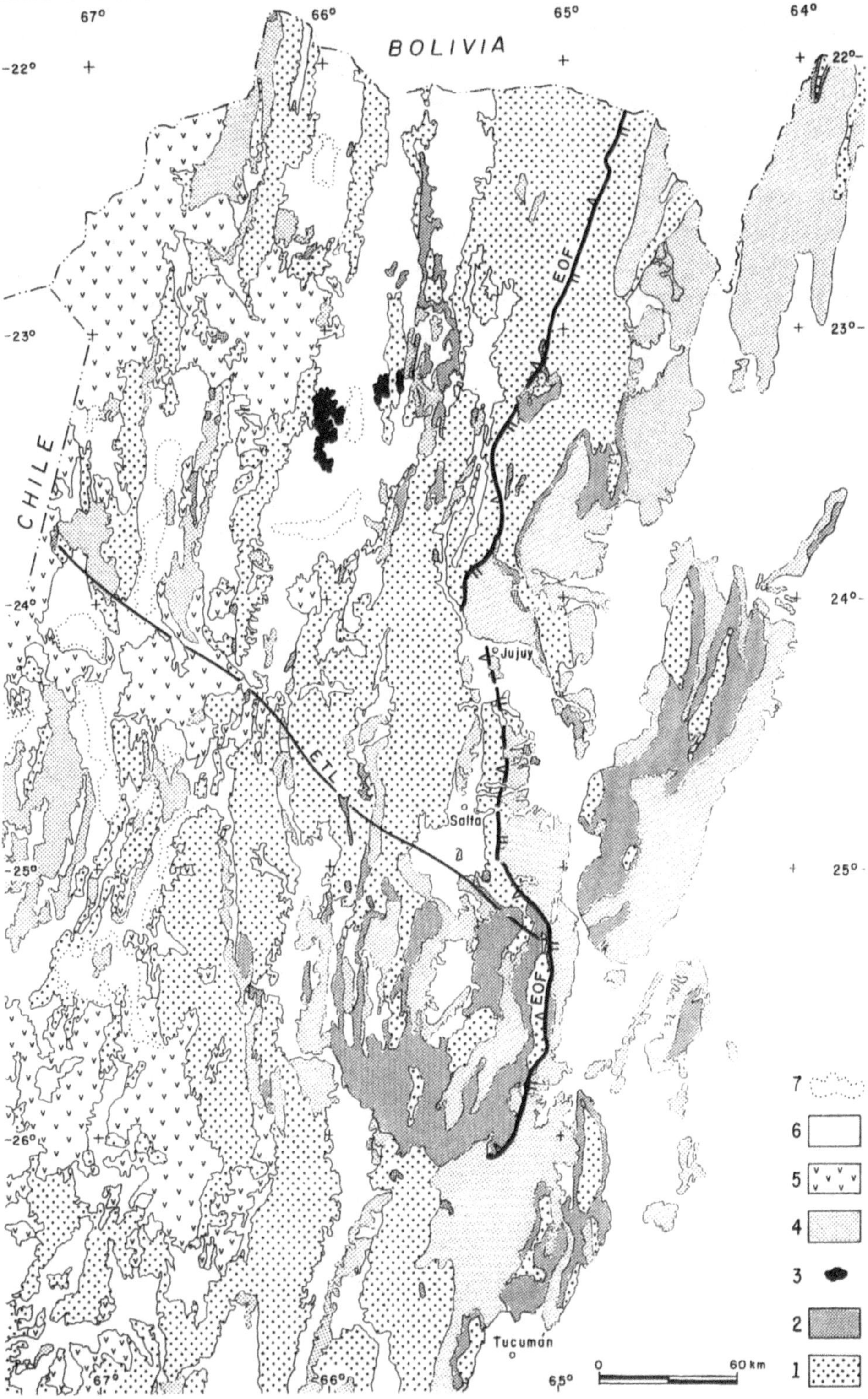

BOLIVIA
CHILE
67°
66°
65°
64°
-22°
-23°
-24°
-25°
-26°
22°
23°
24°
25°
EOF
ETL
Jujuy
Salta
Tucuman
0
60 km
7
6
5
4
3
2
1

Similarly, Ordovician deposits would have been exposed on the Traspampean arch during the Cretaceous. However, it is not known whether the Silurian-Devonian and eventually the Carboniferous- Permian sequences were deposited on the Traspampean arch. At present the Ordovican sequences of the Traspampean arch are covered by Tertiary deposits (Donato, 1987).

To the east of the Ocloyic front various stretches of the Ordovician would have been exposed during the Cretaceous along the Lomas de Olmedo depocenter (Figures 6 and 15). In fact, erosive processes have been inferred for all or part of the Silurian-Devonian deposits, as a consequence of which Ordovician rocks underlie the Cretaceous deposits (Padula et al., 1967; Mingramm and Russo,1972; Carlé et al., 1989). Therefore the Salta Group rests on the Ordovician in those areas in which the Silurian-Devonian was completely eroded (Figure 6).

Silurian-Devonian - For the reasons mentioned before, the Silurian-Devonian deposits also served as a basement for the Cretaceous deposits. The Silurian cropped out along the Ocloyic Fault on the Salta-Jujuy high and in the Lomas de Olmedo depocenter.

In the same way, various stratigraphic levels of the Devonian sequence underlie both synrift and postrift Cretaceous accumulations, though always to the east of the Ocloyic Fault.

The area occupied by the extensive Lomas de Olmedo graben was a positive element, probably from Late Devonian time onward. The strip in which the greatest amount of erosion occurred is precisely that along which almost continuous subsidence processes took place, from Early Cretaceous to Late Tertiary times (Figures 6 and 13).

In turn, the Silurian-Devonian of the northern stretch of the Traspampean arch (Figure 6) would have remained covered by Late Paleozoic deposits, on which the Tertiary sediments rest directly (Donato, 1987).

Late Paleozoic - The Carboniferous-Permian deposits directly related to the Salta Group basin correspond to those found in the Tarija basin (Mingramm and Russo, 1972; Salfity et al., 1987). They formed part of the Michicola arch during the Cretaceous (Figure 6).

The Salta Group overlies Carboniferous deposits to the south of the Cóndor arch (Amengual and Zanettini, 1973) (Figures 6 and 17 A).

However, recent studies (Starck et al., 1992, in press) describe sections in this region in which the Salta Group lies on eolian continental sequences (Tacurú Group, attributed to the Jurassic), which in turn rest on the Carboniferous.

It should be mentioned that to the west of the Michicola arch the Carboniferous underlies the Cuevo Group, the base of which contains eolian deposits (Cangapi Formation) similar to those of the Tacurú Group (Reyes, 1974, 1978). The Vitiacua Formation which overlies the Cangapi Formation is considered to be of Middle to Late Permian-Triassic? by Sempere et al. (1992). Therefore the pre-Cretaceous basement to the south of the

Cóndor arch will be considered to be of Late Paleozoic age, until such time as the exact age of the above-mentioned eolian deposits can be established.

Pre-Maastrichtian Synrift Accumulations: The Pirgua Subgroup

The pre-Maastrichtian deposits in the Salta Group correspond to the Pirgua Subgroup, which represents the synrift stage of basin evolution. The corresponding environmental factors are referred to in a later section of this work.

One of the most complete sections of this subgroup crops out in the Alemanía depocenter. This contains the La Yesera Conglomerate (615 m), the Las Curtiembres Mudstone (1170 m) and the Los Blanquitos Sandstone (1500 m) (Reyes, 1972; Reyes and Salfity, 1973) (Figures 3 and 4).

The evolution of the Pirgua Subgroup can be subdivided into two main episodes, whether in terms of tectosedimentary units (Gómez Omil et al., 1989), or on the basis of the regional distribution of its formations (Figures 7 and 8).

The first episode in the filling of the troughs began with the La Yesera Conglomerate, which has been very well recorded in the Alemanía and Metán depocenters (Reyes and Salfity, 1973; Boso et al., 1974).

Figure 7 has been plotted on the basis of the regional stratigraphic relationships between the three formations, although it has not been possible to draw isopach lines; furthermore, the outline of the northern half of the basin is purely conceptual. This figure illustrates the presence of isolated troughs separated by structural highs; these last providing the sedimentary supply.

The La Yesera Conglomerate contains fragments of the immediate basement on which it was deposited: including Precambrian metamorphic and granitic rocks, Cambrian and Ordovician sediments, Ordovician granites, and Silurian, Devonian, and Carboniferous sedimentary rocks.

This Conglomerate is over 600 m thick in the type section in the Alemanía depocenter. Alternatively, on the Guachipas high (Figure 7) the Pirgua Subgroup is represented by sandstones similar to those of the Los Blanquitos Formation, with a basal conglomerate a few meters thick.

The la Yesera Formation is not so well developed in other depocenters -as for instance in that of Tres Cruces- where the predominant facies are sandstones. It is very possible that

Fig. 3

Cretaceous and Cenozoic correlations in Northwest Argentina. 1 Granitoid, 2 Volcanics, 3 Conglomerate, 4 Limestone, 5 Salt, 6 Subvolcanics, 7 Quaternary, 8 Unconformity, 9 Lithologic transition (at the base of stratigraphic units indicates overlapping), 10 Lateral facies change, 11 Radiometric age. PC-Precambrian, E-Cambrian, O- Ordovician, S-Silurian, D-Devonian, C-Carboniferous, R-Permian, Triassic?. ➤

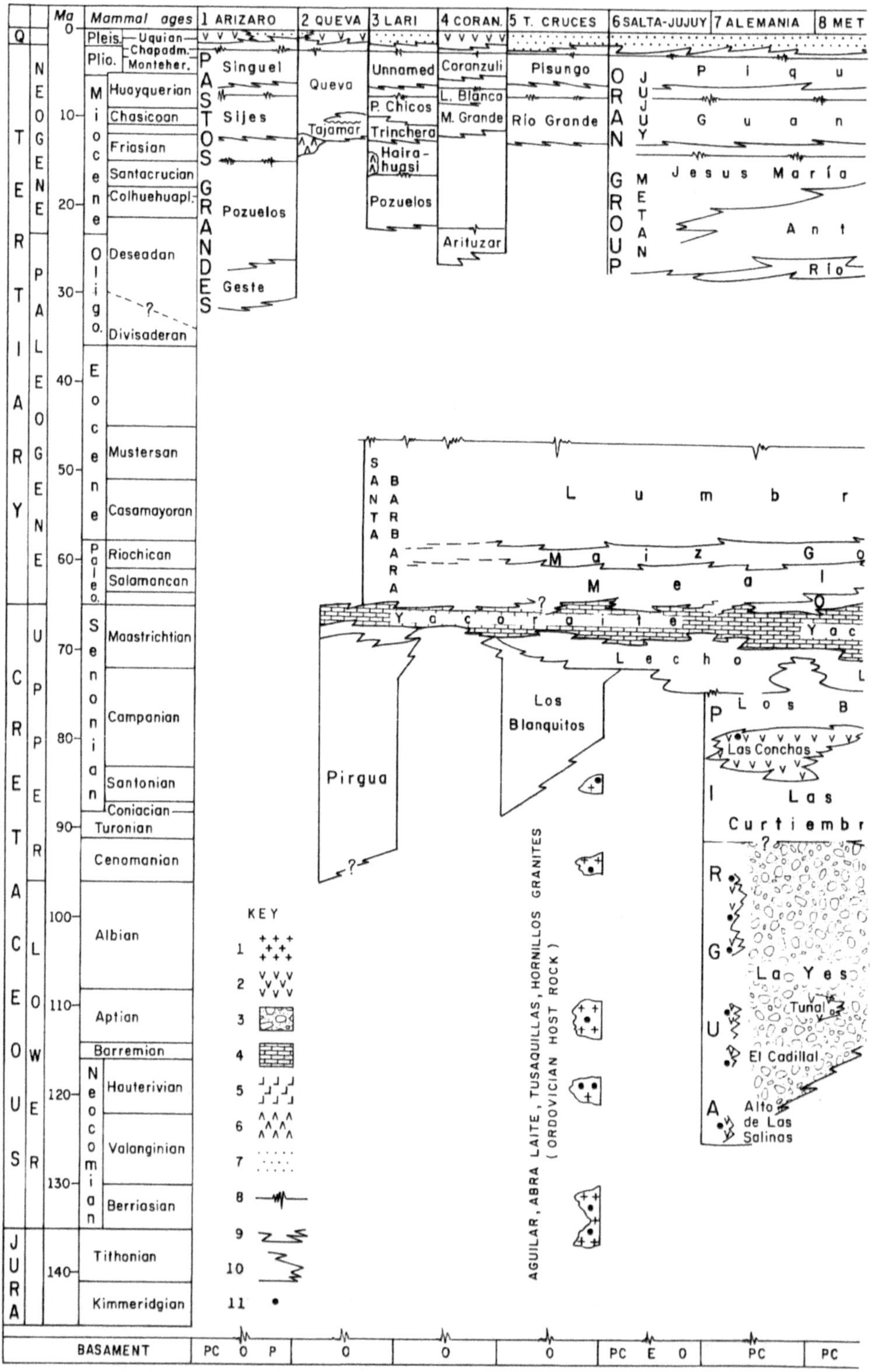
Ma
Mammal ages
1 ARIZARO
2 QUEVA
3 LARI
4 CORAN.
5 T. CRUCES
6 SALTA-JUJUY
7 ALEMANIA
8 MET
Q
TERTIARY
NEOGENE
PALEOGENE
CRETACEOUS
UPPER
LOWER
JURA
Pleis.
Uquian
Chapadm.
Plio.
Monteher.
Miocene
Huayquerian
Chasicoan
Friasian
Santacrucian
Colhuehuapi.
Oligo.
Deseadan
Divisaderan
Eocene
Mustersan
Casamayoran
Paleo.
Riochican
Salamancan
Senonian
Maastrichtian
Campanian
Santonian
Coniacian
Turonian
Cenomanian
Albian
Aptian
Barremian
Neocomian
Hauterivian
Valanginian
Berriasian
Tithonian
Kimmeridgian
PASTOS GRANDES
Singuel
Sijes
Pozuelos
Geste
Queva
Tajamar
Unnamed
P. Chicos
Trinchera
Haira-hugsi
Pozuelos
Coranzuli
L. Blanca
M. Grande
Arituzar
Pisungo
Río Grande
ORAN GROUP
JUJUY
METAN
P i q u
G u a n
Jesus María
A n t
Río
SANTA BARBARA
L u m b r
M a i z G
M e a l
Y a c o r a i t e
Y a c
L e c h o
Los Blanquitos
Pirgua
P
Los B
Las Conchas
Las
Curtiembr
R
G
U
A
La Yes
Tunal
El Cadillal
Alto de Las Salinas
AGUILAR, ABRA LAITE, TUSAQUILLAS, HORNILLOS GRANITES (ORDOVICIAN HOST ROCK)
KEY
1
2
3
4
5
6
7
8
9
10
11
BASAMENT
PC O P
O
O
O
PC E O
PC
PC

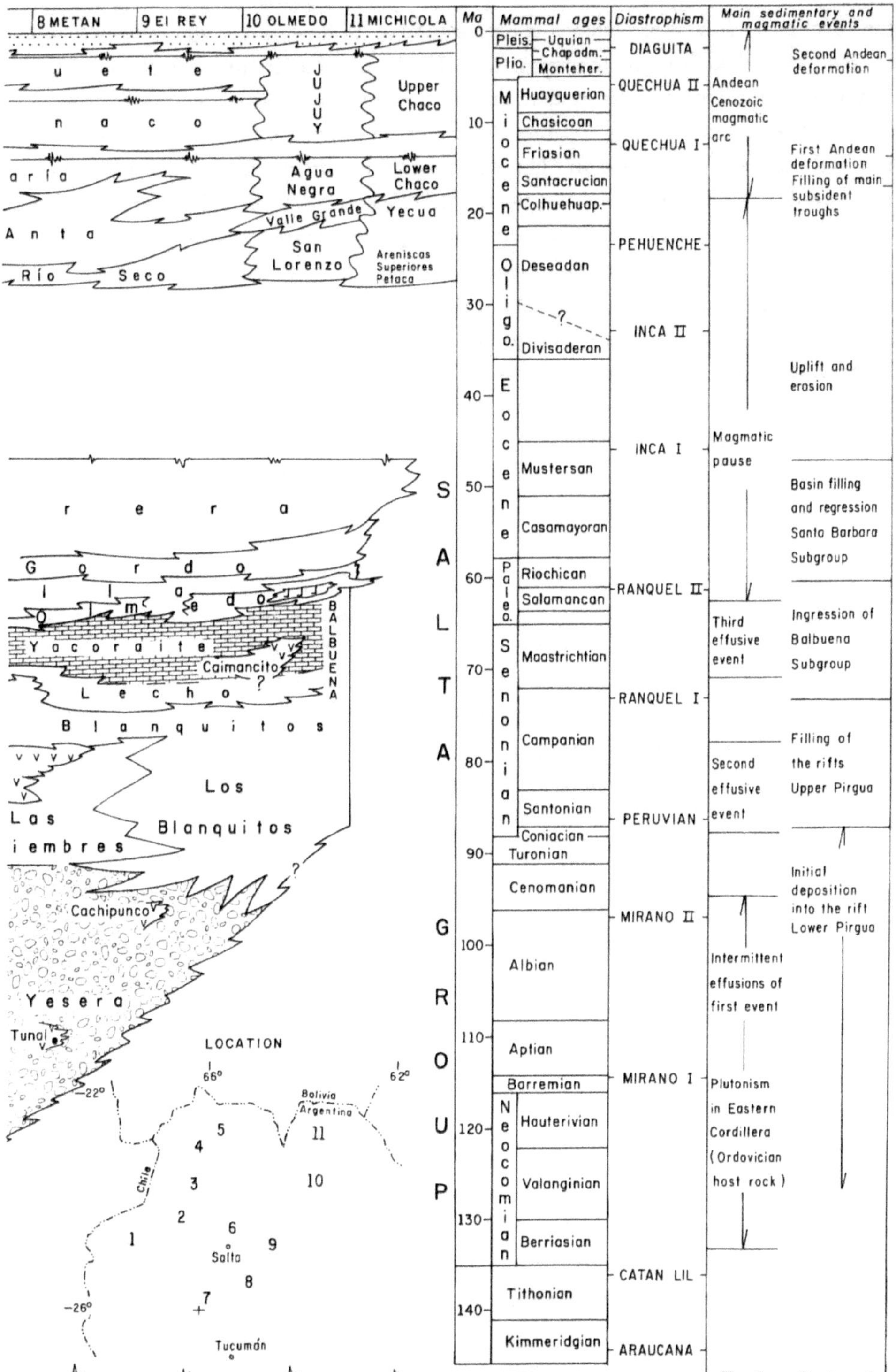

Fig. 3 (Continued).

the basal part of these sandstones is representative, in time, of the whole or part of the La Yesera and Las Curtiembres Formations. The lateral changes in facies observed in the southern part of the basin are evident both in marginal and distal locations (Figure 14).

Toward the end of the conglomerate deposition there was a change of facies between this and the Las Curtiembres and Los Blanquitos Formations (Figure 3). Therefore there is no well-defined limit -either sequential or lithologic- between the first and second synrift episodes. It is possible that this first stage may have come to a close with the basalts of the second extrusive cycle; these form a good local guide level in the Alemanía depocenter (Gómez Omil et al., 1989).

It is possible to infer a lateral change of facies between the Las Curtiembres and Los Blanquitos Formations in the Alemanía depocenter, on the basis of correlations levelled to the top of the second effusive cycle basalts (Figure 14). The same deduction can be made on a regional level (Figure 3).

A second synrift filling episode is embodied in the sandstones of the upper part of the Los Blanquitos Formation. These levelled the depocenters and even accumulated over some of the internal structural highs in the basin (notably the Calete, Cachipunco, Las Víboras and Guachipas uplifts, among others) (compare Figures 7 and 8).

The Pirgua Subgroup only contains age-indicative fossils (Campanian dinosaurs, Bonaparte, 1984) at the top of the Los Blanquitos Formation. The chronology of the rest of the subgroup is based on some radiometric datings of the basalt flows intercalated in the La Yesera Conglomerate and the top of the Las Curtiembres Shale (Figure 3); that is to say, the First and Second Volcanic Cycles.

Up to the present, no regional sequence studies have been carried out on these formations, and therefore there are no regional isotime surfaces plotted for the Pirgua Subgroup stratigraphic column. Only an isopach map for the whole thickness of this subgroup can be shown (Figure 8), with such limited information as is available on the deeper parts of the depocenters.

Fig. 4

Composite stratigraphic columns of the Salta Group. Based on personal data and on Gómez Omil et al. (1989). 1 Conglomerate, 2 Sandstone, 3 Shale, 4 Limestone, 5 Salt, 6 Marl, 7 Volcanics, 8 Unconformity, 9 Base not known. A, B, C, D, E: Depocenters. Pi-Pirgua Subgroup, LY-La Yesera Formation, LC-Las Curtiembres Formation, LB-Los Blanquitos Formation, Le-Lecho Formation, Ya-Yacoraite Formation, Ol+SM- Olmedo Formation plus Saline Member, Tu-Tunal Formation, Me- Mealla Formation, MG-Maíz Gordo Formation, Lu-Lumbrera Formation. ➤

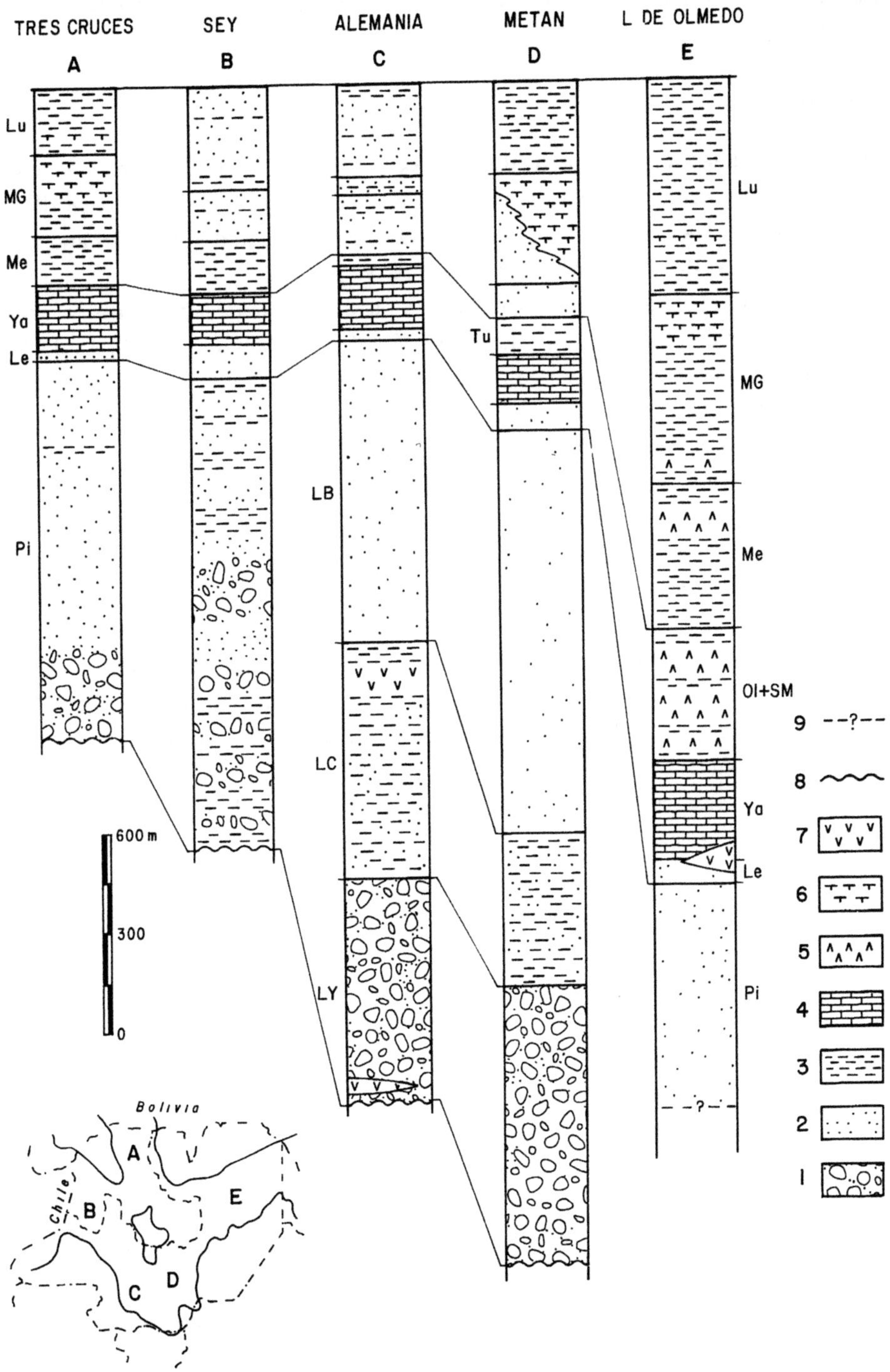
TRES CRUCES
A
SEY
B
ALEMANIA
C
METAN
D
L DE OLMEDO
E
Lu
MG
Me
Ya
Le
Pi
LB
LC
LY
Tu
Ol+SM
600 m
300
0
9 --?--
8
7
6
5
4
3
2
1
Bolivia
Chile

Postrift Maastrichtian-Eocene Accumulations: Balbuena and Santa Barbara Subgroups

The Balbuena Subgroup

The episodes related to the Maastrichtian-Paleocene ingression in the north of Argentina are linked to the eustatic changes which took place at the time of the Ranquel diastrophism (Figure 3). The two pulses of the Ranquel phase are related to the start and finish, respectively, of the Balbuena Subgroup transgressive process.

The Ranquel diastrophism -previously known as the Laramide phase- had a greater structural impact on other basins in the Andes (Salfity and Zambrano, 1990). In the Salta Group basin it is expressed only by the start of the Balbuena Subgroup epeiric flooding, in a frame of tectonic quiescence.

Once the troughs had been silted up with synrift deposits, they were covered by the basal transgressive deposits of the Balbuena Subgroup. The ingression therefore started with the white sandstone and limestone sedimentation of the Lecho and Yacoraite Formations, which accumulated not only over the former synrift troughs but also over the structural highs which bounded them (compare Figures 8 and 9) (Salfity 1982; Marquillas and Salfity, 1989).

Flooding processes were general in the southern cone of South America during de Maastrichtian-Paleocene (Camacho, 1967; Russo et al., 1979; Zambrano, 1987; Riccardi, 1988; Macellari, 1988; Uliana and Biddle, 1988; Salfity and Zambrano, 1990), in concurrence with the global tendency at that time.

These flooding processes were of great paleobiogeographic significance on the American continent. The interchanging of continental herbivorous dinosaur fauna which took place between South America and North America during the Campanian was brought to an end by the Maastrichtian ingressions (Bonaparte, 1984).

These episodes of Campanian continental deposition and later Maastrichtian-Paleocene flooding are very well recorded in the Salta Group basin. The upper part of the Salta Group synrift deposits contains Campanian tetrapods (Bonaparte and Bossi, 1968; Bonaparte and Powell, 1980), and these sequences are covered by Maastrichtian postrift

Fig. 5

Tectonic sketch of the Salta Group and neighbouring basins during pre-Maastrichtian time. 1 Basin edge and high areas, 2 Non-marine Cretaceous depocenters, 3 Present-day lineaments and faults, 4 Distribution of Late Jurassic-Early Cretaceous Serra Geral lava flow, 5 Cretaceous- Eocene magmatic arc, 6 Post-Araucan to Neocomian marine Coastal Basin, 7 Basin name. Lineaments: T Tomasito, LB-Los Blancos, SG- Salinas Grandes, Co-Cobres, C-Calama, O-Olacapato, Cq-Calchaquí, I-Isonza, EB-El Brete, Aq-Aconquija. ➤

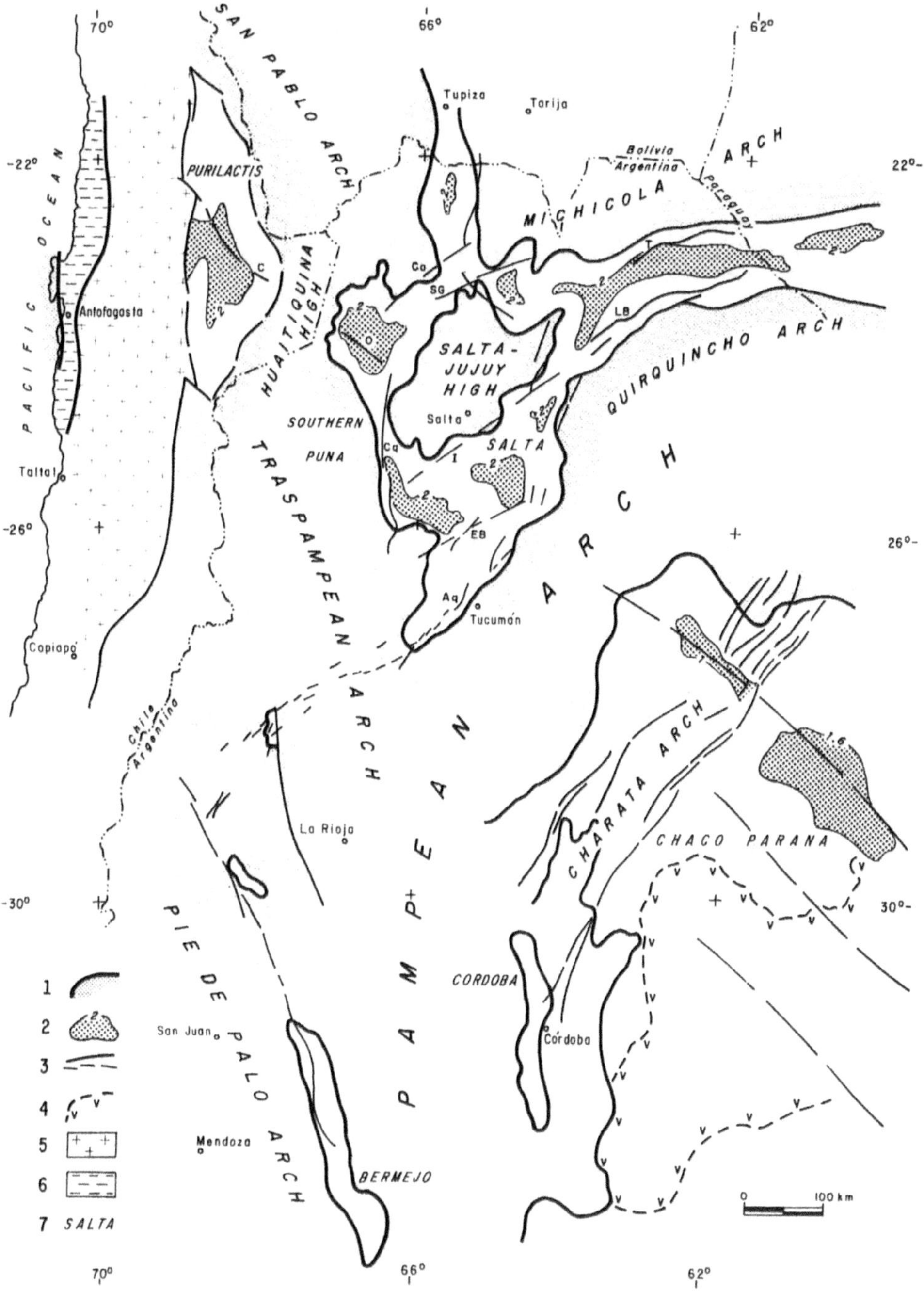
SAN PABLO ARCH
PURILACTIS
HUAITIQUINA HIGH
PACIFIC OCEAN
Antofagasta
Taltal
Copiapo
Tupiza
Tarija
Bolivia
Argentina
Paraguay
MICHICOLA ARCH
QUIRQUINCHO ARCH
SALTA-JUJUY HIGH
Salta
SALTA
SOUTHERN PUNA
TRASPAMPEAN ARCH
PAMPEAN ARCH
Tucuman
Chile
Argentina
La Rioja
CHARATA ARCH
CHACO PARANA
PIE DE PALO ARCH
CORDOBA
Cordoba
San Juan
Mendoza
BERMEJO
0 100 km
1
2
3
4
5
6
7 SALTA

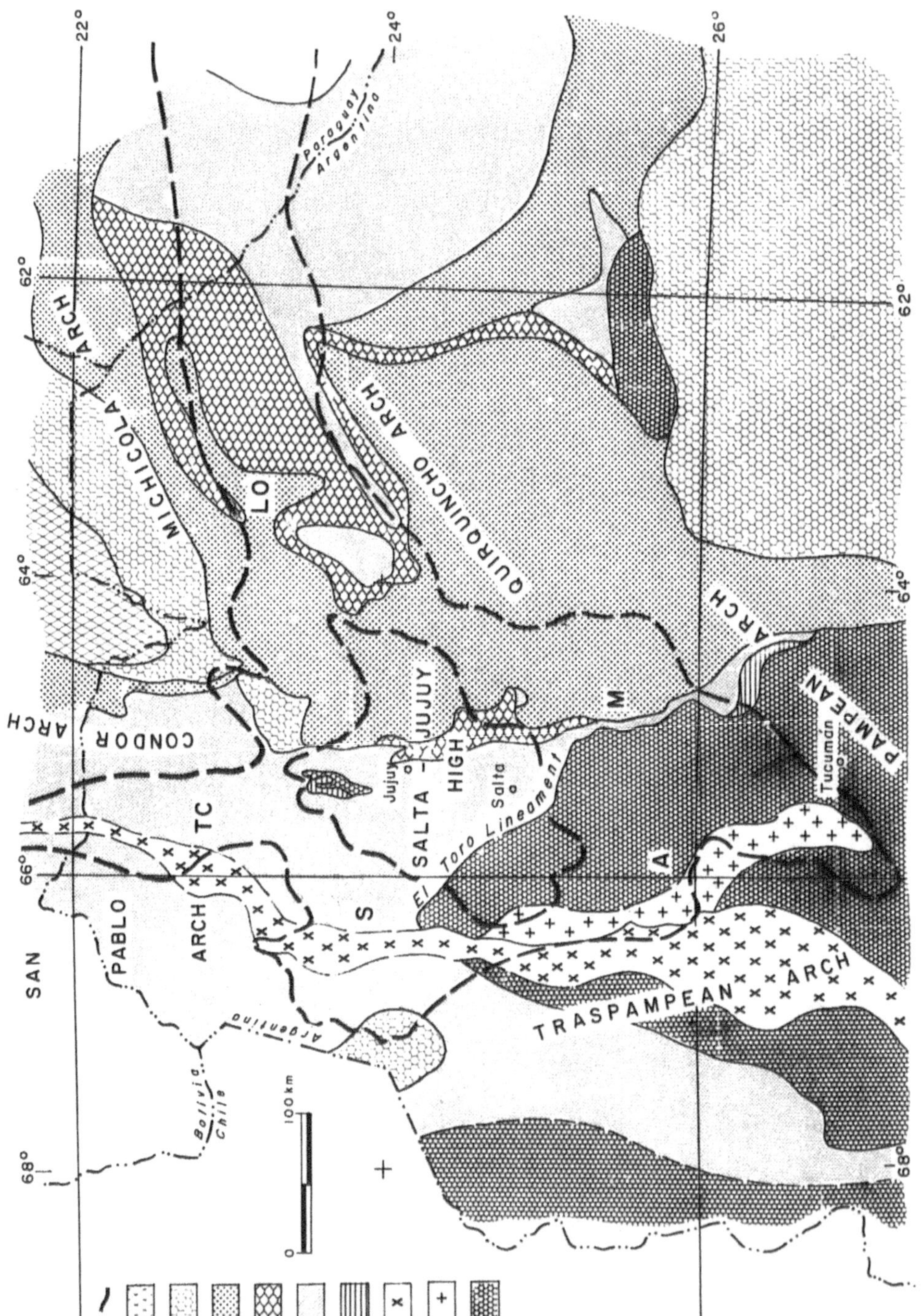
SAN PABLO ARCH
CONDOR ARCH
MICHICOLA ARCH
QUIRQUINCHO ARCH
SALTA - JUJUY HIGH
TRASPAMPEAN ARCH
PAMPEAN ARCH
El Toro Lineament
LO
TC
S
M
A
Jujuy
Salta
Tucumán
Paraguay
Argentina
Bolivia
Chile
100 km
22°
24°
26°
62°
64°
66°
68°
1
2
3
4
5
6
7
8
9
10

Fig. 6

◄ Pre-Cretaceous geology in Northern Argentina. 1 Precambrian, 2 Precambrian granitoids, 3 Early Paleozoic granitoids, 4 Cambrian, 5 Ordovician, 6 Silurian, 7 Devonian, 8 Late Carboniferous-Permian, 9 Permian, Triassic?, 10 Wedge-out of Pirgua Subgroup. Depocenters: TC-Tres Cruces, LO- Lomas de Olmedo, S-Sey, M-Metán, A-Alemanía.

deposits, mainly those of the flooding event embodied in the Yacoraite Formation (Marquillas, 1985).

The tensional tectonics which was active during the synrift accumulations decreased almost abruptly after the Pirgua Subgroup had levelled the basin and the Balbuena Subgroup ingression had begun (Salfity, 1982; Uliana et al., 1988; Gómez Omil et al, 1989; Chiarenza and Ponzoni, 1989) (Figures 8, 16 A and 17E).

The base of the Balbuena Subgroup is made up of a blanket type sandstone, the Lecho Formation (Turner, 1959; Salfity, 1980). This is followed transitionally by the most characteristic unit of the subgroup, the Yacoraite Limestone (Turner, 1959; Marquillas, 1984, 1985, 1986). The top of the subgroup consists of the Olmedo Formation (Moreno, 1970), a black to dark gray shale, which overlies the Yacoraite Formation in the eastern portion of the basin, and is replaced laterally -to the west- by the Tunal Formation (Turner et al., 1979). The pertinent environmental information is given in a later section of this paper.

The Yacoraite Formation is the Balbuena Subgroup unit which achieved the widest areal dispersal. It overlapped the Cóndor, Michicola and Quirquincho arches, and the San Pablo and Salta- Jujuy highs (Figures 9 and 17). In every case the limestone lie on Precambrian or Paleozoic basement.

There was one exception in the southwest sector of the basin, where the Yacoraite Formation deposits had a regressive attitude (Marquillas, 1985). This regression was brought about by the raising of the Traspampean arch and of the southwest portion of the Pirgua Subgroup basin (Figures 9 and 16B), which meant that this region was not covered by the Maastrichtian ingression. Field observations in the southwest corner of the basin show that the base of the Balbuena Subgroup (the Lecho Formation) rests on the Los Blanquitos Formation with a basal conglomerate some 40 m thick, containing Precambrian basement clasts (Lencinas and Salfity, 1973; Grier et al., 1991). This shows the mobility of the provenance areas in the southwest of the basin during the start of the Balbuena Subgroup deposition.

It is in this part of the basin that economically interesting uraniferous mineralizations are known. These are related to the regressive facies of the Yacoraite Formation and to other facies in the Lecho Formation (Figure 9) (Gorustovich, 1988; Gorustovich et al., 1989). Up to the present, these mineralizations form the main uraniferous reserves of northwest Argentina, and are the only ore deposits in the region which have been worked. They are stratiform deposits having a certain regional extension, whose genera-

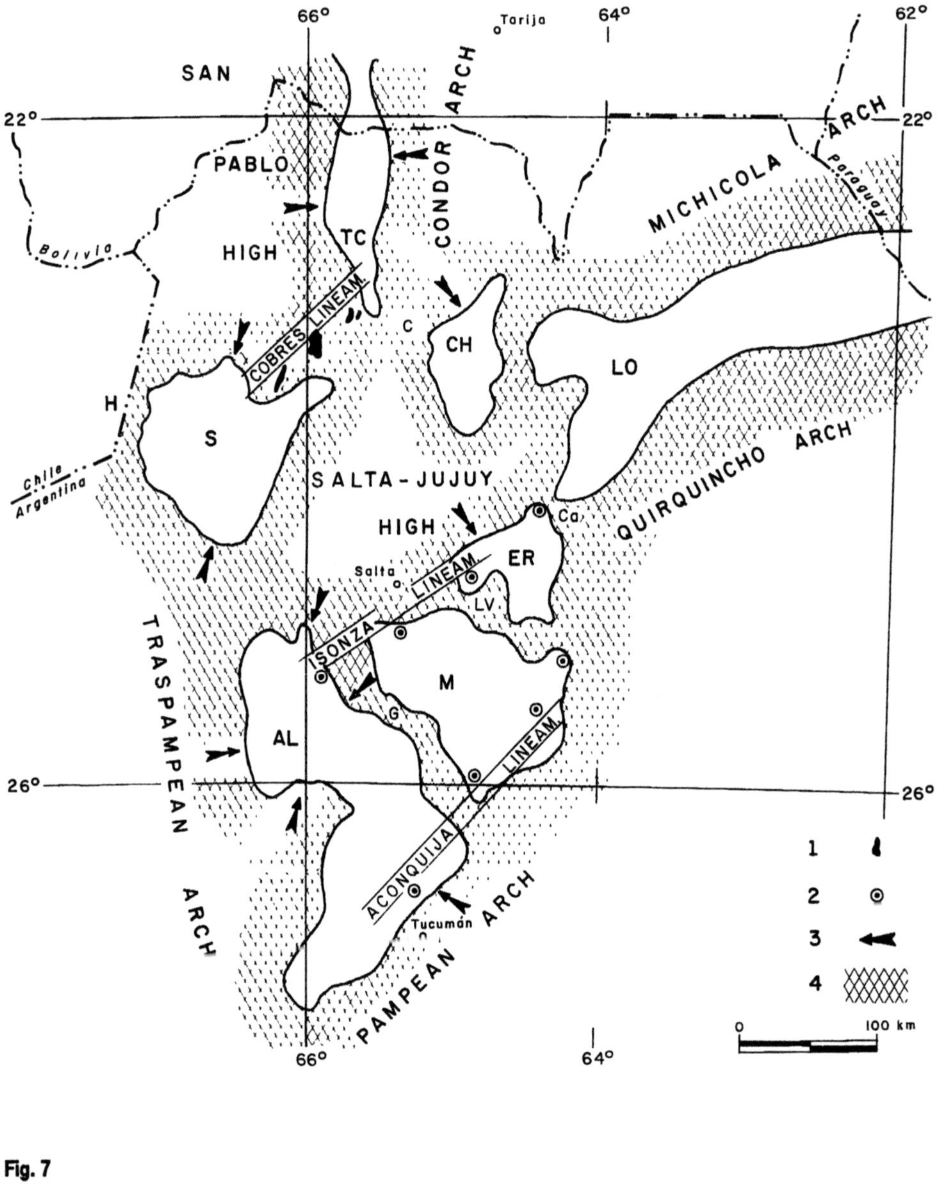

Fig. 7

First synrift sedimentary filling of the lower Pirgua Subgroup (La Yesera Formation and its equivalents). 1 Early Cretaceous granitoids, 2 Early Cretaceous volcanics (First volcanic cycle), 3 Source direction, 4 Structural high. Depocenters: LO-Lomas de Olmedo, CH-Cerro Hermoso, TC-Tres Cruces, S-Sey, ER-El Rey, M-Metán, A-Alemanía. Structural highs: C-Calete, Ca-Cachipunco, LV-Las Víboras, G- Guachipas.

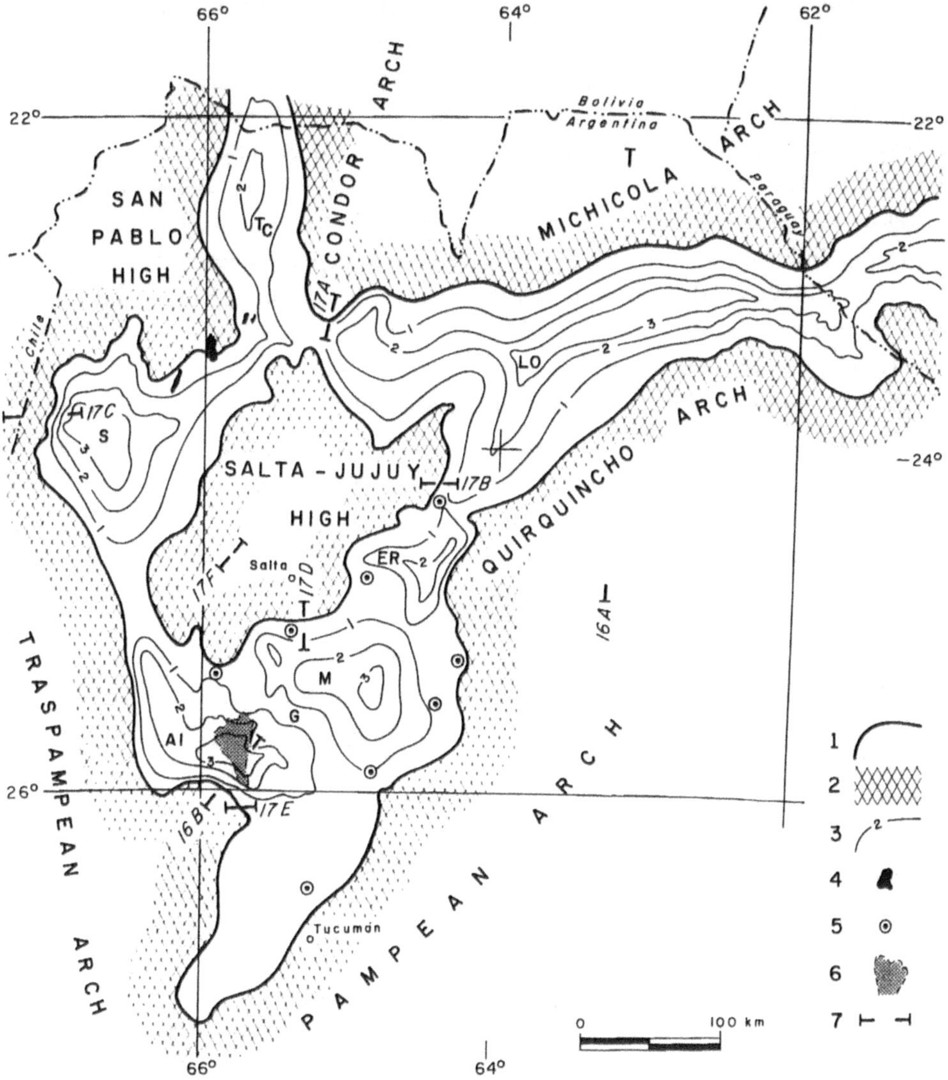

Fig. 8

Synrift sedimentary filling of the Pirgua Subgroup (La Yesera, Las Curtiembres, and Los Blanquitos Formations). 1 Inferred basin edge, 2 Structural high, 3 Isopach line (thickness in km), 4 Early Cretaceous granitoids, 5 Early Cretaceous volcanics (First volcanic cycle), 6 Late Cretaceous volcanics (Second volcanic cycle), 7 Location of the sections shown in Figures 16 and 17. Depocenters: LO-Lomas de Olmedo, CH-Cerro Hermoso, TC-Tres Cruces, S-Sey, ER-El Rey, M- Metán, A-Alemanía. Structural highs: C-Calete, Ca-Cachipunco, LV-Las Víboras, G-Guachipas.

tion was the consequence of syngenetic provenance of the uranil ion from the Precambrian and Paleozoic granite bodies of the Pampean basement; this was then captured by pelitic facies in a reducing environment. Epigenetic remobilization occurred much later, and gave rise to accumulations of phosphates and vanadates in sandy facies, associated with secondary copper minerals (malachite, azurite and some sulphides).

Furthermore, the Yacoraite Formation is without a doubt the main hydrocarbon producing horizon in the Cretaceous Basin (Turic et al., 1987). It serves as a source rock (shales and calcareous mudstones), as a reservoir rock (porous sandstones, fissured and oolitic limestones), and also as a cap rock (impermeable clays and carbonates). Most of the oil and gas fields discovered lie in the northeast portion of the basin (Figures 9 and 15), in which the wells usually reach a depth of 3000 to 4000 meters, and the traps are usually of the structural or combination types.

The Olmedo Formation (Moreno, 1970), also important from an oil bearing point of view, is a unit in the northeastern subsurface of the basin. Its deposition was governes by large-scale subsidence processes in the Lomas de Olmedo depocenter.

Santa Bárbara Subgroup

The second pulse of the Ranquel phase is marked by the ending of the flooding episodes. The surface which marks it consists of the tops of the Olmedo, Tunal and Yacoraite Formations, on which the red pelitic and sandy beds of the Mealla Formation invariably lie. This is the base of the Santa Bárbara Subgroup (Marquillas and Salfity, 1989, 1992) (Figures 3, 4, 12, 15 and 16).

The facies which constitute the Santa Bárbara Subgroup are essentially made up of mudstone, pelitic rock and limestone accumulations of the Mealla, Maíz Gordo and Lumbrera Formations. These have thin tuff levels frequently interbedded in them.

Figure 11 shows a series of idealized schematic drawings of the environmental setting for these formations.

The Santa Bárbara Subgroup starts with the red claystones, mudstones, marls and some sandstones of the Mealla Formation, whose relationship with the underlying units will be discussed in the following section (Figure 12).

The Mealla Formation characteristically contains intercalated deposits of a regional lacustrine episode, known as the "Gray Horizon" (Cazau et al., 1976; Gómez Omil et al., 1989).

The green claystones and siltstones with interbedded limestones and stromatolite limestone of the Maíz Gordo Formation lie on the Mealla Formation. These pelitic and carbonate deposits are those of an extensive stratified meromictic lake, and have aroused a certain amount of interest with regard to oil prospection (del Papa and Marquillas, 1990; del Papa, 1992).

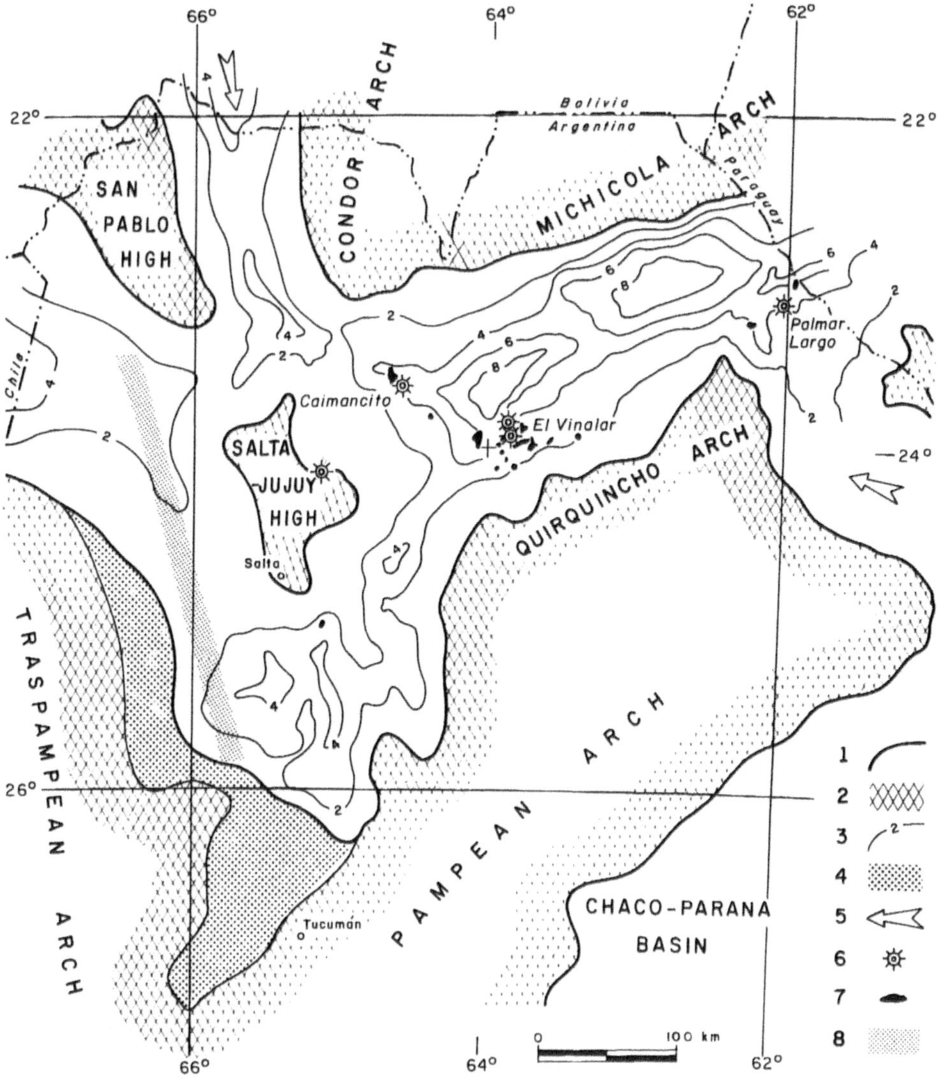

Fig. 9

Postrift sedimentary sequence of the Balbuena Subgroup (Lecho, Yacoraite, and Olmedo-El Tunal Formations, and Saline Member). 1 Inferred basin edge, 2 Structural high, 3 Isopach line (thickness in hundreds of meters), 4 Pirgua Subgroup basin as a part of the Traspampean arch during Late Senonian time, 5 Proposed courses of the Yacoraite Formation ingression from the South and/or from the North, 6 Late Senonian-Paleocene volcanics (Third volcanic cycle), 7 Oil field (Yacoraite Formation as both source rock and reservoir), 8 Uranium belt (Yacoraite Formation as host rock).

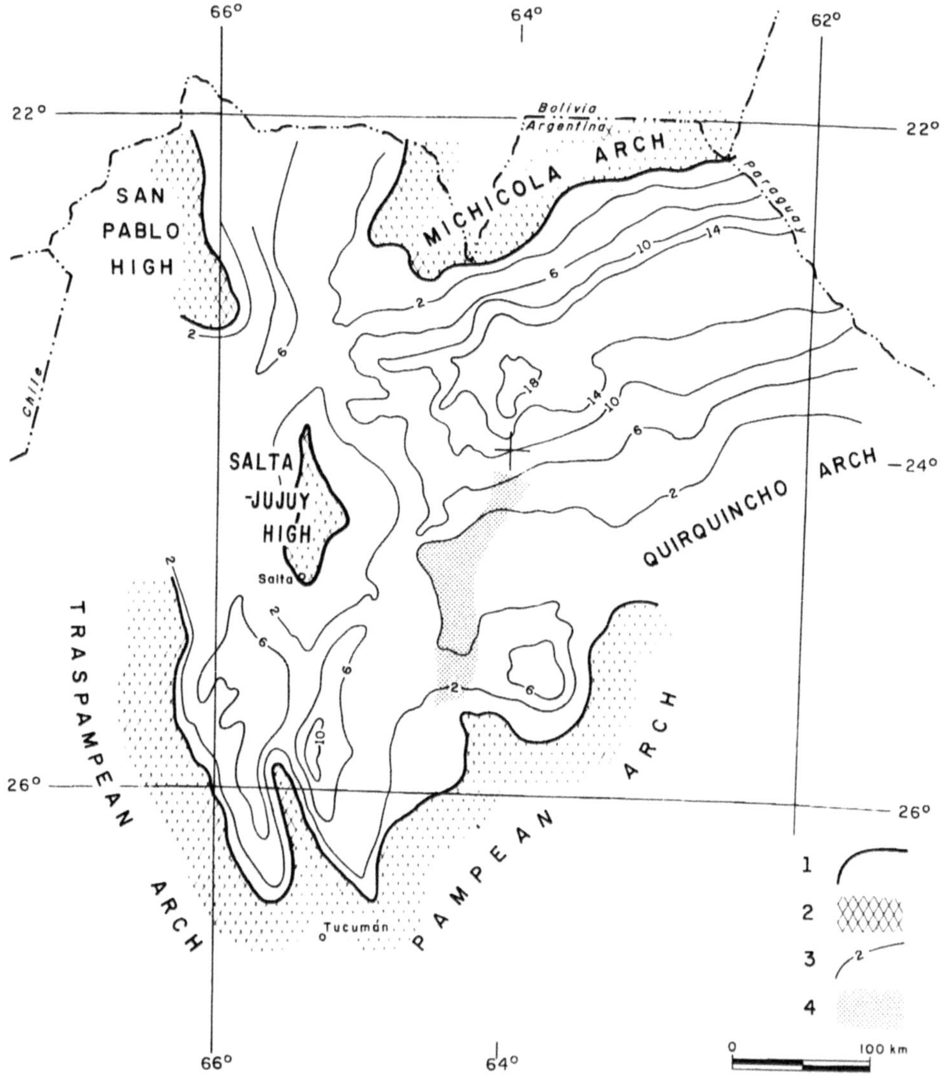

Fig. 10

Final postrift sedimentary sequences of the Santa Bárbara Subgroup (Mealla, Maíz Gordo, and Lumbrera Formations). 1 Inferred basin edge, 2 Structural high, 3 Isopach line (thickness in hundreds of meters), 4 Los Gallos uplift

The Lumbrera Formation consists of red pelitic rocks and subordinate sandstones. It is characterized by a conspicuous green level (the "Green Horizon"), located in its lower third and having wide regional distribution (Cazau et al., 1976; Gómez Omil et al., 1989).

The three formations of the Santa Bárbara Subgroup were mutually overlapping, and the Lumbrera Formation deposition represents the widest expansion registered in the sedimentary record of the Salta Group. In many cases these formations lie directly on the pre-Cretaceous basement over the Cóndor arch (Zanettini, 1973), Michicola arch and Quirquincho arch (Cazau et al., 1976), and round the Salta-Jujuy high (Schlagintweit, 1937; Vilela, 1965) (Figures 10, 15, 16 and 17).

The Santa Bárbara Subgroup contains excellent records of Paleocene-Eocene fossil mammals, of Riochican and Casamayoran mammal ages (Pascual et al., 1981, Pascual, 1984); therefore the age of the Salta Group roof is well identified paleontologically. The youngest stratigraphic levels in which some fossils were found correspond to the "Green Horizon" of the Lumbrera Formation and neighboring strata (Carbajal et al., 1977), in the Alemanía depocenter.

The Santa Bárbara Subgroup also contains fossil remains of reptiles, fish, insects and -particularly- microflora (Quattrocchio and Volkheimer, 1990).

Regional Facies Evolution and Flooding Processes Interbedded in the Red Bed Sequences of the Salta Group

Regional Facies Evolution

Figure 11 has been drawn on the basis of information provided by Gómez Omil et al. (1989), and supplemented with information given by Boso et al. (1984), Marquillas (1984, 1985), Marquillas et al. (1984), Quattrocchio et al. (1986), del Papa and Marquillas (1990), Galli and Marquillas (1990), del Papa (1992), and abundant personal data. The illustration shows an idealized set of environmental models intended to characterize each of the formational units of the Salta Group.

The La Yesera Formation is mainly made up of fluvial bedload facies, deposited in braided rivers and alluvial fans, which form the main fill of the rift troughs. Local eolian and lacustrine sediments are associated with the above facies, as are volcanic rocks. In the type locations of the Alemanía depocenter (Figure 7) the alluvial fans came down off the marginal areas in paleocurrents running crosswise to the rift, whereas the rivers ran both across the rift and longitudinally to it. In general, due to the attitude of the supply areas (Figure 7), the paleocurrent pattern would always tend to be regionally centripetal. This is common in such rift systems (Hubert and Stevens, 1980), because there are always a series of uplifts or highs formed by horsts or volcanic buildups, which bring about a total or partial isolation of the various sub-basins or depocenters.

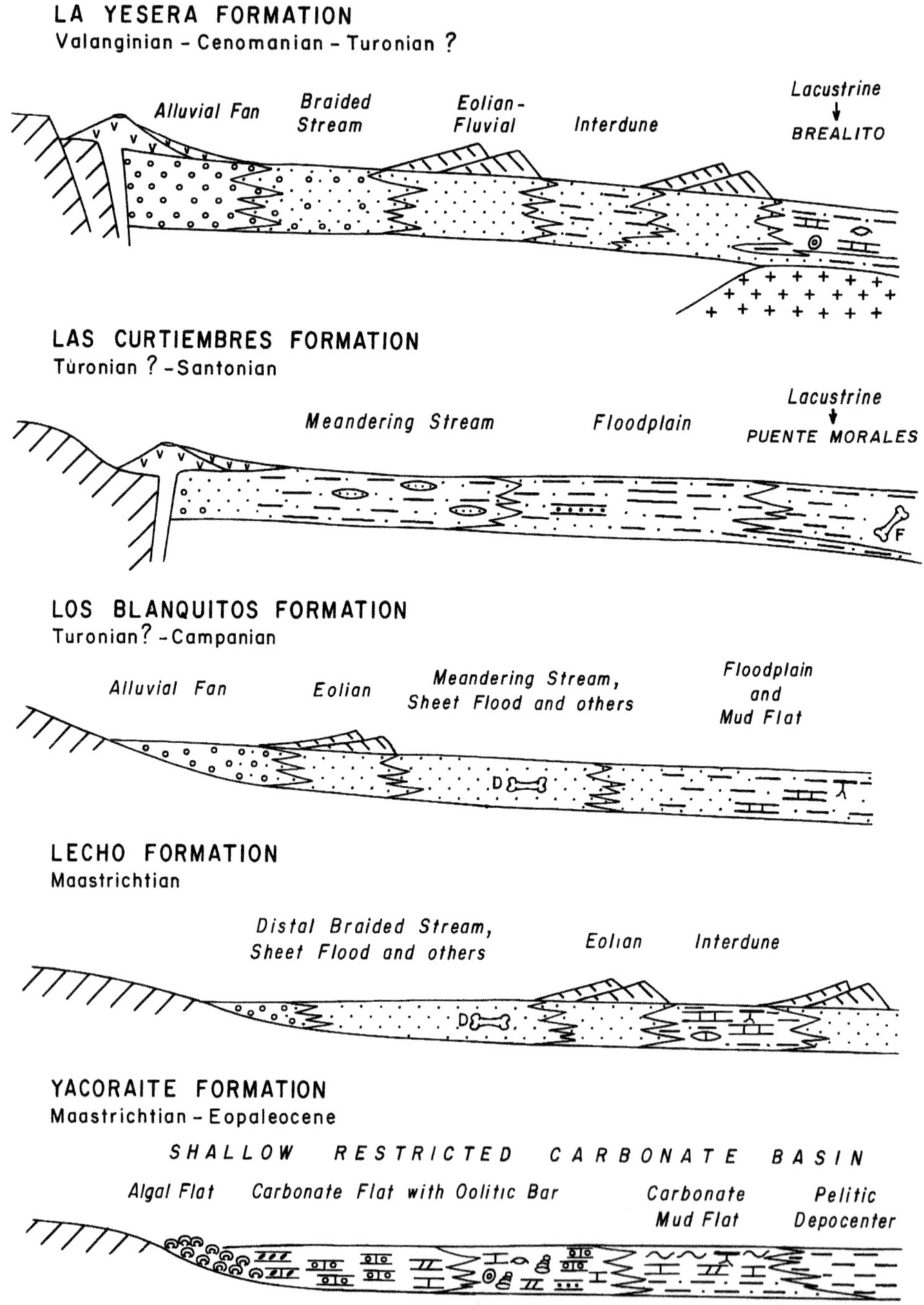

Fig. 11

Idealized environmental evolution of the Salta Group Formations (based on and adapted from Gómez Omil et al., 1989)

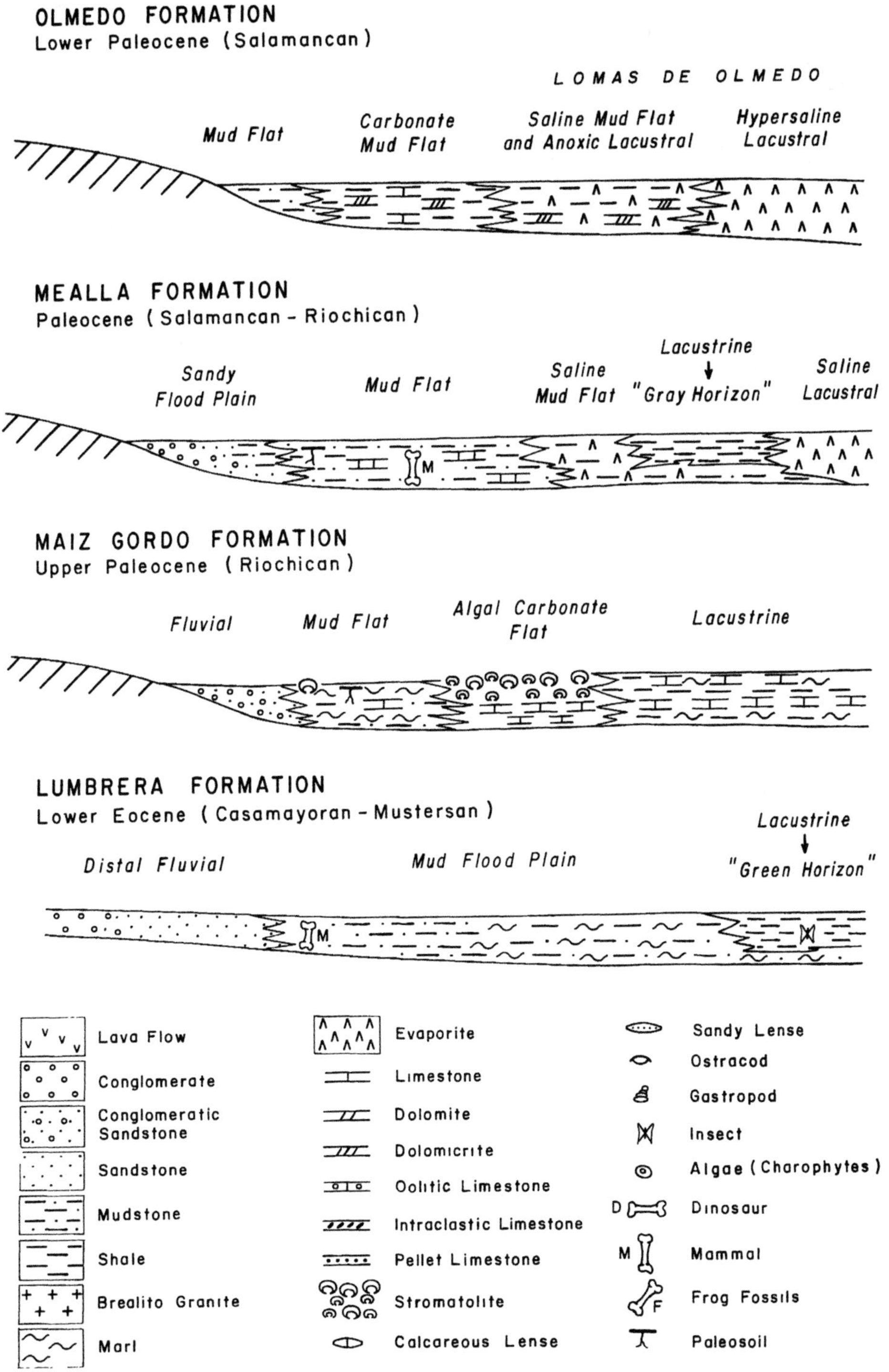

Fig. 11 (Continued).

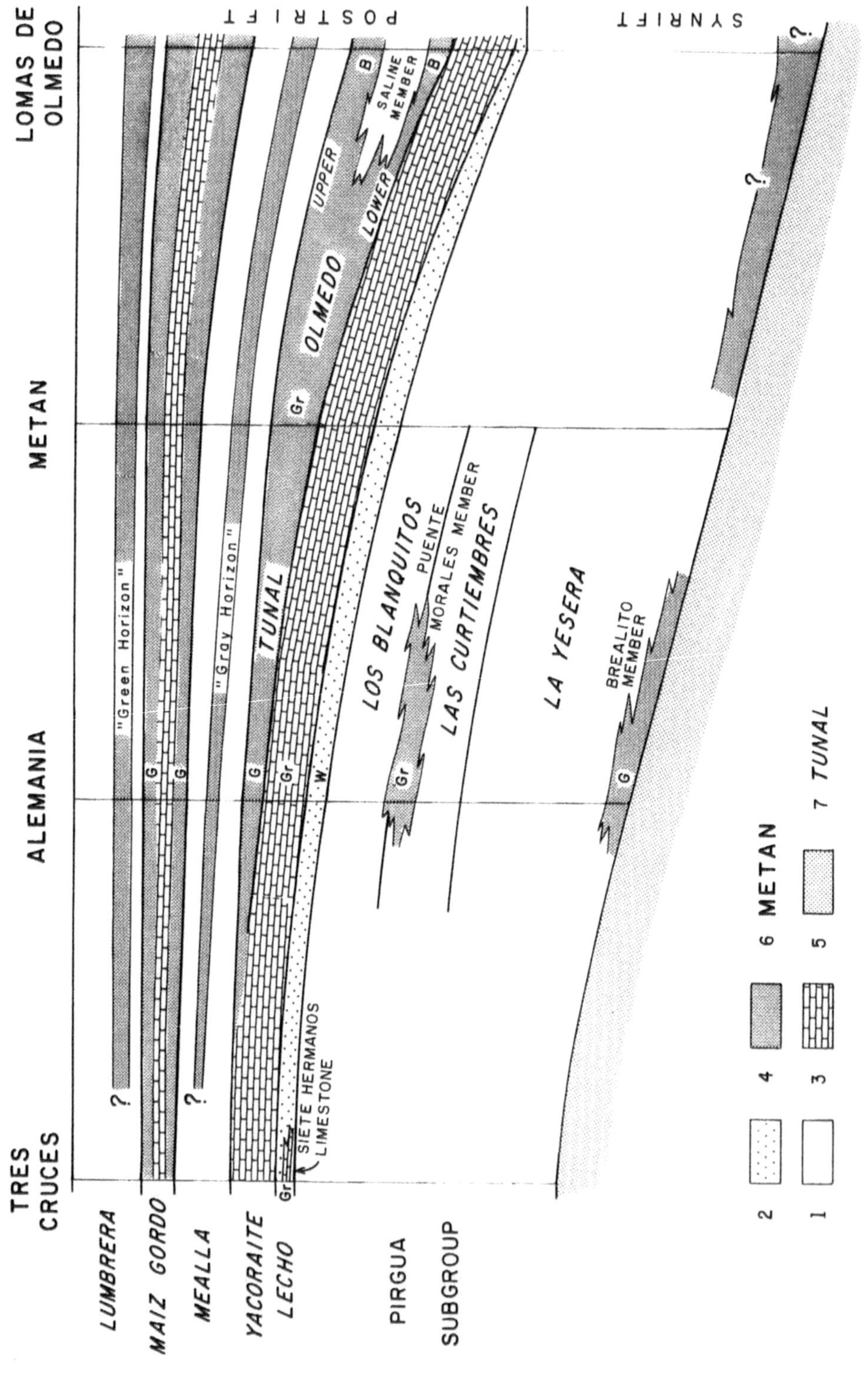
LOMAS DE OLMEDO
METAN
ALEMANIA
TRES CRUCES
POSTRIFT
SYNRIFT
LUMBRERA
MAIZ GORDO
MEALLA
YACORAITE
LECHO
PIRGUA SUBGROUP
"Green Horizon"
"Gray Horizon"
TUNAL
OLMEDO
UPPER
LOWER
SALINE MEMBER
B
Gr
G
W
SIETE HERMANOS LIMESTONE
LOS BLANQUITOS
PUENTE MORALES MEMBER
LAS CURTIEMBRES
LA YESERA
BREALITO MEMBER
?
1
2
3
4
5
6 METAN
7 TUNAL

Fig. 12

◄ Cartoon showing the main flooding events during deposition of the Salta Group red beds (formation thicknesses and distances between depocenters are off scale). 1 Red beds, 2 Sandstone, 3 Limestone, 4 Shale, 5 Pre- Cretaceous basement, 6 Depocenter name, 7 Formation name. B- Black, G-Green, Gr-Gray, W-White.

In the Alemanía depocenter a 300-meter thickness of the La Yesera Formation is recognized as having an 80% conglomeratic lithofacies content and a 20% sandy facies content. Obviously this does not include the volcanic rocks found farther to the south, nor the pelitic rocks present in the southwest portion of the depocenter.

The Las Curtiembres Formation is mainly an accumulation of meandering stream deposits; these streams eventually flooded wide plains and depressed areas. It is presumed that tectonic activity was considerable, as it must have brought about changes in slope (Ouchi, 1985) and consequently favored the flooding episodes. Furthermore, volcanic activity brought about local damming of the waters.

In the Alemanía depocenter the Las Curtiembres Formation, shows the following facies distribution: 64% sandstones, 22% pelitic rocks, 5% conglomerates, and 9% volcanic and pyroclastic rocks, for a total 700 m thickness.

The Los Blanquitos Formation, which covers the Las Curtiembres Formation, is interpreted as having had a very extensive lateral relationship in time with the Las Curtiembres Formation, and it even lies directly on the La Yesera Formation (Salfity,1980) in some parts of the basin (Figure 3).

The Los Blanquitos Formation is made up almost entirely of sandy facies, having areal distribution, which levelled the Pirgua Subgroup basin. A quick regional-scale analysis allows a somewhat complex origin to be supposed for these sandstones, as a consequence of coalescing fluvial systems and of eolian influences, furthered by the development of flood plains and calcareous mud flats in the red and brown sand environments. Nor are the actions of sandy and mixed-load braided systems discarded (Miall, 1977), nor those of unchannelled shallow streams or sheet floods (Mckee et al., 1967), as these probably gave origin to many of the massive or laminated tabular sand bodies having great lateral extension.

In Alemanía the Los Blanquitos Formation reaches a thickness of 1300 meters. This includes 95% sandstones, 3% conglomerates and 2% pelitic rocks.

The Lecho Formation is a blanket-like, commonly calcareous, white sand deposit. Although it marks the limit between the Pirgua Subgroup redbeds and the postrift deposits it shows a region- wide physical continuity with both of these, to a greater or lesser degree (Salfity, 1980). Though only of moderate thickness, it has a wide areal distribution, developed under the influence of distal braided systems, sheet floods and calcareous mud flats -probably of interdune nature- and of accompanying eolian processes. The Lecho sandstone definitely marks the beginning of the transgressive episode embodied in the overlying Yacoraite Formation, and as such can be considered a prior clastic platform.

Within the area here used as a reference -the Alemanía depocenter- the Lecho Formation does not exceed 40 m in thickness, even in atypical facies (Quitilipi Formation; Salfity and Marquillas, 1981). The sandstones here make up 90% of the lithofacies, and the pelitic rocks 10%. In some depocenters, such as Tres Cruces, the calcareous facies are thickly bedded (Marquillas and Salfity, 1990), and in others, such as Metán, they are less abundant but still notable (Galli and Marquillas, 1990).

The Yacoraite Formation is an extensive partly dolomitic tabular limestone. Although this deposit was contemporaneous with the Late Cretaceous (Maastrichtian) global transgressive pulse -to the effects of which it certainly owes its origin- it is very difficult to match it against any of the classic models of platforms described in the literature (among others, by Ahr, 1973; Ginsburg and James, 1974; Wilson, 1975; James 1984a, 1984b; Read, 1982, 1985), despite its having many elements compatible with several of them (Marquillas, 1984, 1985, 1986).

The lack of diagnostic marine fauna is in this regard decisive, when added to the lack or non-preservation of tide marks. In the final event, however, this problem of a suitable model should not seem too strange, if one recalls the large epicontinental platforms covered by epeiric seas which penetrated far inland (Irwin, 1965; Heckel, 1972); as well as other ancient carbonate deposits which evolved during certain moments of the planet's history, and which have no comparable models in present times.

In this Northwest Basin deposit no sustained, nor permanent, nor even periodic, connection with the open sea can be discerned. However, this is not to deny that such a connection may be recognizable in some other part of this extensive basin, which even includes regions of other countries (Salfity et al., 1985; Sempere et al., 1988; Sempere, this volume).

Deposition occurred under shallow water conditions, with frequent subaerial exposure, wave and current action, rapid changes in salinity, and appreciable organic productivity; this last in spite of the lack of diversity in the biotic community which formed its population (Marquillas, 1985). These characteristics obviously correspond to a restricted carbonate basin, which could have been a very internal and restricted platform, or perhaps some type of poorly connected lagoon. In support of the above assertions, mainly those referring to shallow conditions and frequent exposures, signs of advanced diagenetic processes such as dolomitization and early cementing are frequently observed.

The environment was favorable both to stromatolite plains and to the ample development of ooid limestones (grainstones and packstones), associated with other non-skeletal grains (intraclasts, pellets, peloids, pisolites and lumps) and with bioclasts (gastropods, ostracods and pelecypods). Added to this are the shales and calcareous mudstones

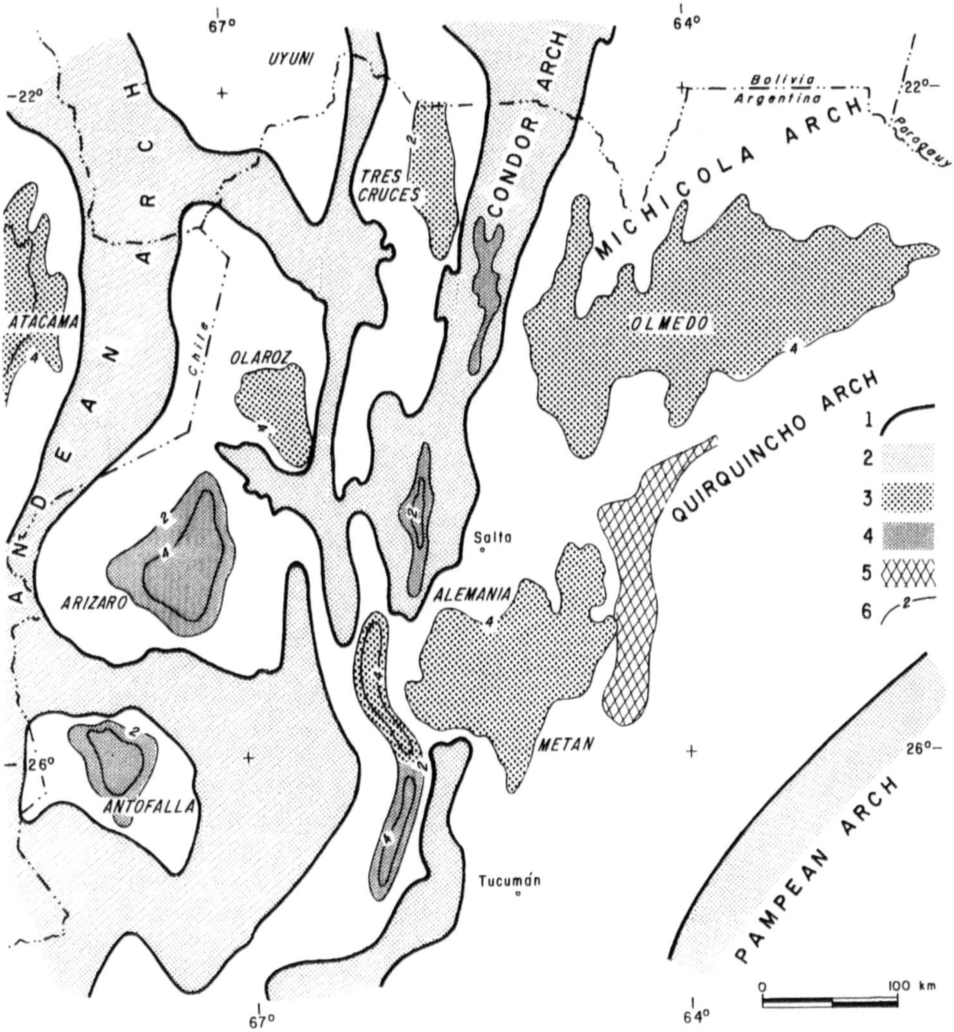

Fig. 13

Tertiary (post Inca and pre Quechua I) sedimentary basins. 1 Inferred basin edge, 2 Structural high (drawn before Late Cenozoic magmatic arc emplacement), 3 Cretaceous plus Tertiary depocenter, 4 Tertiary depocenter, 5 Los Gallos uplift, 6 Isopach line (thickness in km).

which make up the relatively deepest facies -typical of depocenter axes- and which are normally found interbedded among the facies mentioned above; they become slightly predominant in the upper levels of the deposit.

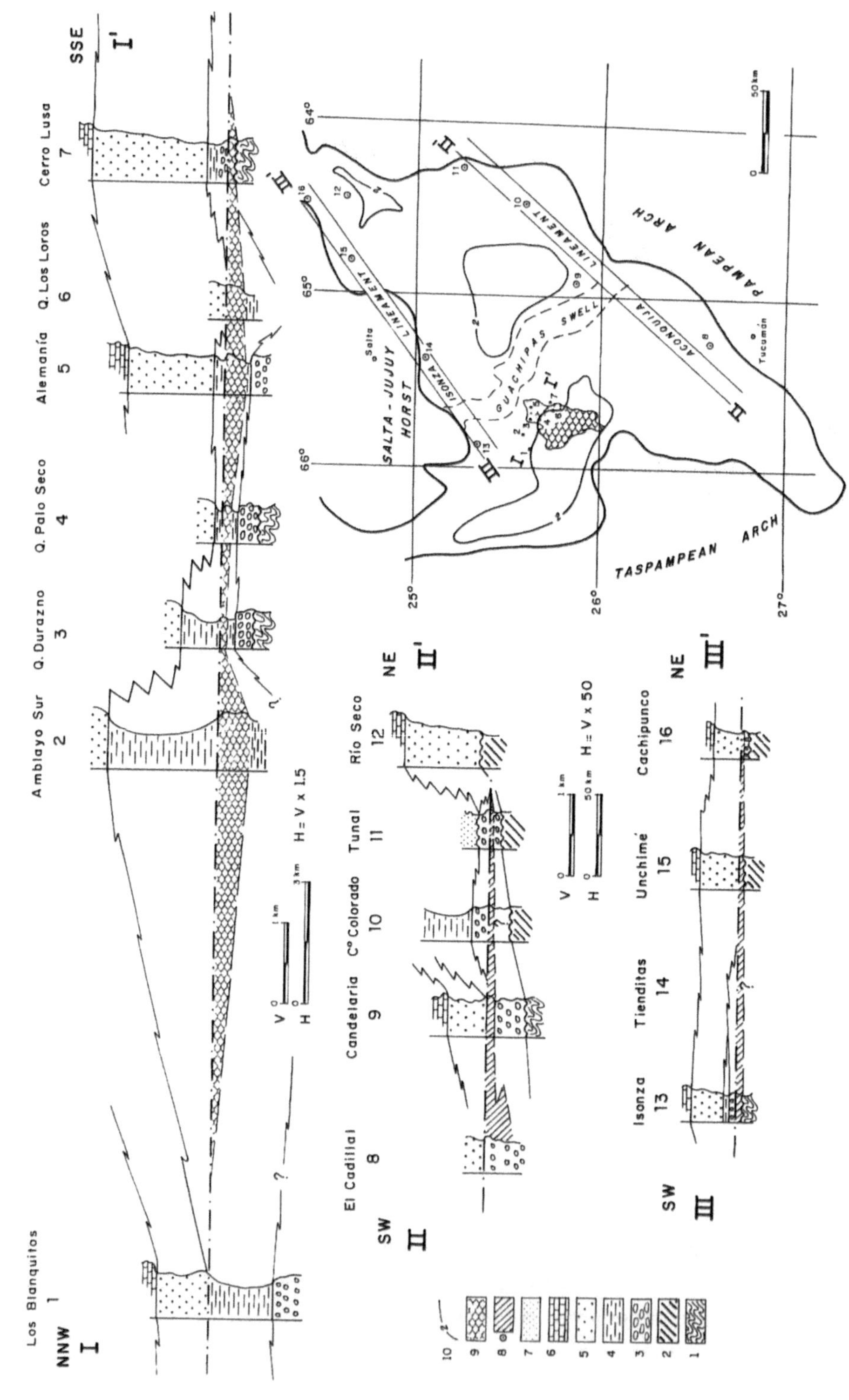

NNW I
Los Blanquitos 1
Amblayo Sur 2
Q. Durazno 3
Q. Palo Seco 4
Alemanía 5
Q. Los Loros 6
Cerro Lusa 7
SSE I'
V 0 1 km
H 0 3 km
H = V x 1.5
SW II
El Cadillal 8
Candelaria 9
Cº Colorado 10
Tunal 11
Río Seco 12
NE II'
V 0 1 km
H 0 50 km
H = V x 50
SW III
Isonza 13
Tienditas 14
Unchimé 15
Cachipunco 16
NE III'
SALTA - JUJUY HORST
ISONZA LINEAMENT
GUACHIPAS SWELL
ACONQUIJA LINEAMENT
PAMPEAN ARCH
TASPAMPEAN ARCH
Salta
Tucumán
64º
65º
66º
25º
26º
27º
0 50 km

Fig. 14

◄ Regional distribution of the First (Isonza Basalt) and Second (Las Conchas Basalt) volcanic cycles of the Pirgua Subgroup. Datum line: Top of each basalt. 1 Precambrian, 2 Paleozoic, 3 La Yesera Formation, 4 Las Curtiembres Formation, 5 Los Blanquitos Formation, 6 Balbuena Subgroup, 7 Santa Bárbara Subgroup, 8 Isonza Basalt, 9 Las Conchas Basalt, 10 Isopach line of the Pirgua Subgroup (thickness in km).

A thickness of 185 m has been recorded in Alemanía for the Yacoraite Formation, with the following lithofacies distribution: 54% limestones, 40% pelitic rocks and 6% sandstones.

The Olmedo Formation (containing black shales and evaporites in the subsurface) is more extensively dealt with in following paragraphs, in which the relationships between red beds and flood deposits are covered. In the locality of Alemanía it is found in the shape of its surface equivalent in this region, the Tunal Formation. This consists of pelitic rocks of various colors, and of scarce calcareous and sandy levels. Its thickness here is 25 m, consisting of 90% pelitic lithofacies, 5% calcareous lithofacies and 5% sandy lithofacies.

Finally, reference should be made to the deposits which ended the postrift process, and which correspond to the Santa Bárbara Subgroup. These are the Mealla, Maíz Gordo and Lumbrera Formations.

These formations are essentially accumulations of muddy, marly and sandy sediments; partly red in color (Mealla and Lumbrera Formations) and partly green, with limestones (Maíz Gordo Formation). In the Lomas de Olmedo depocenter subsurface they also contain evaporites (Moreno, 1970; Gómez Omil et al., 1989; Carlé et al., 1989). They are distal fluvial-lacustrine deposits with ample mud flat development, which accumulated in a tectonically stable environment only affected by periods of gentle subsidence that gave rise to regional flooding episodes.

Despite the overall predominance of suspended loads in the fluvial and other associated mechanisms which took part in the accumulation of these deposits, bedloads were prevalent in this regions proximal to provenance areas (the western border of the basin, for instance -Figure 10).

In the Alemanía depocenter, used herein as a reference for lithofacies distribution, the above-mentioned formations are mainly sandy, and have the following characteristics: Mealla Formation (180 m thick, with 70% sandstones, 20% pelitic rocks + marls, 10% conglomerates); Maíz Gordo Formation (50 m thick, with 50% sandstones, 40% pelitic rocks + marls, 10% conglomerates); Lumbrera Formation (280 m thick, with 67% pelitic rocks + marls, 33% sandstones).

It should be mentioned that fossil soils have been recognized in several of the Salta Group units; the most outstanding of which have been shown in the Figure 11 illustrations. They have been identified in the Los Blanquitos, Lecho, Yacoraite, Mealla and

Maíz Gordo Formations. In some cases these paleosoils are associated with unmistakably phreatophyte levels (Los Blanquitos and Lecho Formations; Galli and Marquillas, 1990), and in other cases with tuff or tuffite levels (Yacoraite Formation; Marquillas, 1985).

Tuffs, chonites and cinerites are conspicuous in the Lecho, Yacoraite, Mealla, Maíz Gordo and Lumbrera Formations deposits, as well as in the Tertiary units of the overlying Orán Group.

Flooding Processes

The sedimentary fill of the Salta Group basin reveals an evolutionary history from Early Cretaceous to Middle Eocene times which is characterized by a diversity of factors, among them the tectonic structuring, which gave rise to a series of different sedimentary environments.

Thus, the stratigraphic column of the Salta Group records the consequences of a series of dissimilar events, which favored the alternative accumulation of red beds and of inundation deposits of varying depth -mostly shallow- and frequently anoxic.

Figure 12 has been specially designed to give a regional picture of the alternation between these two large but different types of deposits, their areal distribution and, where necessary, the proportional fitting of their thickness to a larger space provided by subsidence. This figure has been drawn out of scale, both vertically and horizontally, and is only indicative of the Salta Group deposits in the main depocenters: Tres Cruces in the north and, from SW to NE, Alemanía, Metán and Lomas de Olmedo.

In general, the red clastic deposits -mainly conglomerates, sandstones and mudstones- accumulated in continental fluvial environments and associated settings. This is the case with the La Yesera, Las Curtiembres, Los Blanquitos, Mealla and Lumbrera Formations. In the first three instances eolian events also had a part in their deposition (see also Figure 11).

Meanwhile the deposits linked to locally and regionally flooded environments, consisting of gray, green and black pelitic rocks, limestones, dolomicrites and evaporites, are generally recognized as lacustrine in origin (Figure 11).

Such is the case with the Brealito Member of the La Yesera Formation, the Puente Morales Member of the Las Curtiembres Formation, the Olmedo Formation and its surface equivalent (the Tunal Formation), the "Gray Horizon" in the Mealla Formation, the Maíz Gordo Formation, and the "Green Horizon" in the Lumbrera Formation. On the other hand, the most conspicuous representative of such flooding events, the Yacoraite Formation, defines an extensive shallow carbonate environment of peculiar characteristics.

The Olmedo Formation, in the subsurface of the Lomas de Olmedo depocenter, is perhaps the most important of the lacustrine deposits, due to its hydrocarbon-generating characteristics.

The accumulation of the black-colored pelitic and calcareous deposits of this formation must have been controlled by a sinkage mechanism that brought about the notable differential subsidence in the Lomas de Olmedo depocenter, in which this relatively deeper environment evolved.

In the Lomas de Olmedo depocenter there is a thick salt bed (Moreno, 1970) interstratified in the lower part of the black -commonly bituminous- shales.

It can be supposed that these shales were deposited during the stages of lacustrine expansion (Eugster, 1985), whereas the saline deposits -halite, gypsum and anhydrite- would correspond to a retraction of the water body.

Therefore, in an initial lacustrine expansion stage, due to an increase in subsidence favored by the collapse of central basin blocks (a phenomenon recognized in the area by Chiarenza and Ponzoni, 1989), the Lower Olmedo Formation would have been laid down. This stage would have been followed by a shrinkage stage with the consequent deposition of evaporites. Lacustrine expansion would then have begun again with the deposition of the Upper Olmedo Formation (Figures 11 and 12). Water circulation there were suitable for creating the anoxic situations which characterize these Olmedo Formation deposits.

This flooding event was regionally manifest, though not with the magnitude seen in the Lomas de Olmedo subsurface. It appears in the shape of shallow lake facies with carbonate sediments and of ephemeral lake facies.

Thus, in the Alemanía and Metán depocenters, interbedded between the gray limestones of the Yacoraite Formation and the red mudstones of the Mealla Formation, lies a varicolored mainly green calcareous-pelitic bundle, some tens of meters thick; this is known as the Tunal Formation. This formation must have been deposited at the same time as was the thick Olmedo Formation mentioned above.

Meanwhile, in the Tres Cruces and Sey depocenters in the northern and western areas of the basin, no facies of the Olmedo Formation nor any of its equivalents were laid down during this time. In these two depocenters the red mudstones of the Mealla Formation were deposited directly onto the Yacoraite Formation limestones. In addition to the lack of deposition of the Olmedo Formation facies erosive processes occurred, as can easily be seen by the presence of fragments of Yacoraite Formation rocks in the base of the Mealla Formation.

In the Subandean Ranges -to the east of the Salta-Jujuy high- there are outcrops of the Olmedo Formation, consisting mainly of pelitic rocks, micrite limestones and dark dolomicrite limestones, with some interbedded evaporite levels. This thickness of this outcrop -some 60 m- is far less than recorded for this formation in the Lomas de Ol-

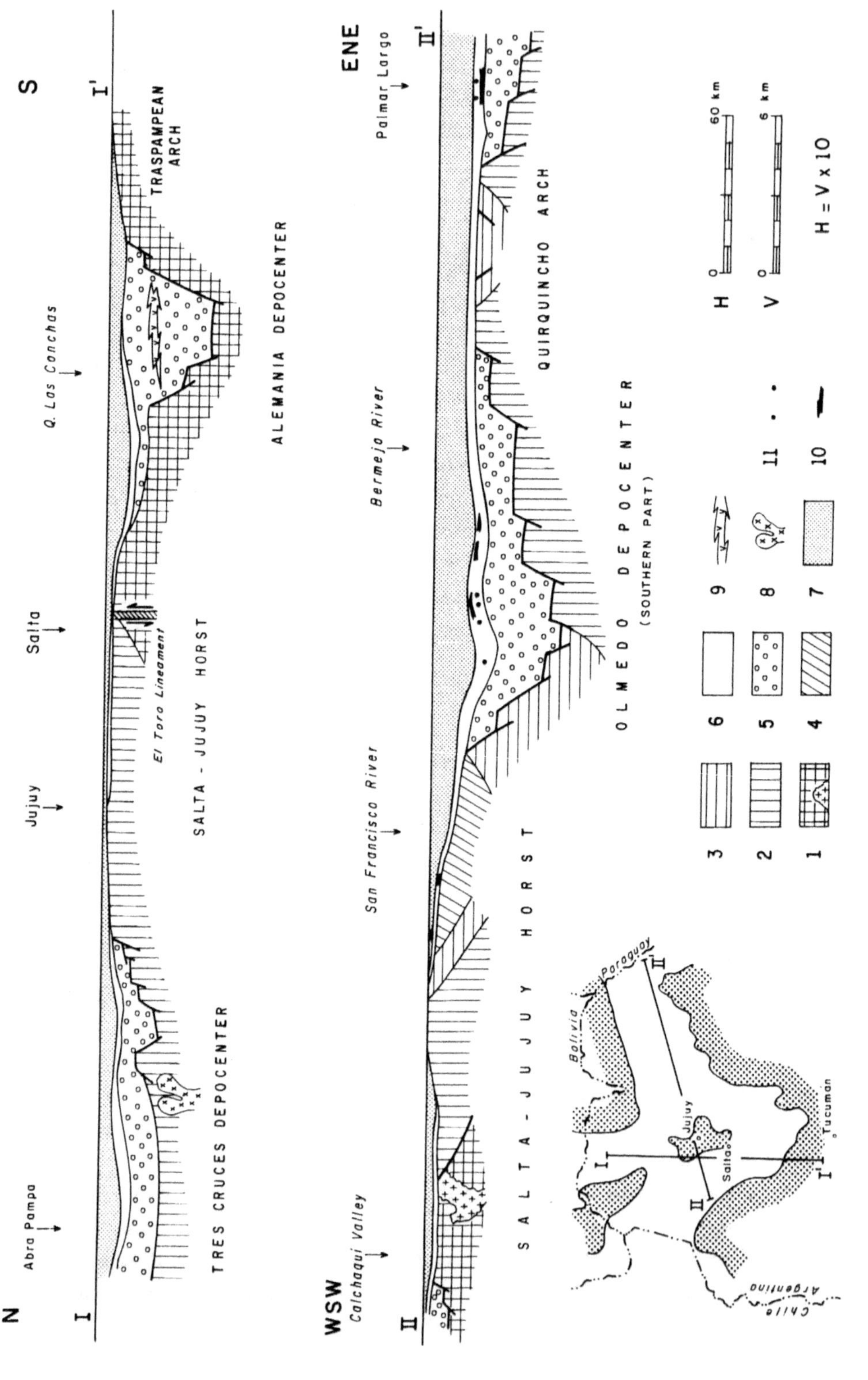
N
S
I
I'
Abra Pampa
Jujuy
Salta
Q. Las Conchas
TRES CRUCES DEPOCENTER
SALTA - JUJUY HORST
El Toro Lineament
ALEMANIA DEPOCENTER
TRASPAMPEAN ARCH
WSW
ENE
II
II'
Calchaqui Valley
San Francisco River
Bermejo River
Palmar Largo
SALTA - JUJUY HORST
OLMEDO DEPOCENTER
(SOUTHERN PART)
QUIRQUINCHO ARCH
1
2
3
4
5
6
7
8
9
10
11
H 0 60 km
V 0 6 km
H = V x 10
Bolivia
Paraguay
Chile
Argentina
Jujuy
Salta
Tucuman

Fig. 15

◄ Regional cross-sections in the Salta Group basin. Datum line: Top of Santa Bárbara Subgroup. 1 Precambrian (metamorphics and granitoids), 2 Ordovician, 3 Silurian, 4 Devonian, 5 Pirgua Subgroup, 6 Balbuena Subgroup, 7 Santa Bárbara Subgroup, 8 Cretaceous granitoid, 9 Second volcanic cycle, 10 Third volcanic cycle, 11 Oil field.

medo depocenter, where Gómez Omil et al. (1989) mention a thickness of over 300 meters.

The remaining regional flooding events have been caused by similar regional mechanisms, and are mainly those already mentioned and which can be seen in Figure 12.

Among the local processes of a certain magnitude, the example of the lacustrine deposit located in the base of the Pirgua Subgroup in the SW portion of the Alemanía depocenter (Boso, et al., 1984) can be cited. This deposit, here called the "Brealito Member" (Figures 11 and 12), is made up of green and red sandy shales and stinkstones with ostracods and freshwater algae, all having parallel horizontal lamination. In some other sections of the Alemanía depocenter pelitic and sandy facies are seen at the base of the Pirgua Subgroup; that is, beneath the La Yesera Conglomerate (Reyes and Salfity, 1973).

No other deposits are known in the same stratigraphic position in other parts of the basin, but the possibility remains that they may be present in the subsurface of the Lomas de Olmedo, with all the generative prospects that this could signify (Uliana et al., 1988; Gómez Omil et al., 1989).

Whitin this same Pirgua Subgroup, but almost in the roof of the intermediate unit -the Las Curtiembres Formation- a thick lacustrine deposit made up of gray, green and red laminated pelitic rocks containing fossil frogs (Báez, 1981) was laid down in the Alemanía depocenter. This deposit, here called the "Puente Morales Member" (Figures 11 and 12), is hypothetically considered to have been generated by local damming phenomena due to contemporary volcanic activity in the area.

Finally, the presence of calcareous levels, probably of mud flat origin, interbedded in the red sandstones of the upper unit of the Pirgua Subgroup -the Los Blanquitos Formation- is mentioned as having been recognized in the Tres Cruces, El Rey and Metán depocenters (Figure 8) (Gómez Omil et al., 1989; Galli and Marquillas, 1990; personal data).

Suggested burial history

As part of the present research, the authors have attempted to reconstruct the burial history of the Salta Group, despite the limitations imposed by the scarcity of available datings, both radiometric and those provided by diagnostic fauna, which makes it difficult to carry out a proper chronostratigraphic reconstruction. For this reason the subsidence rate values given here are considered highly estimative, until such time as a proper adjustment of all the deposition ages can be reached.

For the Alemanía depocenter area, considered herein as a normal reference, the average subsidence values for the whole Salta Group vary between 6.2 cm and 7.0 cm per 1000 years. Meanwhile the Lomas de Olmedo depocenter average subsidence rate varies between 4.0 cm/1000 yrs and 18.0 cm/1000 years.

Thus, in Alemanía an average 6.0 cm/1000 yr subsidence has been calculated for the Pirgua Subgroup deposits, including the levelling of the basin by the Los Blanquitos Formation. An average 8.0 cm/1000 yr subsidence is estimated for the Balbuena Subgroup deposition, during which time the depocenter was levelled for the second time. This includes the reactivation of ancient fractures which favored subsidence, as has also been recognized in other areas of the basin (Chiarenza and Ponzoni, 1989). From that time on a new subsidence stage commenced, allowing the Santa Bárbara Subgroup to accumulate at a subsidence rate falling between the previous two: an average of 6.66 cm/1000 years.

Solely for the purpose of recording the huge variations which occurred during the Salta Group filling of the various depocenters, here follow the subsidence rate values obtained by the authors for the Lomas de Olmedo depocenter. An average 4.0 cm/1000 yr rate was obtained for the Pirgua Subgroup, an average 18.0 cm/1000 yr rate for the Balbuena Subgroup, and an average 14.0 cm/1000 yr rate for the Santa Bárbara Subgroup. These extremely large variations in subsidence rates between the two depocenters considered should not seem too surprising if some of the varied factors in the geologic evolution of the basin already described are borne in mind.

The Eruptive Processes

During the Cretaceous, magmatic events which were limited in scope but of significant structural, stratigraphic and petrologic importance took place (Bossi and Wampler, 1969; Halpern and Latorre, 1973; Reyes et al., 1976; Valencio et al., 1976; Mädel, 1984; Galliski and Viramonte, 1985, 1988; Gait et al., 1989; Avila Salinas, 1989; Toselli, 1992).

Three well-defined magmatic pulses have been distinguished in the history of the Salta Group (Reyes and Salfity, 1973; Reyes et al., 1976; Salfity, 1982; Marquillas and Salfity, 1988). The first was related to the initial fill of the Pirgua Subgroup in the south of the basin and to plutonic manifestations in the north, the second was intercalated in the middle of that fill, and the third was intruded during the first stage of postrift deposition.

The plutonic rocks of the first magmatic cycle are seen in the Aguilar, Abra Laite and Tusaquillas subalkaline granites (Figures 2, 7 and 15), as well as in other alkaline granites such as those of Hornillos, Fundiciones and Rangel (not shown in Figure 2 due to scale limitations) (Halpern and Latorre, 1973; Galliski and Viramonte, 1988; Viramonte and Rapella, 1991; Toselli, 1992; Moya and Salfity, 1992) (Figures 2, 7 and 15).

The host rock for these plutons is mostly Ordovician sedimentary rock in the southern portions of the San Pablo and Cóndor highs. No sections are known in which these plutons intrude the Pirgua Subgroup, or in which this subgroup lies unconformably on these rocks. The basal conglomerates of the Pirgua Subgroup in the Tres Cruces depocenter contain cobbles from the underlying Ordovician formations, but no granite cobbles. Therefore it must be presumed that the plutonic rocks did not outcrop in the Tres Cruces depocenter at the time the Pirgua Subgroup was being laid down (Salfity, 1979). It is very probable that these plutons only cropped out in Late Tertiary time as a consequence of the Andean movements.

The NE-SW-striking lineaments governed both plutonic and volcanic action during the Early Cretaceous (Figure 7) (Salfity, 1985). It is thus that the Cretaceous intrusive rocks are aligned in a NE-SW direction along the Cobres lineament. In the same way, in the southern part of the basin, the Isonza and Aconquija lineaments favored the alkaline effusions of the first extrusive cycle intercalated into the Pirgua Subgroup (Figures 7 and 14). Both lineaments governed synrift sedimentary history in the El Rey, Metán and Alemanía depocenters, and the emplacement of the Salta-Jujuy high and the Pampean arch.

The Pirgua Subgroup contains two interbedded alkaline volcanic episodes, which are well represented in the south of the basin (Figures 3 and 14).

The first such episode (Alto de las Salinas and Isonza Basalts) is interbedded in the La Yesera Formation, and consists of trachytes, basanites and foidites, whose radiometric ages range between 128 and 90 Ma (Bossi and Wampler, 1969; Reyes et al., 1976; Valencio et al.,1976; Galliski and Viramonte, 1988). These volcanic rocks were the result of spot effusions which were mainly intercalated into the base of the La Yesera Formation.

The Second volcanic episode (the Las Conchas Basalt) affecting the Pirgua Subgroup is that located in the middle to upper part of the Las Curtiembres Formation (Figures 3, 8 and 14). This is not structurally related to the lineaments mentioned above but to the Traspampean arch, where this is connected with the Alemanía trough (Figure 16B). The structural controls of this volcanic action were normal faults, mainly of north-south strike, which were reactivated as thrust faults at the time of the Late Cenozoic tectonic inversion (Reyes et al., 1976; Grier et al., 1991).

The Las Conchas Basalt consists of a characteristic set of alkaline rocks, mainly basanite with accompanying mugearite and tephriphonolite, which appear in the shape of Strombolian volcanism with hydromagmatic pulses (Reyes et al., 1976; Galliski and Viramonte, 1985, 1988; Viramonte and Rapella, 1991). Records of upper mantle ultramaphic xenoliths contained in these rocks are also known (Galliski et al., 1990).

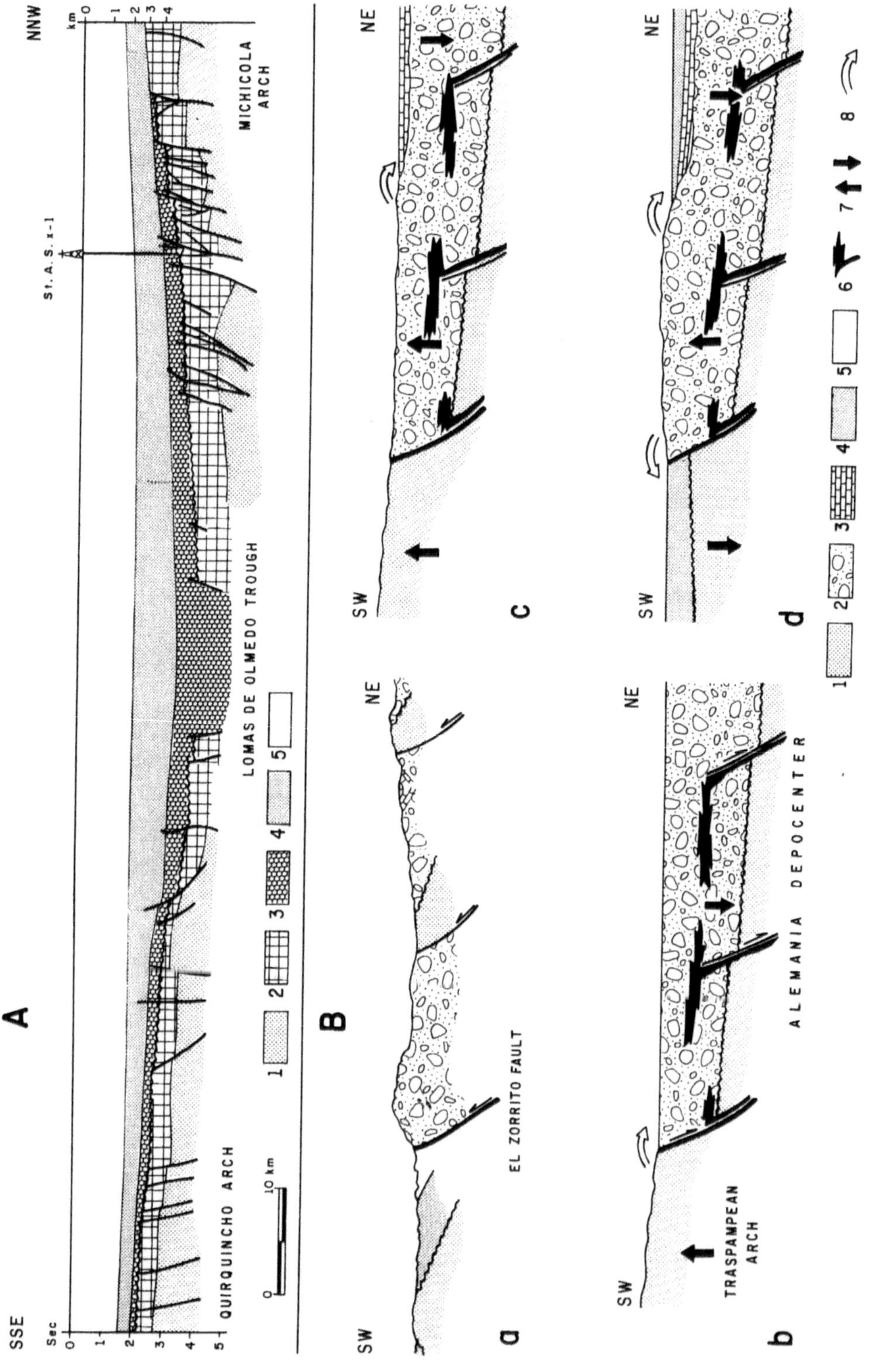
A
SSE
NNW
Sec
km
St. A. S. x-1
QUIRQUINCHO ARCH
LOMAS DE OLMEDO TROUGH
MICHICOLA ARCH
0 10 km
1 2 3 4 5
B
SW
NE
a
EL ZORRITO FAULT
b
TRASPAMPEAN ARCH
ALEMANIA DEPOCENTER
c
d
1 2 3 4 5 6 7 8

Fig. 16

◄ (A) Cross-section in the eastern side of the Lomas de Olmedo depocenter, based on and adapted from a regional seismic line, after Chiarenza and Ponzoni (1989). See location in Figure 8. 1 Crystalline basement, 2 Paleozoic, 3 Pirgua and Balbuena Subgroups, 4 Santa Bárbara Subgroup, 5 Orán Group. St.A.S. x-1: YPF Alto de la Sierra exploration well. (B) Cartoon showing an evoluting section in the southern part of the Salta Group basin, between the Traspampean arch and the Alemanía trough (after Grier et al., 1991). Formation thicknesses and horizontal distances are off scale. See location in Figure 8. 1 Precambrian basement, 2 Pirgua Subgroup, 3 Balbuena Subgroup, 4 Santa Bárbara Subgroup, 5 Late Tertiary sediments, 6 Volcanics, 7 Uplift and subsidence processes, 8 Source direction. (a) Present-day cross-section showing thrusting tectonic style. (b) Pirgua Subgroup deposition during synrift basin stage; First and Second volcanic events are shown. (c) Balbuena Subgroup deposition; note uplift of the southwest portion of the Pirgua basin. (d) Santa Bárbara Subgroup deposition; subsidence occurs both in the Traspampean arch and the Alemanía depocenter.

The radiometric age of the Second volcanic episode -78 to 76 Ma (Reyes et al., 1976; Valencio et al., 1976)- is contemporaneous with Peruvian diastrophism.

Both the First and Second volcanic episodes are characteristic of the southern stretch of the basin; that is, south of the Salta- Jujuy high. In the Lomas de Olmedo, Tres Cruces and Sey troughs no record of them has been mentioned. Mention has also been made of volcanism in the Pirgua Subgroup (Schlagintweit, 1937) in the Cerro Hermoso depocenter.

The Third volcanic episode to occur in the basin took place in the Lomas de Olmedo depocenter (Figure 10), in whose subsurface the basalt flows are intercalated in the Lecho, Yacoraite and Olmedo Formations (Mädel, 1984; Gómez Omil et al., 1989; Gait et al., 1989; Carlé et al., 1989).

The geographic location of some oil fields coincides with the volcanic manifestations referred to (Figures 10 and 13). The Palmar Largo is particularly important in this sense, because the reservoir rock is a pyroclastic facies of the outflow, which has been reworked and weathered, is very porous, and is sealed by the source rock; that is, the Yacoraite Formation (Mädel, 1984; Turic et al., 1987; Gait et al., 1989; Carlé et al., 1989). The Palmar Largo volcanic rocks have a 70±5 Ma age and are located stratigraphically between the top of the Pirgua Subgroup and the base of the Balbuena Subgroup, and even intercalated into the Lecho Formation (Carlé et al., 1989). Therefore they can clearly be correlated with the third extrusive cycle of the Salta Group, recorded in other wells drilled into the Lomas de Olmedo depocenter (Bianucci et al., 1981; Bercowski, 1982, 1987).

Outcrops of lava flows and sills of these basalts are also known on the eastern flank of the Salta-Jujuy high, intercalated into the roof of the Yacoraite Formation and the base of the Mealla Formation (Schlagintweit, 1937; Lyons, 1951). Similarly, a 65-60 Ma sill is present in the north of the Alemanía depocenter (Omarini et al., 1987).

Another volcanic manifestations of the third extrusive cycle is that of the stratified white tuff levels found in the basal portion of the Yacoraite Formation. These normally consist of a fine crystalline white tuff having ample regional distribution in the basin (Tres Cruces, Lomas de Olmedo, El Rey, Metán and Alemanía depocenters). Sometimes there are chonite or tuffite levels. The only radiometric dating available gives a 60$\pm$2 Ma age (Fernández, 1975). These tuff levels have been used as guide levels in regional correlations of the Yacoraite Formation member units (Marquillas, 1985; Marquillas and Salfity, 1989).

In the west of the basin, in Chilean territory, the Lomas Negras Formation (Marinovic and Lahsen, 1984) and the Quebrada Blanca de Poquis Formation (Gardeweg and Ramírez, 1985) crop out on the San Pablo arch, where they lie on a Paleozoic basement. Both formations can be correlated with the Yacoraite Formation (Salfity et al., 1985). The Lomas Negras Formation contains intercalated andesite lavas which can be attributed to the Third volcanic cycle.

Structural Processes in the Basin After Filling

The Salta Group basin underwent lifting and erosion on at least three occasions after being filled: (a) as a result of the Inca phase, during the Late Eocene-Early Oligocene; (b) as a result of the second pulse of the Quechua phase, in Late Miocene time; (c) as a result of the Diaguita phase, in Pleistocene time.

The Inca Phase

The interruption of the Salta Group deposition and the erosion of several levels of the Santa Bárbara Subgroup are attributed to the Inca phase movements which occurred during the Late Eocene- Early Oligocene. These movements were general throughout the whole basin, although their structural intensity was not great. For this reason the angular value of the Inca unconformity is only visible in a few sections, and in general can only be detected through regional analysis.

The Inca unconformity is very well recorded along a stretch of over 100 km on the Los Gallos uplift (Salfity, 1982), where the Salta Group basin underlies Tertiary foreland basin deposits (Figures 3, 10 and 13). The sandy red deposits of the Orán Group base are there found lying on the Lumbrera, Maíz Gordo, Mealla and Yacoraite Formations, showing the different degrees of erosion undergone by the Salta Group basin before the Orán Group deposition (Cazau et al., 1976).

Although the Orán Group sediments are more widely dispersed than those of the Salta Group in the eastern portion of the basin, the general lineaments of the Tertiary basin -at least with regard to its basal unit, the Metán Subgroup- are not much different to those of the Cretaceous basin (Russo and Serraiotto, 1979). Therefore the Inca unconformity in the east of the basin was not sufficiently intense to substantially modify the previous tectonic frame of that region (Figure 13).

Up to the present, the age of the base of the Orán Group is not known (Russo and Serraiotto, 1979). It has been proposed that the sedimentary history of the Orán Group began immediately after the lifting and erosion of the Inca phase; that is, during the Middle-Late Eocene (Gebhard et al., 1974; Salfity, 1982). Because of this, the hiatus made by the Inca phase would have been very brief. However, other authors consider sedimentation to have occurred later, in connection with the second pulse of the Inca phase; that is, during the Late Eocene-Early Oligocene (Jordan and Alonso, 1987). New regional correlations presume an even younger age for the start of sedimentation, approximately during the time the Pehuenche phase occurred (Late Oligocene) (Salfity and Gorustovich, 1992) (Figures 3 and 13).

There are fossil records in the basal portion of the Pastos Grandes Group in the Arizaro basin in the Puna (Pascual, 1983; Alonso, 1986) (Figure 3). The Arizaro basin developed over the Traspampean arch, and therefore outside the Salta Group basin (compare Figures 5 and 13). According to the paleontologic data available, the age of the base of the Pastos Grandes Group may be similar to that of its equivalents in the Subandean basin, which would mean its deposition commenced during the Pehuenche phase (Figure 3).

The basal portion of the Tertiary sequences in the Puna and in the whole of northern Argentina was free of contemporary volcanic outflows. This means there was is a characteristic of the Inca phase in northwest Argentina (Moya and Salfity, 1982).

Figure 13 is symbolic of the paleogeographic situation of the region prior to the emplacement of the Late Miocene-Pleistocene magmatic arc. The host rock of this intracontinental arc would have consisted of Paleozoic or Precambrian rocks forming the Andean arch (Figure 13).

The Andean arc corresponds roughly to the Huaitiquina high and the western flanks of the San Pablo and Traspampean arches which were in place during the Cretaceous (compare Figures 5 and 13). During the Oligocene-Middle Miocene the Andean arch was the structural high which separated the intermontane Neogene Olaroz- Arizaro troughs from the Atacama trough.

In this work it is proposed that the Miocene-Pleistocene volcanic arc of the northern Argentine Andes was emplaced along the structural axis of the Andean arch, in a distensive tectonic setting consisting of Paleozoic or Precambrian host rock.

The magmatic arc was structurally isolated from the Cenozoic troughs of the Puna, but was associated with the Andean arch and with the transverse Paleozoic or Precambrian horsts which formed part of the intermontane basin borders (Salfity et al., 1984; Salfity, 1985). Therefore neither the Cretaceous sequences nor the post-Inca basal Tertiary

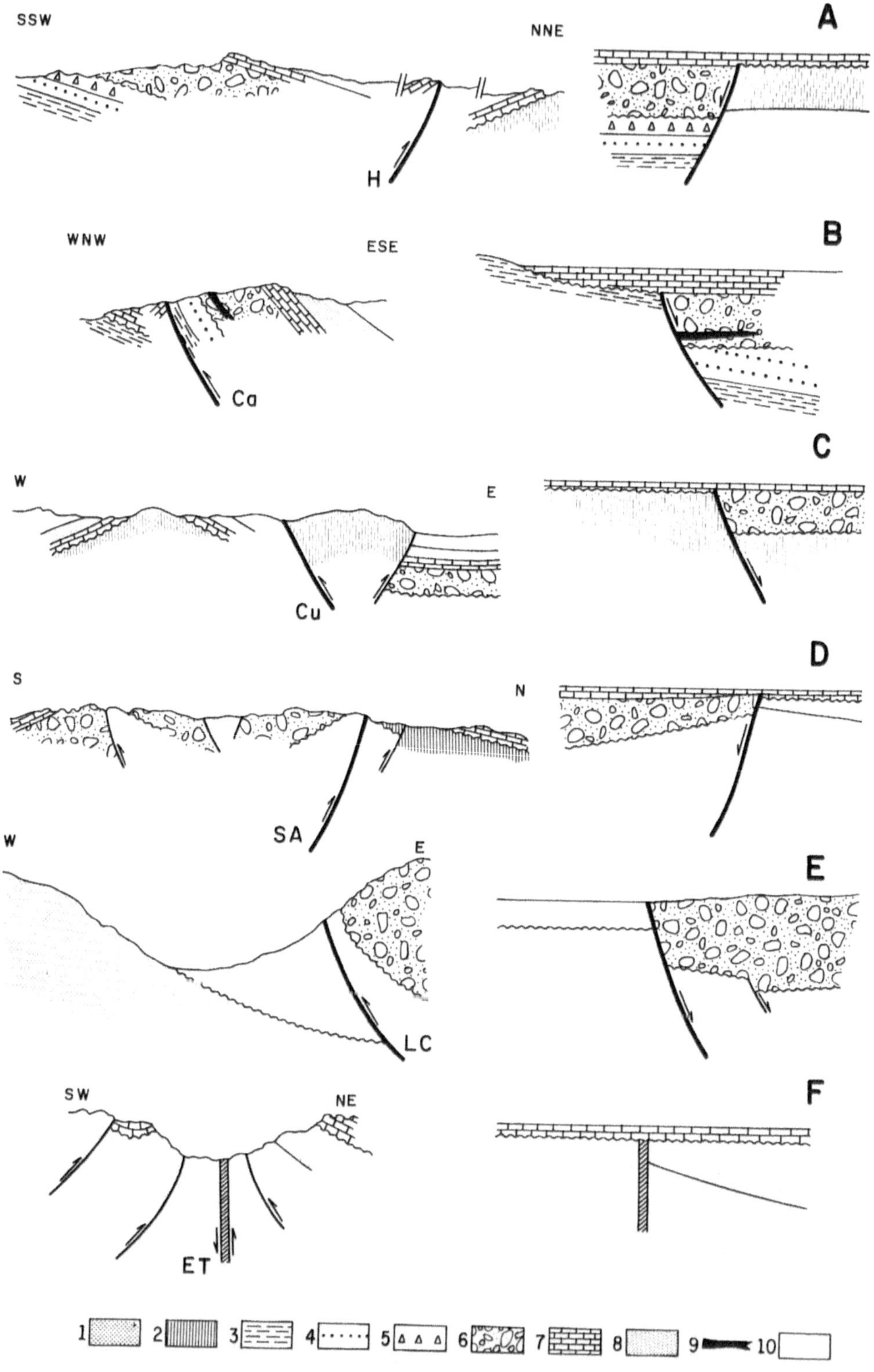
A
SSW
NNE
H
B
WNW
ESE
Ca
C
W
E
Cu
D
S
N
SA
E
W
E
LC
F
SW
NE
ET
1
2
3
4
5
6
7
8
9
10

Fig. 17

◄ Cartoon showing evoluting sections in the Salta Group basin. Formation thicknesses and horizontal distances are off scale. See location of sections in Figure 8. Faults: H-Horconal, Ca-Cachipunco, Cu-Cauchari, SA-San Agustín, LC-Las Chacras; ET-El Toro lineament. 1 Precambrian, 2 Ordovician, 3 Silurian, 4 Devonian, 5 Carboniferous, 6 Pirgua Subgroup, 7 Balbuena Subgroup (mainly Yacoraite Formation), 8 Santa Bárbara Subgroup, 9 Volcanics, 10 Late Cenozoic sediments. The drawings on the left of the Figure show the present structural layout of the Salta Group, its basement, and, in some cases, its Late Cenozoic cover. The drawings on the right show a paleogeographic reconstruction, levelled at the roof of the Yacoraite Formation (except for section E, levelled at the top of the Santa Bárbara Subgroup). Sections A, B, C, D and E show Cretaceous tension faults, reactivated as thrusts faults as a result of the tectonic inversion of the basin during the Late Tertiary. (A) Horconal Fault - Pirgua Subgroup over Carboniferous and Yacoraite Formation over Ordovician, according to data supplied by Amengual and Zanettini (1973). (B) Cachipunco Fault - Pirgua Subgroup over Devonian and Yacoraite Formation over Silurian, according to data supplied by Hagerman (1933) and Arias et al. (1980). (C) Cauchari Fault - Pirgua Subgroup and Yacoraite Formation over Ordovician, according to data supplied by Amengual et al. (1979), Donato (1987) and Donato and Vergani (1987). (D) San Agustín Fault - Pirgua Subgroup over Precambrian and Yacoraite Formation over Ordovician, according to data supplied by Ortiz (1962) and Vergani and Starck (1989). (E) Las Chacras Fault - Pirgua Subgroup over Precambrian and Santa Bárbara Subgroup over Precambrian, according to data supplied by Grier (1990) and Grier et al. (1991). (F) The El Toro lineament in the southwest sector of the Salta-Jujuy high; sealed by the postrif deposits of the Yacoraite Formation, according to data supplied by Vilela (1956).

sequences of the Uyuni, Olaroz, Arizaro and Antofalla basins (Figure 13) were the host rocks of the Miocene-Pleistocene magmatic arc.

The Quechua II Phase

This tectonic inversion stage of the Salta Group basin arose from the folding and erosion of its formations at the end of the Miocene (the second pulse of the Quechua phase).
The Piquete Formation (Late Miocene-Pliocene, Figure 3) conglomerates contain clasts of the Yacoraite Formation limestones and of other formations from the Salta Group (Russo and Serraiotto, 1979). The conglomerates and associated sediments of the Piquete Formation reveal that the basin in which they were formed was of taphrogenic type, with provenance from the inverse Cretaceous basin itself and from the Precambrian-Paleozoic basement. The age of the Piquete Formation is Late Miocene-Pliocene.
It is possible that the first pulse of the Quechua phase also had compressive effects, and that it therefore folded the Salta Group before the Jujuy Subgroup was deposited (Figure 3). It has been seen that some anticline structures could have been formed before the Quechua II and Diaguita movements (Jordan and Alonso, 1987). However, it has not proved possible to determine the age of these folds, which may be even older than estimated; that is, contemporaneous with the Inca or Pehuenche tectonic phases.

Some Salta Group sections are known, in particular those outside the rift troughs -that is not including the Pirgua Subgroup- in which the Balbuena and Santa Bárbara Subgroups are faulted or folded (Figure 17). The folds are invariably related genetically to reverse faults -some of them reactivated- which lay the Paleozoic or Precambrian basement over the Salta Group sequences. The post-Inca Tertiary was not deposited on some of these structures, probably because they acted as positive elements during the Late Tertiary. In other cases the post-Inca Tertiary rests unconformably on the Salta Group.

The Diaguita Phase

The definitive inversion of the Salta Group basin took place with the compressive Diaguita phase, the most important of the Andean movements. These movements folded and faulted the Cretaceous sedimentary column as well as the Cenozoic, in many cases reactivating the pre-Cretaceous basement.

Their effects reached eastward approximately as far as the 64th meridian (Figure 2), to the east of which the Cretaceous basin remains buried under a 3 km average thickness of the Orán Group sequences (Figure 13).

The Diaguita phase caused certain stretches of the western part of the Salta-Jujuy high to lie now at an altitude of 4500 masl; where the Yacoraite Formation rests on a Precambrian basement. Meanwhile, in in the subsurface to the northeast of the Salta- Jujuy high some wells reached the top of the Yacoraite Formation at a burial depth of 6100 meters.

One of the most singular aspects of the Diaguita tectonics which affected the Cretaceous basin, its basement and its Late Tertiary overburden, consisted of the renewed reactivation of the faults which bounded the ancient Cretaceous rift, and which had commenced their compressive action during the Quechua phase.

Some of these faults are classic examples of the Andean reactivation of the Cretaceous rift, such as the El Zorrito fault (Grier, 1990; Grier et al., 1991) (Figure 16B), which bounded the Traspampean arch with the Alemanía depocenter in the south of the basin.

Other examples (Figure 17) are the Horconal fault (Amengual and Zanettini, 1973) at the south of the Cóndor arch; the Cachipunco fault (Hagerman, 1933; Arias et al., 1980) on the eastern border of the Salta-Jujuy high; the Cauchari fault (Amengual et al., 1979); Donato, 1987) on the Huaitiquina high; the San Agustín fault (Ortiz, 1962; Vergani and Starck, 1989) in the south of the Salta-Jujuy high; finally, the Las Chacras fault (Grier, 1990) on the edge of the Traspampean arch in the south of the basin.

Figure 17F shows a Neogene trough located on the south of the Salta-Jujuy high, along the El Toro lineament, which was intensely deformed by the Diaguita phase. However, this lineament was inactive throughout the Cretaceous, and was therefore sealed by Yacoraite Formation deposits. The Yacoraite Formation was also deformed by the Diaguita phase movements.

Acknowledgment

The authors are grateful to Mr. Peter J. Edmunds for translating the Spanish version of this paper into English, and to Mr. Omar Enrique López for drawing the illustrations. Funds for this research were provided by Conicet (National Research Council of Argentina).

References

Ahr, W.M. (1973): The carbonate ramp: an alternative to the shelf model. Gulf Coast Association of Geological Societies, Transactions, v. 23, 221-225.

Alonso, R. N. (1986): Ocurrencia, posición estratigráfica y génesis de los depósitos de boratos de la Puna Argentina. Dr Thesis, Universidad Nacional de Salta, Salta, 196 pp.

Amengual, R., and Zanettini, J.C. (1973): Geología de la comarca de Cianzo y Caspalá (Provincia de Jujuy). Asociación Geológica Argentina, Revista, v. 28, 341-352, Buenos Aires.

Amengual, R., Méndez, V., Navarini, A., Viera, O., and Zanettini, J.C. (1979): Geología de la Región Noroeste, República Argentina, Provincias de Salta y Jujuy, Escala 1:400.000. Buenos Aires, Dirección General de Fabricaciones Militares, two sheets.

Arias, J.E., Chalabe, E., Moreno Espelta, C., and Chávez, A. (1980): Geología del área termal de Cachipunco, departamento Santa Bárbara, provincia de Jujuy. Asociación Geológica Argentina, Revista, v. 35, 557-569, Buenos Aires.

Avila-Salinas, W. (1989): Fases diastróficas y magmatismo jurásico-cretácico en Bolivia. Servicio Geológico de Bolivia, Boletín, Serie A, v. 6, 47-68, La Paz.

Báez, A.M. (1981): Redescription and relationships of *Saltenia ibanezi*, a late Cretaceous pipid frog from northwestern Argentina. Ameghiniana, v. 18, 127-154, Buenos Aires.

Bercowski, F. (1982): Manifestaciones volcánicas en la Formación Yacoraite, Cretácico Superior de la Cuenca del Noroeste, Argentina. Congreso Latinoamericano de Geología, 5th, Resúmenes, p. 122, Buenos Aires.

Bercowski, F. (1987): Colada basáltica en la Formación Yacoraite (Cretácico Superior), pozo Caimancito, Norte Argentino. Simposio PIGC-Proyecto 242 Cretácico de América Latina, 3rd, Resúmenes, Tucumán.

Bianucci, H.A., and Homovc, J.F. (1982): Tectogénesis de un sector de la cuenca del Subgrupo Pirgua. Congreso Latinoamericano de Geología, 5th, Buenos Aires, Actas, 1, 539-546.

Bianucci, H.A., Acevedo, O., and Cerdán, J. (1981): Evolución tectosedimentaria del Grupo Salta en la subcuenca Lomas de Olmedo (provincias de Salta y Formosa). Congreso Geológico Argentino, 8th, San Luis, Actas, 3, 159-172.

Bianucci, H.A., Acevedo, O., and Vásquez, J.R. (1984): La tectónica Varíscica en el noroeste argentino y sus estilos estructurales. Congreso Geológico Argentino, 9th, Bariloche, Actas, 2, 25-30.

Bogdanic, T. (1990): Kontinentale Sedimentation der Kreide und des Alttertiärs im Umfeld des subduktionsbedingten Magmatismus in der chilenischen Präkordillere (21°-23° S). Berliner geowissenschaftliche Abhandlungen, A, 123, 117 pp.

Bonaparte, J.F. (1984): El intercambio faunístico de vertebrados continentales entre América del Sur y del Norte a fines del Cretácico. Congreso Latinoamericano de Paleontología, 3rd, Oaxtepec, Memoria, 438-450.

Bonaparte, J.F., and Bossi, G.E. (1968): Sobre la presencia de dinosaurios en la Formación Pirgua del Grupo Salta y su significado cronológico. Acta Geológica Lilloana, v. 9, 25-44, Tucumán.

Bonaparte, J.F., and Powell, J.E. (1980): A continental assamblage of tetrapods from the Upper Cretaceous beds of El Brete, northwestern Argentina (Sauropoda-Coelurosauria-Carnosauria-Aves). Société Géologique de France, Mémoir, N.S., 139, 19-28.

Bonarelli, G. (1913): Las Sierras Subandinas del Alto y Aguaragüe y los yacimientos petrolíferos del distrito minero de Tartagal, departamento de Orán, provincia de Salta. Ministerio de Agricultura, Sección Geología, Mineralogía y Minas, Buenos Aires, Anales VIII, no 4, 50 p.

Bonarelli, G. (1921): Tercera contribución al conocimiento geológico de las regiones petrolíferas subandinas del norte (provincias de Salta y Jujuy). Ministerio de Agricultura, Sección Geología, Mineralogía y Minas, Buenos Aires, Anales XV, no 1, 96 p.

Boso, M.A., Brandán, E.M., and Salfity, J.A. (1984): Estratigrafía y paleoambientes del Subgrupo Pirgua (Cretácico) en la comarca de Brealito, provincia de Salta. Congreso Geológico Argentino, 9th, Bariloche, Actas, 1, 108-123.
Bossi, G.E., and Wampler, M. (1969): Edad del Complejo Alto de las Salinas y Formación El Cadillal según el método K-Ar. Acta Geológica Lilloana, v. 10, 141-160, Tucumán.
Brackebusch, L. (1883): Estudio sobre la Formación Petrolífera de Jujuy. Academia Nacional de Ciencias, Boletín, v. 5, 137-184.
Brackebusch, L. (1891): Mapa Geológico del interior de la República Argentina, escala 1:1.000.000. Gotha.
Camacho, H. H. (1967): Las transgresiones del Cretácico Superior y Terciario de la Argentina. Asociación Geológica Argentina, Revista, v. 22, 253-280, Buenos Aires.
Carbajal, E., Pascual, R., Pinedo, R., Salfity, J. A., and Vucetich, M. G. (1977): Un nuevo mamífero de la Formación Lumbrera (Grupo Salta, Argentina). Edad y correlaciones. Museo Municipal de Ciencias Naturales de Mar del Plata "L. Scaglia", Publicación, v. 2, 148-163, Mar del Plata.
Carlé, R.J., Di Persia, O.E., and Olivieri, G.A. (1989): Análisis geológico y petrolero del sector noroeste de la provincia de Formosa, República Argentina. Congreso Nacional de Exploración de Hidrocarburos, 1st, Mar del Plata, 1, 163-185.
Cazau, L., Cellini, N., and Oliver G., J. (1976): El Subgrupo Santa Bárbara (Grupo Salta) en la porción oriental de las provincias de Salta y Jujuy. Congreso Geológico Argentino, 6th, Bahía Blanca (1975), Actas, 1, 341-355.
Chiarenza, D.G., and Ponzoni, E. (1989): Contribución al conocimiento geológico de la Cuenca Cretácica en el ámbito oriental de la subcuenca de Olmedo, provincia de Salta, República Argentina. Congreso Nacional de Exploración de Hidrocarburos, 1st, Mar del Plata, 1, 209-228.
Cherroni, C. (1977): El Sistema Cretácico en la parte boliviana de la cuenca andina. Revista Técnica de Yacimientos Petrolíferos Fiscales Bolivianos, v. 6, 5-46, La Paz.
Clebsch, C. (1991): The geologic evolution of the Paraguayan Chaco. Ph.D. Dissertation, Texas Tech University, Lubbock, 185 pp.
del Papa, C.E. (1992): Ciclos de somerización en la secuencia Maíz Gordo (Terciario Inferior), noroeste argentino. Reunión Argentina de Sedimentología, 4th, La Plata, Actas, in press.
del Papa, C.E., and Marquillas, R.A. (1990): Interpretación ambiental de la Formación Maíz Gordo (Terciario) en el arroyo Chamical, Provincia de Salta. Congreso Geológico Argentino, 11th, San Juan, Actas, 2, 73-76.
Donato, E. (1987): Características estructurales del sector occidental de la Puna Salteña. Yacimientos Petrolíferos Fiscales, Boletín de Informaciones Petroleras, Tercera Epoca, v. IV, 12, 89-98, Buenos Aires.
Donato, E., and Vergani, G. (1987): Estratigrafía de la Formación Yacoraite (Cretácico) en Paso Huaytiquina, Salta, Argentina. Congreso Geológico Argentino, 10th, Tucumán, Actas, 1, 263-266.
Eugster, H.P. (1985): Oil shales, evaporites and ore deposits. Geochimica et Cosmochimica Acta, v. 49, 619-635.
Fernández, J. (1975): Peces pulmonados fósiles en la Puna jujeña. Sociedad Científica Argentina, Anales, Serie II, 41, 13-18, Buenos Aires.
Frenguelli, G. (1930): Geosinclinali continentali. Bolletino della Societa Geologica Italiana, v. 49, 1-24.
Frenguelli, J. (1936): Investigaciones geológicas en la zona salteña del valle de Santa María. Universidad Nacional de La Plata, Instituto Museo, Obra del Cincuentenario, 2, 215- 572, La Plata.
Gait, M.A., Cesano, S., Schachner, G., and Jiménez, E. (1989): Vulcanitas de Palmar Largo en el subuselo de Formosa. Yacimientos Petrolíferos Fiscales, Boletín de Informaciones Petroleras, Tercera Epoca, v. VI, 17, 38-56, Buenos Aires.
Galli, C.I., and R.A. Marquillas (1990): Análisis ambiental de las Formaciones Los Blanquitos y Lecho (Cretácico Superior, Grupo Salta) en Cabra Corral, provincia de Salta. Congreso Geológico Argentino, 11th, San Juan, Actas, 2, 165-168.
Galliski, M.A., and Viramonte, J.G. (1985) Un paleorift cretácico en el noroeste argentino. Universidad de Chile, Departamento de Geología y Geofísica, Comunicaciones, 35, 89-91, Santiago.
Galliski, M.A., and Viramonte, J.G. (1988): The Cretaceous paleorift in north-western Argentina: A petrologic approach. Journal of South American Earth Sciences, v. 1, 329- 342.
Galliski, M.A., Aparicio Y., A., Risso, C., Viramonte, J.G., and Araña S., V. (1990): Contribución a la petrología y geoquímica de los xenolitos y basaltos alcalinos cretácicos de la quebrada de Las Conchas, provincia de Salta, República Argentina. Asociación Argentina de Mineralogía, Petrología y Sedimentología, v. 20, 71-87, Buenos Aires.

Gardeweg, M., and Ramírez, C.F. (1985): Hoja Río Zapaleri, II Región de Antofagasta. Servicio Nacional de Geología y Minería, Carta Geológica de Chile, 66, 122 pp., Santiago.

Gebhard, J.A., Giudice, A.R., and Oliver G., J. (1974): Geología de la comarca entre el río Juramento y arroyo Las Tortugas, provincias de Salta y Jujuy, República Argentina. Asociación Geológica Argentina, Revista, v. 39, 359- 375.

Ginsburg, R.N., and James, N.P. (1974): Holocene carbone sediments of continental shelves, *in* The Geology of Continental Margins, (Eds.: Burke, C.A., and Drake, C.L.), New York, Springer-Verlag, 137-155.

Gómez Omil, R.J., Boll, A., and Hernández, R.M. (1989): Cuenca cretácico-terciaria del Noroeste argentino (Grupo Salta), *in* Cuencas Sedimentarias Argentinas, (Eds.: Chebli, G.A., and Spalletti, L.A.), Universidad Nacional de Tucumán, Serie Correlación Geológica, 6, 43-64, Tucumán.

Gorustovich, S.A. (1988): Metalogénesis del uranio en el noroeste de la República Argentina. Dr Thesis, Universidad Nacional de Salta, Salta, 178 pp.

Gorustovich, S.A., Marquillas, R.A., and Salfity, J.A. (1989) Depósitos estratoligados de uranio en el Cretácico Superior del norte argentino, *in* Contribuciones de los Simposios sobre el Cretácico de América Latina, B: Depósitos Minerales del Cretácico de América Latina (Eds.: Brodtkorb, M.K. de, and Schalamuk, I.B.), GSGP-IGCP 242, 61-78, Buenos Aires.

Grier, M.E. (1990): The Influence of the Cretaceous Salta Rift Basin in the Development of Andean Structural Geometries. Ph. D. Thesis, Cornell Universitity, Ithaca, 178 pp.

Grier, M.E., Salfity, J.A, and Allmendinger, R.W. (1991): Andean reactivation of the Cretaceous Salta rift, northwestern Argentina. Journal of South American Earth Sciences, v. 4, 351-372.

Groeber, P. (1953): Andico, *in* Geografía de la República Argentina. Mesozoico, (Groeber, P., Stipanicic, P.N., and Mingramm, A.R.G.), Buenos Aires, Sociedad Argentina de Estudios Geográficos, GAEA, 2, Part 1 (1952), 349-541.

Hagerman, T. (1933): Informe preliminar sobre el levantamiento geológico del Departamento Santa Bárbara en la provincia de Jujuy. Yacimientos Petrolíferos Fiscales, Boletín de Informaciones Petroleras, v. 10, 107, 3-35, Buenos Aires.

Halpern, M., and Latorre, C.O. (1973): Estudio geocronológico inicial de las rocas del noroeste de la República Argentina. Asociación Geológica Argentina, Revista, v. 28, 195-205.

Heckel, P.H. (1972): Recognition of Ancient Shallow Marine Environments, *in* Recognition of Ancient Sedimentary Environments, (Eds.: Rigby, J.K., and Hamblin, W.K.), Society of Economic Paleontologists and Mineralogists, Special Publication 16, 226-286.

Hubert, J.F., and Stevens, R.L. (1980): Alluvial fans, braided rivers and lakes in fault bounded semiarid valley: Sugarloaf Arcose, Newark Supergroup, Deerfield Basin, Massachussets. Northeastern Geology, v. 2, 100-117.

Irwin, M.L. (1965): General theory of epeiric clear water sedimentation. American Association of Petroleum Geologists Bulletin, v. 49, 445-459.

James, N.P. (1984a): Introduction to carbonate facies models, *in* Facies Models, 2nd Edition, (Ed.: Walker, R.G.), Geoscience Canada Reprint Series, 1, 209-211.

James, N.P. (1984b): Shallowing-upward sequences in carbonates, *in* Facies Models, 2nd Edition, (Ed.: Walker, R.G.), Geoscience Canada Reprint Series, 1, 213-228.

Jordan, T.E., and Alonso, R.N. (1987): Cenozoic Stratigraphy and Basin Tectonics of the Andes Mountains, 20°-28° South Latitude. American Association of Petroleum Geologists Bulletin, v. 71, 49-64.

Lencinas, A.N., and Salfity, J.A. (1973): Algunas características de la Formación Yacoraite en el oeste de la cuenca andina, provincias de Salta y Jujuy, República Argentina. Congreso Geológico Argentino, 5th, Carlos Paz (1972), Actas, 3, 253-267.

Lyons, W.A. (1951): El filón-capa basáltico de la Formación Petrolífera de Jujuy. Asociación Geológica Argentina, Revista, v. 6, 114-132, Buenos Aires.

Macellari, C. (1988): Cretaceous paleogeography and depositional cycles of western South America. Journal of South American Earth Sciences, v. 1, 373-418.

Mädel, F. (1984): Estratigrafía del tramo inferior del pozo descubridor Palmar Largo x-1. Yacimientos Petrolíferos Fiscales, Boletín de Informaciones Petroleras, Tercera Epoca, v. 1, 2, p. 109, Buenos Aires.

Marinovic, N., and Lahsen, A. (1984): Hoja Calama, Región de Antofagasta. Servicio Nacional de Geología y Minería, Carta Geológica de Chile, 58, 144 pp., Santiago.

Marquillas, R.A. (1984): La Formación Yacoraite (Cretácico superior) en el Río Juramento, Salta: estratigrafía y ciclicidad. Congreso Geológico Argentino, 9th, Bariloche, Actas, 5, 186-196.

Marquillas, R.A. (1985): Estratigrafía, sedimentología y paleoambientes de la Formación Yacoraite (Cretácico superior) en el tramo austral de la cuenca, Norte argentino. Dr Thesis, Universidad Nacional de Salta, Salta, 139 pp.

Marquillas, R.A. (1986): Ambiente de depósito de la Formación Yacoraite (Grupo Salta, Cretácico-Eoceno), Norte argentino. Simposio PIGC-Proyecto 242 Cretácico de América Latina, 1st, La Paz, 153-169.

Marquillas, R.A., and Salfity, J.A. (1988): Tectonic framework and correlations of the Cretaceous-Eocene Salta Group, Argentine, *in* The Southern Central Andes, (Eds.: Bahlburg, H., Breitkreutz, Ch., and Giese, P.), Berlin- Heidelberg, Springer-Verlag, Lecture Notes in Earth Sciences, v. 17, 119-136.

Marquillas, R.A., and Salfity, J.A. (1989): Dispersión regional de los miembros de la Formación Yacoraite (Cretácico Superior) en el norte argentino, *in* Contribuciones de los Simposios sobre el Cretácico de América Latina, A: Eventos y Registro Sedimentario, (Ed.: Spalletti, L.A.), GSGP-IGCP 242, 253-272, Buenos Aires.

Marquillas, R.A., and Salfity, J.A. (1990): Las facies calcáreas cretácicas del cordón de Los Siete Hermanos, provincia de Jujuy, República Argentina. Congreso Geológico Argentino, 11th, San Juan, Actas, 2, 117-120.

Marquillas, R.A., and Salfity, J.A. (1992): Las relaciones estratigráficas del techo de la Formación Yacoraite (Maastrichtiano-Paleoceno) en el norte argentino. IGCP Projects 301 & 308 Meeting, Abstracts, Punta Arenas.

Marquillas, R.A., Boso, M.A., and Salfity, J.A. (1984): La Formación Yacoraite (Cretácico superior) en el Norte argentino, al sur del paralelo 24°. Congreso Geológico Argentino, 9th, Bariloche, Actas, 2, 300-310.

McKee, E.D., Crosby, E.J., and Berryhill, H.L. (1967): Flood deposits, Bijou Creek, Colorado. Journal of Sedimentary Petrology, v. 37, 829-851.

Méndez, V., Navarini, A., Plaza, D., and Viera, V. (1973): Faja Eruptiva de la Puna Oriental. Congreso Geológico Argentino, 5th, Carlos Paz (1972), Actas, 4, 89-100.

Miall, A.D. (1977): A review of the braided-river depositional environment. Earth-Science Reviews, v. 13, 1-62.

Mingramm, A., and Russo, A. (1972): Sierras Subandinas y Chaco Salteño, *in* Geología Regional Argentina, Primer Simposio (1969), (Dir. and Ed.: Leanza, A.F.), Córdoba, Academia Nacional de Ciencias, 185-221.

Mon, R., and Hongn, F. (1991) The structure of the Precambrian and Lower Paleozoic Basement of the Central Andes between 22° and 32°S.Lat. Geologische Rundschau, v. 80, 745-758.

Moreno, J.A. (1970): Estratigrafía y paleogeografía del Cretácico superior en la cuenca del noroeste argentino, con especial mención de los Subgrupos Balbuena y Santa Bárbara. Asociación Geológica Argentina, Revista, v. 24, 9-44.

Moya, M.C., and Salfity, J.A. (1982): Los ciclos magmáticos en el noroeste Argentino. Congreso Latinoamericano de Geología, 5th, Buenos Aires, Actas, 3, 523-536.

Moya, M.C., and Salfity, J.A. (in press): Herencia de estructuras eopaleozoicas en la evolución de las cuencas cretácicas y terciarias de la Cordillera Oriental argentina. Real Sociedad Española de Historia Natural, Geología, Boletín.

Omarini, R.H., Salfity, J.A., Linares, E., Viramonte, J.G., and Gorustovich, S.A. (1987): Petrología, geoquímica y edad de un filón lamproítico en el Subgrupo Pirgua (Alemanía-Salta). Universidad Nacional de Jujuy, Instituto de Ciencias Geológicas, Revista, 7, 89-99, Jujuy.

Ortiz, A. (1962): Estudio geológico de las Sierras de Castillejo, Departamento Capital, Salta. Dr Thesis, Universidad Nacional de Tucumán, Facultad de Ciencias Naturales de Salta, Salta, 60 pp.

Ouchi, S. (1985): Response of alluvial rivers to slow active tectonic movement. Geological Society of America Bulletin, v. 96,

Padula, E.L, and Mingramm, A.R.G. (1968): Estratigrafía, distribución y cuadro geotectónico-sedimentario del "Triásico" en el subsuelo de la llanura Chaco-Paranense. Jornadas Geológicas Argentinas, 3rd, Comodoro Rivadavia (1966), Actas, 1, 291-331.

Padula, E.L., Rolleri, E.O., Mingramm, A.R.G., Criado Roque, P., Flores, M.A., and Baldis, B.A. (1967): Devonian of Argentina. International Symposium on the Devonian System, Calgary, Proceedings, v. 2, 165-199.

Pascual, R. (1983): Novedosos marsupiales paleógenos de la Formación Pozuelos (Grupo Pastos Grandes) de la Puna, Salta, Argentina. Ameghiniana, v. 20, 265-280, Buenos Aires.

Pascual, R. (1984): La sucesión de las edades- mamífero, de los climas y del diastrofismo sudamericanos durante el Cenozoico: fenómenos concurrentes. Academia Nacional de Ciencias Exactas, Físicas y Naturales, Anales, Buenos Aires, v. 36, 15-36.

Pascual, R., Bond, M., and Vucetich, M.G. (1981): El Subgrupo Santa Bárbara (Grupo Salta) y sus vertebrados. Cronología, Paleoambientes y Paleobiogeografía. Congreso Geológico Argentino, 8th, San Luis, Actas, 3, 743-758.

Quattrocchio, M.E., and Volkheimer, W. (1990): Paleogene paleoenvironmental trends as reflected by palynological assemblage types, Salta Basin, NW Argentina. Neues Jahrbuch für Geologie und Paläontologie Abhandlungen, 181, 377-396.

Quattrocchio, M., R.A. Marquillas, and Volkheimer, W. (1988): Palinología, paleoambientes y edad de la Formación Tunal, cuenca del Grupo Salta (Cretácico-Eoceno), República Argentina. IV Congreso Argentino Paleontología y Bioestratigrafía, 4th, Mendoza (1986), Actas, 3, 95-107.

Read, J.F. (1982): Carbonate platforms of Passive (extensional) continental margins: types, characteristics and evolution. Tectonophysics, v. 81, 195-212.

Read, J.F. (1985) Carbonate Platform Facies Models. American Association of Petroleum Geologists Bulletin, v. 69, 1- 21.

Reyes, F.C. (1972a): Correlaciones en el Cretácico de la cuenca andina de Bolivia, Perú y Chile. Revista Técnica de Yacimientos Petrolíferos Fiscales Bolivianos, v. 1, 101-144, La Paz.

Reyes, F.C. (1972b): On the Carboniferous and Permian of Bolivia and Northwestern Argentina. Anais da Academia Brasileira de Ciências, v. 44 (Suplemento), 261-277, São Paulo.

Reyes, F.C. (1974): Consideraciones sobre el Cretácico de la Cuenca Subandina de Bolivia. Revista Brasileira de Geociências, v. 4, 95-113, São Paulo.

Reyes, F.C. (1978): Algunas consideraciones sobre la posible edad geológica del cambio de pendiente regional en las Sierras Subandinas del Noroeste Argentino y Sudeste de Bolivia. Universidad Nacional de Jujuy, Instituto de Ciencias Geológicas, Revista, 3, 7-25, Jujuy.

Reyes, F.C., and Salfity, J.A. (1973): Consideraciones sobre la estratigrafía del Cretácico (Subgrupo Pirgua) del noroeste argentino. Congreso Geológico Argentino, 5th, Carlos Paz (1972), Actas, 3, 355-385.

Reyes, F.C., Salfity, J.A., Viramonte, J.G., and Gutierrez, W. (1976): Consideraciones sobre el vulcanismo del Subgrupo Pirgua (Cretácico) en el norte argentino. Congreso Geológico Argentino, 6th, Bahía Blanca (1975), Actas, 1, 205-223.

Riccardi, A.C. (1988): The Cretaceous System of Southern South America. Geological Society of America Memoir, 168, 161 pp.

Russo, A., and Serraiotto, A. (1979): Contribución al conocimiento de la estratigrafía terciaria en el noroeste argentino. Congreso Geológico Argentino, 7th, Neuquén (1978), Actas, 1, 715-730.

Russo, A., Ferello, R., and Chebli, G. (1979): Llanura Chaco-Pampeana, *in* Geología Regional Argentina, Segundo Simposio (1976), (Coord.: Turner, J.C.M.), Córdoba, Academia Nacional de Ciencias, 1, 140-183.

Salfity, J.A. (1979): Paleogeología de la cuenca del Grupo Salta (Cretácico-Eogénico) del norte de Argentina. Congreso Geológico Argentino, 7th, Neuquén (1978), Actas, 1, 505- 515.

Salfity, J.A. (1980): Estratigrafía de la Formación Lecho (Cretácico) en la Cuenca Andina del Norte Argentino. Universidad Nacional de Salta, Publicación Especial, Tesis 1, 91 pp., Salta.

Salfity, J.A. (1982): Evolución paleogeográfica del Grupo Salta (Cretácico-Eogénico), Argentina. Congreso Latinoamericano de Geología, 5th, Buenos Aires, Actas, 1, 11-26.

Salfity, J.A. (1985): Lineamientos transversales al rumbo andino en el noroeste argentino. Congreso Geológico Chileno, 4th, Antofagasta, 2 (2), 119-237.

Salfity, J.A., and Gorustovich, S.A. (1992): Cenozoic correlations in the Argentine Puna. Geowissenschaftliches Lateinamerika Kolloquium, 13th, Münster, Abstracts.

Salfity, J.A., and Marquillas, R.A. (1981): Las unidades estratigráficas cretácicas del Norte de la Argentina, *in* Cuencas Sedimentarias del Jurásico y Cretácico de América del Sur, (Eds.: Volkheimer, W., and Musacchio, E.), 1, 303-317, Buenos Aires.

Salfity, J.A., and Marquillas, R.A. (1989): El basamento precretácico de la Argentina septentrional y regiones adyacentes. Universidad de Chile, Departamento de Geología y Geofísica, Comunicaciones, 40, 17-28, Santiago.

Salfity, J.A., and Zambrano, J.J. (1990): Cretácico, *in* Geología de América del Sur (Eds.: J.F. Bonaparte, A.J. Toselli and F.G. Aceñolaza), Universidad Nacional de Tucumán, Serie Correlación Geológica, 2, III, 185-284, Tucumán.

Salfity, J.A., Gorustovich, S.A., Moya, M.C., and Amengual, R. (1984): Marco tectónico de la sedimentación y efusividad cenozoicas en la Puna argentina. Congreso Geológico Argentino, 9th, Bariloche, Actas, 1, 539-554.

Salfity, J.A., Marquillas, R.A., Gardeweg, M., Ramírez, C., and Davidson, J. (1985): Correlaciones en el Cretácico Superior del norte de Argentina y Chile. Congreso Geológico Chileno, 4th, Antofagasta, Actas, 4 (1), 654-667.

Salfity, J.A., C.L. Azcuy, O. López G., D.A. Valencio, J.F. Vilas, A. Cuerda, and G. Laffitte (1987): Cuenca Tarija, *in* El Sistema Carbonífero en la República Argentina (Ed.: S. Archangelsky), SCCS-Project PICG 211, Academia Nacional de Ciencias: 15-39, Córdoba.

Schlagintweit, O. (1937): Observaciones estratigráficas en el norte argentino. Yacimientos Petrolíferos Fiscales, Boletín de Informaciones Petroleras, v. 14, 156, 1-49, Buenos Aires.

Schwab, K. (1984): Contribución al conocimiento del sector occidental de la cuenca sedimentaria del Grupo Salta (Cretácico-Eogénico), en el noroeste argentino. Congreso Geológico Argentino, 9th, Bariloche, Actas, 1, 586-604.

Schwab, K. (1985): Basin formation in a thickening crust-The intermontane basins in the Puna and the Eastern Cordillera of NW-Argentina (Central Andes). Congreso Geológico Chileno, 4th, Antofagasta, Actas, 1 (2), 138-158.

Sempere, T., Oller, J., and Barrios, L. (1988): Evolución tectosedimentaria de Bolivia durante el Cretácico. Congreso Geológico Chileno, 5th, Santiago, Actas, 3 (H), 37-65.

Sempere, T., Aguilera, E., Doubinger, J., Janvier, Ph., Lobo, J., Oller, J., and Wenz, S. (1992) La Formation de Vitiacua (Permien moyen à supérieur-Trias ?inférieur, Bolivie du Sud): stratigraphie, palynologie et paléontologie. Neues Jahrbuch für Geologie und Paläontologie Abhandlungen, 185, 239-253.

Starck, D., Gallardo, E., and Schulz, A.E. (1992): La cuenca de Tarija: Estratigrafía de la porción argentina. Yacimientos Petrolíferos Fiscales, Boletín de Informaciones Petroleras, Tercera Epoca, v. IX, 30, 2-14, Buenos Aires.

Starck, D., Gallardo, E., and Schulz, A.E. (in press): Estratigrafía del Neopaleozoico de las Sierras Subandinas Occidentales y de la Cordillera Oriental de la República Argentina y algunas consideraciones sobre los bordes actuales de la cuenca de Tarija. Carboniferous-Permian Stratigraphy and Geology International Congress, 12th, Buenos Aires (1991).

Toselli, A. (1992): El magmatismo del noroeste argentino. Universidad Nacional de Tucumán, Serie Correlación Geológica, 8, 243 pp., Tucumán.

Turic, M., Aramayo Flores, F., Gómez Omil, R., Pombo, R., Sciutto, J., Robles, D., and Cáceres, A. (1987): Geología de las cuencas petroleras de la Argentina, *in* Evaluación de las Formaciones en la Argentina (Coord.: Felder, B.A.), Buenos Aires, Schlumberger, I/1-I/41.

Turner, J.C.M. (1959): Estratigrafía del cordón de Escaya y de la sierra de Rinconada (Jujuy). Asociación Geológica Argentina, Revista, v. 13 (1958), 15-39, Buenos Aires.

Turner, J.C.M., Méndez, V., Lurgo, C., Amengual, R., and Viera, O. (1979): Geología de la región noroeste, provincia de Salta y Jujuy, República Argentina. Congreso Geológico Argentino, 7th, Neuquén (1978), Actas, 1, 367-387.

Uliana, M., and Biddle, K. (1988): Mesozoic- Cenozoic paleogeographic and geodynamic evolution of southern South America. Revista Brasileira de Geociencias, v. 18, 172-190, Sao Paulo.

Uliana, M., Biddle, K., and Cerdán, J. (1988) Mesozoic extension and formation, Argentine sedimentary basins, *in* Extensional Tectonics and Stratigraphy of the North Atlantic Margins (Eds.: Tankard, A., and Balkwill), American Association of Petroleum Geologists Memoir, 46, 599-614.

Valencio, D.A., Giudice, A., Mendía, J.E., and Oliver G., J. (1976): Paleomagnetismo y edades K/Ar del Subgrupo Pirgua, provincia de Salta, República Argentina. Congreso Geológico Argentino, 6th, Bahía Blanca (1975), Actas, 1, 527-542.

Vergani, G., and Starck, D. (1989): Aspectos estructurales del Valle de Lerma, al Sur de la ciudad de Salta. Yacimientos Petrolíferos Fiscales, Boletín de Informaciones Petroleras, Tercera Epoca, v. VI, 20, 2-9, Buenos Aires.

Vilela, C.R. (1956): Descripción geológica de la Hoja 7 d, Rosario de Lerma (provincia de Salta). Buenos Aires, Dirección Nacional de Minería, Boletín 84, 61 p.

Vilela, C.R. (1965): El petróleo en las cuencas de Orán y Metán. Jornadas Geológicas Argentinas, 2nd, Salta (1962), *in* Acta Geológica Lilloana, v. 7, 425-438, Tucumán.

Viramonte, J.G., and Rapella, C.W. (1991): Cretaceous rift of the northwestern Argentina and the northwestern limit of the Pampean Ranges. Universidad Nacional de Salta, Alfred Wegener Conference, Salta, Field Trip Guide.

Wilson, J.L. (1975): Carbonate Facies in Geologic History. Berlin-Heidelberg-New York, Springer-Verlag, 471 pp.

Zambrano, J.J. (1987): Las cuencas sedimentarias de América del Sur durante el Jurásico y Cretácico: Su relación con la actividad tectónica y magmática, *in* Bioestratigrafía de los Sistemas Regionales del Jurásico y Cretácico de América del Sur, (Ed.: Volkheimer, W.), Comité Sudamericano del Jurásico y Cretácico, 1, 1-48, Mendoza.

Zanettini, J.C.M. (1973) Observaciones sobre la facies proximal del Subgrupo Santa Bárbara en el sector noroeste de la Cordillera Oriental (provincias de Salta y Jujuy). Asociación Geológica Argentina, Revista, v. 28, 219-226, Buenos Aires.

Cretaceous Evolution of the Magallanes Basin

V.A. Ramos
Universidad de Buenos Aires, Departamento de Ciencias Geológicas, Ciudad Universitaria-Pabellón II, (1428) Núñez, Buenos Aires
M.B. Aguirre-Urreta
Universidad de Buenos Aires - CONICET, Departamento de Ciencias Geológicas, Ciudad Universitaria-Pabellón II, (1428) Núñez, Buenos Aires

Abstract

This work presents a summary of the evolution of the Magallanes basin in the Patagonian Andes, between 45°S and 54°S latitude, during the Cretaceous.
A brief description is given of the pre-Mesozoic basement and the Jurassic volcanic sequences which underlie the Tithonian- Cretaceous deposits.
The Basin is analyzed on the basis of the following segments:
a) Lake Fontana-Coyhaique (45°S-46°S). The basin here was of the intra-arc type, with a western volcanic arc and an eastern outer volcanic arc, filled with turbidites, fossiliferous Neocomian shales and deltaic deposits; all covered by Mid- Cretaceous volcanic rocks.
b) Lake Posadas-Lake San Martín (47°S-49°S). The outer volcanic arc is absent. The back-arc basin was filled by intertidal sandy deposits of the Springhill Formation, fossiliferous Neocomian shales and regressive Barremian facies. Deposition came to a close with red beds and pyroclastic sediments. In the south of this segment there are Albian-Cenomanian marine facies.
c) Lake Viedma-Cerro Cazador (49°S-52°S). The Neocomian marine black shale sequences were deposited over the Springhill Formation and over the Jurassic volcanic rocks. The regressive continental rocks are Mid-Cretaceous. The Upper Cretaceous is marine in the south of the segment, and continental in the north.
d) Fueguian Andes (52°S-54°S). The Tithonian-Neocomian deposits correspond to the Yahgan Formation, which in the type area consists of low grade metasediments and metabasalts. This Formation shows differing lithofacies: black shales and tuffaceous beds; partially metamorphosed black shales with sills and dikes; black shales with chert, turbidites and resedimented conglomerates; coarse clastic turbiditic sequences, and turbiditic sandstones and shales.

Biostratigraphic data are given for the assemblage zones of almost all the stages of the Cretaceous System in this basin.
Finally, a transverse reconstruction of the basin and the successive stages of its paleogeographic and structural evolution are analyzed.

Introduction

The purpose of this chapter is to summarize the Cretaceous evolution of the Magallanes Basin, based on previously published contributions, and in the personal field observation of the authors, most of which is still unpublished. Quotations of previous works in the different topics have been selected in order to orientate further readings.

Geographic Setting

The Magallanes Basin encompasses a segment of the Patagonian and Fueguian Cordilleras from the 45°S to the 54°S latitude. The boundary at 45° S latitude is based on the outline of the marine Mesozoic basin, also known as the Austral Basin (Riccardi and Rolleri, 1980; Riccardi, 1988), which does not extend further north, at least not with any important sedimentary fill (Leanza, 1972). So defined, this basin is bounded to the north by the Northern Patagonian Andes, to the east with the table mountains and plains of the Precambrian Deseado Massif, and to the west by the Magmatic Belt which characterizes the western slope of the Patagonian Andes (see figure 1).

Previous Work

The knowledge of this part of the Andes started at the end of the nineteen century with the pioneer work of Hatcher (1897, 1900, 1903), who established the Meso-Cenozoic stratigraphy of the region. Several explorers from the recently founded Museo de La Plata as well as from some foreign institutions, such as Hauthal (1903), Quensel (1910, 1912, 1913), and Halle (1913), continued Hatcher's work along different sectors of the area producing significant contributions on geologic aspects of the cordillera.
The fundamental work of Feruglio (1931, 1938, 1950), and the contributions of Bonarelli and Nágera (1921); Piatnitzky (1938); Frenguelli (1935); Riccardi (1970, 1971, 1979, 1983), and Leanza (1970, 1972) have call the attention on the quality and fossiliferous content of the Cretaceous sections of the Cordillera, being one of the most complete marine sequences cropping out along the Andean Cordillera.
The regular survey of the Cordillera was undertaken during the last decade through the expeditions of the Servicio Geológico Nacional, which significantly contributed to the present knowledge of the Southern Patagonian Andes (Nullo et al., 1978, 1981a,b; Ramos, 1979a,b, 1988, 1989; Ramos et al., 1982).
Recent paleontological work has been presented by Aguirre-Urreta and Riccardi (1988a,b); Riccardi and Aguirre-Urreta (1988, 1989); Riccardi et al. (1987), among others.

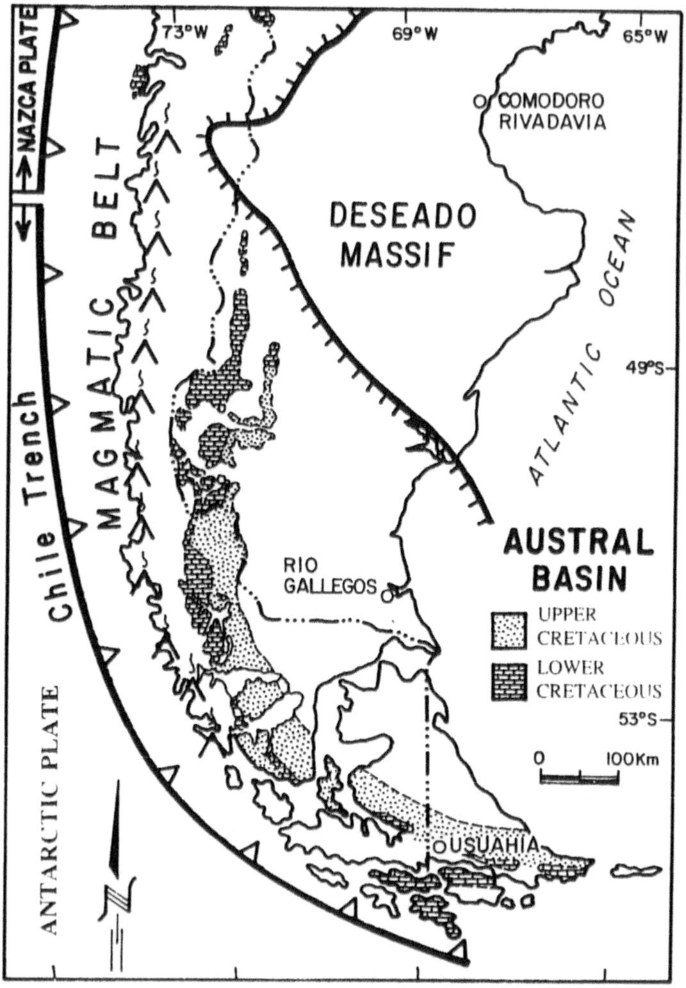

Fig. 1

Cretaceous outcrops of the Magallanes basin in the Southern Patagonian Andes, with the main morphostructural units. Copyright: Episodes.

Geologic Setting

The Meso-Cenozoic sequences of this segment of the Andes rest on highly deformed sedimentary and metasedimentary deposits of Middle to Late Paleozoic age. These rocks constitute the basement of the Cordillera at these latitudes.

A strong angular unconformity, resulted from the gondwanic orogeny (Du Toit, 1927), separates these rocks from volcanic, pyroclastic and volcaniclastic deposits of Jurassic age.

Thick sedimentary sequences overly the previous rocks and constitute the fill of the Magallanes Basin. This basin opens towards the south, where the most complete Late Jurassic and Cretaceous sections were deposited.

Tertiary molassic deposits overly some basaltic rocks in the eastern foothills, and they are covered by younger basalt flows. The basalts are widely distributed throughout the extra-andean region, and are covered by glacial and alluvial deposits of Late Tertiary and Pleistocene age.

Stratigraphy

The Pre-Mesozoic Basement

Thick sequences of flysch deposits that constitute the basement of the cordillera, are represented by metasedimentary and sedimentary facies. The first one is well represented in the Lago San Martín, where crops out the Bahía La Lancha Formation, a marine sequence of graywackes and shales strongly folded and faulted (Nullo et al., 1978). Further north in the Lago Belgrano area, a metasedimentary sequence of metasandstones, slates and phyllites constitutes the Río Lácteo Formation (Ramos, 1979a).
Both units represent a forearc rock assemblage, developed in the eastern margin of the Panthalassa sea, during middle Late Paleozoic times. A few granitoid stocks, such as the Sobral Tonalite of Late Carboniferous age (Ramos, 1983a), represent the scarce remnants of the magmatic arc of this age.

The Jurassic Volcanic Sequence

A widespread volcaniclastic sequence unconformably overlies the Paleozoic deposits. This is represented by acidic tuffs, ignimbritic flows and breccia flows of andesitic to rhyolitic composition. These rocks are irregularly distributed in several units that have been grouped in the El Quemado Volcanic Complex (Riccardi, 1971). It is widely distributed from the 45 to 52°S as pointed out by Quensel (1912). This unit is also known in the southern extremity of the Cordillera in Tierra del Fuego island, as the Serie Tobífera of equivalent age (Thomas, 1949).
The best exposed section of this unit outcrops between the Pueyrredón and Belgrano Lakes. There a 1.000 m thick sequence of alluvial fan conglomerates, andesitic breccia flows, and ignimbritic rhyolites, is slightly tilted and unconformably overlying the Río Lácteo Formation (Ramos, 1979b).
A few isotopic dates in the cordilleran region indicate a Middle Jurassic age for that volcanic section (Ramos, 1981). Based on stratigraphic relationships a Middle to Late Jurassic age was previously assigned (Nullo et al., 1978).
There are not geochemical or petrological studies of the El Quemado Volcanic Complex in the cordilleran region, but based on the geological setting the series is interpreted as a subduction-related arc assemblage, where significant melting of the sialic basement is responsible for the large amount of acidic rocks that characterize the Complex. Eastward, the lower part of the El Quemado Complex, interfingers with extensive acidic

flows and tuffs of the Chon Aike province (Kay et al., 1989). This event represents a generalized Jurassic extension characteristic of the pre-rift stage that preceded the opening of the South Atlantic ocean (Malumián and Ramos, 1984).

The Development of the Magallanes Basin

The development of the basin during Mesozoic times may be subdivided in two marine cycles, one of Late Jurassic-Early Cretaceous age and a second one in the Late Cretaceous. The latter is restricted to the southern portion of the basin, where the most complete sequences of that age are cropping out.

In order to describe the geologic evolution of the Mesozoic, the basin has been split in four segments with different characteristics:

The Lago Fontana-Coyhaique Segment (45°S- 46°S)

This northern segment of the basin is developed as an intra-arc basin, bounded by a western volcanic arc of andesitic to dacitic composition and an eastern outer volcanic arc of bimodal composition (Ramos and Palma, 1983) (Figure 2).

The basin fill began during Late Jurassic times, when turbiditic series of the Tres Lagunas Formation quickly cover the previous volcanic relief. The black shales interbedded with the distal turbidites bear some ammonites like *Aulacosphinctes* sp. which indicates a Tithonian age (Blasco and Ramos, 1978; Leanza, 1981; Ramos, 1981 b).

The sedimentary section is completed by 350 m of black shales of the Katterfeld Formation (see Figure 3) and by several hundred meters of sandstones and shales of the Apeleg Formation (Ramos, 1976; Ploszkiewicz and Ramos, 1977); both constitute a deltaic system that prograded the basin from the eastern side. The black shales have an abundant ammonite fauna of the *Favrella assemblage zone* of late Early to Late Hauterivian age (Riccardi, 1984).

The sedimentary series are covered by thick piles of volcanic and pyroclastic rocks of the Divisadero Group (Ramos, 1981b), which have a highly variable composition, from andesitic in the inner areas to acidic in the eastern foothills of the Cordillera, as seen in Figure 2. Some isotopic data indicate an Aptian-Albian age for these rocks, which are dominant further north (Ramos, 1978; Haller and Lapido, 1982). On this basis it is possible to reconstruct the distinctive paleogeographic setting of this segment that is depicted in Figure 2.

The Lago Posadas-Lago San Martín Segment (47°S-49°S)

The main difference with the northern segment is the absence of the eastern outer volcanic arc. The sedimentary sequence is developed in a retroarc setting, and participation

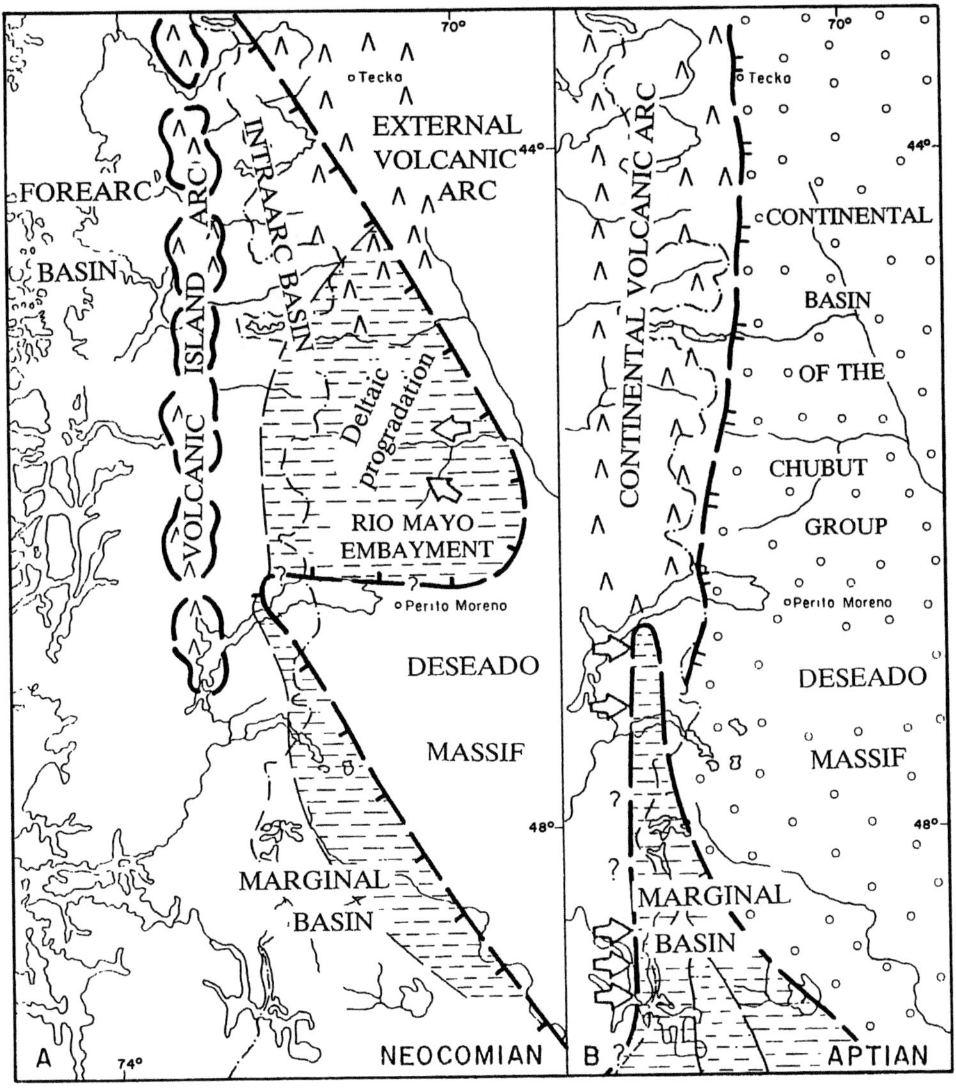

Fig. 2

Paleogeographic setting during Early Cretaceous times for the Lago Fontana -Coyhaique segment (modified from Aguirre-Urreta and Ramos, 1981b).

of volcanic debris is almost negligible, in spite of a well developed magmatic arc further west (Figure 4).

The volcanic relief was significantly eroded, prior to deposition of thick blankets of quartzose sandstone of intertidal facies of the Springhill Formation (Ramos, 1979b). An

Fig. 3

Black shales of the Katterfeld Formation in the Cerro Dedo, Lago Fontana.

erosional unconformity separates the Berriasian-Valanginian Springhill Formation (Blasco et al., 1979; Kielbowicz et al., 1983) from the El Quemado Volcanic Complex (Riccardi, 1971, 1977).

This initial fill is dominant in the eastern flank of the basin, while the inner areas are covered by black shales of basinal facies, corresponding to the Río Mayer Formation (Hatcher, 1897; Favre, 1908). The oldest fossil recorded from these black shales is an ammonite assigned to *Olcostephanus* sp. of Late Valanginian age (Ramos, 1979b).

The black shales of the Río Mayer Formation are widely exposed at the foothills of the Cordillera, reaching thicknesses of more than 750 m (Nullo et al., 1978). Several fossil assemblages are known from this unit (Aguirre-Urreta, 1985, 1991) (Table I).

The regression of the Early Cretaceous sea is diachronicaly registered, as a sand progradation advance from north to south, and from east to west (see Figures 5 and 6) (Aguirre-Urreta and Ramos, 1981a,b; Arbe, 1989).

The dominant facies are represented by bioturbated green sandstones and shales of the Río Belgrano Formation bearing an abundant *Hatchericeras* fauna of Barremian age (Ramos, 1979b). One of the best exposed sections crops out in the Río Roble area (see Figure 7).

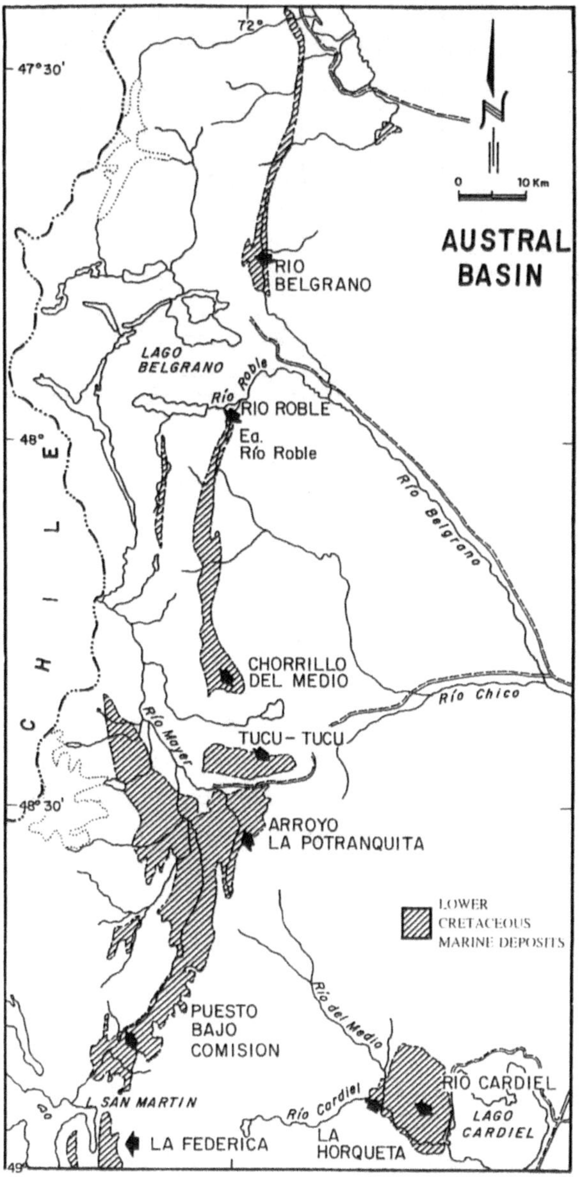

Fig. 4

Extension of the Early Cretaceous deposits of the Lago Posadas - Lago San Martin segment. Copyright: Episodes.

The Cretaceous sequences end with continental deposits of the Río Tarde Formation, composed of red beds and pyroclastic material derived from the active volcanic arc situated at that time farther west. The deposits are a few hundred meters thick, and were formed in an upward-thinning sequence, which varies from alluvial fan and proximal fluvial facies to a more distal fluvial and lacustrine series. Abundant plant remains and

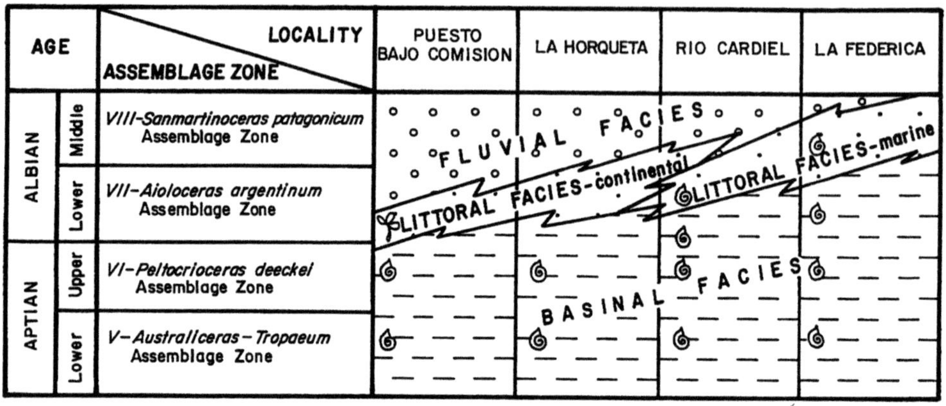

Fig. 5

Paleontological control of the diachronism of the east-west Early Cretaceous regression. See localities in Figure 4 (based on Aguirre-Urreta, 1991). Copyright: Episodes.

microflora indicate an Aptian-Albian age for this sequence (Ramos and Baldoni, 1981; Archangelsky et al., 1981).

The abundant fossil occurrences, mainly consisting of ammonites, bivalves and decapods, permitted recognition for the first time in the basin of a *Colchidites* fauna (Blasco et al., 1980), which indicates a Late Barremian age for the Río Mayer Formation at Loma Pelada, north of the Lago Belgrano region. This fauna is also associated with *Sanmartinoceras africanum* (Aguirre-Urreta and Klinger, 1986), and some decapods (Aguirre-Urreta, 1983, 1989).

At the southern end of this segment, sedimentation took place in marine facies up to Albian-Cenomanian times (see Figure 8).

Interfingering of continental and marine facies is noticed in the Albian and slightly younger deposits, as recorded in the Kachaike Formation (see Figure 9).

The main fossil assemblages of Table 1 and the units recognized for this segment are depicted in the block diagram in Figure 9, where progradation of the continental deposits from north to south can be seen. The set of paleogeographic maps in Figure 10 illustrates the beginning of the marine transgression in this segment (Figure 10 A), the maximum extension of the sea (Figure 10 B), the progradation of the continental deposits (Figure 10 C), and the final regression during Albian times (Figure 10 D), after Aguirre-Urreta and Ramos (1981b), Ramos (1982a), and Aguirre-Urreta (1991).

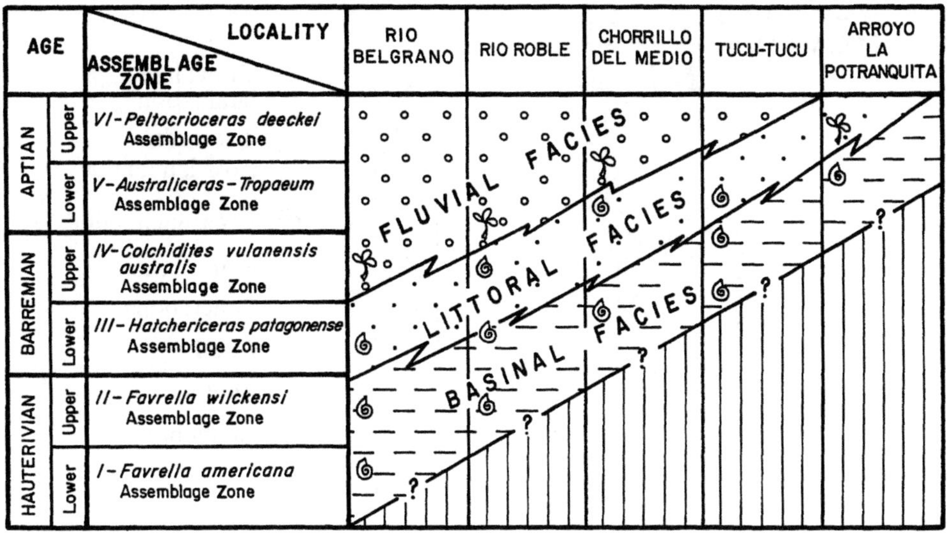

Fig. 6

Paleontological control of the diachronism of the north-south Early Cretaceous regression. See localities in Figure 4 (based on Aguirre-Urreta, 1991). Copyright: Episodes.

The Lago Viedma-Cerro Cazador Segment (49°S- 52°S)

The Cretaceous fill of this segment begins with a conglomeratic sequence of variable thickness, that in some areas lies unconformably over the volcanic rocks of the El Quemado Volcanic Complex, while in others it is interfingered with them. At the base some layers of tuffaceous deposits, produced by recycling of the pyroclastic material, are recognized as the Springhill Formation (Thomas, 1949; Cecioni, 1955; Katz, 1963; Leanza, 1972; Riccardi, 1976 and Blasco et al., 1979).

This basal unit has a homogeneous lithology, mainly in the subsurface of the eastern flank of the basin, where it constitutes the Springhill platform, which has been extensively studied because it is the main reservoir for hydrocarbon exploration (Robles, 1982).

The age of the Springhill Formation varies from Tithonian in Lago Argentino (Blasco et al., 1979) to Early Valanginian at Lago San Martín (Kielbowicz et al., 1983).

The marine sequence continues with a homogeneous lithology of black shales and fine-grained sandstones of the Río Mayer Formation (Hatcher, 1897; Riccardi, 1971; Nullo et al., 1978, 1981a). The base of this unit grades to the conglomeratic and coarse-grained sandstones of the Springhill Formation in some areas (Blasco et al., 1979), while in others

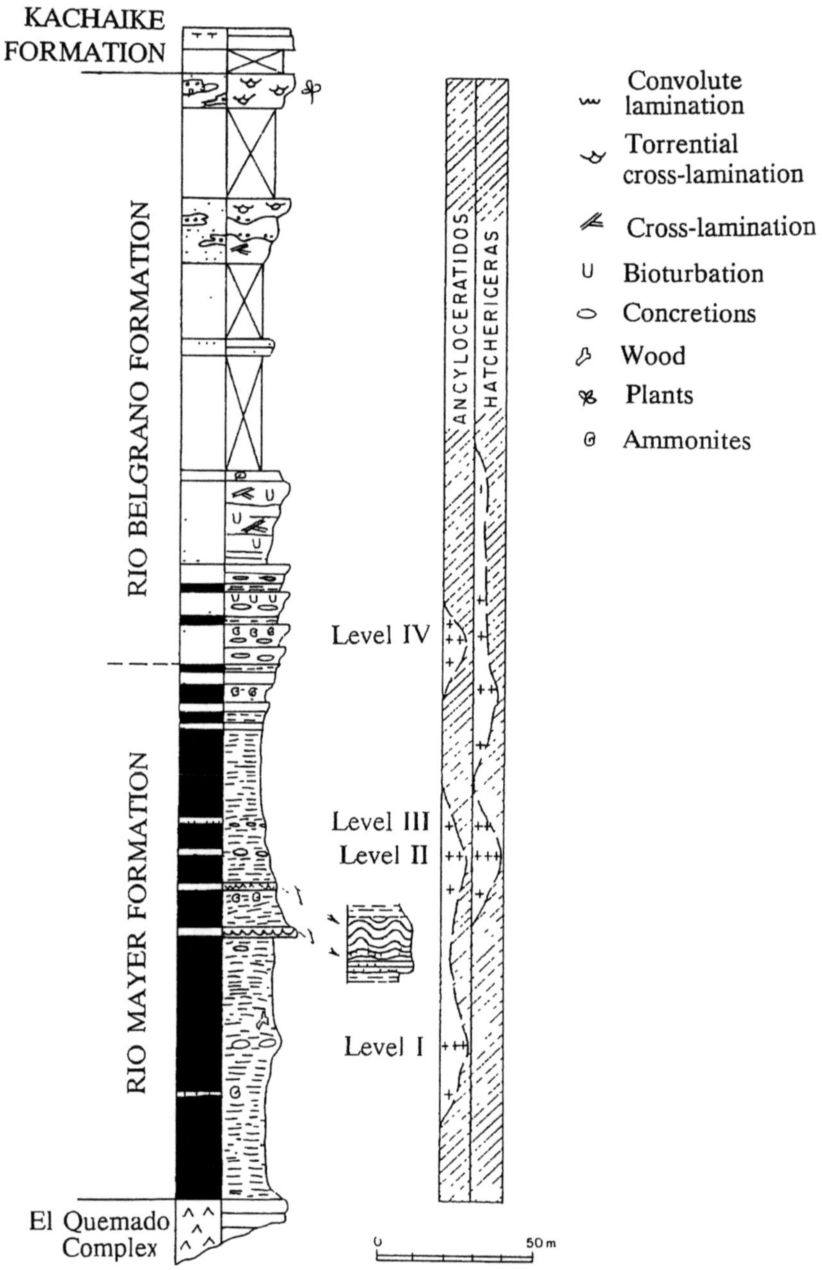

Fig. 7

The classic section of Río Roble, south of the Lago Belgrano region, with the development of the Early Cretaceous sequences (based on Ramos, 1982b).

Fig. 8

Typical basinal facies of the Río Mayer Formation at La Federica, Lago San Martín.

it lies unconformably over the El Quemado Volcanic Complex (Nullo et al., 1981a). The dominant lithology indicates a distal platform environment, although in the southern areas some slope facies are recognized (Arbe and Hechem, 1984a). The base of this unit is diachronous, but the top -in this segment- reaches a Late Albian age (Arbe and Hechem, 1984a; Arbe, 1989). However, some authors have proposed either an Early Turonian (Nullo et al., 1981a) or a Cenomanian age (Malumián et al., 1983).

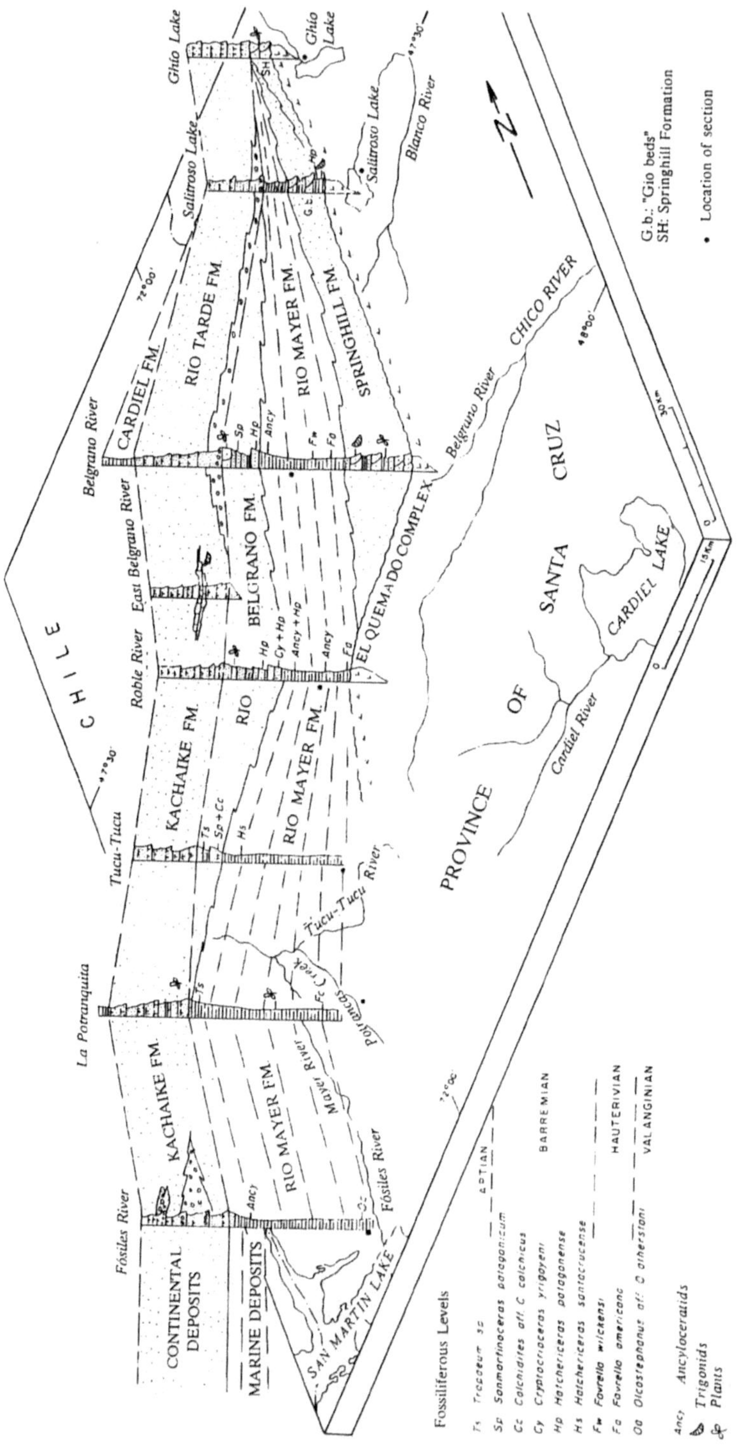

Fig. 9
Stratigraphic correlation of the Early Cretaceous deposits between Lakes Pueyrredón and San Martín. Based on Aguirre-Urreta and Ramos, 1981b.

Table 1: Correlation chart and fossil control of the Early Cretaceous sedimentary sequence of the Magallanes Basin. Copyright: Episodes.

AGE		ASSEMBLAGE ZONE	FAUNAL COMPONENTS / LOCALITY	RIO BELGRANO	RIO ROBLE	CHORRILLO DEL MEDIO	TUCU-TUCU	ARROYO LA POTRANQUITA	PUESTO BAJO COMISION	LA HORQUETA	RIO CARDIEL	LA FEDERICA
ALBIAN	Middle	VIII- *Sanmartinoceras patagonicum* Assemblage Zone	*Sanmartinoceras patagonicum*									●
			"Feruglioceras piatnitzkyi"								●	●
	Lower	VII- *Aioloceras argentinum* Assemblage Zone	*Aioloceras argentinum*									●
			Sinzovia leanzai									●
			Rossalites imlayi									●
			Beudanticeras rolleri								●	●
			Cleoniceras santacrucense								●	
APTIAN	Upper	VI- *Peltocrioceras deeckei* Assemblage Zone	*Peltocrioceras deeckei*						●	●	●	●
			Helicancylus patagonicum						●		●	
			Lithancylus guanacoense								●	●
			Sanmartinoceras cardielense								●	
			Sinzovia piatnitzkyi								●	●
			Ptychoceras sp									●
			Feruglioceras piatnitzkyi								●	●
	Lower	V- *Australiceras* – *Tropaeum* Assemblage Zone	*Australiceras cardielense*						●	●	●	
			Tropaeum (Tropaeum) sp			●	●					
			Tropaeum (Tropaeum) inflatum							●	●	
			Tropaeum (Australotropaeum) magnum					●				
			Sanmartinoceras walshense						●		●	●
			Sanmartinoceras africanum cf africanum						●			
			Toxoceratoides nagerai						●	●	●	
			Helicancylus bonarellii									●
BARREMIAN	Upper	IV- *Colchidites vulanensis australis* Assemblage Zone	*Colchidites vulanensis australis*		●?	●	●					
			Heteroceras elegans				●					
			Sanmartinoceras africanum insignicostatum		●	●	●					
	Lower	III- *Hatchericeras patagonense* Assemblage Zone	*Hatchericeras patagonense*	●	●		●					
			Hatchericeras semilaeve	●	●		●					
			Cryptocrioceras yrigoyeni		●							
			Hemihoplites varicostatus	●	●							
HAUTERIVIAN	Upper	II- *Favrella wilckensi* Assemblage Zone	*Favrella wilckensi*	●	●?							
			Protaconeceras patagoniense	●								
			Hemihoplites varicostatus	●								
	Lower	I- *Favrella americana* Assemblage Zone	*Favrella americana*	●								
			Hemihoplites ploszkiewiczi	●								
			Aegocrioceras sp	●								

In this segment the regression of the sea is also younger to the south, due to a progradation of the clastic continental facies similar to that observed in the northern segments. This fact is represented by a sandy unit recognized as the Piedra Clavada Formation, which ranges in age from Late Albian to Cenomanian north of Lake Viedma (Leanza, 1970; Nullo et al., 1981a).

The Upper Cretaceous marine deposits of this segment crop out south of Lago San Martín. Further north, most of the deposition took place in continental environments.

The Early Turonian-Maastrichtian span is represented by several lithosomes mainly dated on their ammonite fauna (see Table 2). Various authors have essayed a paleogeographic analyses of this area such as Vilela and Csaky (1968), Riccardi and Rolleri (1980), Nullo et al. (1981a,b), Malumián et al. (1983), Arbe and Hechem (1984a,b), Macellari (1988), and Macellari et al., (1989).

The marine deposits of the Río Mayer Formation are overlaid north and south of Lago Viedma by the sandstones and shales of the Puesto El Alamo (see Figure 11) and Mata Amarilla Formations.

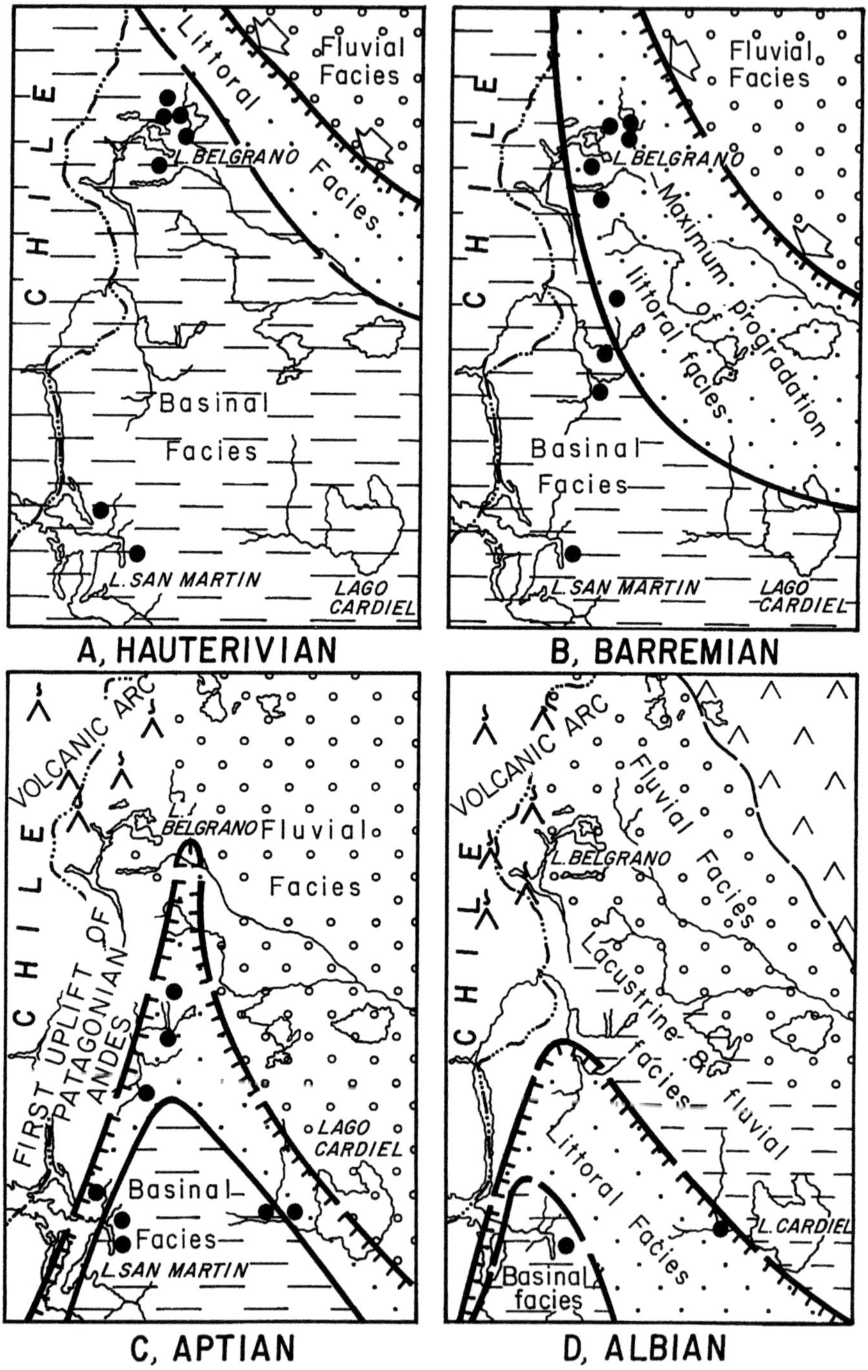

Fig. 10

Paleogeographic setting of the Early Cretaceous deposits between Lakes Pueyrredón and San Martin. Based on Aguirre-Urreta and Ramos 1981b and Aguirre- Urreta, 1991. Copyright: Episodes.

Table 2: Upper Cretaceous fossil assemblages of the Magallanes Basin

AGE		ASSEMBLAGE ZONE	FAUNAL COMPONENTS
MAASTRICH-TIAN	Lower	*XI-Gunnarites kalika* Assemblage zone	*Gunnarites kalika* *Gunnarites antarcticus* *Maorites spp.* *Diplomoceras australe* *Grossouvrites gemmatus*
CAMPANIAN	Upper	*VIII-Hoplitoplacenticeras plasticum* Assemblage zone	*Hoplitoplacenticeras plasticum* *Saghalinites kingianus* *Gaudryceras varagurense* *Hypophylloceras nera* *Pseudophyllites peregrinus* Baculites duharti
	Lower	*VII-Karapadites centinelaensis* Assemblage zone	*Karapadites centinelaensis* *Natalites hauthali* *Neograhamites morenoi*
SANTONIAN		*VI-Polyptychoceras Faunule*	*Polyptychoceras sp.* *Baculites sp.* *Anagaudryceras sp.*
CONIACIAN		*V-Gauthiericeras Faunule*	*Gauthiericeras santacrucense*
TURONIAN	Middle	*IV-Anapachydiscus steinmanni* Assemblage zone	*"Anapachydiscus" steinmanni* *"Anapachydiscus" patagonicus* *"Anapachydiscus" hauthali* *"Patagiosites" (?) amarus* *Parabinneyites paynensis* *Parapuzosia magellanica* *Placenticeras viedmaense* *Placenticeras santacrucense* *Placenticeras washbournei* *Placenticeras patagonicum* *"Canadoceras" megasiphon* *Argentoscaphites mutantibus* *Sciponoceras santacrucense*
	Lower		
CENO-MANIAN	Upp	*III-Calycoceras Faunule*	*Calycoceras sp.*
	Mid	*II-Desmoceras Faunule*	*Desmoceras floresi*
	Low	*I-Hypoturrilites Faunule*	*Hypoturrilites cf. gravesianus* *Sciponoceras sp.*

Fig. 11
The type locality of Puesto El Alamo Formation, north of Lago Viedma.

The Cerro Toro Formation, towards the south of the segment, was characteristic at this time of deeper environments. These overlying units are themselves covered by marine neritic and continental deposits of the Anita and Chorrillo Formations (see Figure 12).
The Cerro Toro Formation (Cecioni, 1955; Riccardi and Rolleri, 1980) is characterized by a thick sequence of turbiditic shales and sandstones, typically exposed south of Lago Viedma. Its base has been considered younger than Cenomanian (Nullo et al., 1981a).
The upper boundary of the Cerro Toro marine sequence was placed in the Campanian by Riccardi and Rolleri (1980) south of Lago Argentino, as well as on the Chilean side (see Figure 13). An alternative age has been proposed by Blasco et al. (1981).
Several units with similar lithologic characteristics overlie the Cerro Toro sequences, marking the continentalization of the region. These units are known as the Puesto El Alamo-Anita Formation and the Chorrillo-Cerro Fortaleza Formation between Lakes Viedma and Argentino, or as the Anita and Chorrillo Formations between Lago Argentino and Cerro Bagual. To the south, in the Río Turbio area, they are known as the Cerro Cazador and Dorotea Formations (Paülcke, 1907; Hünicken, 1965; Nullo et al., 1981a,b; Macellari et al., 1989).

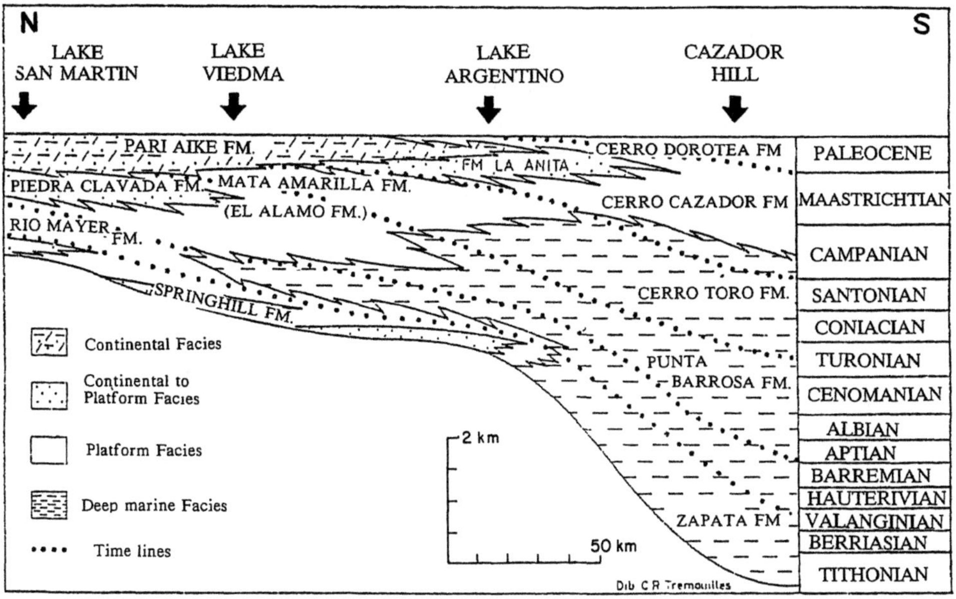

Fig. 12

Interpretative diagram showing the stratigraphic relationships of the Cretaceous deposits between Lago San Martín and Cerro Cazador (based on Riccardi and Aguirre-Urreta, 1988).

The Campanian-Maastrichtian span is dominated by a rapid continentalization of the basin, indicated in some areas by the Lago Sofía conglomerates (Late Turonian-Early Campanian), which represent submarine canyon deposits developed on the slope margin of the basin (Winn and Dott, 1979). This important clastic supply is related to the first uplift of the Patagonian Cordillera. Paleogeographic evolution for the Late Cretaceous was depicted by Arbe (1989) and Macellari et al. (1989) (see Figure 14).

The Fueguian Andes Segment (52°S-54°S)

The Fueguian Andes are located south of 52° S latitude, where the Andean chain shows an east-west strike at the southern end of the continent (see Figure 1). This segment is therefore the link between the Andean Cordillera and the northern segment of the Scotia Arc.

The Cordillera at these latitudes encompasses a mountain chain, mainly developed in the Island of Tierra del Fuego and in a series of smaller islands, such as Navarino, Los Estados, and many others, which constitute the Fueguian Archipelago.

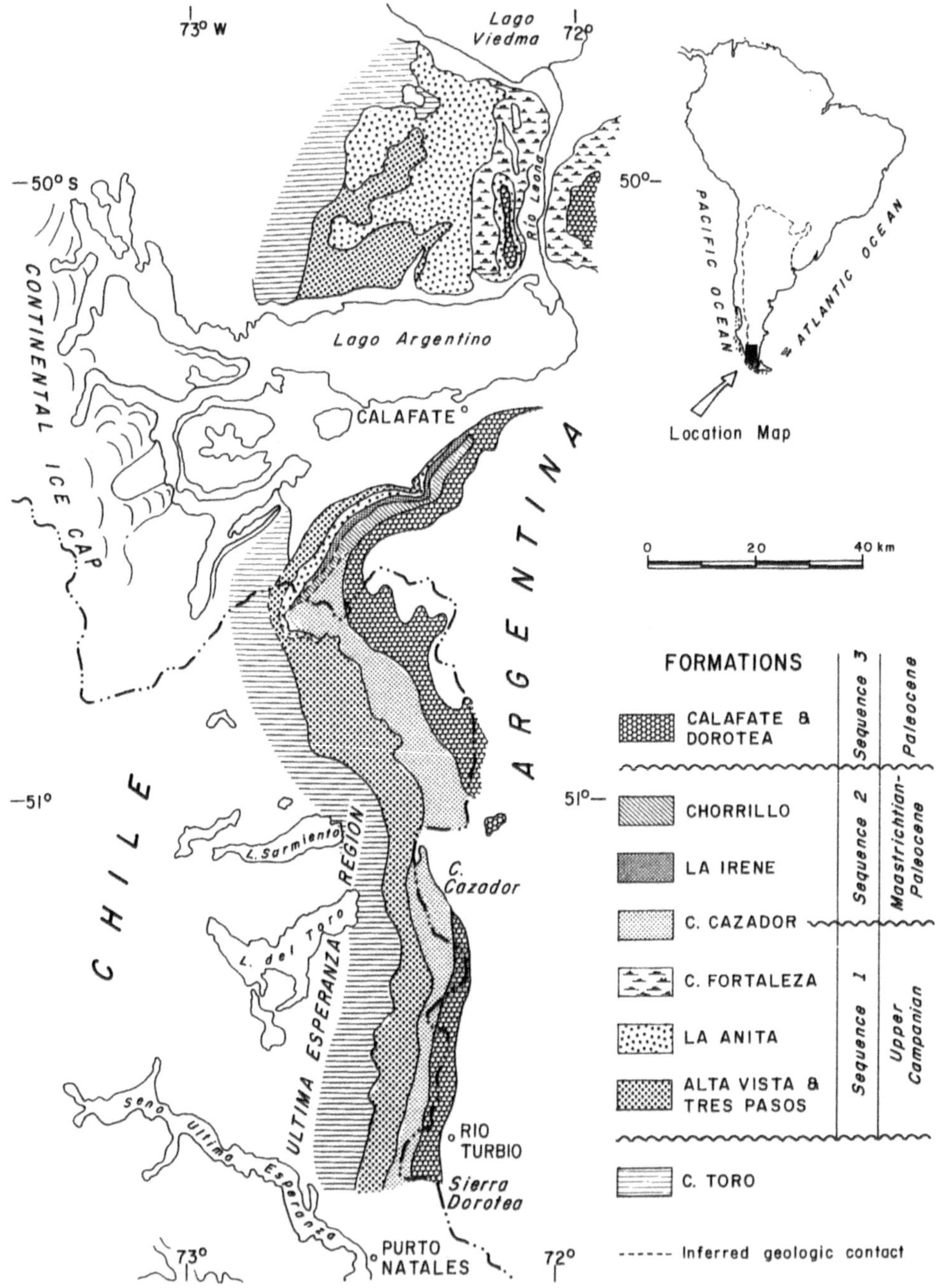

Fig. 13

Geologic sketch of the distribution of Late Cretaceous units between Lago Viedma and Río Turbio (based on Macellari et al., 1989).

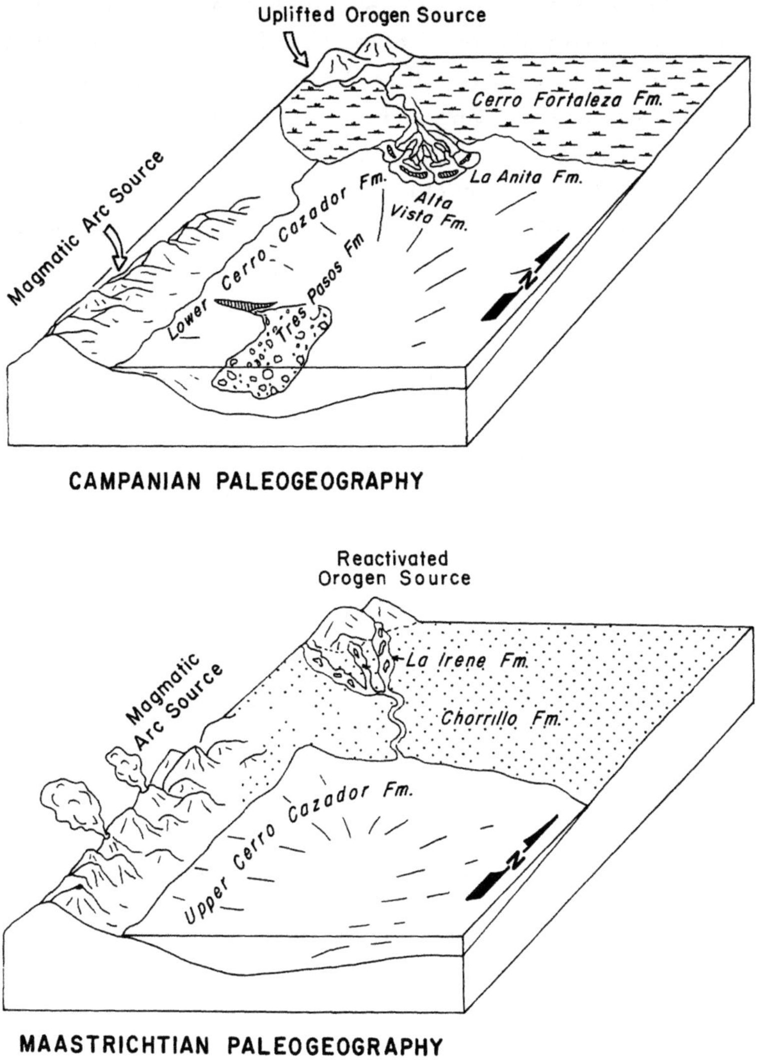

Fig. 14

Block diagrams showing proposed paleoenvironmental reconstruction of the southern segment of the Magallanes Basin (after Macellari et al., 1989).

The first survey of the area was carried out by Darwin, who identified the first Cretaceous outcrops and collected invertebrate fossils from the Brunswick Peninsula (Darwin, 1846).

The area was later studied through the efforts of several international expeditions, such as the American expedition headed by Dana (1839-1842), the French polar expedition carried out by Dumont-D'Urvielle (1837-1840) and reported by D'Orbigny (1848), and the more recent Swedish expeditions (Nordensjold 1897, 1905; Quensel 1910, 1912, 1913). Later, outstanding contributions were those of Bonarelli (1917), Kranck (1932), Feruglio (1936, 1950) and Leanza (1967, 1969), and the recent ones of Dalziel et al. (1974 and others), Suárez and Pettigrew (1976), Suárez (1978 and others), Caminos (1980), and Caminos et al. (1981). The basement of the Mesozoic sequences along both sides of the Cordillera de Darwin is made up of a series of phyllites, micaschists and gneisses of low to high grade metamorphism that are intruded by Late Triassic granitoids. An Rb/Sr numerical age for this basement indicates 240 +/- 40 Ma for the metamorphic rocks at Pluschöw Bay (Hervé et al., 1981). These rocks are known on the Argentine side as the Lapataia Metamorphics (Borrello, 1968, 1969), and are composed of thinly laminated phyllites (Caminos et al., 1981). There are some metabasaltic rocks represented by green schists and low-rank amphibolites. On the Argentine side the dominant metamorphism is of prehnite pumpellite to low green schist grade (Caminos, 1980).

The presence of acidic volcanic rocks, widely extended through out most of the northern slope of the Fueguian Cordillera has been known since the early work of Quensel (1910, 1912) . It was Kranck (1932) who identified these rocks at Monte Buckland and classified their characteristic lithology. They are there unconformably layed over Late Paleozoic metamorphic rocks, and a conglomerate is widely developed at the base of the sequence. On the Argentine side, the volcanics are represented by the Lemaire Formation (Borrello, 1969; Caminos, 1976), which is widely developed at the Sierras de Alvear and Lucio López and in the Los Estados island further east. It is composed of thick sequences of ignimbritic tuffs, rhyolitic breccias and some interbedded volcaniclastic deposits. The volcanic and volcaniclastic sequences are also highly deformed, and are presently represented by low grade metamorphics of the prehnite pumpellite facies (Caminos, 1978).

There are no suitable radiometric ages for this unit, since most of the dates obtained are minimum ages related to deformational events, such as the ones presented by Ramos et al. (1986). Based on regional correlation and the age of the overlying units, a Late Jurassic age is accepted for the Serie Tobífera and the Lemaire Formation.

The Cretaceous deposits were known as the "Clay-slate" Formation since the first survey made by Darwin (1846) in the Fueguian Andes. These deposits were included in the Yahgan Formation by Kranck (1932), who first classified this unit. He recognized two formations: one in the western segment of the Cordillera, in the surroundings of the Buckland Peninsula -the Monte Buckland Series or Formation- and one to the east, along both sides of the Beagle, the Yahgan Formation. Most of the work done in the area has referred to this latter unit.

The type area of the Yahgan Formation outcrops on the northern side of the Beagle Channel, between Yendegaia and Ushuaia. In this area a series of low grade metasediments and metabasalts, highly deformed, are well exposed along the southern slope of the Fueguian Cordillera. Several facies have been recognized in the Yahgan Formation, according to Caminos et al. (1981) and Suárez et al. (1985a,b).

a) The Sierra de Alvear lithofacies: a basal section of the Yahgan Formation, which consists of a series of intercalated black shales and white tuffaceous beds, thinly laminated. This facies is developed in transition to the Lemaire Formation, and is indicative of an active volcanism during its deposition in a restricted environment. It is well exposed south of Lake Fagnano, on the northern flank of the Sierra Valdivieso, where it is several tens of meters thick.

b) The Paso Garibaldi lithofacies: this is also a basal section of the Yahgan Formation, which is composed of basalt sills and dikes, emplaced in black shales, partially metamorphosed to cloritic-sericitic slates, thinly laminated. In this facies the mafic material, represented by green schists, is dominant. The metabasalts have a vesicular to amygdaloid texture, which indicates their effusive subaerial character. They are well exposed east and south of Lago Fagnano. In some areas, such as the vicinity of El Tunel, along the northern shore of the Beagle Channel, a few kilometers east of Ushuaia, some ultramafic rocks have been identified (pyroxenite bodies, according to Genini; pers. com., 1979).

c) The Canal de Beagle lithofacies: this is mainly composed of black shales, with some chert levels and radiolarites, with frequent *Chondrites* sp. traces. There are also some minor basalts in sills and dikes, but not as frequent as in the former lithofacies. Along the northern shore of the Beagle Channel, typical turbiditic facies have been identified. Resedimented conglomerates, as well as pebble mudstones, are widely exposed. On the northern shore of the Beagle Channel there are frequent turbiditic layers, with angular clasts of acidic volcanic rocks of the Lemaire Formation.

d) The Peninsula Mitre lithofacies: this is a sequence of black shales, with conspicuous fissility parallel to the bedding planes, that contains no chert nor radiolarites, such as does the previously described facies. Turbiditic facies are not present, and their dynamic metamorphism is absent or poorly developed.

e) The Bahia Douglas Lithofacies: this consist of a coarse clastic sequence exposed south of the Beagle Channel. These deposits are interpreted as turbidites of braided channels on suprafan lobes, associated with debris flows and volcanic breccias and resedimented conglomerates (Suárez et al., 1985a,b). The main difference with the northern facies is the provenance of conglomerates and sandstones, where abundant mesosilicic rocks and subordinate gabbros and diorites are present in the clasts.

f) The Wulaia Lithofacies: this is characterized by thinly laminated sandstones and shales, interfingered with more massive sandstones, which also crop out in the southern part of the Beagle Channel, but north of the previous facies. These deposits have been interpreted as mid-fan turbiditic facies, locally associated with suprafan lobe channels (Suárez et al., 1985a,b).

Based on the previous descriptions it is possible to reconstruct a transverse section of the basin (see Figure 15), which has striking differences from north to south.

Although turbiditic sequences are dominant in both areas, the provenance and the associated facies are different. To the north, acidic tuffs are interbedded with the black shales, while in the southern sector breccia flows of andesitic composition are frequent. Clast provenance in the north is mainly of rhyolites and basement rocks, while in the south volcanic and plutonic rocks derived from a magmatic arc are dominant.

In the north it is possible to recognize distal platform facies, absent in the southern sectors, and a rapid transition to deep basinal facies associated with radiolarites, *Chondrites*, and chert levels.

Some garnet-bearing basement clasts in the southern areas indicate erosion of a possible continental margin, where a much older Precambrian basement was exposed, similar to the Cabo Belgrano (Cape Meredith) metamorphic complex of about 1,000 Ma (Cingolani and Varela, 1976). Figure 14 summarizes the relationships among the different facies, as well as the paleogeographic setting of the area during the Yahgan Formation deposition.

Based on the occurrence in several localities on the southern shore of the Beagle Channel of *Belemnopsis* spp. and *Favrella americana* (Favre), a Tithonian to Early Cretaceous age has been assigned (Suárez et al., 1979, Aguirre-Urreta and Suárez, 1985) to the Yahgan Formation.

Minor outcrops of granitic rocks are found on the southern slope of the Cordillera. For example, some granitic porphyries have been found at the southern end of the Ushuaia Peninsula, just south of the airstrip. These dikes unconformably cut the deformed Yahgan Formation, and yielded K/Ar ages of 77 +/- 3 Ma (Ramos et al., 1986).

Farther east of Ushuaia, half way to Estancia El Túnel, a small granite stock is exposed along the Beagle Channel, with similar facies to the one outcropping on the southern shore of the Channel at Navarino Island. Ultramafic rocks, mainly pyroxenites, are also exposed in this area (Caminos et al. 1981).

Geologic Evolution

The evolution of the Cretaceous deposits in the Magallanes basin exposed along the southern Patagonian Andes and the Fueguian Cordillera exhibit a series of well-defined stages.

Basin development in Latest Jurassic and Early Cretaceous times occurred during a regional crustal extension. The initial development coincided with the pre-rift facies of the Chon-Aike province in extra-Andean Patagonia. Further south, in Ultima Esperanza, the silicic volcanism of the Serie Tobífera was a precursor of the formation of a deep marine trough (Wilson, 1991). Differential subsidence of the trough suggests that more extreme

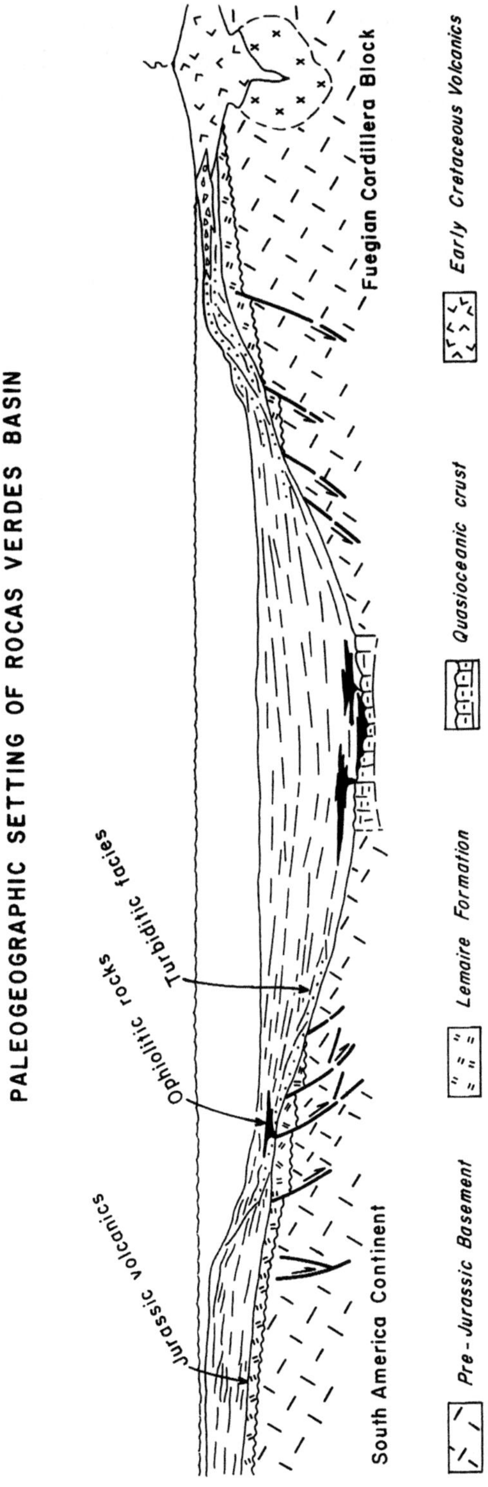

Fig. 15
Paleogeographic setting of the Yahgan Formation in southeastern Tierra del Fuego (based on Dalziel and Ramos, 1988).

extreme crustal thinning by normal faulting was localized along the margin, governed by its underlying thin basement of accreted Gondwanide fore-arc material (Ramos, 1983a). Subsidence further south was extreme, developing the "Rocas Verdes" back-arc basin (Dalziel et al., 1974). These ophiolitic rocks were generated from Cordillera Sarmiento (52°S) to the Tortuga Complex (55°S). Farther north, this stage was characterized by minor subsidence, and a continuous supply of terrigenous debris from the cratonic side. North of 47°S, the retro-arc basin graded to an intra-arc basin.

A fundamental change in basin geometry occurred during Late Albian to Cenomanian times. This change corresponds to the closure of the back-arc basin, the obduction of the ophiolitic material, and the beginning of compressional deformation. This setting contrasts with the evolution of the northern segments where subsidence was mild, and an ocean floor was not formed. A series of deltaic systems from the northwestern side (Arbe, 1989), as well as a magmatic source to the west, controlled the sedimentation and paleogeography of the Late Cretaceous (Macellari et al., 1989).

As deformation prograded eastward, a series of foredeeps were formed, causing a total subsidence of more than 6 kilometers in the main western depocenter of the Magallanes basin. This latter stage of the foreland basin persisted up to Latest Cenozoic times, which shaped the present geometry of the basin.

Acknowledgements

The authors are grateful to the Servicio Geológico Nacional for many years of financial support, and to many colleagues specially Dr. Miguel Palma and Graciela Marín for many fruitful discussions, and assistance in the field.

References

Aguirre-Urreta, M.B. (1983): Crustáceos decápodos barremianos de la región del Tucu-Tucu, Provincia de Santa Cruz. Ameghiniana, v. 19, 303-317, Buenos Aires.

Aguirre-Urreta, M.B. (1985): Ancyloceràtidos (Ammonoidea) Aptianos de la Cordillera Patagónica Austral, Provincia de Santa Cruz, Argentina. Academia Nacional de Ciencias, Boletín, v. 56, 135-235, Córdoba.

Aguirre-Urreta, M.B. (1989): The Cretaceous Decapod Crustacea of Argentina and the Antarctic Peninsula. Palaeontology, v. 32, 499-552.

Aguirre-Urreta, M.B. (1991): Paleogeography and Biostratigraphy of the Austral basin (Argentina-Chile): an appeal for sound systematics. Episodes, v. 13, 247-255.

Aguirre-Urreta, M.B., and Klinger, H.C. (1986): Upper Barremian Heteroceratinae (Cephalopoda:Ammonoidea) from Patagonia and Zululand, with comments on the systematics of the subfamily. South African Museum, Annals, v. 96, 315-358, Cape Town.

Aguirre-Urreta, M.B., and Ramos, V.A. (1981a): Crustáceos decápodos del Cretácico inferior de la Cuenca Austral, provincia de Santa Cruz, Argentina, *in* Cuencas Sedimentarias del Jurásico y Cretácico de América del Sur, (Eds.: Volkheimer, W., and Musacchio, E.), 2, 599-623, Buenos Aires.

Aguirre-Urreta, M.B., and Ramos, V.A. (1981b): Estratigrafía y Paleontología de la Alta Cuenca del río Roble, Cordillera Patagónica. Congreso Geológico Argentino, 8th, San Luis, Actas, 3, 101-138.

Aguirre-Urreta, M.B., and Riccardi, A.C. (1988a): Albian Heteromorph Ammonoids from Southern Patagonia. Journal of Paleontology, v. 62, 598-614.

Aguirre-Urreta, M.B., and Riccardi, A.C. (1988b): El género Lithancylus Casey (Ammonoidea) en el Aptiano superior de Patagonia. Notas Museo de La Plata 21, Paleontología, 107, 195- 207, La Plata.

Aguirre-Urreta, M.B., and Suárez, M. (1985): Belemnites de una secuencia volcaniclástica de la Formación Yahgan, Titoniano-Cretácico inferior del extremo sur de Chile. Congreso Geológico Chileno, 4th, Antofagasta, Actas, 1 (1), 1-16.

Arbe, H.A. (1989): Estratigrafía, discontinuidades y evolución sedimentaria del Cretácico en la Cuenca Austral, provincia de Santa Cruz, *in* Cuencas Sedimentarias Argentinas, (Eds.: Chebli, G.A., and Spalletti, L.A.), Universidad Nacional de Tucumán, Serie Correlación Geológica, 6, 419-442, Tucumán.

Arbe, H.A., and Hechem, J. (1984a): Estratigrafía y facies de depósitos marinos profundos del Cretácico Superior, lago Argentino, provincia de Santa Cruz. Congreso Geológico Argentino, 9th, Bariloche, Actas, 5, 7-41.

Arbe, H.A., and Hechem, J. (1984b): Estratigrafía y facies de depósitos continentales, litorales y marinos del Cretácico Superior, lago Argentino, provincia de Santa Cruz. Congreso Geológico Argentino, 9th, Bariloche, Actas, 7, 124-158.

Archangelsky, S., Baldoni, A., Gamerro, J.C., Palamarczuk, S., and Seiler, J. (1981): Palinología estratigráfica del Cretácico de Argentina Austral. Diagramas de grupos polínicos del suroeste de Chubut y noroeste de Santa Cruz. Congreso Geológico Argentino, 8th, San Luis, Actas, 4, 719-742.

Blasco, G., and Ramos, V.A. (1978): El Titoniano de arroyo Pedregoso, Lago Fontana, Provincia del Chubut. Congreso Argentino de Paleontología y Bioestratigrafía, 2nd, and Congreso Latinoamericano de Paleontología, 1st, Buenos Aires, Resúmenes, p. 32, Buenos Aires.

Blasco, G., Nullo, F.E., and Proserpio, C. (1979): Aspidoceras en Cuenca Austral, Lago Argentino, provincia de Santa Cruz. Asociación Geológica Argentina, Revista, v. 34, 282-293, Buenos Aires.

Blasco, G., Nullo, F.E., and Ploszkiewicz, J.V. (1980): El género Colchidites Djánelidzé, 1926 y la posición estratigráfica del género Hatchericeras Stanton, 1901 en la Estancia Tucu-Tucu, provincia de Santa Cruz. Asociación Geológica Argentina, Revista, v. 35, 41-58, Buenos Aires.

Blasco, G., Nullo, F.E., and Proserpio, C. (1981): Santoniano-Campaniano: estratigrafía y contenido amonitífero, Cuenca Austral. Asociación Geológica Argentina, Revista, v. 35 (1980), 467-493, Buenos Aires.

Bonarelli, G. (1917): Tierra del Fuego y sus turberas. Ministerio de Agricultura, Anales, Sección Geología, XII, 3, 1-66, Buenos Aires.

Bonarelli, G., and Nágera, J.J. (1921): Observaciones geológicas en las inmediaciones del Lago San Martín (Territorio de Santa Cruz). Dirección General de Minas, Boletín 27 B, 1-39, Buenos Aires.

Borrello, A.V. (1968): Sobre el vulcanismo embrionario neoídico de los Andes Patagónicos-Fueguinos. Academia Nacional de Ciencias, Boletín, v. 47, 21-26, Córdoba.

Borrello, A.V. (1969): Los Geosinclinales de la Argentina. Dirección Nacional de Geología y Minería, Anales, 14, 1-136, Buenos Aires.

Caminos, R. (1976): Tobas y pórfidos dinamometamorfizados de la Isla de Los Estados, Tierra del Fuego. Congreso Geológico Argentino, 6th, Bahía Blanca, Actas, 2, 9-24.

Caminos, R. (1980): Cordillera Fueguina *in* Geología Regional Argentina, Segundo Simposio, (Coord.: Turner, J.C.M.), Córdoba (1976), Academia Nacional Ciencias, 2, 1463-1501, Córdoba.

Caminos, R., Haller, M.A., Lapido, O., Lizuaín, A., Page, R., and Ramos, V.A. (1981): Reconocimiento geológico de los Andes Fueguinos. Territorio Nacional de Tierra del Fuego. Congreso Geológico Argentino, 8th, San Luis, Actas, 3, 759-786.

Cecioni, G. (1955): Edad y Facies del Grupo Springhill en Tierra del Fuego. Universidad de Chile, Facultad de Ciencias Exactas, Físicas y Naturales, Anales, 12, 243-256, Santiago.

Cingolani, C., and Varela, R. (1976): Investigaciones geológicas y geocronológicas en el extremo sur de la isla Gran Malvina, sector cabo Belgrano (Cabo Meredith), Islas Malvinas. Congreso Geológico Argentino, 6th, Bahía Blanca, Actas, 1, 457-474.

Dalziel, I.W.D., and Ramos. V.A. (1988): Eastern Tierra del Fuego and Islas de los Estados. International Geological Congress, 28th, Washington, Fieldtrip Guide-Book T 180, Tectonics of the Scotia Arc, 168-178.

Dalziel, I.W.D, de Wit, M.F. and Palmer, K.F. (1974): Fossil marginal basin in the southern Andes. Nature, v. 250 (5464), 291-294.

Dana, J.D. (1848): Geology, *in* U.S. Exploring Expedition 1839-1842 under Ch. Wilkes, Chap. 15-16, New York.

Darwin, Ch. (1846): Geological observations of South America, being the third part of the Geology of the voyage of the Beagle during 1832-1836. London, Smith, Elder, vii + 279 pp.
Du Toit, A.L. (1927): A geological comparisson of South America with South Africa. Publications Carnegie Institute, 381, Washington.
Favre, F., (1908): Die Ammoniten der untern Kreide Patagoniens. Neues Jahrbuch für Mineralogie, Geologie und Paläontologie, 25, 601-647.
Feruglio, E. (1931): Nuevas observaciones geológicas en la Patagonia Central. Contribución YPF Primera Semana de la Geografía, 3-24, Buenos Aires.
Feruglio, E. (1936): Paleontographia Patagónica. Memoria Instituto Geológico, 11-12, 1-384, Padova.
Feruglio, E. (1938): El Cretácico superior del lago San Martín y de las regiones adyacentes. Physis, v. 12, 293-342, Buenos Aires.
Feruglio, E. (1950): Descripción Geológica de la Patagonia. Dirección General Y.P.F., I-III, Buenos Aires.
Frenguelli, J. (1935): "Ptilophyllum hislopi" (Oldham) en los "Mayer River Beds" del lago San Martín. Museo La Plata, Notas 1, Paleontología, 3, 71-83, La Plata.
Halle, T.G. (1913): Some Mesozoic plant-bearing deposits in Patagonia and Tierra del Fuego and their floras. Kungl. Svenska Vetenskapsakademiens Handlingar, 51, 3-58, Stockholm.
Haller, M., and Lapido, O. (1982): The Jurassic- Cretaceous volcanism in the Septentrional Patagonian Andes, *in* Magmatic Evolution of the Andes, (Eds.: Linares, E., Cordani, U.G., and Munizaga, F.), Earth-Science Reviews, v. 18, 395-410.
Hatcher, J.B. (1897): On the Geology of Southern Patagonia. American Journal of Science, (4) 4 (23), 327-354.
Hatcher, J.B. (1900): Sedimentary rocks of Southern Patagonia. American Journal of Science (4) 9 (50), 85-108.
Hatcher, J.B. (1903): Narrative and geography. Report Princeton University Expedition to Patagonia, 1, 1-314.
Hauthal, R. (1903): Mitteilungen uber den heutigen Stand der geologischen Erforschung Argentiniens. Congress Geologique Internationale, 9th, Vienne, Comptes-Rendus 1903, 649- 656.
Hervé, F., Nelson, E., Kawashita, K., and Suárez, M. (1981): New isotopic ages and the timing of orogenic events in the Cordillera Darwin, southernmost Chilean Andes. Earth and Planetary Science Letters, v. 55, 257-265.
Hünicken, M. (1965): Algunos cefalópodos supracretácicos del Río Turbio, Santa Cruz. Universidad Nacional de Córdoba, Facultad de Ciencias Exactas, Físicas y Naturales, Revista, Serie Ciencias Naturales, 26, 49-99, Córdoba.
Katz, H.R. (1963): Revision of Cretaceous Stratigraphy in Patagonian Cordillera of Ultima Esperanza, Magallanes Province, Chile. American Association of Petroleum Geologists Bulletin, v. 47, 506-524.
Kay, S.M., Ramos, V.A., Mpodozis, C., and Sruoga, P. (1989): Late Paleozoic to Jurassic silicic magmatism at the Gondwanaland margin: analogy to the Middle Proterozoic in North America? Geology, v. 17, 324-328.
Kielbowicz, A.A., Ronchi, D.I., and Stach, N.H. (1983): Foraminíferos y ostrácodos valanginianos de la Formación Springhill, Patagonia Austral. Asociación Geológica Argentina, Revista, v. 38, 313-339.
Kranck, E.H. (1932): Geological investigations in the Cordillera of Tierra del Fuego. Acta Geographicae (Geographic Society of Finland Bulletin), v. 4, 2, 1-231, Helsinki.
Leanza, A.F. (1967): Descripción de la fauna de Placenticeras del Cretácico superior de la Patagonia Austral con consideraciones acerca de su posición estratigráfica. Academia Nacional de Ciencias, Boletín, v. 46, 5-28, Córdoba.
Leanza, A.F. (1969): Sobre el descubrimiento del piso Coniaciano en Patagonia Austral y descripción de una especie de ammonites (Peronioceras santacrucense n.sp.). Academia Nacional de Ciencias, Boletín, v. 47, 5-20, Córdoba.
Leanza, A.F. (1970): Amonites nuevos o poco conocidos del Aptiano, Albiano y Cenomaniano de los Andes Australes con notas acerca de su posición estratigráfica. Asociación Geológica Argentina, Revista, v. 25, 197-261, Buenos Aires.
Leanza, A.F. (1972): Andes Patagónicos Australes, *in* Geología Regional Argentina, Primer Simposio, (Dir. and Ed.: Leanza, A.F.), Córdoba (1969), Academia Nacional de Ciencias, 689-706, Córdoba.
Leanza, H.A. (1981): Faunas de ammonites del Jurásico superior y del Cretácico inferior de América del Sur, con especial consideración de la Argentina, *in* Cuencas Sedimentarias del Jurásico y Cretácico de América del Sur, (Eds.: Volkheimer, W., and Musacchio, E.), 2, 559-597, Buenos Aires.

Macellari, C., Barrio, C.A., and Manassero, M.J. (1989): Upper Cretaceous to Paleocene depositional sequences and sandstone petrography of Southwestern Patagonia (Argentina and Chile). Journal South American Earth Sciences, v. 2, 223-240.

Malumián, N., and Ramos, V.A. (1984): Magmatic intervals, transgression-regression cycles and oceanic events in the Cretaceous and Tertiary of Southern South America. Earth and Planetary Science Letters, v. 67, 228-237.

Malumián, N., Nullo, F.E. and Ramos, V.A. (1983): The Cretaceous of Argentina, Chile, Paraguay and Uruguay, *in* The Phanerozoic Geology of the World, (Eds.: Moullade, M., and Nairn, A.E.M.), II, The Mesozoic, Amsterdam, Elsevier Sci. Pub., B, 265- 304.

Nordenskjöld, O. (1897): Notes on Tierra del Fuego. An account of the Swedish Expedition of 1895-1897. Scottish Geographical Magazine, v. 13, 393-399, Edinburgh.

Nordenskjöld, O. (1905): Die kristallinen Gesteine der Magallansländer. Wiss. Ergebn. Schwed. Exp. Magellasländer, v.1, 6, 175-240, Stockholm.

Nullo, F.E., Proserpio, C.A., and Ramos, V.A. (1978): Estratigrafía y tectónica de la vertiente este del Hielo Continental Patagónico, Argentina-Chile. Congreso Geológico Argentino, 7th, Neuquén, Actas, 1, 455-470.

Nullo, F.E., Proserpio, C.A., and Blasco de Nullo, G. (1981a): Estratigrafía del Cretácico superior en el cerro Indice y alrededores, provincia de Santa Cruz. Congreso Geológico Argentino, 8th, San Luis, Actas, 3, 373-387, Buenos Aires.

Nullo, F.E., Proserpio, C.A., and Blasco de Nullo, G. (1981b): El Cretácico de la Cuenca Austral entre el lago San Martín y Río Turbio, *in* Cuencas Sedimentarias del Jurásico y Cretácico de América del Sur, (Eds.: Volkheimer, W., and Musacchio, E.), 1, 181-216, Buenos Aires.

Orbigny, A. d'(1848): Géologie, Minéralogie et Géographie Physique du Voyage. Voyage au Pole Sud et dans l'Océanie sur les corvettes l'Astrolabe et la Zélée pendant les années 1837-38-39-40 sous le commandement de M.J. Dumont d'Urville.

Paülcke, W. (1907): Die Cephalopoden der oberen Kreide Südpatagoniens. Naturforsch Gessellschaft Freiburg, v. 15, 167-248, Berlin.

Piatnitzky, A. (1938): Observaciones geológicas en el oeste de Santa Cruz. Yacimientos Petrolíferos Fiscales, Boletín de Informaciones Petroleras, Segunda Epoca, 165, 45-85, Buenos Aires.

Ploszkiewicz, J.A., and Ramos, V.A. (1977): Estratigrafía y Tectónica de la Sierra de Payaniyeu. Asociación Geológica Argentina, Revista, v. 32, 209-226, Buenos Aires.

Quensel, P.D. (1910): On the igneous rocks of the Patagonian Cordillera. Congress Géologique Internationale, 11th, 2, 905-907, Stockholm.

Quensel, P.D. (1912): Geologische-petrographische Studien in der patagonischen Cordillera. Geological Institute of Uppsala Bulletin, v. 11, 1-114.

Quensel, P.D. (1913): Die Quarzporphyr und Porphyroidformation in Südpatagonien und Feuerland. Geological Institute of Uppsala Bulletín, v. 12, 9-40.

Ramos, V.A. (1976): Estratigrafía de los lagos La Plata y Fontana, Provincia del Chubut, República Argentina. Congreso Geológico Chileno, 1st, Santiago, Actas, 1 (A), 43-64.

Ramos, V.A. (1978): Descripción Geológico-Económica de la Hoja 55 c, Meseta de la Muerte, provincia de Santa Curz. Servicio Geológico Nacional, unpublished, Buenos Aires.

Ramos, V.A. (1979a): El vulcanismo del Cretácico inferior de la Cordillera Patagónica de Argentina y Chile. Congreso Geológico Argentino, 7th, Neuquén, Actas, 1, 423-436.

Ramos, V.A. (1979b): Tectónica de la región del Río y Lago Belgrano, Cordillera Patagónica, Argentina. Congreso Geológico Chileno, 2nd, Arica, Actas, 1 (B), 1-32.

Ramos, V.A. (1981a): Descripción Geológica de la Hoja 47 a-b, Lago Fontana, Provincia del Chubut. Servicio Geológico Nacional, Boletín, 183, 1-130, Buenos Aires.

Ramos, V.A. (1981b): Descripción Geológico-Económica de la Hoja 55 a-b, Sierra de Sangra, provincia de Santa Cruz. Servicio Geológico Nacional, unpublished, Buenos Aires.

Ramos, V.A. (1982a): Geología de la región del lago Cardiel, provincia de Santa Cruz. Asociación Geológica Argentina, Revista, v. 37, 23-49, Buenos Aires.

Ramos, V.A. (1982b): Descripción Geológico-Económica de las Hojas 53 a, Cerro San Lorenzo y 53b Meseta Belgrano, provincia de Santa Cruz. Servicio Geológico Nacional, unpublished, Buenos Aires.

Ramos, V.A. (1983a): Evolución tectónica y metalogénesis de la Cordillera Patagónica. Congreso Nacional de Geología Económica, 2nd, San Juan, Argentina, Actas, 1, 108-124.

Ramos, V.A. (1983b): Descripción geológica de la Hoja 53 ab Cerro San Lorenzo y Meseta Belgrano, provincia de Santa Cruz. Servicio Geológico Nacional, unpublished, Buenos Aires.

Ramos, V.A. (1988): La estructura de la Cordillera Patagónica (47-49° S) de Argentina y Chile. Congreso Geológico Chileno, 5th, Santiago, Actas, 1 (A), 101-114.

Ramos, V.A. (1989): Foothills structure in Northern Magallanes Basin, Argentina. American Association of Petroleum Geologists Bulletin, v. 73, 887-903.

Ramos, V.A., and A. Baldoni. (1981): Nuevas localidades con plantas fósiles cretácicas en la Cordillera Patagónica, provincia de Santa Cruz. Congreso Geológico Argentino, 8th, San Luis, Actas, 4, 743-759.

Ramos, V.A., and Palma, M.A. (1983): Las lutitas pizarreñas del cerro Dedo, Lago La Plata, provincia del Chubut. Asociaciacón Geológica Argentina, Revista, v. 38, 148-160, Buenos Aires.

Ramos, V.A., H. Niemeyer, J. Skarmeta, and J. Muñoz. (1982): The magmatic evolution of the Austral Patagonian Andes, *in* Magmatic Evolution of the Andes, (Eds.: Linares, E., Cordani, U.G., and Munizaga, F.), Earth-Science Reviews, v. 18, 411-443.

Ramos, V.A., Haller, M.J. and Butrón, F. (1986): Geología y Evolución tectónica de las Islas Barnevelt: Atlántico Sur. Asociación Geológica Argentina, Revista, v. 41, 137-154, Buenos Aires.

Riccardi, A.C. (1970): Favrella R. Douvillé, 1909 (Ammonitina: Cretácico Inferior): Edad y distribución. Ameghiniana, v. 7, 110-138, Buenos Aires.

Riccardi, A.C. (1971): Estratigrafía en el oriente de la Bahía de La Lancha, lago San Martín, Santa Cruz, Argentina. Museo de La Plata, Revista, 7, Geología, 61, 245-318, La Plata.

Riccardi, A.C. (1976): Paleontología y edad de la Formación Springhill. Congreso Geológico Chileno, 1st, Santiago, Actas, 1 (C), 41-56.

Riccardi, A.C. (1977): Berriasian invertebrate fauna from the Springhill Formation of Southern Patagonia. Neues Jahrbuch für Geologie und Paläontologie, Abhandlungen 155, 216- 252.

Riccardi, A.C. (1979): El género Calicoceras Hyatt (Ammonitina, Cretácico superior) en la Patagonia Austral. Museo de la Plata, Obra del Centenario, 5, 63-72, La Plata.

Riccardi, A.C. (1983): Kossmaticeratidae (Ammonitina) y nomenclatura estratigráfica del Cretácico tardío en Lago Argentino, Santa Cruz, Argentina. Ameghiniana, v. 20, 317-345, Buenos Aires.

Riccardi, A.C. (1984): Las asociaciones de amonitas del Jurásico y Cretácico de la Agentina. Congreso Geológico Argentino, 9th, Bariloche, Actas, 4, 559-595.

Riccardi, A.C. (1988): The Cretaceous System of Southern South America. Geological Society of America Memoir, 168, 1-161.

Riccardi, A.C., and Aguirre-Urreta, M.B. (1988): Bioestratigrafía del Cenomaniano-Santoniano de la Patagonia Argentina. Congreso Geológico Chileno, 5th, Santiago, Actas, 2 (C), 375-394.

Riccardi, A.C., and Aguirre-Urreta, M.B. (1989): Hemihoplitidae (Ammonoidea) from the Hauterivian-Barremian of Southern Patagonia. Palaeontology, v. 32, 447-462.

Riccardi, A.C., and Rolleri, E.O. (1980): Cordillera Patagónica Austral, *in* Geología Regional Argentina, Segundo Simposio, (Coord.: Turner, J.C.M.), Córdoba (1976), Academia Nacional Ciencias, 2, 1173-1306, Córdoba.

Riccardi, A.C., Aguirre-Urreta, M.B., and Medina, F. (1987): Aconeceratidae (Ammonitina) from the Hauterivian-Albian of Southern Patagonia. Palaeontoqraphica, A 196, 105-185.

Robles, D. (1982): El desarrollo de la Formación Springhill en la cuenca de Magallanes. Congreso Nacional de Hidrocarburos, 1st, Buenos Aires, Exploración, 293-312.

Stanton, T.W. (1901): The marine Cretaceous Invertebrates. Report Princeton Expedition Patagonia, 4, 1, 1-43, Princeton.

Suárez, M. (1978): Geología de la región al sur del Canal de Beagle, Chile. Instituto de Investigaciones Geológicas, Carta Geológica de Chile, 36, 48 pp., Santiago.

Suárez, M., and Pettigrew, T.H. (1976): An upper Mesozoic Island arc back-arc system in the Southern Andes and South Georgia. Geological Magazine, v. 113, 305-400.

Suárez, M., Puig, A., Hervé, M., Piracés, R., and Cepeda, A. (1979): Geología de la región al sur de los canales Beagle y Ballenero, Andes del Sur, Chile, Nota Preliminar. Congreso Geológico Chileno, 2nd, Arica, Actas, 4 (J), 19-28.

Suárez, M., Puig, A., and Hervé, M. (1985a): Depósitos de un abanico submarino de tras-arco del Mesozoico superior: Formación Yahgán, Islas Hoste y Navarino. Congreso Geológico Chileno, 4th, Antofagasta, Actas, 1 (1), 534-545.

Suárez, M., Hervé, M., and Puig, A. (1985b): Plutonismo diapírico del Cretácico en Isla Navarino. Congreso Geológico Chileno, 4th, Antofagasta, Actas, 3 (4), 549-563.

Thomas, C.R. (1949): Geology and Petroleum Exploration in Magallanes Province, Chile. American Asocíation of Petroleum Geologists Bulletin, v. 33, 1553-1578.

Vilela, C.R., and Csaky, A. (1968): Las turbiditas en los sedimentos cretácicos de la región de lago Argentino (Provincia de Santa Cruz). Jornadas Geológicas Argentinas, 3rd, Comodoro Rivadavia, Actas, I, 209-225.

Wilson, T.J. (1991): Transition from back-arc to foreland basin development in the southernmost Andes: stratigraphic record from the Ultima Esperanza District, Chile. Geological Society of America Bulletin, v. 103, 98-111.

Winn, R.D., and Dott, R.H. (1979): Submarine-fan turbidites and resedimented conglomerates in a Mesozoic arc-rear marginal basin in Southern South America, *in* Sedimentation in submarine canyons, fans and trenches, (Eds.: Stanley, D.J., and Kelling, G.), Stroudsberg, Pennsylvania, Dowden, Hutchinson & Ross, 362-373.

Rhenohercynian and Subvariscan Fold Belts

Edited by R. A. Gayer, R. O. Greiling, and A. K. Vogel.

1993. vi, 390 pages. (Earth Evolution Sciences; edited by Andreas Vogel) Hardcover
ISBN 3-528-06488-9

The volume provides a case for example for the tectonic evolution of a foreland fold-and-thrust belt from early extension and basin formation to basin inversion and various aspects of deformation and modern structural techniques, based on the Variscan northern margin from Ireland in the west through Wales and SW England to central Europe including Poland in the east. The papers are selected from the conference on Rhenohercynian and Subvariscan Fold Belts organized by the Editorial Group of Earth Evolution Sciences (EES) and sponsored by the Tectonic Studies Group (TSG) of the Geological Society of London. Although the text is directed primarily towards researchers both in academia and in the industry it gives also case examples suitable for advanced of postgraduate students.

Vieweg Publishing · P.O.Box 58 29 · D-65048 Wiesbaden

vieweg

Geophysical Data Interpretation by Inverse Modelling

Proceedings of the Ninth International Seminar on Model Optimization in Exploration Geophysics, Berlin 1991

Edited by Andreas Vogel, Rudolf Gorenflo, Wigor Webers, and Roman S. Rutmann (Eds.)

1993. viii, 217 pages. (Theory and Practice of Applied Geophysics, Vol. 6; edited by Andreas Vogel) Hardcover
ISBN 3-528-06523-0

This monograph contains a selection of papers presented at the Ninth International Seminar on Model Optimization in Exploration Geophysics, held at the Free University of Berlin, Feb. 18-23, 1991. The book covers a wide spectrum of inverse modelling from basic mathematical theory to practical Applications in the different branches of geophysics, such as gravity, magnetics, seismics, geoelectrics and geothermics. The requirements arising from practical applications stimulate the development of new analytical and numerical methods, in particular the construction of approximate solutions and their stabilization which is necessary in the case of insufficient data sets. In general it is obvious that the type of the governing partial differential equations essentially influences how to treat the inverse problem both on a global and regional scale as well as in applied geophysics.